This textbook deals with the composition and evolution of material objects in the universe, from terrestrial and moon rocks to quasars: what are their compositions and how do they arise? There are two main themes: chemical processes responsible for the abundances we observe, and nuclear processes in which chemical elements originate.

The author presents a self-contained and interdisciplinary synthesis, aimed at the advanced undergraduate/graduate student. The book has problem sets to aid understanding and an extensive bibliography.

AN INTRODUCTION TO COSMOCHEMISTRY

AN INTRODUCTION TO COSMOCHEMISTRY

CHARLES R. COWLEY
Professor of Astronomy, University of Michigan

Published by the Press Syndicate of the University of Cambridge
The Pitt Building, Trumpington Street, Cambridge CB2 1RP
40 West 20th Street, New York, NY 10011-4211, USA
10 Stamford Road, Oakleigh, Melbourne 3166, Australia

First published 1995

A catalogue record for this book is available from the British Library

Library of Congress cataloguing in publication data available

Cowley, Charles R., 1934–
An introduction to cosmochemistry / Charles R. Cowley.
p. cm.
Includes bibliographical references.
ISBN 0 521 41538 1
1. Cosmochemistry. 2. Astrogeology. 3. Astrophysics. I. Title.
QB450.C68 1994 93-14079
523′.02–dc20 CIP

ISBN 0 521 41538 1 hardback

ISBN 0 521 45920 6 paperback

Transferred to digital printing 2004

TAG

Contents

Foreword

My career as a professional astronomer was some 15 years old when it first became necessary for me to learn something of the new developments in the solar system. At that time, in the mid-1970's, I was about as ignorant of the solar system as one trained in astronomy could possibly be. Worse than that, I had an attitude typical of many astronomers today. Because the field was old, I thought it was dull and uninteresting! Nevertheless, when it became necessary for me to give an introductory course in solar system astronomy, I thought I must try to understand what all the fuss over moon rocks was all about.

Moon rocks are not so terribly different from terrestrial rocks, and so I began to read an introductory geology text. Soon, I was making trips to the building next door to visit the Geology Department. I became an amateur geologist, and a rockhound. On automobile trips I would stop at various rock formations, and bash off samples with a rock hammer. These samples were typically shown to a geologist, sometimes in a nearby university or college, sometimes back at Michigan.

The experience of becoming an amateur geologist was immensely broadening. Not only did I become a great fan of planetary science, but I began to be interested in other areas of astronomy that had never particularly appealed to me. Eventually, I began to realize that there was a single theme behind all of these endeavors – the history of matter. In due course, I will explain how astronomers typically focused on nuclear aspects of the history of matter while geologists were largely concerned with chemical and mineralogical changes. Strategically positioned with respect to both groups are those marvelous people who study the composition – and history – of meteorites. It is a great shame that one cannot just stop at a road cut, and bash off meteorite samples!

There is a great advantage in being able to see one's research as a piece of a large fabric. It is not merely psychologically fulfilling; there are practical advantages. Techniques are often developed and applied in one area of science well ahead of another. Individuals with a broad perspective have a chance to make progress in their own fields by adapting techniques developed in another.

This book is an attempt to present the broad picture of the chemical and nuclear history of matter as we understand it today. It would have been quite impossible for me to write it without enormous amounts of help from friends and colleagues. There is no way for me to adequately express my thanks; I wish there were some way to be sure to include everyone in this inadequate acknowledgement.

Many people have graciously let me use figures from their books and articles. The sources of these figures are cited in the text. M. W. Chase, Lawrence Grossman, Hartmut Holweger, Nancy Houk, Francesca Matteucci, Harry McSween, Charles Pouchert, and Scott Sandford were kind enough to supply original drawings. Saul Perlstein at Brookhaven patiently taught me how to use their nuclear data base. In the process of making requests of James Kaler (Illinois), Francesca Matteucci (Max Planck Institute), and Scott Sandford (NASA-Ames) for permission to use some of their material, I managed to persuade them to read drafts of Chapters 14, 15 and 16. Their help is very gratefully acknowledged. I appreciate the advice of Leroy Doggett of the Naval Observatory concerning the system of astronomical constants. I thank Greg Bothun and Jim Schombert for helping me draw one of the figures.

Geologists at Michigan, Paul Cloke, Charles DeWolf, Eric Essene, Michael Gurnis, William Heinrich, Donald Peacor, Stuart Weinstein, and Youxye Zhang, have helped with the present book, or answered endless questions about rock samples or general geology. John Valley, now a Professor of Geology at the University of Wisconsin, was especially generous of his time when he was a graduate student here at Michigan. Michigan chemists, Seyhan Eğe, Henry Griffin and Larry Lohr suggested references, answered many questions, and commented on sections from early drafts. Physicists Fred Becchetti, Michael Bretz, K. T. Hecht, and Joachim Janecke are not to be blamed for errors that may linger in my discussion of nuclear or surface physics. They have been patient and extraordinarily helpful. Physicists at the US National Bureau of Standards (now NIST) have provided data and advice relevant to this book. I am especially grateful to J. R. Fuhr, W. C. Martin, and J. Sugar. I was most fortunate to have had the advice and help of John Barker and Sushil Atreya of the Michigan Department of Atmospheric and Oceanic Sciences. I thank Thomas Donahue who kindly supplied some key numerical values. Astronomical colleagues at Michigan, Donald Bord, Beth Brown, Gordon MacAlpine, and Guy Worthy, have commented on drafts, and answered numerous questions. Special thanks are due to my friend and colleague of three decades, Richard Sears. He has helped me in more ways than could possibly be set down, and I do not think I could do astronomy without him.

1
Overview

1.1 The Scope of Cosmochemistry

The cosmochemist has two basic tasks. The first is to determine the chemical composition of matter in the material universe. The form of this matter ranges from such mundane materials as terrestrial rocks to distant galaxies. The second task is to understand the reasons for the compositions that are found. While the first of these tasks is basically a matter of analytical chemistry, the second has important evolutionary aspects.

From a logical and historical point of view, cosmochemistry is an extension of the well-established discipline of geochemistry. Victor Goldschmidt, one of the pioneers of modern geochemistry, had a keen interest in abundances of the chemical elements in meteorites and the sun and stars. Goldschmidt (1937) was an early compiler of what became known as the *cosmic* abundances, in which the analyses of extraterrestrial sources played important roles.

The logical complement to the term geochemistry, in the astronomical domain, would be *astrochemistry*. This word is frequently in the literature, but typically with a more restricted meaning, such as the formation of molecules in cool interstellar clouds. Solar system astronomers have used terms such as planetary geology or lunar geochemistry rather than astrogeology or astrochemistry.

For many years, astronomers thought of cosmochemistry primarily in terms of the nuclear history of matter, and the search for a standard abundance distribution (SAD). The modern aspects of this work began with Goldschmidt, and were continued by Hans Suess and H. C. Urey (1956). The team of Margaret and Geoffrey Burbidge, Nobel Laureate William A. Fowler, and Fred Hoyle (1957) are known to astronomers as B^2FH. Along with A. G. W. Cameron (1957), they founded the modern theory of the origin of the chemical elements. Their pioneering studies were based on the elemental abundance compilation of Suess and Urey. Chemistry *per se* played a minor role in most astrophysical studies dealing with the origin and history of the chemical elements.

While this *nuclear astrophysics* was growing into a mature discipline, radio astronomers began to explore the extraordinarily rich observational domain of interstellar molecules. As the sensitivity of receivers in the *microwave region* (about 10^9–10^{11} Hz) improved, the astronomical literature swelled with articles dealing directly with chemically bonded atoms. For the first time, chemical processes beyond the solar system became a subject of intense study. In the present work, we shall integrate interstellar molecules into the broad notion of cosmochemistry.

The most distant objects known in the universe are the quasars. They may be analyzed with the help of techniques developed for emission-line regions in our own Galaxy. Quasars, and their congeners, the *active galactic nuclei*, are discussed in Chapters 15 and 16. Distant objects, especially the quasars, allow us to extend the domain of cosmochemistry not only in space, but also backward in time, when the universe was a fraction of its present age of 10 to 20 billion (10^9) years.

A few elements and isotopic species originate in the big bang itself. Enough is now known about *cosmological nucleosynthesis* that details of the structure of the present universe can be fixed from a knowledge of the abundances of deuterium, ^{3}He, and lithium, beryllium and boron. Thus, the discipline of cosmochemistry covers our universe, in space and time.

1.2 Cosmochemistry and the Four Physical Sciences

It will be useful to place cosmochemistry within the framework of the four branches of physical science: *astronomy, geology, chemistry*, and *physics*. The first two disciplines have far more in common than is often recognized, and they are separated from the latter two by a vast gulf. In physics and chemistry, one can do experiments; astronomical and geological investigations are largely observational. The astronomer and geologist must analyze experiments done for them by nature. There are usually no means to exercise direct control over the phenomena studied. Fortunately, the observer may palliate this shortcoming by the clever selection of objects to study, and the tools used to study them.

In observational science only partial information is typically available; usually, there is no way to return to the laboratory and constrain conditions more closely, or make new measurements. A good indication that the astronomer or geologist is dealing with partial information is analysis by the method of modeling. Here, the numerical consequences of specific assumptions are worked out and compared with observations. Modern computing facilities make this technique increasingly popular. It is employed in physics and chemistry too, though not as extensively as in the observational sciences. The modeling technique always carries with it the question of uniqueness. Would another model fit the observations as well? More often than not, there is no competing model, and it is only too easy to forget the question of uniqueness.

The raw materials of the physicist and chemist are electrons, protons, and quarks, elements, compounds, and atoms. These basic entities, we hope, are the same today as they were 10^9 or more years ago. But the raw materials of the geologist, astronomer, and cosmochemist are landforms, planets, stars, nebulae and galaxies. And we know that in general these were *not* the same many times 10^9 years ago as they are today. Indeed, an understanding of the *evolution* of these objects, of their history, has become one of the primary goals of astronomers and geologists. In this sense, these applied disciplines deal with an aspect of time that is and must be missing from physics and chemistry.

With a very few specialized exceptions, physics and chemistry deal with laws of nature or approximations thereto. It is assumed that these laws do not change with time; if they did, one could not confirm yesterday's experiment in the laboratory today, and most scientific endeavor would be impossible.

Cosmochemistry is a branch of astronomy and geology for which physics and chemistry are ancillary disciplines. It is therefore subject to the limitations of an observational science. It is enriched by concerns of history and evolution.

1.3 A Standard Abundance Distribution

For many years, the main task of the cosmochemist was to discover the composition of the gas cloud from which the sun and planetary system were formed. These hypothetical abundances, or approximations to them, have been known for historical reasons as cosmical abundances. Prior to the modern work on the nuclear history of matter (B^2FH, Cameron) it was thought that the solar composition was similar to that of virtually all stars in the universe. Of course, it had been known from the spectroscopy of stars that some of them had very peculiar compositions. Still, stars with palpably peculiar spectra were relatively few in number, and it was possible as recently as the middle of the twentieth century to argue that they might be accounted for by special processes.

The underlying concept of a universal composition is central in the introduction to Suess and Urey's (1956) celebrated study "Abundances of the Elements." The basis for this composition was to be sought in some primordial synthesis, possibly related to that made famous by Gamow and his collaborators in their work on the early universe (Alpher, Bethe, and Gamow 1948). But discoveries within the last several decades have steadily undermined the conceptual basis for *a* primitive composition. Gordon Goles, a cosmochemist at the University of Oregon, has compared attempts to find the original composition of the solar system to the search for the Holy Grail.

The Suess–Urey work did not lead to the Holy Grail. Rather, it provided the empirical background for the theories of stellar nucleosynthesis. But if the chemical elements were built up in stars, why, then, are cosmic abundances so similar? The general explanation, details of which we shall come to in due course (Chapter 16), is that most nucleosynthesis took place rapidly, perhaps during the first few hundred million years of the history of galaxies. Most stars were born after this initial burst of star formation and nucleosynthesis.

The bulk composition of the sun represents a fundamental, though not universal, abundance sample. The best set of experimental values is still pursued, but with a more realistic purpose than previously.

Figure 1.1 gives the distribution of elemental abundances in the primitive solar nebula (SAD) according to Anders and Grevesse (1989). According to the practice among astronomers, the abundance of hydrogen is assumed to be 10^{12} by number, and values for other elements are given *relative* to it. Geochemists use a scale where silicon is fixed at 10^6. We shall comment on the unusual shape of this curve at various points in the text. For the present, we confine ourselves to the following remarks. First, the nearly universal sawtooth pattern is readily understood in terms of the stability of nuclei with even numbers of protons (§7.4 and Chapter 9). In abundance patterns, it is called the odd–even effect. Other properties of the curve,

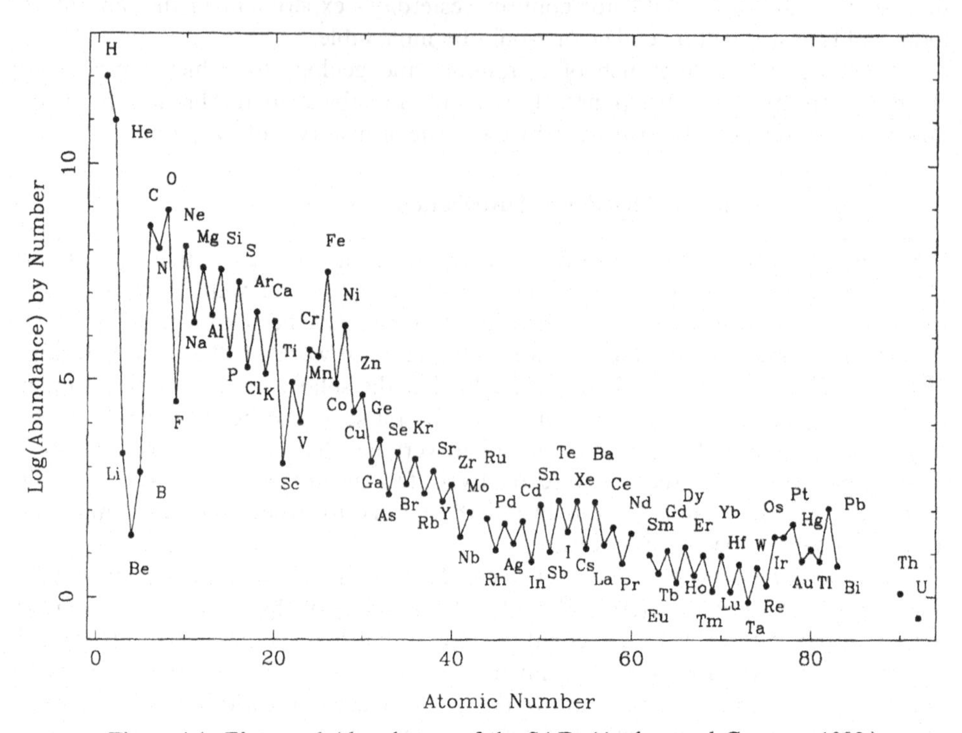

Figure 1.1. Elemental Abundances of the SAD. (Anders and Grevesse 1989.)

such as the maximum at $Z = 26$ (iron), can also be explained in terms of nuclear astrophysics (Chapter 10).

It was believed until ca. 1970 that the material forming the sun and the planets was at one time thoroughly homogenized. This was because significant isotopic abundance variations among meteoritic and terrestrial samples were not observed (see Reynolds 1967, pp. 255f). The uniformity was considered quite overwhelming. (We must exclude, of course, mass-dependent fractionations as well as variations that can be attributed to known radioactivities.) The skeptical and cautious Suess (1965) wrote that the well-mixed primordial nebula was "one of the very few assumptions...which can be considered well justified and firmly established."

In the face of such pronouncements, the avalanche of data on isotopic anomalies that began at the end of the 1960s can only make the student of science wonder about the meaning of a "scientific fact."

In spite of ample evidence for heterogeneities in the early solar system, there still remains the possibility that such anomalies are confined to a relatively small mass fraction, and/or that the fractional variations involved are small. Suppose, for example, the isotopic anomalies could be explained by the admixture of one part in 10^2 to 10^4 of "foreign" material, and the bulk of the mass was well mixed, and uniform in composition. Under these circumstances, it is still useful to adopt a "reference" composition that may be appropriate to most of the primordial solar system. The reference composition would also be, in some sense, primitive. Audoze and Tinsley (1976) use the term "standard abundance distribution," or SAD.

Throughout this work we shall refer to the SAD. By this we shall mean one or another of the abundance compilations that attempt to represent the original composition of the solar nebula. Usually there are fewer than half a dozen current compilations. For the most part, they are very similar, and we need not distinguish among them.

The sun itself may represent the most satisfactory sample of the primitive composition, as its mass comprises more than 99.9% of the present solar system. Traditionally, that part of the material that did not become a part of the sun has been known as the *solar nebula* (see Kuiper 1951, Magisos 1985), although the distinction is not particularly helpful if we wish to use the present composition of the sun as a guide to the composition of the residual gas from which the planets formed. It is generally thought that material from the original nebula was driven off by a violent, early solar wind. It is not known how much mass was driven away. Ideas from the experts have ranged from a minimum of about 0.03 to several solar masses.

Practical difficulties in determination of the solar composition (see Chapter 12) have made other materials more useful for fixing the SAD. It is still generally agreed that the meteorites known as carbonaceous chondrites are most useful for this purpose.

Some qualifications must be made, of course. The carbonaceous chondrites are solid substances containing only minimal portions of hydrogen and helium, the principal constituents of the sun. These meteorites are also strongly depleted in noble gases and other *volatile elements.* However, the abundances of nonvolatile elements are considered as the most securely known in the SAD. The reasoning is involved, and, to some extent, circular. We shall take it up in Chapter 7.

Solid materials which can be analyzed in terrestrial laboratories have great advantages for the student of the nuclear history of matter over spectroscopic analyses of stars and cosmic plasmas. Not only is the accuracy much higher, but isotopic abundances are available. By contrast, isotopic abundances of stellar atmospheres and gaseous nebulae may be obtained only in special cases. Until the microwave investigations of molecular gases, we knew virtually nothing about the isotopic composition of external galaxies. There are 83 stable or long-lived chemical elements and 264 stable nuclei. For this reason alone, there is more information available from a meteoritic sample than from the spectrum of a star or nebula.

Thus the study of vagabond fragments of solid solar system material, the meteorites, has played a dominating role in the study of cosmochemistry, quite out of proportion to the fractional mass that, as far as we know, is represented by such matter.

1.4 The Chemical History of Planetary Material

Most of the data on planetary chemistry concerns earth materials. We shall limit ourselves here to specialized aspects of terrestrial and lunar chemistry. We must bypass such matters as metamorphism or hydrothermal processes, ore and petroleum geology. A temporary excuse is that such matters have little application beyond the earth. The surficial layers of the earth, where such processes take place, represent only a few tenths of a percent of earth's mass.

Future discussions of cosmochemistry may not be able to dismiss such specialized chemistry. Complicated reactions involving liquids *other* than those of the magmatic processes discussed here are certain to be required for an understanding of the materials to be returned from the surfaces of Mars and Venus. Even now, certain aspects of the chemistry and mineralogy of the carbonaceous chondrites seem inexplicable in the absence of liquid water (Chapter 7).

The canonical picture of the formation of the planets was presented in Urey's (1952) *The Planets.* Further work by Lewis, Larimer, Grossman and others (see Chapter 5) pushed this theory to the limits imposed by uncertainties of thermodynamic data and models (cf. homogeneous vs. inhomogeneous accretion). In this scheme, solid materials condense from a gas with the composition of the SAD. The condensed matter is then chemically *differentiated* or fractionated from the primitive or parental gas. The bulk compositions of the planets can then be understood in terms of the temperature in the solar nebula where they formed. Roughly, dense materials (metal and rock) form solids at higher temperatures than low-density materials (ices). In a nebula where the temperature decreases outward from the sun, we would then expect a decrease in mean planetary (decompressed!) density, as found for the planets from Mercury through Saturn.

This theory is at present in a state of considerable flux because we do not yet know *how* significant the overall inhomogeneities within the solar nebula may have been. In addition, it is difficult to assess the contributions to planetary chemistry of bombarding material.

We know that comets enter the inner part of the solar system from a kind of reservoir known as the Oort Cloud. These clouds are thought to exist some tens of thousands of astronomical units (AU) from the sun. Some of them have been swept up, for example, by the earth and other planets. What fraction of the abundances of trace species and volatiles may the earth have gained in this way?

If we assume that the planets accreted as approximately homogeneous bodies, then it is clear from the present core–mantle structure of the earth that further chemical differentiation took place. The current crust of the earth is chemically distinct from the mantle. Its composition can be generally understood in terms of fractionation of mantle material. Because of the activity of the earth, with its extensive oceans and plate tectonics, the chemical differences between the crust and mantle are profound. The chemical history of the Moon (probably also Mercury and Mars) has been much simpler.

A formidable problem in the history of the solar system is the fate of the residual material from planetary formation. According to the canonical theory, only a relatively small fraction of the SAD went into the formation of the terrestrial planets. The residual material must somehow be accounted for, but unfortunately for us, it is unavailable for study. We must content ourselves with the usual hand-waving argument that an active, early solar wind drove the material away.

The traditional field of geochemistry is a rather more developed one than that of the chemical history of the Galaxy. This may be understood simply in terms of the far greater amounts and reliability of analyses and the larger number of active research workers. The cosmochemist has, therefore, much to learn from a study of the general approach of the geochemist.

1.5 Abundances Beyond the Solar System

Stellar, nebular, and galactic abundances have been of great importance in the development of current ideas about the history of matter. Stellar compositions are closely related to both the dynamical and kinematical properties of the stars and their ages. Abundance variations in galactic and planetary nebulae (Chapter 15) may be generally understood within a theoretical framework that includes the abundances in stars.

It is now believed that the majority of the chemical elements were synthesized in the interiors of stars. Principally hydrogen and helium emerged from the cosmic fireball or "big bang." The heavier elements, through iron, may be made as a part of the consumption by the stars of nuclear fuels. As the fuels are consumed, the "ashes," the heavier nuclides, become energetically more and more tightly bound, and this process continues until the most tightly bound nuclei are reached. This is the *iron peak,* so called because there is a local maximum of the SAD in the neighborhood of these nuclei. Trans-iron-peak elements show clear evidence that they were created by neutron addition processes. Such neutron buildup can take place quiescently, in the interior of a red giant star, or it may happen very rapidly in a stellar explosion. When a stellar explosion or supernova occurs, some of the material may remain locked up in a remnant, such as a neutron star or a black hole. The remainder of the material is returned, chemically enriched, to the interstellar medium.

The surface chemistry of lower main-sequence stars must reflect the composition of the interstellar gas from which they formed. Red giants often have surface compositions palpably different from lower main-sequence dwarfs, and this is generally attributed to mixing of the products of nucleosynthesis from the stellar interiors to their surfaces by convective currents or to mass transfer from evolved binary companions.

Upper main-sequence stars exhibit a bizarre and variegated chemistry. It is generally believed to be the result of a chemical separation (§13.7). In the more massive stars, we see instances where the outer layers have been stripped away by prodigious winds. This mass loss is sometimes so great that the results of interior nuclear burning can be seen at the stellar surfaces (Michaud and Tutukov 1991).

Space experiments have made it possible to measure relative abundances in particles that have come from the solar atmosphere, either in the solar wind or in the more energetic solar cosmic rays. These abundances show strong evidence for fractionation which may be the result of plasma processes. The still more energetic galactic cosmic rays show different fractionations that can be accounted for, at least in part, by spallations resulting from encounters with the interstellar gas. Solar and galactic cosmic rays will be discussed in §§7.3, 10.5 and 10.6

The chemistry of diffuse matter in space has been studied by a variety of techniques ranging from radio and millimeter wavelengths to γ-rays. In our own Galaxy it is thought that a sizable fraction of the heavier elements are in the solid phase in the form of grains.

The great preponderance of luminous matter in the universe occurs in stars. Some ten percent or less of the luminous universe is in the form of gas that may be observed at optical or radio frequencies. Modern astronomers are now very much concerned with *dark* or *missing matter* that is inferred from dynamical considerations in extragalactic systems. This material in principle lies within the domain of cosmochemistry, and current ideas recognize the possibility that it may comprise the major fraction of the mass of the universe. At present, there are mainly theoretical speculations concerning the nature of this material upon which we shall touch only briefly (see Chapter 16).

1.6 Two Approaches to Cosmochemistry

The present discussion will focus on the chemical history of material objects in the universe. We will begin with the materials closest at hand, those found in the earth's surface, and proceed to discuss the Moon rocks, meteorites, and planets. Our sun, a typical star, is unique in the power of analytical tools that may be employed in the study of its chemistry. We shall discuss other stars and stellar systems in turn, and devote some attention to such matters as the cosmic rays and interstellar matter.

It is often useful to focus on the individual chemical elements, and to discuss each of them in various cosmical settings. This was the approach taken by Victor Goldschmidt (1954) in his great study *Geochemistry*, as well as by Mason (1971) in his *Handbook of Elemental Abundances in Meteorites.*

In principle, a complete discussion by either method would contain the same material, but in practice what is often wanted is so widely scattered throughout a work that one is grateful for reference books taking the other approach.

A few of the books in the references use the element-by-element technique. The standard work for astronomical spectroscopy of Merrill (1956) is sorely in need of revision. Fortunately, a new work by C. and M. Jaschek (1995) should fill this gap.

2

Minerals: An Introduction to the Nomenclature and Chemistry

2.1 Introduction

The chemical composition of matter is often the result of factors which, at first, are not at all obvious. Consider the following examples. In certain very stable stellar atmospheres a separation of the chemical elements can take place. One might think that the heavier elements would be the first to sink with respect to the abundant hydrogen, which forms the bulk of (most) stellar matter. In the earth's upper atmosphere, for example, there is a region where the heavy species sink. The number density of a given species roughly follows the law for an isothermal atmosphere, $N \propto \exp(-z/H)$, where z is the altitude, and H the *scale height*, $H = \mathscr{R}T/g\mu$. (See the Index for the meaning of symbols not explained in the text.) Thus molecules such as O_2 and N_2 are concentrated at low altitudes relative to atomic hydrogen and helium.

For the stars in question, the situation is not so simple. There is a competing, upward force due to radiation pressure that can overwhelm gravity. Given time, exotic heavy elements such as mercury or platinum can be pushed up from the envelope of a star and concentrated in the photosphere, where they may be revealed by spectroscopy. In these stars, the abundant, light elements have a tendency to sink! We shall discuss this counterintuitive process in some detail in Chapter 13, since much of the writer's own research has been concerned with its observational consequences.

We now mention some peculiarities of the chemistry of the earth that are also counterintuitive. In order to understand the phenomena, we must learn something of the physical and chemical properties of minerals.

The gross divisions of the earth are the core, on the one hand, and the mantle and crust on the other. These may be understood entirely on the basis of the high density of the nickel–iron mixture that forms the core, and the rocky materials of the mantle and crust. During an early phase of the earth's history, the dense nickel–iron sank to form the present core. If one looks at the detailed chemistry, the situation is much more complex.

In the earth's crust, the element potassium, for example, is much more abundant than it is in the SAD. This unusual abundance may be attributed to the *size* of the potassium ion (K^+), rather than the density of the element. The ion has a relatively large size that prevents it from being easily incorporated into the crystalline structures that form the bulk of the earth's mantle. Thus whenever some of the mantle material is partially melted (§2.6), the resulting magma becomes enriched in potassium because the ion is more loosely bound than some of the others (e.g., Fe^{2+}, Mg^{2+}) that are more typical of mantle minerals. Magma is usually carried upward, and the potassium is eventually frozen. It typically goes into the mineral orthoclase ($KAlSi_3O_8$).

All of the noble gases are depleted in the earth, but there is much more of the isotope ^{40}Ar than one would expect. This ^{40}Ar has come primarily from the decay of radioactive potassium, ^{40}K. It is therefore impossible to understand why the earth has its present content of argon without knowing something of the chemical history of potassium. The abundance and distribution of argon in the earth are determined, in part, by the characteristics of a completely different element.

The location of trace elements in the earth's crust depends primarily on the ability of the ions to be accommodated within the crystalline structures of major minerals. On the earth, as well as on the Moon and terrestrial planets, the radioactive elements uranium and thorium are enriched in the crust. This is because they are squeezed out of the major minerals of the mantle, just as potassium was. Indeed, it is typical to find enhancements of uranium and thorium in minerals and rocks that are also rich in potassium. The detailed history of these and other trace elements depends very much on the mineralogical properties of their hosts.

It is clear that we cannot get a fundamental understanding of the chemical history of matter unless we know something of the properties of individual chemical compounds, especially the minerals.

There are many instances in cosmochemistry where the mineralogy is unknown or uncertain, even though something is known of the chemical composition. The significance of this lack of information will become clear as we proceed. One of the most frustrating cases concerns the composition of the Martian surface. The X-ray fluorescence spectrometers carried by the Viking landers did not allow a direct determination of the mineralogy. Compositions that one may find, for example, in textbooks, for the Martian surface were derived largely through modeling techniques – blunt analytical tools compared with the methods used in the analyses of meteorites and Moon rocks.

It would be useful to have cosmochemical references that focused on the individual chemical compounds and minerals. Such a study would list the minerals one by one, and describe their properties and natural occurrences. (We described a similar work for the elements in §1.6.) This is not generally done. Secondary sources, especially, have a proclivity to report the chemical analysis and leave the reader in the dark about the possible influence of the mineralogy. Useful references which do focus on minerals are Deer, Howie, and Zussman's (1966) *An Introduction to the Rock-Forming Minerals*, and Frondel's (1975) *Lunar Mineralogy*. The *Lunar Sourcebook* (Heiken, Vaniman, and French 1991) contains an excellent article on lunar minerals by Papike, Taylor, and Simon (1991).

We end this section with a final example of chemistry and mineralogy in cosmochemistry. It has been known since the early work of H. N. Russell (1934) that the very high dissociation energy of the CO molecule would cause it to dominate the molecular chemistry of cool stars. Whether a star becomes a member of the carbon (R, N) or oxygen (M) sequences can depend on very subtle changes in the carbon-to-oxygen ratio. If the abundance of carbon is slightly less than that of oxygen, the CO molecule may be assumed to tie up all of the carbon, leaving only oxygen to form other molecules, such as TiO, or VO. When carbon is even slightly more abundant than oxygen, these molecules may not be seen at all, while the stellar spectra may be rich in bands of CN and C_2.

In the formation of solid particles, the C/O ratio will also play a dominating role. If oxygen is more abundant than carbon, we may expect oxides and silicates to form, as in the models of the condensing gas of the solar nebula. However, if carbon is more abundant the mineralogy is expected to be completely different, with carbides forming in abundance.

2.2 Mineralogy

Minerals are naturally occurring solid substances. The qualifications are usually added that they must be formed by inorganic processes, have a definite chemistry, and an ordered atomic arrangement. We shall not stress the latter qualifications. Cosmochemists are certainly concerned with some amorphous materials occurring in meteorites, and it is a minor concern whether such matter technically qualifies as a mineral. There is, of course, the practical consideration that we are unlikely to gather much information about amorphous materials in a standard mineralogy text!

The language of mineralogy may be unnecessarily turgid, but we need not deal with its full complexity here. All together, there are several thousand known minerals, but *exactly* how many there are depends as much on the nomenclature as nature itself. Nickel and Nichols' (1991) useful reference lists some 3800 mineral names along with brief descriptions. Minerals are typically grouped into *families*, and it is always useful to know if a given mineral name is a general one (e.g., oxide, or silicate), or more specific (e.g., corundum or forsterite, see below).

We present a simplified classification here, of minerals that are important for an understanding of the bulk chemistry of the earth's mantle, Moon rocks, and meteorites. Refinements will be introduced later as they are needed. The classification is outlined in Table 2.2.

A number of the physical properties of minerals are worth mentioning. A field geologist may make a preliminary identification of a mineral in a rock sample by considering the hardness or the color of the mineral. A diagnostic technique often used is to make a scratch with the mineral on a piece of porcelain and note the color of the "streak." Likewise, one may examine the "luster," a word that is used to describe the reflected light from the mineral. These properties are discussed in standard works on mineralogy (see Berry and Mason 1959).

Definitive mineralogical analyses are made with the help of *thin sections* and an optical microscope. A thin slice of a rock or mineral is made with a diamond saw, and the sample is then mounted on a glass slide. It is ground and eventually polished with successively finer abrasive materials until it is transparent. The resulting thin sections are then examined with the help of a special microscope fitted with polarizing optics. One skilled in this technique, called *optical mineralogy*, may then identify the minerals and determine their relative percentages in a rock sample.

The diagnostic techniques of mineralogy will be a minor concern for us. However, we will be keenly interested in the density, melting points, and condensation temperatures of minerals.

2.3 Some Useful Concepts from Crystallography

Before we describe the minerals in more detail, it will be useful to introduce some terminology from that branch of solid state physics known as crystallography. Only a brief discussion will be given here. Additional details may be found in physical chemistry texts, such as that of Castellan (1983), or works on solid state physics, e.g., Kittel's (1971) text.

Let us first define what is meant by a lattice. A lattice is a collection of points in space with a specific relationship to one another. Let **a, b**, and **c** be three linearly independent vectors, whose origin is some common, arbitrary point (see Figure 2.1(a)). We shall call them *basis* vectors. In general, they will not be orthogonal, and will have unequal lengths. The lattice will be defined as the collection of all points $n_1\mathbf{a} + n_2\mathbf{b} + n_3\mathbf{c}$, where n_1, n_2, and n_3 are integers, which may be positive, negative, or zero. The set of all such points will define an infinite lattice. Obviously, crystals in nature are not infinite in size. Nevertheless, in any macroscopic sample, the most elementary patterns are repeated so many times that it is a useful idealization to imagine them to be infinite.

A collection of lattice points is not sufficient to define the structure of a crystal. In order to do this we must in general associate additional points with the vectors **a, b**, and **c**, that is, we need points other than the origin and end points of these vectors. These additional points will specify the locations of the ion centers that make up the crystal. We shall call this collection of ion centers a *basis*.

If there are $\mathcal{N}$ ions in a basis, their locations may be specified by vectors of the form $x_i\mathbf{a} + y_i\mathbf{b} + z_i\mathbf{c}$, where the subscript i runs from 1 to $\mathcal{N}$. We shall assume that it is always possible to describe the crystal structure with coefficients x_i, y_i, and z_i that are always between zero and unity, so that no ion of the basis extends beyond the parallelepiped defined by the basis vectors **a, b**, and **c**. The crystal structure is now built up by iterating the basis in space, that is, by choosing all possible integers n_1, n_2, and n_3 to reach all of the lattice points (see Fig. 2.1(b)).

In general, there may be any number of positive and negative ions in a basis, but for the minerals with which we shall be most frequently concerned, the number will often be less than a dozen or so. There will commonly be different numbers of positive and negative ions, and in these cases, charge neutrality can be preserved because the ions represent different valence states.

By definition, the crystalline pattern would be unchanged by the linear displacement of all of the lattice points by $n\mathbf{e}_k$. Here, n could be any integer, and $\mathbf{e}_k$ could be **a, b**, or **c**. The lattice structure is said to be *invariant* to such translations. The set of all such transformations to which a given (idealized, infinite) lattice structure is invariant forms a mathematical group. For example, the transformations $n\mathbf{e}_1$ form a one-to-one correspondence with the group whose elements are the positive and

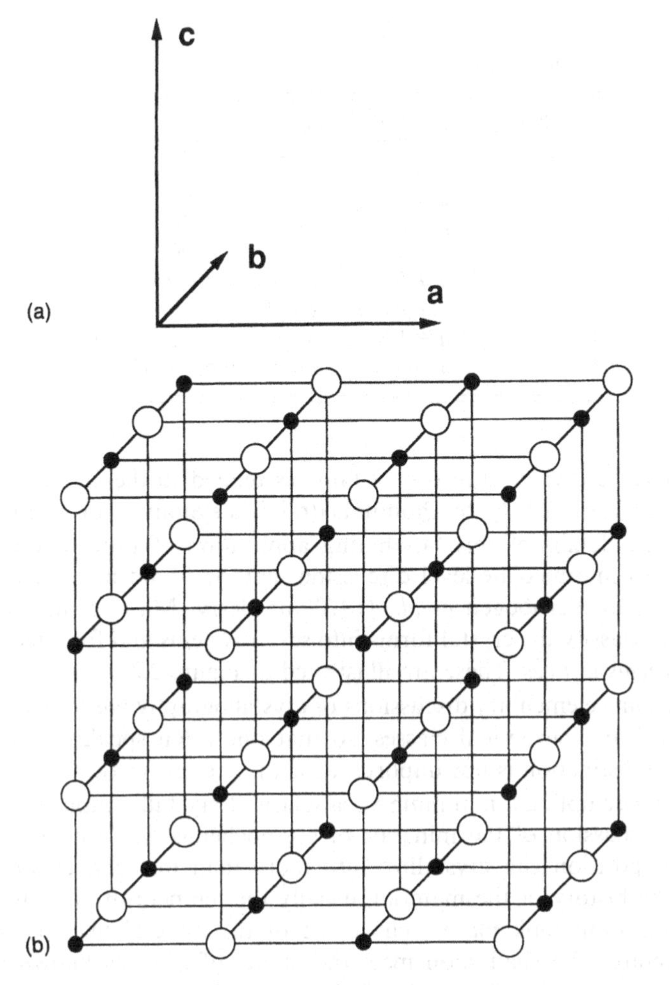

Figure 2.1. (a) Basis Vectors of a Lattice. (b) The Halite Lattice (NaCl).

negative integers including (the identity) element zero where the rule for combination is addition.

Rotations and reflections may also be applied to a given lattice structure. Such transformations leave at least one point fixed, so they are called *point transformations*, and the associated groups are therefore called *point groups*. It is relatively easy to show, although we shall not do so here, that among the purely rotational point groups, only those with 1, 2, 3, 4, and 6 members are relevant for crystal structure.

One may classify the lattice structure of minerals according to the different groups of transformations to which the pattern is invariant. If we restrict ourselves to cases which obey the *law of rational indices* (see Hammermesh 1962), only 32 point groups are of interest. Essentially, this "law" restricts us to the case of crystal structures that can be built up by the iteration of a basic structure using integers n_1, n_2, n_3. The

Table 2.1. *The Crystal Systems*

System	Number of lattice types	Description of sides	angles
Cubic	3	$a = b = c$	$\alpha = \beta = \gamma = 90°$
Tetragonal	2	$a = b \neq c$	$\alpha = \beta = \gamma = 90°$
Orthorhombic	4	$a \neq b \neq c$	$\alpha = \beta = \gamma = 90°$
Monoclinic	2	$a \neq b \neq c$	$\alpha = \gamma = 90° \neq \beta$
Triclinic	1	$a \neq b \neq c$	$\alpha \neq \beta \neq \gamma$
Trigonal	1	$a = b = c$	$90° \neq \alpha = \beta = \gamma < 120°$
Hexagonal	1	$a = b \neq c$	$\alpha = \beta = 90°, \gamma = 120°$

32 point groups may be divided into 6 or 7 families related to the crystal systems discussed below. (See Table 2.1. Note 'rhombohedral' is a synonym for 'trigonal.')

We must leave the details of this fascinating application of pure mathematics to the references. An intermediate-level discussion may be found in the advanced general geology text by Verhoogen *et al.* (1970) or Berry, Mason, and Dietrich (1983). It is usual to classify the crystal forms into seven systems and fourteen lattice types known as *Bravais lattices.* These are illustrated in Figure 2.2.

It is common in some elementary discussions of crystallography *not* to distinguish between the trigonal and hexagonal classes, so that one often speaks of only six crystal systems. The distinction is not important for the present work.

We have finished assembling a minimum vocabulary of crystallography necessary to allow a simple discussion of the minerals of cosmochemistry. Minerals with a given chemical composition can crystallize into more than one crystal structure. This depends on the history of the material, usually the temperature and pressure under which solidification took place. The name of the mineral may be related to the crystal structure. We shall soon meet the family of silicates known as the pyroxenes. One often sees orthopyroxene or clinopyroxene, depending upon whether the crystalline structure is orthorhombic or monoclinic. Similarly, there is a family of pyroxenoid silicates, which have analogous chemistry to the pyroxenes, but which belong to the triclinic family of crystals.

2.4 A Simplified Mineral Classification

We shall follow Berry and Mason's (1959) classification of the minerals into eight classes. We give only a brief description of the classes here. Additional material for mineral classes of importance for cosmochemistry is given in the following section. The classic work on minerals is that of Dana (1985); as of this writing, the work is in its twentieth edition.

In the list below, the subclasses that are of lesser importance for cosmochemistry are given in parentheses.

1. *Native elements*: Although they are not pure elements, the Nickel-iron alloys are

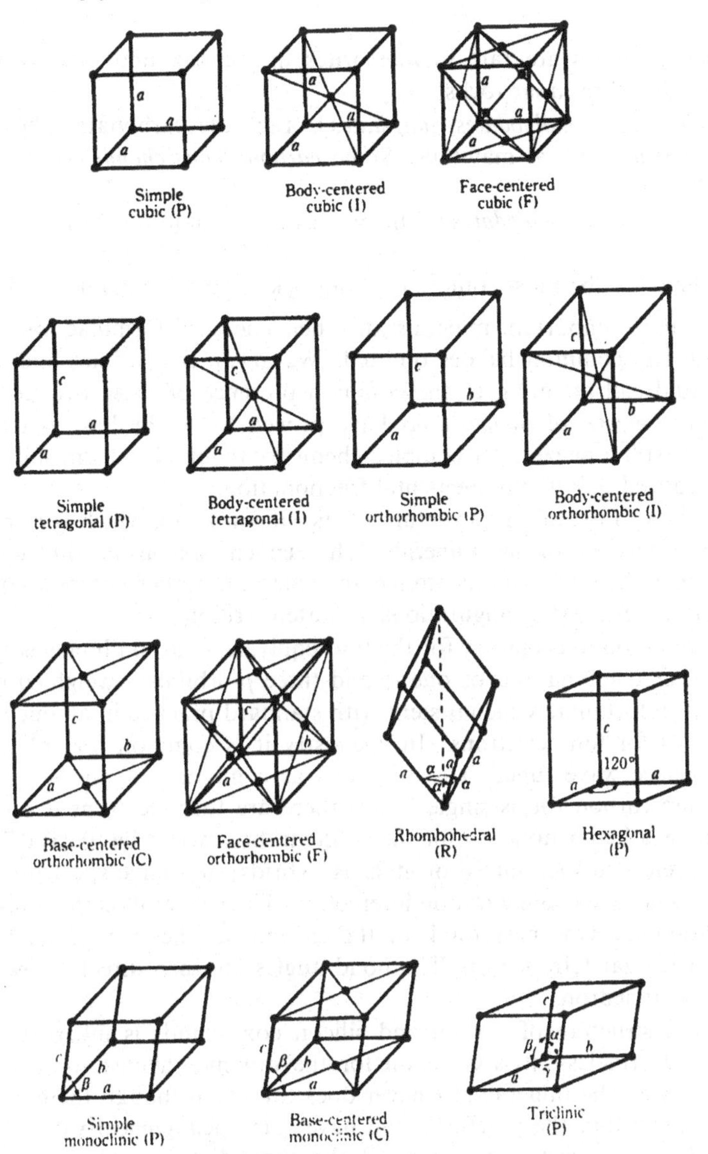

Figure 2.2. The Seven Crystal Systems and Fourteen Bravais Lattices.

included by Berry and Mason among the native elements. These minerals are common in meteorites.

2. *Sulfides.* FeS or troilite is an important mineral in meteorites.
3. *Oxides (and hydroxides).* Several important oxides will be mentioned in the following section. Hydroxides are of minor importance in extraterrestrial materials.
4. *Halides.* Halides are relatively unimportant in extraterrestrial materials, as far as we know.
5. *Carbonates, (Nitrates, borates, and iodates).* Certain meteorites contain carbon-

ates. The carbonates are rare in meteorites in general, and only very small amounts are found in lunar rocks.

6. *Sulfates, (Chromates, molybdates, tungstates).* Like the carbonates, the sulfates are rare in extraterrestrial materials. Some *carbonaceous chondrites* (see below) contain sulfate minerals.
7. *Phosphates, (arsenates, vanadates).* Phosphates are common in Moon rocks and meteorites.
8. *Silicates.* These are the most common of the rock-forming minerals.

Some of the more important minerals and their chemical formulae are listed in Table 2.2, with less common halides omitted. We group the sulfates and carbonates together and call attention to the common presence of these two families in meteorites whose *elemental compositions* have proven to be the best guide to the SAD (§1.3). It is still a puzzle how complex chemistry required to form sulfates and carbonates produced only minor elemental fractionation.

Let us now turn to the atomic structure of the silicon atom, which is responsible for the complex class of silicate minerals. The element silicon lies directly below carbon in the periodic table, and its atomic and chemical properties are accordingly similar. Both outer electron configurations are often written s^2p^2.

A common notation in chemistry for the four equivalent carbon bonds is sp^3. The orbitals are linear combinations of one *s*- and three *p*-orbitals – weighted equally. The notation is unfortunately inconsistent with standard practice in atomic physics, where sp^3 stands for four electrons. In chemistry it is common for sp^3 to mean *one* electron, whose wave function is described as a mixture of one *s*- and three *p*-orbitals. When carbon forms single bonds there are four electrons described in this way. Perhaps a better notation for the configuration would be $(0.25s\ 0.75p)^4$.

The chemist refers to such mixed orbitals as hybrids; in atomic spectroscopy the phenomenon is known as configuration interaction. The net result is the well-known equiangular bonding. One may think of the carbon or silicon atom as being at the center of a regular tetrahedron. The bond angles are then directed toward the vertices of the tetrahedron.

The molecular structure of carbon and silicon compounds is inherently three-dimensional. Nevertheless, it is common for the organic chemist to write "flat" structural formulae. The mineralogist never does this, even though there are times when it would seem that the practice could have some pedagogical value. Instead, one typically finds perspective drawings of the three-dimensional structures (see Fig. 2.3).

The silicates may be classified into six groups, according to the scheme outlined in Table 2.3 (below). It is common that a single mineral name may apply to a limited range of compositions. This is illustrated by the rather complex formula shown in the footnote to Table 2.3 for the mica biotite.

2.5 The Common Minerals of Cosmochemistry

In this section we describe additional minerals that are common in cosmic materials, or which are stressed, for one reason or another, in the literature on cosmochemistry.

Table 2.2. *Some Mineral Classes of Importance for Cosmochemistry*

Class	Examples		Remarks
	Name	Formula	
Native elements	kamacite taenite	Fe(-Ni) Ni-Fe	Alloys: reduced phases of meteorites
Oxides	corundum ilmenite perovskite	Al_2O_3 $FeTiO_3$ $CaTiO_3$	"refractory" oxides are among the first to condense from the gas phase
Sulfides	troilite	FeS	This dense mineral may control the bulk density of some terrestrial planets
Carbonates, sulfates, hydroxides	calcite epsomite	$CaCO_3$ $MgSO_4 \cdot 7H_2O$	The appearance of these minerals is an indication of a volatile-rich past.
Phosphates	apatite whitlockite	$Ca_5(PO_4)_3Cl$ $Ca_3(PO_4)_2$	These minerals are frequently enriched in lanthanides and the actinides uranium and thorium.
Silicates	silica forsterite	SiO_2 Mg_2SiO_4	The most common of the rock-forming minerals

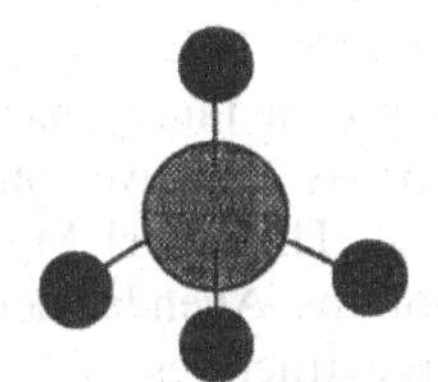

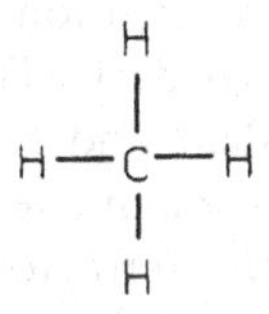

Figure 2.3. Tetrahedral Bonds and Flat Structural Formulae. On the left, the structure could represent SiO_4 or methane, CH_4. In mineralogy books, one typically sees perspectives of tetrahedra for SiO_4. This kind of bonding is often called sp^3.

While this list may seem something of a burden to one who spurns memory work, the rewards are worth the effort. We discuss first several oxides, and then three silicate families, namely, the olivines, the pyroxenes, and the feldspars. Two other silicate families will be mentioned, the amphiboles and the micas. The latter minerals are common in the earth's crust, but they are rare in lunar rocks. They are produced by geochemical processes that are virtually absent on the Moon. We briefly discuss some of the more complex mineralogy that is found in terrestrial materials exposed to liquid water, or more generally, to weathering.

Table 2.3. *The Silicates*

Name	Description	Unit	Example: Mineral	Example: Formula
Nesosilicates	Isolated tetrahedra	SiO_4	Olivine	Mg_2SiO_4
Sorosilicates	Two tetrahedra sharing one oxygen	Si_2O_7	Melilite	$Ca_2MgSi_2O_7$
Cyclosilicates	Rings of tetrahedra sharing two oxygens	$(SiO_3)_n$ n=3,4,6	Beryl Benitoite	$Be_3Al_2(SiO_3)_6$ $BaTi(SiO_3)_3$
Inosilicates	Single chains or double chains	$(SiO_3)_\infty$ Si_4O_{11}	Pyroxenes Amphiboles	$MgSiO_3$ $Mg_7Si_8O_{22}(OH)_2$
Phyllosilicates	Sheets of tetrahedra	Si_4O_{10}	Micas	[a]
Tektosilicates	Three-dimensional framework	SiO_2	Quartz Feldspars	SiO_2 $CaAl_2Si_2O_8$

[a]biotite mica: $K_2(Mg, Fe^{2+})_{6-4}(Fe^{3+}, Al, Ti)_{0-2}[(Si, Al)Si_5Al_2O_{20}](OH, F)_4$

2.5.1 *Oxides*

Corundum, Al_2O_3, is a common refractory oxide, that is, it is one of the first solids to form when a gas with the solar composition cools. Other refractory oxides are perovskite, $CaTiO_3$, and the spinels. Perovskite is a family name for minerals with the same structure as $CaTiO_3$. A magnesian–silicon perovskite, which has recently been the subject of much attention (Hemley and Cohen 1992), should *not* be confused with the refractory oxide $CaTiO_3$. Spinel is also a family name for an oxide that contains one divalent and two trivalent cations (positive ions). A very common terrestrial spinel is magnetite, $Fe^{2+}(Fe^{3+})_2O_4$. The spinel $MgAl_2O_4$ is a major mineral in some of the refractory inclusions of the Allende meteorite (see below). These inclusions often contain relatively large fractions of "foreign material." By this, we mean matter whose isotopic composition is distinct from that of the bulk of solar system material. Ilmenite, $FeTiO_3$, is one of the major minerals of certain of the lunar rocks; it is far less common on the earth.

Silicon dioxide, SiO_2, or silica is an extremely common terrestrial mineral family, comprising the common quartz, tridymite, cristobalite, and several other minerals including the amorphous silica glass. Tridymite and cristobalite are high-pressure forms of silica; small fragments are found in lunar rocks and they are almost certainly the result of shock heating during meteoroid impact.

Euhedral (beautiful-angles) or crystalline quartz is commonly found in terrestrial museums. In the usual grubby rocks that a casual collector can find, the quartz may often be recognized because of its lack of reflecting surfaces. Fractures of quartz are often conchoidal, that is, curved like the geometrical figure (conchoid: $r = b \pm a \cdot \sec(\theta)$).

In general, the percentage of silica is a measure of the complexity of the history of a terrestrial rock. Materials that have been repeatedly melted or partially melted tend to be rich in quartz. Quartz is rare in Moon rocks, meteorites and primitive materials.

2.5.2 The Olivines

The most important olivines are fayalite, Fe_2SiO_4, and forsterite, Mg_2SiO_4. These are dense minerals, with relatively high melting temperatures. They may also be considered refractory, although they do not condense at as high a temperature as some of the oxides. The olivines are common in "primitive" solids, such as unweathered igneous rocks that originated deep within the earth, or in some of the "Holy Grail"-candidate meteorites.

Fayalite and forsterite form a solid solution. One rarely finds a mineral that is either pure Mg_2SiO_4 (Fo) or pure Fe_2SiO_4 (Fa). It is becoming common to describe the composition of an olivine by giving the percentage of Fa or Fo. Those who enjoy nomenclature may delight in the special names for ranges of composition between the two "end members" of this series. Within this scheme, one would refer to a mineral whose composition was 90–100% Fa as fayalite, but if the composition range were 70–90% Fa the term ferrohortonolite could be used.

Many geologists now feel that the proliferation of names in mineralogy and petrology has been counterproductive, and they strongly advocate descriptions in terms of percentages of end members rather than burdensome names which have little or no mnemonic value. However, especially in the case of the pyroxenes (see below), some of the names which describe specific composition ranges are in such common use that it is helpful to learn some of them. We will even find that these special names have some mnemonic value, since some of the meteorite type names are derived from the special composition range of the pyroxene which they contain.

2.5.3 The Pyroxenes

The solid solution of fayalite and forsterite may be described by drawing a horizontal line, with Fa at one extreme and Fo at the other. It is useful to introduce the pyroxenes with the help of a two-dimensional extension of this sort of scheme. The ternary diagram shown in Figure 2.4 is from Berry and Mason's (1959) *Mineralogy*. Of the three end members shown, only enstatite ($MgSiO_3$) and ferrosilite ($FeSiO_3$) are pyroxenes. The upper part of the triangle (ca. 50%), with the vertex at Wollastonite ($CaSiO_3$) belongs to the family of pyroxenoids. The distinction of mineral family names arises from the crystal structure, which is very different for the two families. The pyroxenoids crystallize in the triclinic system, while the pyroxenes may crystallize in either the monoclinic or the orthorhombic system.

The chemical formula for enstatite may be written as either $MgSiO_3$ or $Mg_2Si_2O_6$. The latter form is useful as a mnemonic, since a modest change gives the formula for diopside, $CaMgSi_2O_6$. Moreover, one may remember the formula for enstatite by thinking of adding SiO_2 to the formula for olivine.

There are only certain ranges of the pyroxene field in which miscible, solid solutions form, and this depends on the cooling history of a parent melt. There is in general a miscibility gap between the calcium-rich and calcium-poor pyroxenes.

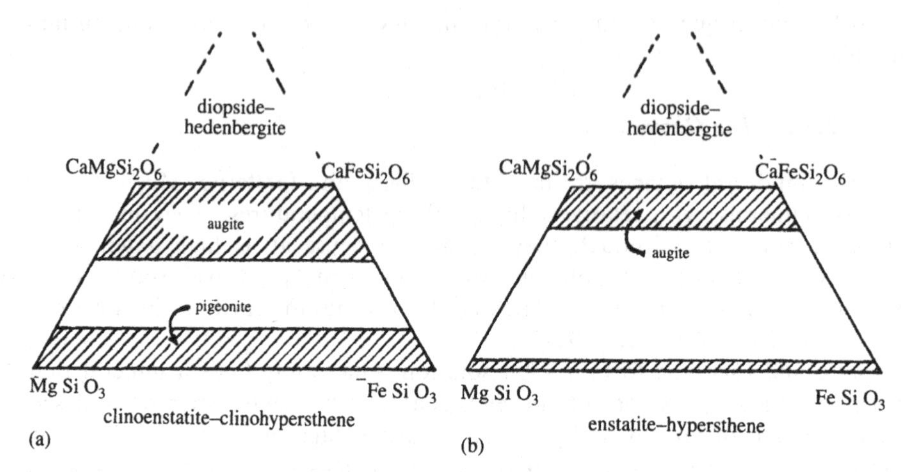

Figure 2.4. The Pyroxenes. From Berry and Mason's *Mineralogy* ©1959 W. H. Freeman and Company. (a) High temperature. (b) Medium temperature.

This gap is wider if the minerals cooled relatively slowly. The calcium-rich pyroxenes are monoclinic, while the calcium-poor ones are orthorhombic. However, when a melt with a composition near the $MgSiO_3$–$FeSiO_3$ range cools rapidly from a temperature above some 1325 K, the solid crystals will belong to the monoclinic class. It is common to indicate this in the mineral name, by referring to *clinoenstatite.*

When the pyroxenes have cooled rapidly from higher temperatures, the miscibility gap is somewhat smaller than in the case where the calcium-rich and calcium-poor pyroxenes crystallize in different systems. These relationships show how the mineralogy of a rock (or meteorite) can give some insight into its history.

If material with the composition $Fe_2Si_2O_6$ cools, it will, on solidification, form a mixture of SiO_2 and Fe_2SiO_4, silica and fayalite. Consequently, the miscible solid solution of the calcium-poor pyroxenes does not extend all the way to ferrosilite. However, some lunar rocks have been found with compositions very close to the ferrosilite vertex. The material has been called pyroxferroite, and is sometimes described as a new mineral, discovered in the Moon rocks.

Augite is perhaps the most common of the terrestrial pyroxenes. If there is a dark pyroxene in a terrestrial igneous rock, the chances are excellent that it is augite. This pyroxene is much less common among the lunar rocks and meteorites, for reasons that are not clear.

Not all pyroxenes fall within the fields illustrated by Figure 2.4. Other possible cations are Mn, Cr, Al, Na, Li, and even Y. The nomenclature may be avoided, although it is sometimes amusing. (We have enjoyed the name spodumene: $LiAlSi_2O_6$.) A rare pyroxene name has often been used in discussions of the mineralogy of meteorites with isotopic anomalies. We must mention it here, fassaite, $Ca(Mg, Al, Ti)(Si, Al)_2O_6$. The parenthesized elements indicate that fassaite may be thought of as occupying a field in a multi-dimensional space, whose extent depends on the percentage of the parenthesized elements.

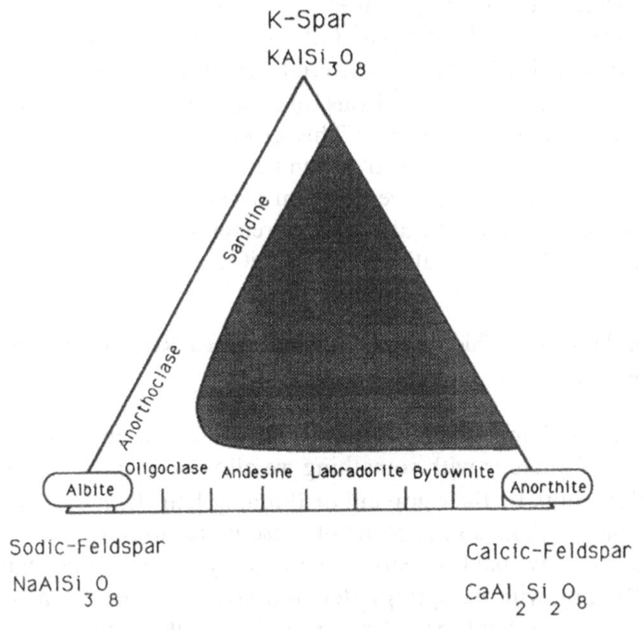

Figure 2.5. The Feldspar Ternary Diagram (schematic).

2.5.4 *The Feldspars*

The feldspars may again be depicted on a ternary diagram, as shown in Figure 2.5. They are the most abundant of all minerals in the earth's crust. One of the end members, anorthite, $CaAl_2Si_2O_8$, is the predominant mineral in rocks of the lunar highlands, and it remains a major mineral in virtually all lunar samples. Anorthite has a higher melting temperature than some of the pyroxenes, and it is one of the early-forming solids in the condensation sequences for the solar nebula.

Albite ($NaAlSi_3O_8$) forms a continuous miscible solid solution with anorthite, and the field at the base of the ternary diagram of Figure 2.5 is given the name plagioclase. Some of the specialized names given by mineralogists to the compositions between albite and anorthite are shown. The more modern way to specify these compositions is by giving the percentage of one of the end members. Thus, for example, albite itself might be Ab_{100} to Ab_{90}, and bytownite Ab_{30} to Ab_{10}.

The plagioclase feldspars can often be recognized in freshly broken hand specimens by the "striations" on the crystal faces. These striations are the result of a particular kind of intergrowth of the feldspar crystals known as twinning. The Apollo 15 astronauts recognized plagioclase feldspar as the dominant mineral in the so-called "Genesis Rock" by the striations. The name arose because it was recognized that it would be primarily highland material that should be very old. In fact, this rock solidified roughly 0.5×10^9 years after the time we think the Moon formed.

While the plagioclase feldspars mix freely, the same is not true of mixtures of albite and orthoclase ($KAlSi_3O_8$, often called K-feldspar). The albite and K-feldspar will often exsolve, forming a striped pattern known as perthite. Note that anorthite and K-feldspar do not form solid solutions. Indeed a large portion of the feldspar ternary diagram is unoccupied by natural materials (shaded area of Fig. 2.5).

A useful mnemonic, at least for the silicons and the oxygens of the feldspars, is to think of adding SiO_2 to the pyroxenes. This gives us the Si_3O_8 that we need for the K-feldspar and albite. We have to remember that the numbers of silicons and aluminums have to change when we go from albite to anorthite, in order to accommodate the divalent calcium. We also have to remember to write the pyroxenes with two cations, that is, $Mg_2Si_2O_6$ rather than $MgSiO_3$.

2.6 Mineralogy as a Clue to History: The Bowen Reaction Series; Paragenesis

The mineralogy of natural materials provides a flawed but useful guide to their history. We draw a brief analogy with something we all understand – human aging.

We recognize older adults by their graying or thinning hair, their increased girth, crow's feet around the eyes, and so on. None of these indications are perfect guides – some people become gray, bald, or stout prematurely. The septuagenarian US President, Ronald Reagan, with his springy step and non-gray, hirsute splendor was exceptional. However, typical bank presidents do not look like high school athletes, and most of us take such things into account in our daily interactions with other people.

The cosmochemist has similar guides. Rocks which originate from the deep earth materials are typically composed of characteristic minerals such as olivines and pyroxenes. As a result of "typical" processes that affect such rocks, the mineralogy changes. Partial melting (see below) and weathering, due to the exposure to water and its dissolved corrosive substances, are the main causes of these modifications.

There is a typical pattern followed by the mineralogy that was recognized by the American geochemist N. L. Bowen (1928), whose *The Evolution of Igneous Rocks* is a classic. Bowen discussed what he called a *reaction principle*, which governs the typical sequences of minerals that appear in the normal evolution of rocks. Characteristic minerals along this sequence constitute what is called the *Bowen reaction series*.

We may start with a consideration of olivine, which is thought to be a major constituent of the earth's mantle. Rocks that are extruded through faults or fissures in the earth's crust (see Chapter 3) are often dominated by olivine. Olivine will react with a liquid rich in SiO_2 to form the pyroxene (clino)enstatite. Crystals of olivine are often found in a background rock matrix, because the relatively high melting point of olivine (forsterite) causes early crystallization. Such crystals, called *phenocrysts* (§3.3), are often *zoned*, because the outer rims of the olivine crystals react with the residual melt. A typical rim of an olivine crystal would contain a layer of pyroxene. Bowen called the olivine forsterite and clinoenstatite a *reaction pair*.

Weather also causes the olivines to alter to a mineral family known as serpentines. The pyroxenes typically weather to amphiboles, and the amphiboles further alter to micas.

Eventually, these and similar minerals are broken down by weathering processes to form clay minerals. At this point the degradation of the rock-forming minerals is essentially complete. The clay minerals, along with organic matter, are the principal constituent of soils. Their chemistry is complex, and we will not write down formulae. However, it is worth noting several names because of speculation that the Martian soils may be dominated by iron-rich clay minerals. Berry and Mason (1959, p.504) give four main families: the kaolin, montmorillonite, clay mica, and chlorite groups. Nontronites of the montmorillonite or smectite group are clay minerals often suggested for the Martian soils.

The sequence of minerals leading from the olivines and pyroxenes through quartz is known as the *discontinuous* Bowen series. This is because the various minerals are all distinct from one another. In a rock, it is possible to point to crystals of olivine that are distinct from those of pyroxene, etc. The olivines, pyroxenes, amphiboles, and micas do *not* form solid solutions. On the other hand, the feldspars *do* form solid solutions, and there results a *continuous* Bowen series which starts with the calcium-rich feldspar, anorthite, passes through albite, and ends with K-feldspar.

The feldspars again degrade into clay minerals, so that both series are shown to end in such materials in Figure 2.6. Bowen's original series ended with quartz. He put it at the junction of both series to indicate an increase in the SiO_2 percentage in rocks composed primarily of "later" minerals in both series. It does not imply, for example, that micas weather to quartz. Bowen's series did not include clay minerals. The present author takes responsibility for this generalization.

The lunar rocks are virtually unweathered, at least by water, and anorthite is one of their major minerals. Calcium-rich feldspar is also typical in those terrestrial rocks that show the least evidence of chemical alteration. However, if we examine the feldspars of rocks that show increasingly greater signs of geochemical alteration, either by weather or partial melting, we will find that the feldspar types become richer first in sodium, and finally in potassium.

We have encountered the term partial melting before, but it is now time to point out that a change in chemistry as well as a change of state, from solid to liquid, is usually implied. In general, if we melt some portion of a rock or meteorite, the composition of both the liquid and the remaining rock will be changed from that of the original rock. Minerals which melt easily will preferentially enter the liquid phase. Whatever chemical elements are predominant in the easily melted mineral will then be enriched in the melt, and depleted in the remaining rock, relative to the original.

In typical processes that take place on the earth, after partial melting, the liquid and the residual solid become separated from one another. The liquid may be forced upward through a fissure in the earth, and erupt as lava, while the solid will be left behind. In this manner, the chemistry of the materials becomes changed or *differentiated* with respect to the original composition. One may refer to magmatic differentiation, for example, as opposed to weathering. Both processes change the chemistry of cosmic substances, and both are important in the Bowen series.

The Bowen reaction series is perhaps the most common illustration of the relationships among aging cosmic materials. These relations are described by the general term *paragenesis*. Deer, Howie, and Zussman's (1966) discussion of the rock-forming

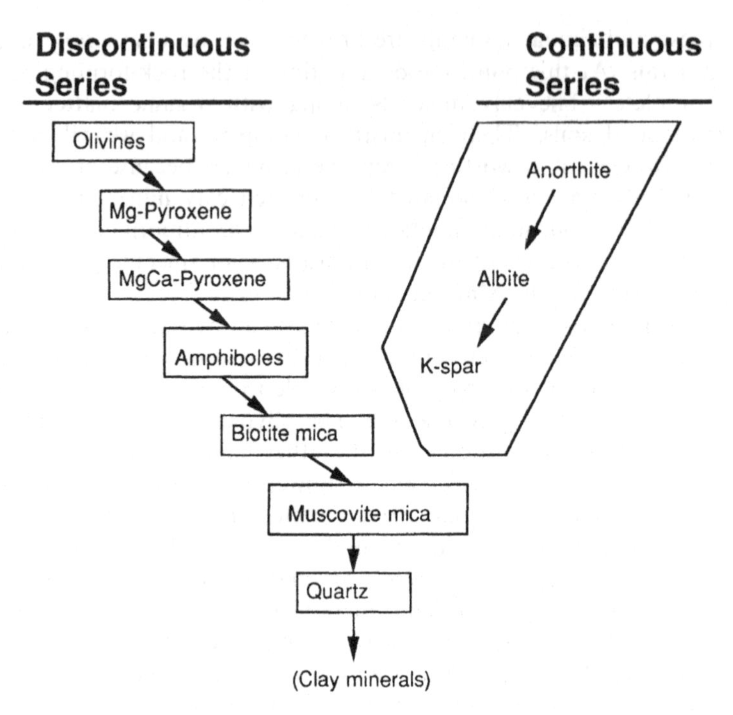

Figure 2.6. The Reaction Series.

minerals includes a separate section on the paragenesis of individual minerals. When we learn that a certain type of meteorite is dominated by the mineral serpentine, it is of considerable interest to read in Deer, Howie, and Zussman, that on the earth, the "principal occurrences of serpentine minerals are in altered [weathered] ultrabasic rocks, e.g., dunites, pyroxenites, and peridotites." Dunites, pyroxenites, and peridotites are (see Chapter 3) terrestrial rocks dominated by the olivines and pyroxenes. The fact that weathering is the principal factor in the transformation to serpentines provides a powerful insight into the history of these particular meteorites. We may understand why most meteoriticists think these meteorites *must* at some time in the past have been in the presence of liquid water.

We now illustrate one possibility for chemical differentiation with the help of a phase diagram for plagioclase feldspar (Figure 2.7). Temperature is plotted as an ordinate, composition as an abscissa.

Consider a mineral fragment with the composition A′. This material is a solid solution of the two end members, and its properties are not identical to those of either pure albite or pure anorthite. If such a fragment is heated, the solid will not melt until it reaches a temperature indicated by the intersection of the line A′–A′ and the curve marked solidus. At this point, the solid still has the composition A′, but the first liquid has a composition enriched in albite, the mineral with the intrinsically lower melting temperature. The composition of the liquid is indicated by the letter A on the curve labeled liquidus. As soon as a finite portion of the

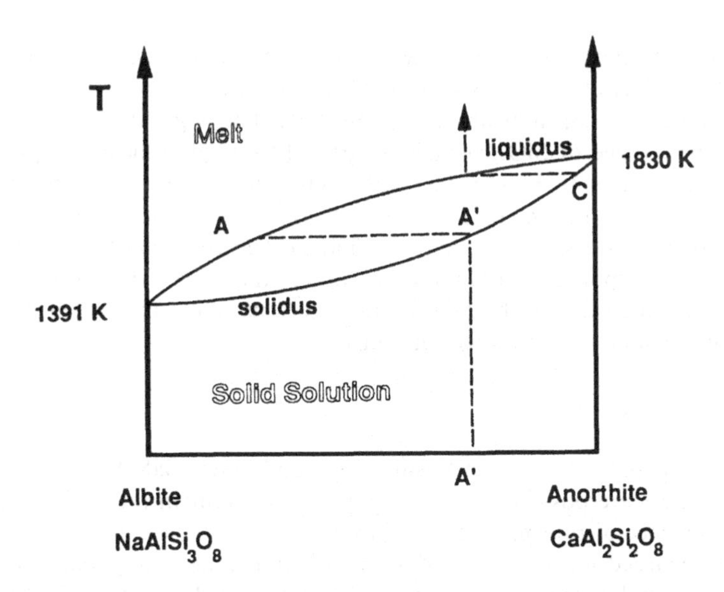

Figure 2.7. Phase Diagram for Plagioclase Feldspar (schematic).

solid is melted, the net composition of the solid must necessarily change. As the temperature rises, and more of the solid is melted, the composition of the residual moves up the solidus curve, indicating enrichment in anorthite. The liquid, provided it remains in contact with the solid, will also become enriched in anorthite. Its composition will be given by a point that moves upward and to the right along the liquidus curve. If one continues raising the temperature, with the solid and liquid always in contact, the last solid to melt will have a composition indicated by the letter C, while the liquid will have the original composition.

If the solid is only partially melted, and the liquid is removed and eventually crystallizes, the new material will be enriched in albite. This is a simple example of how the first steps down the continuous Bowen series may be taken.

Chemical differentiation, by magmatic processes or by weathering, thus leaves its trace in cosmic materials which we may be able to read. Profound inferences are sometimes drawn from quite fragmentary observations. For example, the level of radioactivity measured by the Venera space probes at the surface of Venus indicated to the scientists that the planet was chemically differentiated, like the Moon or the earth. The reasoning went as follows. During partial melting, ions with large radii are preferentially partitioned into the liquid (see the discussion for potassium above). The radioactive ions have large radii, and would be enriched in (liquid) magma. Since these liquids typically have lower densities than their parent rocks, they work their way upward toward the surface of the planet, carrying an enhanced level of radioactivity. The Venera probes measured a greater radioactivity than would be expected from the surface of an undifferentiated planet, and from this it was straightforward to assume the action of magmatic differentiation.

Among terrestrial rocks there are a number of chemical and mineralogical signposts that indicate the complexity of the geochemical history. It will be useful to mention some of them here, along with the caveat that there are important exceptions that will have to be dealt with on an individual basis.

One of the most common indicators of a material that is geochemically mature is the abundant presence of SiO_2, usually quartz. This is of sufficient importance that we have indicated it in Figure 2.6. The chemistry of major elements is also suggested by this figure; the major elements will follow the major minerals. Thus, in geochemically mature rocks, we expect to find diminishing quantities of Fe and Mg, and increasing proportions of Na and K. Among the trace elements, those with large ion sizes will increase in abundance, and the ions that can hide in minerals such as the olivines and pyroxenes will diminish.

2.7 Problems

1. What are the valences of sulfur in troilite (FeS) and barite $BaSO_4$? Explain how these two valences are possible using the position of sulfur in the periodic table.
2. What is the valence of phosphorus in apatite?
3. A negatively charged ion or group of atoms is often called an *anion*, since in a solution, they would migrate toward the positively charged anode. In the formula for biotite mica (Table 2.3) the negative or anionic group $[(Si, Al)Si_5Al_2O_{20}]$ occurs. What are the possible values of the charge of this group?
4. Write an electrically neutral formula for biotite that involves the Ti^{4+} ion, and another using Fe^{3+}.

3
A Brief Introduction to Petrology

3.1 Preliminary Remarks

Petrology is the study of rocks. The prefix, *petro*, derives from a Greek word meaning rock, or stone. Those unacquainted with geology may be surprised to find that this subdiscipline of geology is huge. All aspects of the study of rocks are included. Not only are various rock types, their mineralogy and chemistry a part of petrology, but also the question of which rocks are where on the surface of the earth, and *why*. Cosmochemists, of course, must be concerned with lunar rocks and meteorites as well as their terrestrial congeners. Fortunately, lunar and meteoritic petrology are comparatively simple. A fundamental grasp of terrestrial *igneous petrology* will provide a good introduction to the study of extraterrestrial solids.

The three major divisions of terrestrial rocks are igneous, metamorphic, and sedimentary. Like all classifications these divisions have shortcomings. The boundaries are often ill-defined, and sometimes it is not particularly useful to place a rock in one category rather than another. Igneous rocks are those that have crystallized from a melt or *magma*. Geologists use the term magma to mean primarily liquid material, which may include some already solidified or not yet melted matter as well as dissolved or imprisoned gases. Lava, on the other hand, may be flowing, or long since frozen magma. Magma becomes lava when it erupts on the surface of the earth.

The signature of an igneous rock is usually evident from a freshly broken surface. It is unmistakable to one skilled in optical mineralogy. When the first lunar samples were examined from the Apollo 11 mission, terrestrial petrologists knew a great deal almost immediately about the processes that had gone on on the surface of the Moon. Speculations that the lunar materials might be pristine or SAD-like in composition were quickly extinguished.

Sedimentary rocks have been deposited from a fluid. The fluid is usually water, and the rocks typically occur in layers that may be easily recognized in highway road cuts. The most characteristic sedimentary rocks are limestones and sandstones. There are certain volcanic materials that are called tuffs, which, at least to the novice, can closely resemble limestone layers. The material was blown into the air by volcanic explosions. It eventually settled out, was compacted and *lithified* into the rocky layers where we can find it. While such rocks may contain the calcium carbonate of limestones, the more common minerals are those of the igneous rocks.

One may distinguish such rocks with the help of a drop of weak hydrochloric acid – the acid test! If there is a fizz, the rock is a limestone. Layers of tuff are not usually called sedimentary, although it would be logical to do so.

Igneous or sedimentary rocks that have been substantially modified by temperature and/or pressure are called metamorphic. It is a matter of convenience when one chooses to stop calling a material igneous or sedimentary and start calling it metamorphic. Virtually all of the lunar samples (and many meteorites) have been modified by a process known as *shock metamorphism.* This results from the constant bombardment of the surface of the Moon by meteoroids. Two common metamorphic terrestrial rocks are marbles and gneisses. Marbles are metamorphosed limestones, while gneisses are metamorphosed volcanic rocks.

For the most part, we will not be concerned with the complex petrology of metamorphic and sedimentary rock types. Those wishing to understand the nature of certain meteorites, as well as future returns of rocks from Mars and Venus, may wish to consider metamorphosed terrestrial rocks as a clue to these cosmic samples. A few remarks have already been made concerning the presence of hydrated clay minerals and serpentines in carbonaceous meteorites (see §2.6). We shall discuss such matters again in Chapter 7. We turn now to a discussion of igneous rocks.

3.2 The Classification of Igneous Rocks

The igneous-rock classifications that we shall introduce are based on the mineralogy, chemistry, and texture. Rocks are also described by words that tell something of their history. We have already met an example of the latter; to call a rock a tuff is to imply that it was compacted from material blown out in a volcanic explosion.

These words often have meanings that are interconnected. The history of the material found in a rock will obviously determine its texture and chemistry. As we shall see, many terms have historical overtones that a cautious worker should avoid. The modern trend is to make use of purely descriptive terms when the historical background is uncertain.

The words *mafic* and *felsic* are used to describe the major mineralogy, and the associated chemistry of rocks. Mafic derives from "ma" for magnesium, and "fe" for iron. Mafic rocks are therefore rich in ferromagnesian minerals such as the olivines and pyroxenes. "Felsic" is a word which combines "fel" for feldspar, and "si" for silica, SiO_2. With the help of these two words we can describe most major igneous rock types, and simultaneously say something about their history. If we think of the Bowen series, the mafic rocks are made up of materials that occur early in the series, while felsic ones are made from later minerals. Consequently we may infer that the geochemical history of felsic rocks is more complex than that of mafic rocks.

Ultramafic rocks contain neither feldspar nor quartz. In general, the mineral types change smoothly from mafic to felsic in a way that is illustrated in Figure 3.1. The figure is intended for heuristic rather than definitive purposes.

In the early history of geochemistry, it was thought that the silicates were salts of silicic acid, H_4SiO_4. This concept was not particularly useful, since the formation of silicates is unrelated either to this weak acid, or to other insoluble silicic acids. Nevertheless, it is still common to refer to rocks with a high content of SiO_2 (>66%)

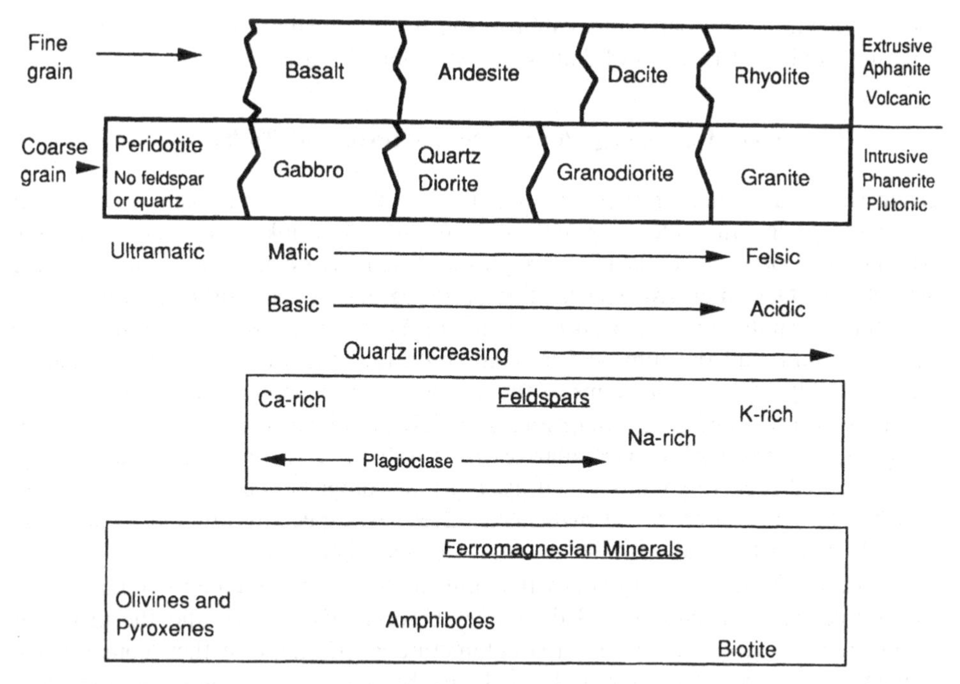

Figure 3.1. A Classification of the Igneous Rocks.

as *acidic*, and those with a lower SiO_2 content (45–52%) as basic. For our purposes, there will be no reasons to discriminate between the words acidic and felsic on the one hand and basic and mafic on the other.

The igneous-rock classification given in the upper portion of Figure 3.1 is partly two-dimensional. The vertical axis gives a measure of the grain size. The size of the grains in an igneous rock is a clue to the cooling rate of the parent liquid. Generally, large grain size implies a slow cooling time.

The cooling time of a lava or magma depends upon the circumstances. Lavas, which have been extruded upon the surface of the earth in thin layers, will cool very quickly. A magma may be intruded into sedimentary or volcanic layers of solid rock, and cool very slowly. In the first instance, we will expect fine-grained rocks, and coarse-grained ones in the second.

Coarse-grained rocks are typically intrusive, while fine-grained ones are typically extrusive. These words clearly imply something about the history of the material and are not descriptions of the material *per se*. Intrusive igneous rocks are sometimes called plutonic, and extrusive ones volcanic. The words plutonic and volcanic require careful attention to the context in which they are used.

Purely descriptive words concerning grain size are phanerite for a fine grained rock and aphanite for a coarse-grained one.

The division between coarse- and fine-grained rocks is generally made on the basis of whether the grains may be seen with the help of a hand lens. Rocks with the composition of basalts and gabbros and an intermediate grain size are called dolerites in Great Britain, and diabases in the United States.

3.3 Some Terminology Related to the Texture of Rocks

Grain size is a matter of rock texture. Rocks of granitic compositions but with very large grains are called pegmatites. Whitten and Brooks (1972) state that the crystals must be 3 cm or larger, though other sources do not make such specific restrictions. One may also speak of gabbro-pegmatites etc., although this is less common. Granites are often enriched in exotic species such as the actinide rare earths uranium and thorium, and in pegmatites, the enrichment is typically higher. This is readily explained in terms of the geochemistry of trace elements (§5.7).

Breccia is a word of great importance in lunar and meteoritic petrology. It is used to describe a rock with a particular texture. A breccia is made up of compacted and cemented rock fragments called *clasts* by petrologists. The adjective is clastic. Volcanic tuffs are often called *pyroclastic*. Breccias, especially lunar breccias, may be made of geochemically similar materials, but not necessarily so.

On the earth, sedimentary rocks that are made up of cemented, rounded pebbles are called conglomerates. Sedimentary breccias also occur, and the principal difference between a breccia and a conglomerate is the shape of their constituents. Conglomerates are made of rounded pebbles, breccias of more angular fragments or clasts. Often, the pebbles of conglomerates have been transported some distance by water.

The lunar rocks are typically brecciated. Many are porphyritic. A porphyry is a rock in which large crystals may be found within a fine-grained matrix. Basalts are frequently porphyritic because certain of the minerals in the magma crystallize at much higher temperatures than the bulk of the melt. A chemical and mineralogical distinction is thus to be expected between the large crystals and the background matrix. There can also be a gradient in the composition of the crystals themselves, since the outer layers are formed later than the centers, at a time when the composition of the magma has changed.

The large grains within a porphyry are called phenocrysts ("pheno" derives from a Greek word meaning to shine or to appear: cf. phenomenology). These phenocrysts are typically euhedral, meaning they have well-defined, "beautiful" angles. When there is a chemical gradient from the inside to the outside of a phenocryst, it is said to be zoned. The amount of zoning is an indication of the cooling history of the parent magma. Lunar rocks are often brecciated and porphyritic, with the phenocrysts zoned.

In igneous rocks, the last material to solidify fills in the spaces between the crystals. This material may be in the form of a glass or it may form the outermost zones of previously crystallized minerals. It is called the *mesostasis*.

Basaltic rocks frequently include rounded cavities known as *vesicles*, formed by gas inclusions in the freezing magma. The degree of vesicularity can vary enormously. Highly vesicular rocks may be called lava or scoria. The word lava may apply to solid or liquid materials, while magma always refers to a melt.

Many of the returned samples from the lunar maria contain vesicles. At the Viking lander sites on Mars, especially that of Viking II in Utopia, the rocks appear to be vesicular. Apparently because the samples have not been investigated in the laboratory, the professional papers describe these holes in the Martian rocks as pits or vugs rather than vesicles.

3.4 Basalts, Gabbros, and Anorthosites

An important rock type, especially in cosmochemistry, is the anorthosite. They do not appear specifically in Figure 3.1, although they may arguably be considered a kind of gabbro (see below). They are primarily (about 90%) plagioclase feldspar. When other major minerals than plagioclase are present in a rock, one may call it a gabbroic anorthosite or an anorthositic gabbro, thereby indicating successively greater amounts of other minerals. Anorthosites are coarse-grained igneous rocks. They dominate the lunar highland materials for reasons that seem to be reasonably well understood (§3.5 below). The origin of terrestrial anorthosites is a much more contentious subject, but since the terrestrial anorthosites are relatively rare, we may leave this problem to the specialist (see Carmichael, Turner, and Verhoogen 1974, henceforth, CTV).

Terrestrial basalts may be divided into four main categories (CTV p. 32), but it is unnecessary for us to do so here. Unfortunately, some lunar gabbros have been described, even in elementary textbooks, as *norites* and *troctolites*, terms that are uncommon in all but advanced discussions of terrestrial petrology (CTV p. 38). Since anyone reading about lunar rocks will surely encounter these terms, we give an abbreviated classification of gabbros in Table 3.1, adapted from Whitten and Brooks (1972). The classification is based on the composition of the plagioclase, the richness of the pyroxene in calcium (augite is richer in calcium than orthopyroxenes such as hypersthene or enstatite), and finally upon whether olivine is present. For simplicity, we omit types that are richest in anorthite with olivine. The interested reader may find these types in Whitten and Brooks.

The terminology of terrestrial petrology is often applied to meteorites with subtle and sometimes confusing changes. For example, in Chapter 7 we shall discuss a group of meteorites known as eucrites, whose reflectance spectrum closely resembles that of the minor planet Vesta. These meteorites are basaltic rather than gabbroic in texture, and presumably have a cooling history commensurate with extrusive rather than intrusive rocks.

3.5 Phase Diagrams

The phase diagram for the system albite–anorthite was discussed in §2.6 (see Figure 2.7). In the following paragraphs, we shall take up more complicated examples, including materials for which solid solutions are not possible. Under these circumstances, the solids are composed of several minerals as in naturally occurring rocks.

We first discuss a binary mixture with a minimum melting temperature. We simplify the situation described by Bowen and Tuttle (1950) for a sodic–K-feldspar

Table 3.1. *A Partial Classification of Gabbros*

Pyroxene	*Plagioclase*		
	Labradorite		*Bytownite–Anorthite*
	No Olivine	*With Olivine*	*No Olivine*
Augite	Orthogabbro	Olivine gabbro	Eucrite
Augite and orthopyroxene	Hypersthene gabbro	Olivine–hypersthene gabbro	Eucrite
Orthopyroxene	Norite	Olivine norite	Hypersthene eucrite
No pyroxene	(Anorthosite)	Troctolite	(Anorthosite)

mixture, following Krauskopf (1979). The real mixture behaves in a complicated way near the solidification of pure K-feldspar. These details are omitted in Figure 3.2. The relations shown were determined experimentally under a water vapor pressure of 2 atmospheres. The phase relationships for a dry mixture are more complicated.

If we ignore for the moment the region below the solidus line, the situation may be considered as two simple phase diagrams of the form already considered in Figure 2.7. We make a conceptual division of the figure at about 70% albite, and consider this composition to be an "end member." This composition is designated 'M' on the figure. We then have two systems, one being K-feldspar and M, the other consisting of M and albite. The interpretation of the solidus and liquidus curves follows the description of §2.6 for the separate systems. In a liquid with less albite than M, the first solid to form on cooling (e.g., $A \rightarrow A'$ on the figure) would be enriched in K-feldspar (e.g., the point B on the solidus). If the percentage of albite were greater than that of M, a solid with composition D would melt to produce a liquid depleted in albite (e.g., $D \rightarrow C'$).

The shaded area within the solid field of the figure indicates that albite and K-feldspar form a stable solid solution only for temperatures above about 650 °C. At lower temperatures, the sodic and potassiac feldspars will slowly exsolve to form separate phases. It is possible to see this exsolution as lineations in certain feldspars called *perthites*.

Solid feldspar with the composition M would melt to form a liquid with exactly the same composition. This is called *congruent* melting.

We next consider a system with a eutectic point (E in Figure 3.3). The data for anorthite and diopside are from N. L. Bowen's (1928) classic work. These two minerals do not form a solid solution, so the situation is different from that of our two previous diagrams (we ignore the exsolution of albite and K-feldspar). A eutectic mixture of diopside and anorthite would melt to form a liquid with that same composition, but for any other composition, the melting history is more complicated.

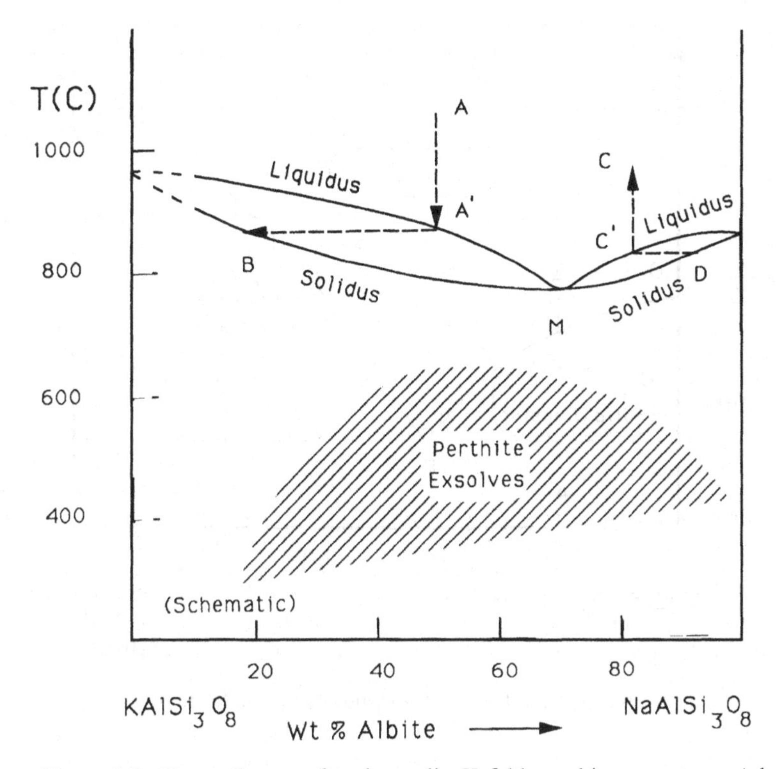

Figure 3.2. Phase diagram for the sodic–K-feldspar binary system. Adapted from Bowen and Tuttle (1950).

Consider the melting of a solid with the composition A, 20% anorthite by weight. When the temperature is raised above 1270 °C, liquid of composition E will appear. Liquid with the eutectic composition will continue to melt "at" 1270 °C until the anorthite is exhausted. If the temperature is raised further, essentially pure diopside must melt, and the liquid must become richer in diopside. When the liquid reaches the composition A, the last solid phase will vanish.

The curves DE and EF represent the composition of the liquid. Since a solid solution does not form, there is no corresponding solidus. The liquidus curve corresponds to the composition of the liquid *at* a temperature where a phase change occurs, for example, the solidification of *either* diopside *or* anorthite for any composition other than a eutectic mixture. For the eutectic mixture, the two minerals crystallize simultaneously.

Many compounds decompose rather than undergo a simple phase change. Bowen (1928) describes the melting of clinoenstatite $Mg_2Si_2O_6$ (or $MgSiO_3$). This compound decomposes at a temperature of 1557 °C. Solid forsterite Mg_2SiO_4 will remain until a temperature slightly below 1600 °C is reached. The corresponding liquid is roughly 52% SiO_2 and 48% Mg_2SiO_4. The melting is illustrated by the following equation:

$$52Mg_2Si_2O_6(c) \rightarrow 4Mg_2SiO_4(c) + 48Mg_2SiO_4(l) + 52SiO_2(l) \qquad (3.1)$$

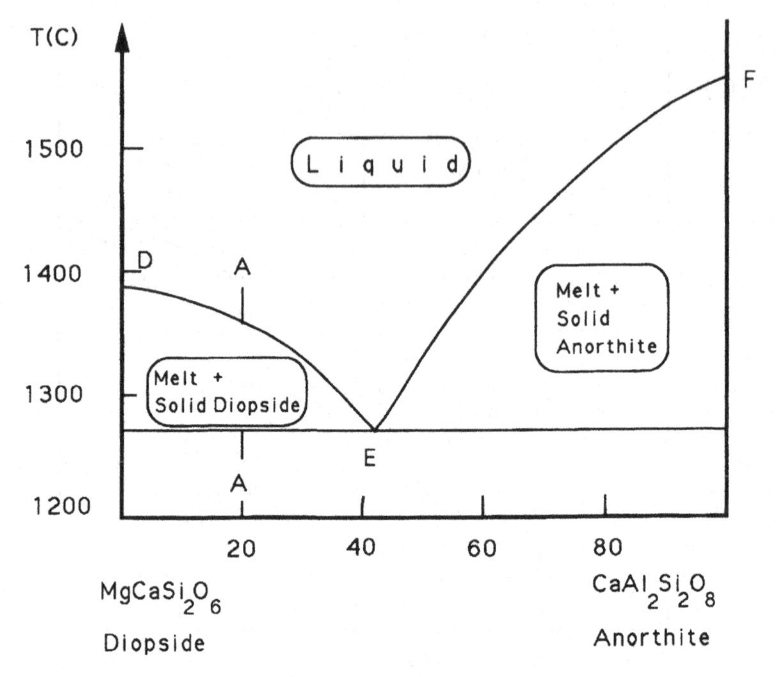

Figure 3.3. Phase Diagram for the Anorthite–Diopside System.

where (c) stands for crystalline, and (l), liquid. The approximate mole percentages of liquid quartz and forsterite are indicated. Note that this liquid is somewhat richer in SiO_2 than clinoenstatite, as it must be if some forsterite is left in the solid state.

The phase diagram for the forsterite–silica system illustrating the incongruent melting of clinoenstatite is given by Bowen (1928) and Krauskopf (1979). We shall not reproduce it here, but pass to a discussion of a ternary system, silica–forsterite–anorthite, again following Krauskopf.

One must pretend to be looking down upon a three-dimensional version of the previous figures. The dimension corresponding to the temperature would be out of the paper. We are, so to speak, looking down on hills and valleys. The valleys are shown as heavy lines within the ternary field. The surface that we see in projection would give the temperature corresponding to a given composition of the liquid, at which some phase change takes place.

Corresponding to eutectic points in a binary mixture, we have minima of the surface whose vertical extent we must imagine. In Figure 3.4, these valleys are indicated by heavy lines. The point marked E at the lowest point in these valleys corresponds to a eutectic mixture; a solid with this composition will melt to give a liquid with this same composition.

A liquid with a composition different from one of the valleys would, as the temperature drops, eventually reach a temperature corresponding to the surface, at which a phase change occurs – one of the three solids would begin to form, and the composition of the liquid would change to reflect loss of the solid. The minerals

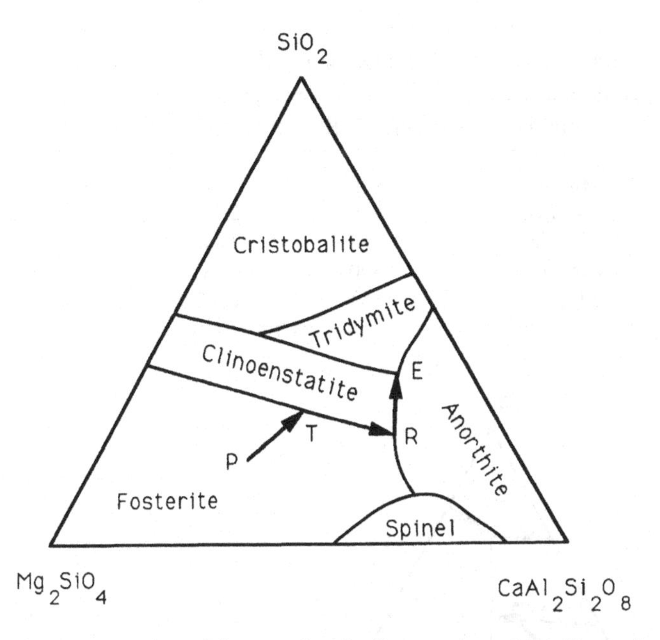

Figure 3.4. Phase Diagram for the System Anorthite–Forsterite–Silica.

written within the fields are the ones which crystallize. For example, liquid with the composition indicated by the letter P would begin to form solid forsterite at a temperature of about 1500 °C. On further cooling, additional forsterite would form, and the composition of the liquid would move toward the point T on the valley between forsterite and clinoenstatite.

Subsequent cooling would result in the presence of both crystalline clinoenstatite and forsterite. The liquid composition would move along the valley to the point R. At R, solid anorthite would appear, and further cooling would move the composition of the liquid along the line R → E, the eutectic. Once the eutectic is reached, solid anorthite, clinoenstatite, and the SiO_2-mineral tridymite would form until the liquid becomes exhausted.

We have omitted a number of complications. Petrologists refer to Figure 3.4 as a *pseudoternary* diagram, because spinel ($MgAl_2O_4$) cannot be made from mixtures of the end members in proportions indicated by the position of the field. This pseudoternary diagram is discussed in most petrology and geochemistry texts (see Ehlers 1987, Chapter 5). A similar construction will be used in the section that follows.

3.6 Formation of the Lunar Highlands: A Petrological Exercise

The lunar highlands are dominated by anorthositic rocks. The type is unusual, though well known on the earth. These rocks derive their name from the calcic feldspar anorthite, which appears at the top of the continuous Bowen series (Fig-

Table 3.2. *Properties of Major Lunar Minerals*

Mineral	Formula	Melting T(K)	ρ (gm/cm^3)
clinoenstatite	$MgSiO_3$	1830	3.19
anorthite	$CaAl_2Si_2O_8$	1830	2.76
forsterite	$MgSiO_4$	2163	3.21
fayalite	$FeSiO_4$	1490	4.34
quartz	SiO_2	~1883	~2.65

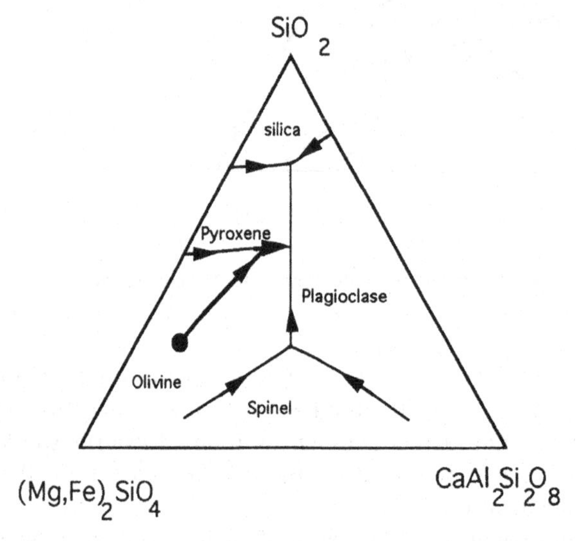

Figure 3.5. Phase Diagram for Formation of Lunar Highlands. The filled circle represents the approximate composition of a molten lunar mantle. The text describes solidification steps relevant to the formation of the anorthositic highlands.

ure 2.6). Since the highlands comprise the majority of the lunar surface, it was at first surprising to find them dominated by an unusual rock type.

There is currently a general agreement on the broad picture of the formation of the highlands, although details remain to be worked out. We shall concentrate on the former here, following Vaniman *et al.* (1991).

Table 3.2 gives melting temperatures and densities of minerals that may be taken as representative of the bulk composition of the Moon. The data are from Robie, Hemingway, and Fisher (1978) supplemented by the *Handbook of Chemistry and Physics*, (henceforth *HCP*, see Lide 1992). The entries for SiO_2 are for quartz, but should approximate the melting temperature and density of the appropriate form of SiO_2 at the relevant pressures. We may plot an estimated lunar bulk composition on a pseudoternary diagram similar to that of Figure 3.4. In the present case, we

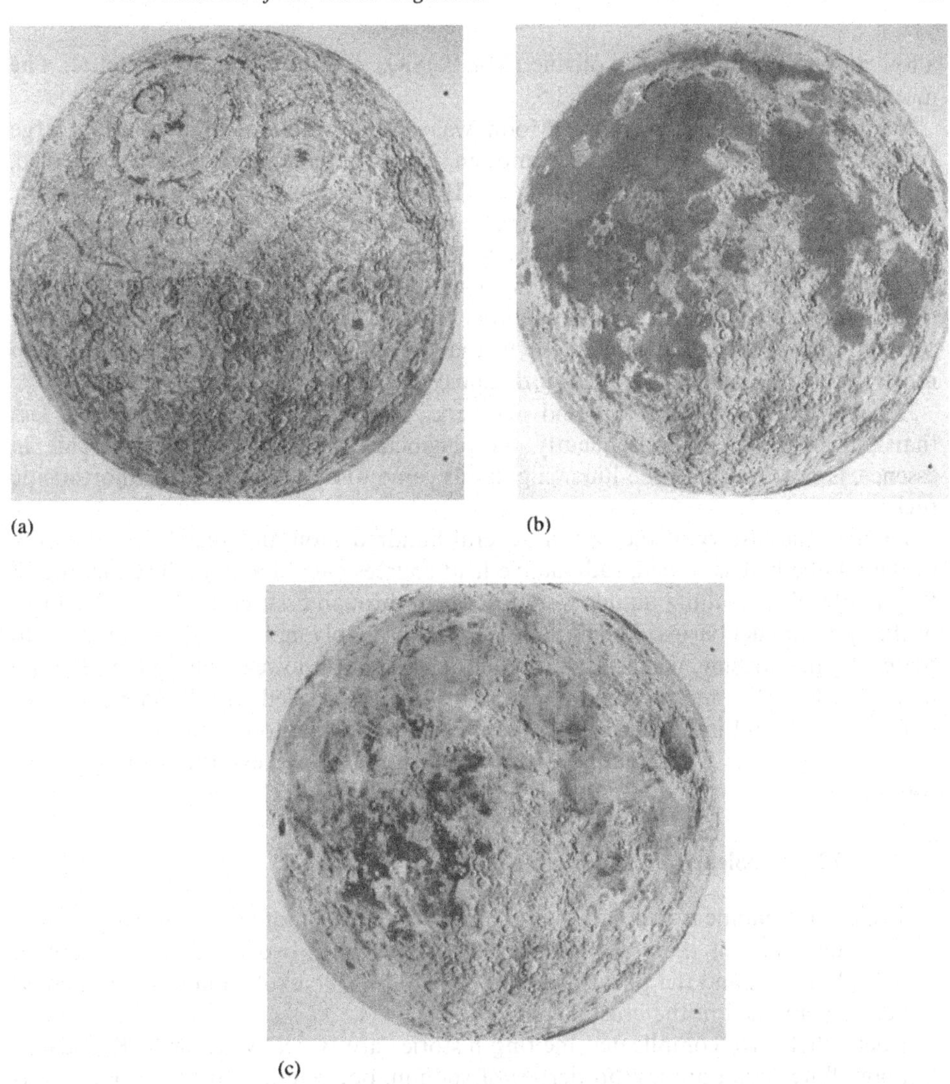

Figure 3.6. Artist's conception of the formation of impact basins and maria. In (a), the highlands and most major basins apart from Imbrium have formed. At (b), Imbrium has formed, and basalt flooding has filled it and the other major basins, as well as many craters. The present Moon (c) has many craters that were formed after the flooding episode. These often have rough floors, with central peaks; the youngest have light-colored, radial rays. Courtesy of the US Geological Survey, drawings by Donald E. Davis.

allow an end member that is olivine $(Mg, Fe)SiO_4$ rather than pure forsterite. The modified plot is shown in Figure 3.5.

We consider a time when the Moon was entirely molten, or at least a large fraction of the lunar mantle was molten. The bulk composition is indicated, roughly, following Longhi and Boudreau (1979). In a liquid with that composition, solid olivine would begin to form at a temperature somewhat below the melting temperature of the olivine end member. Because of the density of olivine, the precipitate would slowly *sink.* In the meantime, the composition of the liquid would move away from the olivine corner, toward the pyroxene field. Subsequent precipitation of olivine and pyroxene would cause the composition of the liquid to move toward the plagioclase field until calcic plagioclase began to crystallize.

The new solid, unlike olivine and pyroxene, would have a density somewhat less than that of the melt. Consequently, the plagioclase would float upwards. This, in essence, is how we think the lunar highlands came to be dominated by anorthositic rocks.

Later in the history of the Moon, several hundred thousand years after the bulk of the mantle had solidified, radioactive heat sources caused a secondary melting of the mantle. The resulting liquids were forced up through fissures created at the birth of the giant impact basins. The molten material probably had a composition close to the pyroxene diopside, which has a higher density, but a lower melting temperature than anorthite. The multi-ringed lunar basins were then filled with flood basalts not unlike those found in northern Idaho and eastern Washington state.

Figure 3.6 shows a fascinating artist's conception of the evolution of the lunar surface.

3.7 Problems

1. The earth's mantle is thought to contain considerably more calcium than sodium or potassium, yet the *crust* is enriched in the latter two. Use the concept of partial melting and the phase diagram of Figure 2.7 to explain how sodium could become enriched in the crustal rocks.
2. Rocks that have complicated melting histories are typically richer in K-feldspar than albite. What are the properties of sodium, potassium, and their compounds that could account for this?

4
A Résumé of Thermodynamics and Statistical Mechanics

4.1 Introductory Remarks

In this chapter, it will be necessary for us to go rather briskly over introductory thermodynamic concepts, and take the practical view that our aim is to learn to *use* the theory. We can only attempt to provide a sufficient understanding of the subject to make our subsequent discussion intelligible. Courses in physical chemistry usually cover the concepts needed for this book.

There are many excellent texts with introductions to thermodynamics. We cite the physical chemistry texts by Atkins (1990), Castellan (1983), and Prutton and Maron (1965). There are many others. Krauskopf's (1979) *Introduction to Geochemistry* is a standard work. The attractive geochemistry text by Richardson and McSween (1989) has a definite cosmochemical bias (see their Chapter 13).

4.2 Systems, States, and Variables

Practical thermodynamics is largely a matter of finding out the direction in which some chemical reaction will proceed. Reactions are described by chemical equations such as

$$A \longrightarrow B. \tag{4.1}$$

Both A and B may stand for several "substances." The substances on the left-hand side are traditionally called reactants, and those on the right, products. The reactants, for the moment, may be thought of as being in a beaker in some laboratory. Assume the beaker is unsealed, so the reactants experience normal atmospheric pressure.

We may refer to the reactants as the *system*, and to the beaker, the laboratory room, and the atmosphere, etc. as the environment, and occasionally as the "bath." It is often useful to think of the environment as another system, in contact with the first one. In order to determine what will happen to our system, we will first endeavor to describe its state in terms of certain state functions or state variables. We could attempt to define these terms, system, state, and state function, in sentences of the form: a state function is .... We shall not do so here. Instead, we provide examples, and hope that this will enable us to use the words in a manner that will be constructive. The philosopher of science may recognize in this approach the notion of undefined terms, analogous to mass, length, and time in classical mechanics. The latter words are defined operationally, by the way things are measured, but not in terms of more words.

Temperature, pressure, and volume are considered state variables, as are energy and density. We assume the reader will be familiar with these terms. The amounts (e.g., concentrations, or mole fractions) of different chemicals in the system must also be considered state functions. We say that the state of a system is specified by a "sufficient" number of state variables.

How many variables are enough? In general, this will depend on the nature of the system as well as its environment. For the present, we *assume* that the state of a pure system will be specified when, for a given amount of substance, two (additional) state variables are known. If our system is an ideal gas, then the system is specified when the number of moles, n, is known along with two other state variables, such as temperature and pressure.

In the general case two state variables would not describe *any* system. One might need to specify the local magnetic or gravitational field. One might even need to specify the individual particle velocities. For the present, we shall not concern ourselves with such complications. In the use of thermodynamics in practice, one typically deals with systems that *can* be described by a relatively small number of these state variables.

Often the concentration variables are considered known, or "understood." One might say then that only two state variables determine the system – with the understanding that the concentration variables are known. It is traditional in chemistry to specify amounts in moles; a mole of H_2 molecules, for example, means Avogadro's number of H_2 molecules, or approximately 6.02×10^{23} molecules. We use the symbol $\mathcal{N}_a$ for Avogadro's number.

Once a system is completely specified, any further description will be redundant. This means that we should be able to determine a state function from any two others. Since the state functions are things that we measure at a specific time, it follows that once the state functions are specified, the history of the system is *irrelevant*. For example, a mole of H_2 gas at 273 K and 1 atmosphere is in the same "state," whether it just finished cooling to 273 K, or was heated to 273 K.

The state functions are given values by measurements that are made on a system. We measure the temperature, pressure, volume, and so on. It is *assumed* that the results of these measurements obey certain relations, such as

$$P = P(T, V), \tag{4.2}$$

which says that P is a mathematical function of two variables T and V. The property of mathematical functions that is analogous to the independence of the state functions from history, is independence of path. We write

$$\oint dV = 0, \tag{4.3}$$

where V might be the volume, or *any* state variable, and the loop any closed path in the space of two other state functions which completely specify V. When

Equation 4.3 holds for the volume, one may write, for example,

$$dV = \left(\frac{\partial V}{\partial P}\right)_T dP + \left(\frac{\partial V}{\partial T}\right)_P dT. \tag{4.4}$$

We assume the reader is familiar with these purely mathematical relationships (see, for example Kaplan's (1991) *Advanced Calculus* or a similar text). The variables held constant in the partial derivatives are often indicated specifically. For example, in the first derivative on the right above, T is held constant, and P is constant in the second derivative.

This independence of path is of great importance in practical thermodynamics, because it provides a certain freedom both in the laboratory and in the mind. It is common to specify a state by giving the temperature and pressure. Suppose we wish to find the change in some other state function when the system passes from the state described by the values P_1, T_1 to P_2, T_2. Independence of path (for the mathematical quantities) and independence of history (for the physical entities) tells us that we may find this change by holding P constant at P_1, and letting T go from T_1 to T_2, and then holding T constant, while letting P go from P_1 to P_2. Alternatively, we could hold first T constant, and then P. The point is that it is often inconvenient analytically, and sometimes experimentally, to change both variables simultaneously. Our assumptions make this complication unnecessary. The great success of thermodynamics justifies these assumptions.

An equation of the form 4.2 is called an equation of state. It is rarely possible to write an exact relation that will describe the state of an actual physical system, but for gases at low pressures, the ideal gas law,

$$PV = n\mathscr{R}T, \tag{4.5}$$

is a very useful form of an equation of state. Here, n is the number of moles of gas, and $\mathscr{R}$ is the gas constant 8.314×10^7 ergs/mole/deg, or 8.314 joules/mole/deg).

The chemists distinguish among intrinsic and extrinsic properties of a system. Extrinsic properties, such as energy and volume, depend on the amount of substance, while intrinsic properties, e.g., temperature and density, do not (we assume a uniform substance). The notation that we shall use here is to let upper case letters such as E or V stand for the total energy or total volume of the system. Chemists often use upper-case letters belonging to a smaller font to denote quantities *per mole*. The energy and volume per mole are intrinsic properties. We shall not introduce these molar quantities here (with the exception of heat capacities, see below).

4.3 Additional State Functions and the Laws of Thermodynamics

The first law of thermodynamics states that if a quantity of heat is added to a system, the result will be that the energy of the system will change, and the system may do some work on its surroundings. We write this

$$\delta q = dE + \delta w, \tag{4.6}$$

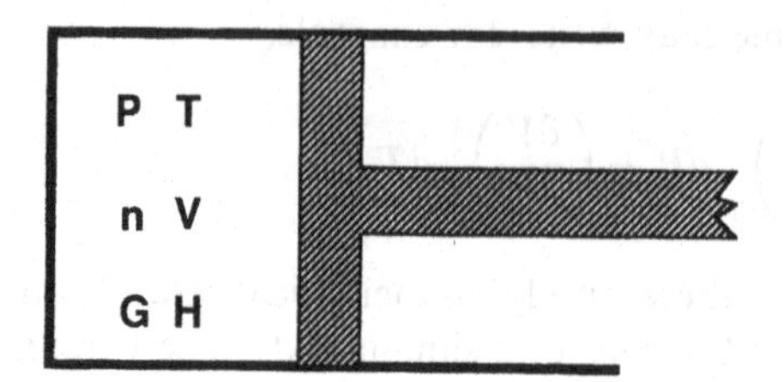

Figure 4.1. A Simple Physical System.

where δq stands for an increment of heat, and δw for an increment of work. The convention adopted here is that work is considered positive if it is done *by* the system. Note that the opposite convention is frequently adopted. Pitzer and Brewer (see Lewis and Randall 1961), for example, write $dE = \delta q + \delta w$. The 'δ' symbols are often written instead of the *d*'s of differential calculus. This is to remind the reader that the amount of heat added to a system, and the amount of work a system does are not state functions. Thus the mathematical symbols should not imply that the incremental quantities of heat and work are differential quantities in the sense of independence of path (Equation 4.3). The incremental change in energy is written dE, since the energy of a system *is* a state function.

In the following discussion, it will be useful to have in mind the specific system of Figure 4.1, an ideal gas enclosed in a cylinder that has been fitted with a piston. For this system, if the pressure outside the piston is maintained at a constant value P, the work increment done by the system is PdV, and the first law becomes

$$\delta q = dE + PdV. \tag{4.7}$$

In this equation, P is the pressure "outside" the piston. We shall often make the assumption that the changes are such that the pressure inside the cylinder differs from it only infinitesimally, so we may assume the pressures inside and outside are the same. Our notation does not reflect this subtlety, so the reader must be aware that there can be a vast difference between the inside and outside pressures.

While physical systems may perform other kinds of work than P–V work – electrical work for example – the simple relation 4.7 will be sufficient for many of our applications.

It is useful to discriminate between cases in which heat is added reversibly (δq_r) and irreversibly. In the former case the heat increment is added to the system so slowly that the entire process could be turned around. The same increment δq could then be extracted from the system which would then return to its original state. Loeb (1961) gives a helpful discussion of the reversible and irreversible compression of an ideal gas.

We may define additional state functions in terms of the ones with which we are already familiar. It is useful to define the heat content, or enthalpy, H, by the relation

$$H = E + PV. \tag{4.8}$$

For a change that takes place at constant pressure, we see from the first law

(Equation 4.7) that

$$\delta q = dH. \tag{4.9}$$

Consequently, the *heat of reaction* corresponding to Equation 4.1 would be

$$\Delta H = H_B - H_A. \tag{4.10}$$

The next state function that we shall discuss is the entropy. We eschew a general definition here, assuming that this is not the first time the reader has encountered this word. We hope the ensuing discussion will at least provide some operational meaning, that is, it will indicate how the entropy might be measured or calculated, at least for simple systems. The concept of entropy may be usefully approached with the help of statistical mechanics and a relation known as Boltzmann's postulate, which we shall discuss below.

Let us define the entropy S of a system in such a way that the change in the entropy is calculated from the relation

$$dS = \delta q / T. \tag{4.11}$$

This relation is sufficient to allow us to calculate the change in the entropy for many physical systems. It does not give the entropy itself. Absolute entropies may often be calculated with the help of the *third law of thermodynamics*. Boltzmann's postulate (see below) provides one of the most general definitions of entropy.

It is not necessary that heat be added to a system for the entropy to increase. But since we postulate that the entropy is a state function, we can still calculate the entropy change using Equation 4.11. A simple example is given as a problem at the end of this chapter.

Much useful thermodynamics can be done even if one restricts one's attention to entropy changes, that is without considering absolute entropies. The second law of thermodynamics postulates that for any change of state that occurs spontaneously in nature, the entropy increases. Unlike the first law, the second may not seem intuitively obvious. Its justification is the success of applications based upon it.

Spontaneous changes are assumed to involve the production of entropy. A spontaneous change, then, cannot take place in an isolated system that is in equilibrium. Therefore, some disequilibrium is always present whenever spontaneous changes occur or, equivalently, when entropy is produced. Let dS_S be the entropy change of a system within which we assume some irreversible change is taking place. Let dS_B be the entropy changes of the surroundings or bath, assumed to be reversible. For this case, write the second law as

$$dS_S + dS_B = dS_{\mathrm{Irr}} > 0, \tag{4.12}$$

where dS_{Irr} denotes the irreversible production of entropy. In the idealized case of reversible processes in both the system and the bath, the right-hand side of Equation 4.12 is zero.

Spontaneous incremental changes are "allowed" by the second law whenever the entropy of a system increases. In actuality, these changes may take a very long time. It is important to note that the discipline of thermodynamics does not deal with the *rates* of processes at all. Reaction rates, which can be of great significance in cosmochemistry, can only be approached by going outside the domain of thermodynamics. In physical chemistry, rates are treated under the heading reaction kinetics, which we shall discuss in Chapter 14.

With the help of the entropy, we now define two new thermodynamic functions, the Helmholtz and Gibbs energies respectively:

$$A = E - TS, \tag{4.13}$$

$$G = H - TS. \tag{4.14}$$

In the older literature, these quantities are called Helmholtz and Gibbs *free* energies, respectively. In the 72nd edition of the *Handbook of Chemistry and Physics* (Lide 1992) these two quantities are also called Helmholtz and Gibbs *functions.*

The Helmholtz energy, A, is also called the work function by the chemists, since it is possible to show that A gives the maximum work that can be obtained from a system, in the idealized case of a reversible process. The function A would be useful for studying equilibrium relations for enclosed systems that are maintained at a constant temperature. Such systems are less useful than those for which both T and P are constant. For these it is valuable to consider the possible changes in the Gibbs energy, G.

Many chemistry texts and references (for example, older editions of the *HCP*), use the symbol F for the Gibbs energy. Physics texts may use F for the Helmholtz energy! In an interdisciplinary field such as cosmochemistry, one must constantly deal with notational ambiguities.

4.4 How to Tell the Direction of a Chemical Reaction

Let us define a system and a bath in the following way. The boundary between the system and the bath is taken such that all irreversible changes are contained entirely within the system. This means that the work that the system does on the bath is assumed to be reversible. Further, the bath is sufficiently large that it can supply heat to or remove heat from the system *without changing its own temperature or pressure.*

A specific example of a system might be the contents of a beaker in a laboratory. The beaker, the (heavy) laboratory bench, the room and the rest of the earth (if necessary) would be the bath.

If the chemicals in the beaker react, they will generally release heat. This heat will eventually pass to the bath, whose temperature by definition remains fixed. Eventually, the system will reach room temperature too, and we ask if the reaction, say Equation 4.1, went to completion. That is to say, after all is said and done, and T and P in the beaker return to the values of the bath, does the beaker contain A or B? To answer this question, we write the entropy change in the system as

$$dS = \delta q/T = \delta q_r/T + dS_{\rm Irr}. \tag{4.15}$$

According to the second law, if dS is positive (throughout the process), A $\rightarrow$ B. We shall now show that positive dS, for processes that take place at constant temperature and pressure, is equivalent to an incremental *decrease* in the Gibbs energy.

The incremental heat change δq for the system is not assumed to take place reversibly. By hypothesis, or equivalently, by the *second law*, $dS_{\rm Irr} > 0$ if the process goes spontaneously. In the limiting case, $dS_{\rm Irr} = 0, \delta q$ becomes δq_r.

Multiply Equation 4.15 by T, and use the first law, Equation 4.6. Then

$$TdS - dE - \delta w = TdS_{\rm Irr} > 0. \tag{4.16}$$

If we restrict ourselves to the case where the only work done is reversible, P–V work,

$$\delta w = PdV. \tag{4.17}$$

Put 4.17 into 4.16, and note that the pressure is constant. The left-hand side of 4.16 is then *minus* $dG_{P,T}$, that is, minus the change in G with *both* P and T constant. Thus

$$-(dG)_{P,T} = TdS_{\rm Irr} > 0. \tag{4.18}$$

For a given temperature and pressure, we thus have the rule that spontaneous processes will occur with a decrease in the Gibbs energy. We may formulate a simple rule to see whether the reaction symbolized by Equation 4.1 will occur. Calculate the Gibbs energy for A and B, individually, subtract, and see if

$$\Delta G \equiv G_{\rm B} - G_{\rm A} \tag{4.19}$$

is less or greater than zero. If $\Delta G < 0$ the reaction will proceed spontaneously from left to right. If $\Delta G > 0$, it will go right to left. The null case, $\Delta G = 0$, corresponds to the important case of (chemical) equilibrium.

Note the chemists' convention that all "Δ-quantities" are defined in the sense of products minus reactants, or right-hand side minus left-hand side. The upper case 'Δ' must not be confused with the 'δ' used to designate an increment of some non-state variable (e.g., δw).

Our problem now is to be able to find or to calculate values of G for the reactions of interest to us. The "reactions" may be interpreted very liberally. The generality of the laws of thermodynamics is such that purely physical and phase changes as well as chemical reactions are included. Indeed the domain of validity of these concepts is very broad, but we must confine ourselves here to a limited task.

Let us postulate a constant T and V for our system rather than T and P. Then it is easy to see from 4.17 that $\delta w = 0$. From 4.16 and 4.13 it follows that $TdS_{\rm Irr} = -dA$. For constant volume processes, the Helmholtz free energy is the useful state variable to examine to find the direction of a reaction. The second law is in essence a statement that $dS > 0$ for spontaneous processes taking place at constant energy and volume.

The natural processes of most interest are typically ones that take place at constant T and P, which is why the Gibbs energy is of such practical importance.

4.5 Calculation of ΔG; Standard States

Gibbs energies may be calculated with the help of various thermochemical ΔG tabulations, such as the *JANAF Thermochemical Tables* (Chase *et al.* 1986), the extensive compilation of Barin (1989), or the *HCP*. For minerals, Robie, Hemingway, and Fisher's (1978) *Thermodynamic Properties of Minerals and Related Substances at 298.15 K and 1 Bar (10^5 Pascals) Pressure and at Higher Temperatures* is a standard reference.

In order to use these tables, it is necessary to have some understanding of what is meant by standard states of substances. For solids and liquids, the standard states are defined to be the most stable forms at a fixed or *standard pressure.* Until recently, that pressure was one atmosphere, slightly more than a million dynes/cm^2, or 101 325 newtons/meter^2 or pascals. This is still the standard pressure used in the 72nd edition of the *Handbook of Chemistry and Physics* (Lide 1992), but the third edition of the *JANAF Thermochemical Tables* adopts 100 000 pascals as the standard pressure. This distinction will usually be unimportant for our purposes.

There is a standard state of a substance for every temperature. When phase changes are possible, the state can change profoundly. For example, the standard state of H_2O is solid below 273 K, vapor above 373 K, and liquid for intermediate temperatures. We shall discuss standard states for real gases below. For the present, we note that an *ideal* gas at the standard pressure (usually, one atmosphere) is in its standard state by definition.

The reason standard states prove to be useful is primarily the fact that differences in the state variables (e.g., ΔG's) play such a key role in thermochemistry. One can measure, directly or indirectly, the differences (ΔG's, say) when a system changes from one of these standard states to other nonstandard states. From the values so obtained, one can calculate the appropriate differences (ΔG's) when the system changes from one nonstandard state to another. This is possible for quantities such as ΔG or ΔH which are state variables.

For any chemical compound, we may define the standard *heat of formation* ΔH_f^0 as the change in enthalpy when the compound is prepared at constant pressure from the chemical elements in their standard states. The superscript '0' implies that the compound is in its standard state; the '*f*' means "of formation." The quantities ΔH may be measured in the laboratory, or calculated from measured quantities. The chemical elements may be brought to their standard states before the reaction is allowed to proceed. The standard heats of formation of the elements themselves, at any temperature, are defined to be zero, if the elements are in their standard states.

Analogous to ΔH_f^0 one may define a ΔG_f^0 for a given compound. It is the change in the Gibbs energy when the compound (in its standard state) is prepared from the chemical elements (in their standard states). Again, the ΔG_f^0's are defined to be zero for the elements in their standard states.

One may calculate ΔG from ΔH if ΔS is known. Values of ΔS may be obtained in a variety of ways, including *ab initio* calculations and thermochemical measurements. The latter involve an invocation of the third law of thermodynamics, which we state here for completeness: *the entropy of pure crystalline solids at absolute zero Kelvin*

(K) is assumed to be zero. We must leave the details of calculations of ΔS as well as its absolute value to the references (see also §4.10).

It is often possible to combine chemical equations in a way to yield values of ΔH or ΔG for a compound for which a tabular entry does not appear directly. Consider the reactions

$$\mathrm{A} \rightarrow \mathrm{B}, \tag{4.20}$$

and

$$\mathrm{B} \rightarrow \mathrm{C}. \tag{4.21}$$

Upon addition, we obtain A → C. *Hess's law of summation* states that the ΔH for A → C is the sum of the ΔH's for Equations 4.20 and 4.21. The law also holds for ΔG and ΔS. It is possible to argue that Hess's law follows from the fact that H, G, and S are state functions.

The *Handbook of Chemistry and Physics* gives tables of ΔH_f^0 and ΔG_f^0 in kilocalories per mole for a temperature of 25 °C, or more precisely, 298.15 K. The older notation for the Gibbs energy is often used so the tabular heading may be ΔF_f^0. One (thermochemical) calorie is 4.184 joules. A kilocalorie is thus 4.184×10^{10} ergs. Most recent chemistry texts have adopted SI units, but a few important references have not yet succumbed to this abomination.

For a reaction that might take place at one atmosphere of pressure and 25 °C, we find ΔG by subtracting the tabulated ΔG's of the reactants from those of the products. If two moles of A react to form one mole of B, we must multiply the tabulated ΔG by two before subtracting from that for B. When the resultant ΔG is negative, the reaction will proceed.

The calculation of ΔG for other temperatures is relatively easy if we can find tabulated values of what we shall call, following Pitzer and Brewer (see Lewis and Randall 1961), the *free energy function*. The third edition of the *JANAF* tables calls it the *Gibbs energy function*. We write it

$$\mathscr{F} = \left(G^0 - H_{298}^0\right)/T. \tag{4.22}$$

This free energy function typically varies slowly with temperature, and is better for interpolation than ΔG^0. It is not usually given a symbol. We find it convenient to do so, and introduce $\mathscr{F}$ here. The reader must be aware that our $\mathscr{F}$ will not be found in other works.

The *JANAF* tables and their supplements, for example, give $\mathscr{F}$ for many substances for a wide range of temperatures. The appropriate columns are headed by the right-hand side of 4.22 rather than the left! The calculation of these functions involves a knowledge of absolute entropies. Details of these calculations are given in the preface to the *JANAF* tables.

In order to calculate ΔG using $\mathscr{F}$, we rewrite 4.22 as

$$G^0 = H_{298}^0 - \mathscr{F}T. \tag{4.23}$$

Consequently, for any reaction of the form A → B, we have

$$\Delta G = \Delta H_{298}^0 - T\Delta\mathscr{F}. \tag{4.24}$$

The ΔH^0_{298} and $\Delta\mathscr{F}$ are formed by subtracting values for the products from those for the reactants, as before. If any of the products or reactants are *elements* in their standard states, the ΔH^0_{298} for them is defined to be zero.

The *JANAF* tables contain a number of other useful entries, such as heat capacities, absolute entropies, and equilibrium constants. The order of the entries follows the *modified Hill indexing system*, discussed in Section 9.2 of the Third Edition of the *JANAF* Tables. Basically, one rewrites the chemical formula for a compound so that the elements appear in alphabetical order. Thus $MgAl_2O_4$ would be rewritten Al_2MgO_4, and the corresponding tabulation filed under Al, *following* those compounds containing *one* atom of aluminum. There is an exception to the alphabetizing for the organic (carbon) compounds, which appear together under C.

In Figure 4.2, we reproduce data for liquid sulfur taken from a forthcoming edition of the *JANAF* tables. In addition to the free energy function (our $\mathscr{F}$), various other thermodynamic quantities are tabulated, including the heat capacity at constant pressure and the absolute entropy. All of the entries refer to a pressure of one atmosphere. The horizontal lines indicate phase changes. As may be seen from the entries, sulfur melts at 388.36 K, and boils at 717.83 K. At a temperature of 882.117 K, the reaction $S(l) \to \frac{1}{2}\,S_2(g,\, f = 1.0)$ is in equilibrium ($\Delta G = 0$). The 'g' means *gas*, and the $f = 1$ means the diatomic gaseous sulfur is in its standard state. The *fugacity*, f, is defined in §4.6.

From 388.36 K to 882.117 K, liquid sulfur is the most stable form of the element, and according to convention, the standard heats and free energies of formation are defined to be zero. In the *JANAF* tables, solid horizontal lines (not shown for sulfur) indicate crystalline phase changes.

We note that while the usual convention is that the standard states for the elements at a given temperature are *usually* the most stable forms at a pressure of one atmosphere, this convention is not always followed. In the *JANAF* tables, the reference states for oxygen and nitrogen, for example, are O_2, and N_2 at *all* temperatures. Users of these tables must be aware of this convention.

4.6 Free Energy Changes At Arbitrary Pressures

It is straightforward to calculate the change in the Gibbs energy of an ideal gas when the pressure changes with the temperature held constant. We have from Equation 4.14

$$dG = dH - TdS. \tag{4.25}$$

Combining Equation 4.8 with the first law of thermodynamics, and the assumption that the only work is reversible P–V work, we readily find

$$\int (dG)_T = \int V dP. \tag{4.26}$$

This equation may easily be integrated between limits P_1 and P_2 using the ideal gas law. We find

$$G_2 - G_1 = n\mathscr{R}T \ln\left(P_2/P_1\right). \tag{4.27}$$

SULFUR (S) LIQUID A_r = 32.06

$S°(298.15\ K) = [36.825]\ J\ K^{-1}\ mol^{-1}$
$T_{trs} = 432.02 \pm 0.2\ K$
$T_{fus} = 388.36 \pm 0.2\ K$

$\Delta_f H°(298.15\ K) = [1.854]\ kJ{\cdot}mol^{-1}$
$\Delta_{trs} H° = 0.0\ kJ{\cdot}mol^{-1}$
$\Delta_{fus} H° = 1.721 \pm 0.003\ kJ{\cdot}mol^{-1}$

Enthalpy of Formation

The enthalpy of formation of liquid sulfur at 298.15 K is calculated from that of monoclinic sulfur by adding $\Delta_{fus}H°$ and the difference in enthalpy, $H°(388.6\ K) - H°(298.15\ K)$, between the monoclinic crystal and the liquid.

Heat Capacity and Entropy

The adopted heat capacity values are based on the studies by Montgomery (1, 405.79 – 433.31 K) and West (2, 373 – 678 K). Liquid sulfur undergoes a second order transition with a maximum reported at 432.02 ± 0.20 K (1) and 432.25 ± 0.30 K (2); this has been attributed to the depolymerization of S_8 molecules (3). We adopt the tabulated heat capacity values of Montgomery (1) up to 434 K and those of West (2) above 434 K. The heat capacity is assumed to be constant at 7.568 cal K^{-1} mol^{-1} above 810 K. Below T_{fus} = 388.36 K, the heat capacity values are obtained by linear extrapolation using the slope of the values in the region T_{fus} to 420 K. The entropy at 298.15 K is calculated in a manner analogous to that used for the enthalpy of formation.

Vaporization Data

The normal boiling point of sulfur, T_{vap} = 717.834 K (1 atm) is a secondary reference on the International Practical Temperature Scale of 1968. At this temperature, equilibrium sulfur vapor contains monomeric and several polymeric sulfur species: $S_2(g)$ and $S(g)$ predominate above 1000 K while $S_6(g)$, $S_7(g)$ and $S_8(g)$ dominate at T_{vap}.

Since our reference state for sulfur is arbitrarily defined to involve 0.5 $S_2(g)$ as the vapor species, we calculate $\Delta_{vap}H°(T_{vap}) = [10.839]\ K\ cal^{-1}\ mol^{-1}$ for the process $S(l) = 0.5\ S_2(g)$. The brackets are used to indicate the arbitrary nature of this value.

References

1. R. L. Montgomery, Ph.D. Dissertation, Oklahoma State University, 1976.
2. E. D. West, J. Am. Chem. Soc. **81**, 29 (1959).
3. B. Meyer, Chem. Rev. **76**, 367 (1976).

Sulfur (S) $S_1(l)$

Enthalpy Reference Temperature = T_r = 298.15K Standard State Pressure = $p°$ = 0.1 MPa

T/K	$C_p°$ (J K⁻¹mol⁻¹)	$S°$ (J K⁻¹mol⁻¹)	$-[G° - H°(T_r)]/T$ (J K⁻¹mol⁻¹)	$H° - H°(T_r)$ (kJ·mol⁻¹)	$\Delta_f H°$ (kJ·mol⁻¹)	$\Delta_f G°$ (kJ·mol⁻¹)	Log K_f
0							
100							
200							
250							
298.15	22.531	36.825	36.825	0.	1.854	0.432	−0.076
300	22.707	36.965	36.826	0.042	1.854	0.423	−0.074
350	27.434	40.821	37.120	1.295	1.941	0.180	−0.027
388.360	31.058	43.859	37.635	2.417	―― BETA ↔ LIQUID ――		
400	32.162	44.793	37.829	2.785	0.	0.	0.
432.020	53.829	47.431	38.442	3.884	―― Cp LAMBDA MAXIMUM TRANSITION ――		
432.020	53.808	47.431	38.442	3.884			
450	43.046	49.308	38.840	4.711	0.	0.	0.
500	37.986	53.532	40.106	6.713	0.	0.	0.
550	35.708	57.037	41.490	8.551	0.	0.	0.
600	34.308	60.078	42.915	10.298	0.	0.	0.
700	32.681	65.241	45.748	13.645	0.	0.	0.
800	31.699	69.530	48.460	16.856	0.	0.	0.
882.117	31.665	72.624	50.568	19.456	―― FUGACITY = 1 bar ――		
900	31.665	73.260	51.012	20.023	−53.090	1.079	−0.063
1000	31.665	76.596	53.407	23.189	−51.780	7.028	−0.367
1100	31.665	79.614	55.654	26.356	−50.485	12.846	−0.610
1200	31.665	82.369	57.767	29.522	−49.205	18.546	−0.807
1300	31.665	84.904	59.758	32.689	−47.941	24.141	−0.970
1400	31.665	87.250	61.639	35.855	−46.693	29.639	−1.106
1500	31.665	89.435	63.420	39.021	−45.460	35.048	−1.220

PREVIOUS: December 1965 CURRENT: September 1977

Sulfur (S) $S_1(l)$

Figure 4.2. Thermodynamic Data for Liquid Sulfur. Courtesy of M. W. Chase.

For any gas phase reaction, we may use Equation 4.27 in conjunction with the tabulations previously cited to calculate ΔG for various temperatures and pressures within the range of the tabulations, and for low pressures where the ideal gas law is valid.

The ideal gas law is a useful approximation in a wide range of cosmochemical applications, but we must also deal with solids and liquids. This is done with the help of two new state functions, called the fugacity, f, and the activity, a.

The fugacity is defined so that for any two pressures P_1 and P_2, Equation 4.27 is valid provided the pressures are replaced by the fugacities. Thus we *define* the fugacities by the relation

$$G_2 - G_1 = n\mathscr{R}T\,[\ln(f_2) - \ln(f_1)]\,. \tag{4.28}$$

We can be sure that the f's are state functions, because they are defined in terms of other state functions, G and T. For any non-ideal gas, we may write

$$G_2 - G_1 = \int_{P_1}^{P_2} \left[V - n\mathscr{R}T/P\right] dP + n\mathscr{R}T \ln\left(P_2/P_1\right) = n\mathscr{R}T \ln\left(f_2/f_1\right). \tag{4.29}$$

The term containing the pressures that has been subtracted from V under the integral sign is precisely canceled by the corresponding integrated form. However, we now let P_1 approach zero, and assume any gas will approach ideal behavior. Then f_1 will equal P_1. We may combine the two logarithmic terms so that the arguments of the logarithms are respectively P_2/f_2 and P_1/f_1. Then in the limit as $P_1 \to 0$, the second ratio becomes unity; since $\ln(1) = 0$, the term with P_1/f_1 will disappear from the relation 4.29. This provides a means of calculating the fugacity of a real gas when V is known as a function of P. If we assume $P_1 \to 0$, and drop the subscripts '2' from P and f, we may write

$$\ln(f/P) = (1/n\mathscr{R}T) \int_0^P [V - n\mathscr{R}T/P]dP. \tag{4.30}$$

For mild departures from ideality, one may use the van der Waals coefficients tabulated, for example, in the *HCP*. If we assume that first-order expansions are valid, we may show that (see Problem 2 below)

$$\ln(f/P) = (1/n\mathscr{R}T) \int_0^P n\left(b - \frac{a}{\mathscr{R}T}\right) dP. \tag{4.31}$$

holds approximately. Here a and b are the van der Waals constants. The van der Waals a must not be confused with the *activity* a, to be defined below.

It is argued that 4.30 is valid for any substance, since we can expect vaporization, at zero pressure. Astronauts know that this vaporization does not happen rapidly – their space crafts do not vaporize! The point is not of practical interest, because fugacities are rarely needed for solids or liquids. Instead, one makes use of the activity.

We define the activity a by the relation

$$a = f/f^0, \tag{4.32}$$

where f^0 is the fugacity of the substance when it is in its standard state. Let us take the subscript '2' in Equation 4.28 to represent a system at some arbitrary pressure, and '1' to represent a system of solids or liquids in their standard states at one atmosphere of pressure. Combining 4.28 and 4.32, we see that

$$G - G^0 = n\mathcal{R}T \ln(a). \tag{4.33}$$

As P approaches one atmosphere of pressure, G approaches G^0, and we must therefore have $a \to 1.0$. This will be true for gases as well as for solids and liquids provided we define their standard states to be such that $f = 1$. Thus a real gas is not in its standard state when its partial pressure is 1 atm. One may in principle use Equation 4.30 to find the partial pressure corresponding to the standard state for a gas ($f = 1$).

In most applications that we shall discuss, departures of gases from ideality are small with respect to the overall uncertainties, e.g., of physical and chemical conditions, or thermodynamic data. Thus we may usefully think of the activity, the fugacity, and the pressure as all being the same thing, provided the units are atmospheres.

It is readily shown that the free energy change with pressure of a gas is very much larger than that of a comparable number of moles of a solid or liquid. Consider, for example, the free energy change of an ideal gas when the pressure decreases from 1 atm to 0.01 atm. For 1 mole, Equation 4.27 gives $\Delta G = -2\mathcal{R}T$. For one mole of a solid, we integrate Equation 4.26 between the appropriate limits, to find $\Delta G = -V(0.99)$. Here, V is the volume of one mole of a solid or liquid, which we may assume, for the present purposes, to be constant with pressure. The volume of one mole is μ/ρ, where μ is the molecular weight, and ρ is the density. If we consider water, μ/ρ is about 18 cm^3/mole, since the density of water is very close to unity. Convert the $P_2 - P_1 = 0.99$ atm to dynes/cm^2 by multiplying by 10^6 (1 atm $= 1.013 \times 10^6$ dynes/cm^2). Then, using (beloved) cgs units we find a ΔG of some $18 \times 10^6 \times 0.99$, or roughly 2×10^7 ergs in absolute value for solid or liquid water. If we assume a T of 300 K, the *corresponding* ΔG for an ideal gas undergoing a pressure change from 0.01 to 1 atm is 1 mole $\times$ 8.314 $\times 10^7$ erg/deg/mole $\times$ 300 deg $\times$ 4.61. The value 4.61 is the natural logarithm of 1 atm/0.01 atm. The product gives a value some 10^{11} ergs.

This calculation is representative. In most instances the pressure-dependent Gibbs energy changes of solids or liquids will be negligible with respect to those involving gases. We therefore obtain the very useful simplification that *the activities of solids or liquids may be assumed to be unity for reactions involving gas-phase constituents.*

In geochemistry, there are, certainly, important cases where a gas phase is not involved in a reaction. Here, to find the free energy change, one integrates Equation 4.26 with the help of compressibility data.

To recapitulate, when we need the Gibbs energy for some arbitrary pressure, we first find G^0 with the help of the *JANAF* or comparable tables. This is done for the temperature in question. We then use Equations 4.33 or 4.26 to find the effect of the pressure change.

The new quantities that we have introduced, the fugacity and the activity, are state functions in the same sense as pressure or entropy. Because of the way in which standard states have been defined, it is usually convenient to work with them in units of atmospheres. A useful energy unit is then the liter-atmosphere, and the gas constant $\mathscr{R}$ is often given as 0.082 06 liter-atm/mole/deg.

4.7 Heat Capacities

We define here, for future reference, the heat capacities at constant pressure, C_P, and constant volume, C_V. We have

$$C_P = \left(\frac{\partial H}{\partial T}\right)_P, \tag{4.34}$$

and

$$C_V = \left(\frac{\partial E}{\partial T}\right)_V. \tag{4.35}$$

For an ideal gas, $C_V = (3/2)\mathscr{R}$ per mole. while $C_P = (5/2)\mathscr{R}$ per mole. The *specific heat* of a substance is the ratio of its heat capacity to that of water at 15 °C. The ratio of the specific heats, C_P/C_V, is often designated by the Greek letter gamma (γ).

4.8 Chemical Equilibrium

Consider a general equation of the form

$$\alpha A + \beta B + \cdots \rightarrow \gamma C + \delta D + \cdots. \tag{4.36}$$

The equation states that α moles of a substance A reacts with β moles of B and possibly other things, to produce γ moles of C, etc. With Equation 4.33, we readily find

$$\Delta G = \Delta G^0 + \ln\left(\frac{a_{\mathrm{C}}^{\gamma}\, a_{\mathrm{D}}^{\delta}\cdots}{a_{\mathrm{A}}^{\alpha}\, a_{\mathrm{B}}^{\beta}\cdots}\right). \tag{4.37}$$

Let us make use of Equation 4.37 to investigate a reaction of great cosmochemical significance. The mineral enstatite may react with CO_2 gas to produce calcite and silica, viz.

$$\mathrm{MgSiO_3(c) + CO_2(g) \rightarrow MgCO_3(c) + SiO_2(gl)}. \tag{4.38}$$

The parenthesized letters indicate crystalline, gas, and amorphous glass, for SiO_2. If we take the activities of the solids as unity, Equation 4.37 simplifies in this case to

$$\Delta G = \Delta G^0 - \mathscr{R}T \ln\left[a\,(\mathrm{CO_2})\right]. \tag{4.39}$$

We now investigate the conditions under which the reaction 4.38 will be in equilibrium. According to the discussion in Section 4.4, we put ΔG equal to zero, and obtain

$$\log_{10}\left[a\,(\mathrm{CO_2})\right] = \frac{\Delta G^0}{2.303\mathscr{R}T}. \tag{4.40}$$

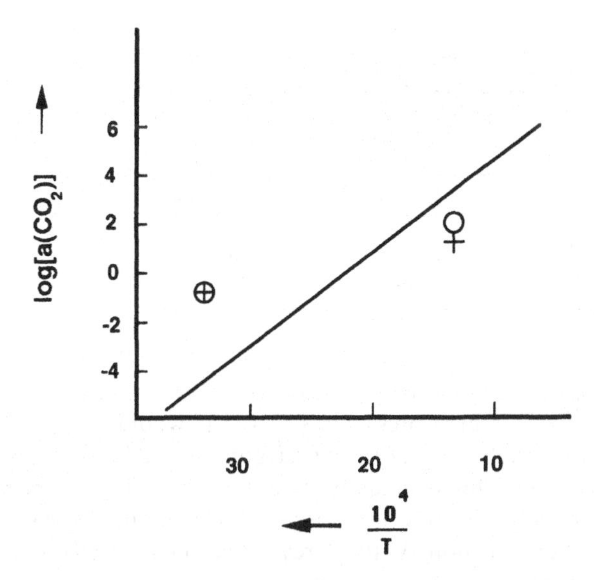

Figure 4.3. The Equilibrium Line. Gaseous CO_2 plus silicates on the right, carbonates plus silica (SiO_2) on the left. Symbols for the earth and Venus are placed according to the surface temperatures and partial pressures. Logarithms are base 10.

It is common in chemistry to convert to base-ten logarithms, so the numerical constant $2.303 = \log_{10}(e)$ is often seen in such relations.

The standard free energy change ΔG^0 is usually a rather slow function of the temperature, so that for heuristic purposes, we may assume it is constant. We then find that the equilibrium relation 4.40 is a straight line in a plot of $\log[a(CO_2)]$ vs. $(10^4/T)$, as is illustrated in Figure 4.3. To the left of the equilibrium line in Figure 4.3, the reaction 4.38 proceeds to the right, implying the production of carbonates and SiO_2 from the pyroxene and CO_2 gas. For temperatures and CO_2 pressures belonging to the right of the line, the reaction would proceed right to left, giving silicates and CO_2 gas.

Equation 4.38 appears in Harold Urey's (1952) celebrated book *The Planets.* He used it to explain why Venus has a dense CO_2 atmosphere, while on the earth, the CO_2 is primarily in the form of carbonate rocks. Urey's original argument is unnecessarily complicated because the high surface temperature of Venus was unknown at the time. Urey wrote that "Venus owes its large nonequilibrium carbon dioxide atmosphere to the slowness of the chemical reactions and the lack of erosion, both the consequence of the absence of water." We now know there is no need to assume non-equilibrium.

The data in Table 4.1 are taken from Robie and Waldbaum (1968) who use calories rather than joules. From the table, we may calculate $\Delta G^0 = -5765$ cal at 298 K, and +10 030 at 700 K. Thus, if the CO_2 pressure on the earth and Venus were 1 atmosphere, we would expect the formation of carbonates on the earth, and the decomposition of carbonates on Venus.

Table 4.1. *Thermodynamic Data for the Urey Reaction*

	$\mathscr{F}$(cal/deg)		
Compound	298 K	700 K	$\Delta H^0_{f,298}$(cal)
$MgSiO_3$(s)	16.22	22.46	−370 140
CO_2(g)	51.06	53.83	− 94 051
$MgCO_3$(s)	15.70	21.71	−266 081
SiO_2(s)	11.33	14.88	−215 870

In actuality, the partial pressure of CO_2 on Venus is some 100 atmospheres, while on the earth, it is roughly 10^{-4}. A high pressure of CO_2 gas would be expected to force the Urey reaction to the right (formation of carbonates). We must therefore take the pressures into account. This is actually done in Figure 4.3, where points have been plotted approximately for the atmospheres of Venus and the earth. We can see that the sense of the direction of the Urey equation is unchanged after adjustment for the CO_2 partial pressures.

Urey's overall analysis, though incomplete because of a lack of knowledge of the surface temperature of Venus, was nevertheless brilliant! Especially important is the fact that various catalytic processes are necessary to speed the reaction of Equation 4.38 from left to right (to form carbonates). No self-respecting pyroxene is going to decompose overnight because of a little CO_2 gas. On the earth, the silicates decomposed because of two basic processes. The CO_2 dissolved in water to form the weak acid H_2CO_3 (carbonic acid). In this form, the silicates could be more efficiently attacked by carbon dioxide, although the process is still slow on the scale of human experience. Animal life has also made a major contribution to the formation of carbonates on the earth, through fossilized bones and shells.

4.9 Statistical Mechanics: Distributions

We must now discuss certain aspects of thermal physics that deal with the distribution of particles and momenta. Thus far, our discussion of state variables has dealt with mean values: of temperature, energy, pressure, etc. For matter containing a large number of particles, such mean values are often sufficient. However, there are instances where mean values do not provide enough information. We may need to know, for example, not merely the mean energy per particle $E/\mathscr{N}$, but the fraction of particles with energies between certain limits, say E_1 and E_2. The discipline that deals with such matters is known as statistical mechanics.

The distribution of particle energies turns out to depend, in general, on the quantum mechanical nature of the particles. The formalism that takes these properties into account is known as *quantum statistics.* There is, of course, a broad domain of practical importance in which the quantum nature of particles is unimportant. We can derive the distribution of particles within this domain by means of *classical statistics*, ignoring all considerations of the quantum nature of matter. On the other hand, these same relations can be derived from quantum statistics, *in the classical limit*, and it is the latter approach that we must take here. The equilibrium relations derived in §4.8 avoid this issue only because we assumed the availability of the G's. As we shall see, if these must be calculated from first principles, we need to consider quantum phenomena.

All particles that obey the laws of quantum mechanics may be divided into two categories called fermions and bosons. For reasons that we do not pursue here, fermions have half-integral spin ($\frac{1}{2}; \frac{3}{2}; \frac{5}{2}; ...$), while bosons have integral spin ($0, 1, 2, 3, ...$). Thus electrons, protons, and neutrons are fermions. The least exotic particle with integral spin is the photon, with spin 1. Spins refer to a particle's internal angular momentum in units of Planck's constant h divided by 2π, or $\hbar = h/2\pi$. The existence of these two "kinds" of particles has been known since the early days of quantum mechanics. We shall only enumerate the properties of fermions and bosons necessary for our purposes, and refer to more profound discussions (see Landau and Lifshitz 1958, Reif 1965).

Both bosons and fermions are said to be indistinguishable . This property manifests itself whenever two or more identical particles are described, and is therefore not encountered in the most elementary problems of quantum mechanics, such as the particle in a box (§4.11), or the harmonic oscillator, which deal with single particles. Indistinguishability requires that the properties of states of two or more identical particles be invariant to an exchange of particles.

For fermions, there is an additional restriction, known as the *Pauli exclusion principle*, that no two particles can occupy the same elementary state. This property of fermions may be described in various ways; for example, in the atom, we say that no two electrons can have the same four quantum numbers. For a "free" particle, there is an analogous limitation, to which we shall come shortly.

We shall now derive the distribution function for fermion energy states. Imagine a large box in thermal equilibrium containing $\mathcal{N}$ fermions. In general, these particles will not all have the same energies, but there will be a distribution in which, say, $\mathcal{N}_1$ particles have the energies E_1, $\mathcal{N}_2$, have E_2, etc. We suppose that the energies take discrete values rather than continuous ones. We shall see that this is correct if the particles are confined, or in a "box." For the present, we may imagine these energies to be very closely spaced, so the distinction between the discrete values and a continuum is unimportant.

We assume that the particles will take their most probable distribution among these energies, subject to the following restrictions:

1. the total number of particles must be constant, that is

$$\mathcal{N} = \sum \mathcal{N}_i; \tag{4.41}$$

2. the total energy must be constant; thus

$$E = \sum \mathcal{N}_i E_i. \tag{4.42}$$

The different energy states may now be thought of as compartments or boxes, but in an abstract energy space that must not be confused with the box that contains the $\mathcal{N}$ total particles. Each compartment has a certain statistical weight that gives the maximum number of particles allowed by the Pauli exclusion principle. In an atom, these statistical weights for an energy level are usually equal to $2J + 1$, where J is the total angular momentum, orbit plus spin (see Chapters 9 and 11). We shall use the symbol G_i for the weight of the ith compartment. Thus, we imagine that there are G_i different substates, all with the same energy, in which no two particles share the same four quantum numbers. Such substates are said to be degenerate.

We compute the probability of a given distribution of particles by counting the number of ways in which the distribution may come about. For example, with $\mathcal{N}_i$ particles in a box with G_i slots, we have G_i choices for the first particle, $G_i - 1$ for the second, and so on, with $G_i - \mathcal{N}_i + 1$ choices for the last. If we multiply these possibilities out, the product may be written $G_i!/(G_i - \mathcal{N}_i)!$. We have, of course, assumed that $G_i > \mathcal{N}_i$; if the inequality went in the opposite direction, the Pauli principle would be violated.

There is only one more consideration that must be added to the results thus far assembled. In order to take account of the indistinguishability of the fermions, we divide from the above quotient those $\mathcal{N}_i!$ arrangements of the particles that merely represent permutations of the identical particles.

We obtain the *thermodynamic probability* W on taking the product of the number of "ways" (that we have decided to count!) for each partition. Thus

$$W = \prod_i \frac{G_i!}{\mathcal{N}_i!(G_i - \mathcal{N}_i)!}. \tag{4.43}$$

The quantity W is not a true probability, because it has not been normalized to unity for all possible values of the $\mathcal{N}_i$'s. However, it is proportional to a probability, and we now *assume* that the true distribution of $\mathcal{N}_i$'s will be that one for which W takes its maximum value, subject to the two side conditions (1) and (2). Rather than W, we work with its more amenable natural logarithm, $\ln(W)$, which will shortly be related to the entropy of the system.

Formally, we may say that $\ln(W)$ is a function of the multiple variables $\mathcal{N}_i$, and we wish to find a maximum of the function with two side conditions. This is a classical problem in mathematics that may be solved with the method of Lagrange multipliers, which is described by almost any text in advanced calculus (see Kaplan 1991, Arfken 1985). To find the maximum of $\ln(W)$, we let the variables undergo arbitrary variations, $\delta\mathcal{N}_i$. There will be a maximum of $\ln(W)$ when the resulting change, $\delta \ln(W)$, is zero. We include the two side conditions by multiplication by arbitrary constants, say λ and μ, and taking the variations of 4.41 and 4.42. Altogether, we have

$$\delta \ln(W) + \lambda \sum_i \delta\mathcal{N}_i + \mu \sum_i E_i \delta\mathcal{N}_i = 0. \tag{4.44}$$

It is sufficient to evaluate $\delta \ln(W)$ using Stirling's approximation in its most rudimentary form, that is

$$\ln(n!) \simeq n \ln(n) - n. \tag{4.45}$$

When this recipe is used in 4.44, a tiresome but elementary reduction leads to

$$\sum_i [-\ln(\mathcal{N}_i) + \ln(G_i - \mathcal{N}_i) + \lambda + \mu E_i]\,\delta\mathcal{N}_i = 0. \tag{4.46}$$

We now argue that the variations $\delta\mathcal{N}_i$ must be assumed arbitrary, that is, we wish the sum in 4.46 to vanish for all possible combinations of them. This means all may be zero except for $i = 1$, all zero except for $i = 2$, and so on. The sum will only vanish under these circumstances if the summand vanishes. We thus find

$$\frac{\mathcal{N}_i}{(G_i - \mathcal{N}_i)} = \exp(\lambda + \mu E_i), \tag{4.47}$$

or

$$\mathcal{N}_i = \frac{G_i}{\exp(-\lambda - \mu E_i) + 1}. \tag{4.48}$$

We next have the problem of identification of the Lagrange multipliers. We do this for the case where $\exp(-\lambda - \mu E_i)$ is a very large number. In this case the unity in the denominator of 4.48 may be neglected. Under this assumption, which we shall justify below, 4.41 leads to

$$\mathcal{N} = \sum_i \mathcal{N}_i = \exp(\lambda) \cdot \sum_i G_i \exp(\mu E_i). \tag{4.49}$$

The neglect of unity with respect to $\exp(-\lambda - \mu E_i)$ results in an expression for $\mathcal{N}_i$ that may be obtained from classical as opposed to quantum statistics.

We define the *partition function*, Q, by the relation

$$Q = \sum_i G_i \exp(\mu E_i) = \sum_i G_i \exp(-E_i/kT), \tag{4.50}$$

where in the last term we have anticipated the identification of the second multiplier (see below).

4.10 Boltzmann's postulate: the Connection Between W and the Entropy S

A system that is left to reach equilibrium in the thermodynamic sense may reasonably be assumed to achieve that distribution of energies among its constituents that is, in some sense, the most probable. Since the second law of thermodynamics also tells us that the equilibrium state is the state of maximum entropy, it is natural to assume some connection between the entropy of a state and its thermodynamic probability.

Consider two systems with entropies S_1 and S_2 and thermodynamic probabilities W_1 and W_2. Because the entropy is an extrinsic property of systems, the combined entropy of the two systems is $S_1 + S_2$. On the other hand, since probabilities of independent events are multiplicative, the thermodynamic probability of the combined system is $W_1 \cdot W_2$. It follows that if there is any connection between S and W, the relation must be a logarithmic (or exponential) one. Boltzmann *assumed* that one may write

$$S = k \cdot \ln(W), \tag{4.51}$$

where k is the familiar Boltzmann's constant.

Let us now make use of Equations 4.7 and 4.11, the former of which holds when the only work done by the system is P–V work. We have then

$$dS(E,V) = \frac{1}{T} \cdot dE + \frac{P}{T} \cdot dV, \tag{4.52}$$

from which it follows from the rules of (total) differentiation that

$$\left(\frac{\partial S}{\partial E}\right)_V = \frac{1}{T}. \tag{4.53}$$

If we use Equation 4.45 to evaluate the factorials, we obtain from 4.43

$$\ln(W) = -\sum \left[G_i \ln\left(\frac{G_i - \mathcal{N}_i}{G_i}\right) + \mathcal{N}_i \ln\left(\frac{\mathcal{N}}{G_i - \mathcal{N}_i}\right)\right]. \tag{4.54}$$

Now if we use 4.48 through 4.50, we obtain

$$\frac{\mathcal{N}_i}{G_i} = \frac{\mathcal{N}}{Q} \cdot \exp(\mu E_i). \tag{4.55}$$

We substitute this into 4.54 after making the following simplifications. First, we assume that $G_i \gg \mathcal{N}_i$, which is tantamount to assuming the validity of classical statistics. Then

$$\ln\left(\frac{G_i - \mathcal{N}_i}{G_i}\right) \simeq -\frac{\mathcal{N}_i}{G_i}, \tag{4.56}$$

and

$$\ln\left(\frac{\mathcal{N}_i}{G_i - \mathcal{N}_i}\right) \simeq \ln(Q/\mathcal{N}) + \mu E_i. \tag{4.57}$$

Inserting these into Boltzmann's postulate 4.51, and using 4.41 and 4.42, we readily obtain a useful result, valid under the assumption of classical statistics:

$$S = k \cdot \left[\mathcal{N}\left(1 + \ln(Q/\mathcal{N}) - \mu E\right)\right]. \tag{4.58}$$

It appears from 4.58 that

$$\left(\frac{\partial S}{\partial E}\right)_V = \frac{1}{T} = -k\mu, \tag{4.59}$$

which justifies the identification in 4.50, and allows us to write the Boltzmann formula

$$\frac{\mathcal{N}_i}{\mathcal{N}} = \frac{G_i}{Q} \cdot \exp\left(-E_i/kT\right). \tag{4.60}$$

4.11 The Particle in a Box: Quantum Cells for Free Particles

Quantum phenomena enter into ionization and dissociation in a subtle way that is related to the counting of energy states. While the Boltzmann formula 4.60 can be derived entirely from classical concepts, this is not possible for the ionization or dissociation equations. If we wish to apply 4.60 to an atom, the energies E_i, obtained primarily from experiment, may be looked up in tabulations such as those of Moore to be discussed in Chapter 11. These tabulations also contain information from which the statistical weights, G_i (usually equal to $2J+1$) may be calculated.

When an atom undergoes ionization, the electron goes into energy states in the continuum, and it is obviously not practical to tabulate them. How should the corresponding statistical weights be assigned? In order to answer these questions we use an artifice that is often useful. We consider a situation in which the energy states are quantized, namely, when the particles are confined to a box. We then ask what will happen as we let the volume of the box approach infinity. Under these circumstances, the allowed energy levels approach one another very closely, and approximate a continuum.

Schrödinger's time-independent equation may be obtained by applying the principle of classical analogy. In most instances, we may simply write the classical expression for the Hamiltonian, and equate it to the constant, W. Thus if the particle momenta are p_x, etc., and the potential energy is $V(x, y, z)$,

$$\frac{1}{2m}\left(p_x^2+p_y^2+p_z^2\right)+V(x,y,z)=W. \tag{4.61}$$

We next make the substitutions

$$p_x \rightarrow -i\hbar\frac{\partial}{\partial x} \tag{4.62}$$

etc., and multiply on the right by the wave function $\psi(x, y, z)$.

For the particle in a box, the potential energy is zero within the box. We require the wave function to vanish at the walls, which is tantamount to assuming a potential function that becomes infinite there. The wave equation is

$$-\frac{\hbar^2}{2m}\left(\frac{\partial^2\psi}{\partial x^2}+\frac{\partial^2\psi}{\partial y^2}+\frac{\partial^2\psi}{\partial z^2}\right)=E\psi. \tag{4.63}$$

The trial solution $\psi = X(x)Y(y)Z(z)$ allows separation of the variables, and we obtain solutions usually familiar from introductory courses in quantum mechanics:

$$X(x)=A_x\sin\left(\sqrt{\frac{2mE_x}{\hbar^2}}\cdot x\right)+B_x, \tag{4.64}$$

where A_x and B_x are constants of integration. The requirement that $X(x)$ vanishes at the walls of the box gives $B_x = 0$, and the quantization of the energies in the x-direction. Thus if the walls of the box are located at $x = 0$ and a, $y = 0$ and b, and $z = 0$ and c, we have

$$E_x=n_x^2h^2/\left(8a^2m\right),\quad n_x=1,2,3\ldots, \tag{4.65}$$

with similar equations for y and z. The requirement that the integral over the box be unity determines the coefficients of the sine functions, thus

$$A_x = \sqrt{2/a}, \ A_y = \sqrt{2/b}, \ A_z = \sqrt{2/c}. \tag{4.66}$$

Let us now work out the partition function for the particle in a box. We shall call this Q_{tr} because in our classical model, we have assumed that only translational energy states are involved. In general, of course, internal energy states are also possible.

$$Q_{tr} = \sum_{n_x n_y n_z} \exp\left[-\frac{h^2}{8mkT} \cdot \left(\frac{n_x^2}{a^2} + \frac{n_y^2}{b^2} + \frac{n_z^2}{c^2}\right)\right]. \tag{4.67}$$

We now imagine the sides of the box to recede to infinity. This makes the energy level spacing very small, and the sum may then be approximated by an integral, viz.

$$Q_{tr} = \frac{1}{8}\int\int\int \exp\left[-\frac{h^2}{8mkT} \cdot \left(\frac{n_x^2}{a^2} + \frac{n_y^2}{b^2} + \frac{n_z^2}{c^2}\right)\right] dn_x dn_y dn_z. \tag{4.68}$$

The limits of all three integrals are $+\infty$ and $-\infty$, and the factor of $1/8$ appears because the previous values of n were all positive, covering only an octant in 3-space. The integrals in 4.68 factor into three error-type integrals, leading to the result

$$Q_{tr} = \frac{(2\pi mkT)^{3/2}}{h^3} \cdot V, \tag{4.69}$$

where $V = abc$, the volume of the box.

It is instructive to change the variables of integration in 4.68 from n_x, n_y, and n_z to the corresponding momenta p_x, p_y, p_z, which may be done with the help of the quantum conditions represented by 4.65. We have, then,

$$\frac{1}{8} \cdot dn_x dn_y dn_z = \frac{dp_x dp_y dp_z}{h^3} \cdot V. \tag{4.70}$$

This relation is of fundamental importance. Assign a free particle, without internal degrees of freedom such as spin, to a cell in six-dimensional phase space (x, y, z, p_x, p_y, p_z) of size h^3. This is equivalent to assigning the three quantum numbers n_x, n_y, and n_z to a bound particle. Alternatively, we may say that quantum weights, analogous to our quantities G_i, should be assigned to volumes of phase space, by dividing these volumes by h^3. Note that the dimensions of Planck's constant, erg-sec, are also those of momentum $\times$ length.

For free particles that have internal degrees of freedom, the complete partition function may be written

$$Q = Q_{int}\frac{(2\pi mkT)^{3/2}}{h^3} \cdot V. \tag{4.71}$$

For free electrons, protons, or other particles of spin $\frac{1}{2}, Q_{int} = 2$, for the two spin states. Atoms and molecules have more complicated internal degrees of freedom. The internal partition functions for atoms are traditionally designated by u, or $u(T)$ to indicate their dependence on temperature. We shall discuss these partition functions and their calculation in Chapter 11, after we have introduced the necessary spectroscopic nomenclature.

4.12 Chemical Equilibrium and Statistical Mechanics

With the help of Boltzmann's postulate, and the developments of §4.10, we may write the entropy in the form

$$S = k[\mathcal{N}(1 + \ln(Q/\mathcal{N}) + (E/kT))]. \tag{4.72}$$

We may then write the Gibbs energy

$$G = H - TS = E + PV - kT\mathcal{N} - kT\mathcal{N}\ln(Q/\mathcal{N}) - E. \tag{4.73}$$

When the ideal gas law is valid, this simplifies to

$$G = kT\mathcal{N} \cdot \ln(\mathcal{N}/Q). \tag{4.74}$$

Now consider the general reaction 4.36, for which we have

$$\begin{aligned} \Delta G = \mathcal{R}T \big(&\gamma \ln(\mathcal{N}/Q)_C + \delta \ln(\mathcal{N}/Q)_D \\ + \quad &\ldots - \alpha \ln(\mathcal{N}/Q)_A + -\beta \ln(\mathcal{N}/Q)_B - \ldots\big). \end{aligned} \tag{4.75}$$

by setting $\Delta G = 0$, we obtain the equilibrium relations

$$\frac{\mathcal{N}_C^\gamma \mathcal{N}_D^\delta \cdots}{\mathcal{N}_A^\alpha \mathcal{N}_B^\beta \cdots} = \frac{Q_C^\gamma Q_D^\delta \cdots}{Q_A^\alpha Q_B^\beta \cdots}. \tag{4.76}$$

The relevant partition functions are given by 4.71 for each species.

Let us consider two cases of great practical interest: the ionization of an atom, and the dissociation of a diatomic molecule. Astronomers call the ionization equation the *Saha* equation. For either ionization or dissociation of a diatomic molecule, the reaction simplifies to A $\rightarrow$ C + D. In each of these cases, there are only three species to be considered, and the exponents α, γ, and δ are each unity. Each partition function of translation contains a volume V, which may be divided into the corresponding (absolute) number $\mathcal{N}$ to yield number densities, N. Two of the factors $(2\pi kT/h^2)^{3/2}$ cancel, while the term containing the masses, $m_\mathrm{C}m_\mathrm{D}/m_\mathrm{A}$, may be written with the help of the reduced mass $(1/\mu = 1/m_\mathrm{C} + 1/m_\mathrm{D})$, since (to a very high order of precision!) $m_\mathrm{A} = m_\mathrm{C} + m_\mathrm{D}$.

Altogether, we find, for ionization or dissociation,

$$\frac{N(\mathrm{C})N(\mathrm{D})}{N(\mathrm{A})} = \frac{Q_\mathrm{int}(\mathrm{C})Q_\mathrm{int}(\mathrm{D})}{Q_\mathrm{int}(\mathrm{A})} \cdot \left(2\pi\mu kT/h^2\right)^{3/2} \cdot \exp(-E/kT). \tag{4.77}$$

The partition functions are now all internal since the translational parts have been accounted for separately. We have written the exponential term with an energy E, which could be either the *ionization energy* χ or the *dissociation energy, D*.

The $\exp(-E/kT)$ arises because it is usual to refer the zero point for the internal partition functions to the ground energies for all of the respective atoms or molecules. However, this energy is not consistent. For specificity, let us consider ionization of an atom. The same principles apply to molecular dissociation.

A consistent set of zero points for atom, electron, and ion would be the following. The electron is "free" with zero kinetic energy; the atom and the ion likewise have zero kinetic energy. The "just" freed electron plus the ion corresponds to the neutral atom with an amount of energy χ_I, namely the ionization energy of the atom. If we call the levels of the neutral atom referred to the ground state E_i, then referred to the common zero point, the energies are $E_i - \chi_I$; they are negative. Consider the *partition function* for the neutral atom, e.g., Equation 4.50. For consistency with the ion and free electron energies, we must replace the E_i that appears there by $E_i - \chi_I$. The $\exp(\chi_I/kT)$ is a constant factor for all of the levels of the neutral atom, and can be taken outside the sum.

In Equation 4.77, the constant factor has been taken out of the internal partition function for the neutral atom; it now appears in the numerator as $\exp(-E/kT)$. We used the symbol E rather than χ_I, to suggest energy. For molecules, E must be the dissociation energy (usually designated D).

4.13 Problems

1. Consider an ideal gas enclosed by a piston. Let the piston move out very rapidly, so that the gas expands, essentially into a vacuum, to a new volume that is twice that of the original volume. Calculate the corresponding entropy change in two ways. First, calculate the $\int(\delta q/T) = \int P\,dV$, for an *isothermal*, reversible expansion. State carefully why one may assume the expansion to be isothermal. Second, use Boltzmann's postulate.
2. The *van der Waals* equation of state for non-ideal gases is discussed in physical chemistry texts. It is

$$\left(P + \frac{n^2 a}{V^2}\right)(V - nb) = n\mathscr{R}T, \tag{4.78}$$

   where a and b are constants, tabulated for common gases in the *HCP* and other references. Derive Equation 4.31 under the assumption that the departures from the ideal gas law are *small*.
3. The Maxwell–Boltzmann distribution of particle velocities follows directly from Equation 4.60 if the prescription of §4.11 is used to pass from discretely defined energy levels to continuously defined ones. Use the concepts of §4.11 along with appropriate changes of variables to obtain

$$f(v_x, v_y, v_z) = \sqrt{\frac{m}{2\pi kT}} \cdot \exp\left[-\frac{m}{2kT} \cdot (v_x^2 + v_y^2 + v_z^2)\right]. \tag{4.79}$$

   Note that the function f is normalized so that its integral over all velocities is unity.
4. Show that the entropy change when an ideal gas is heated from T to $2T$ is $(3/2)n\mathscr{R} \cdot \ln(2)$.

5. Show that for an ideal gas, Equation 4.58 may be written

$$S = k\mathcal{N}[(5/2) + \ln(2\pi mkT/h^2)^{3/2}]. \tag{4.80}$$

This relation is known as the Sackur–Tetrode equation.

6. Show that the entropy changes calculated in Problems 3 and 4 are the same as those that would be obtained from the Sackur–Tetrode equation.

5

Condensation Sequences and the Geochemical Classification of the Elements

5.1 The Concept of Geochemical Classification

The Norwegian geochemist Victor Goldschmidt is the father of the notion of geochemical classifications of the chemical elements. Goldschmidt's (1954) posthumous work *Geochemistry* is still of great value. His basic aim was to divide the elements into groups which might be identified with the major divisions of the earth during its cooling history. He thought there might be three separate liquid phases, one metal, one silicate, and one primarily iron sulfide. These would be surrounded by a gaseous phase. He classified the elements from their association with, or preference for one or the other of these phases.

Let us begin with a consideration of the chemistry of meteorites and the earth. The earth may be divided into a core, a mantle and a crust. The chemistry of the core must be largely inferred, and this is essentially true for most of the mantle (Ringwood 1975, Pasteris 1984). The upper continental crust is relatively well sampled (Taylor and McLennan 1985), but it is not representative. Meteorites, on the other hand, have been repeatedly and thoroughly analyzed in the laboratory. Moreover, they are thought to be pieces of a broken-up planet, not unlike the earth (see, e.g., McSween 1987). Because of this they have been used to infer the chemistry of the earth as well as of much of the cosmos.

The meteorites may be very roughly classified as stones, stony irons, and irons, and there is a rough correspondence between the chemistry of the irons and the earth's core on the one hand, and the stones and the mantle and crust on the other. We now know that the pressure–density structure of the earth's core can be rather closely matched by numerical models if it is assumed the composition is iron and nickel with an admixture of some light element, possibly silicon, sulfur, or oxygen (Jeanloz 1990). Even before the rather sophisticated laboratory experiments supporting this conclusion, it was generally inferred that the earth's core was an iron–nickel mixture *because* of the iron meteorites.

It was argued that the meteorites probably originated in a planet or planetoid. The body differentiated chemically into a core and mantle, and was later broken by an interplanetary collision. This general hypothesis, involving the notion of meteoritic parent bodies, is widely accepted today (McSween 1987, and Chapter 7 below). Naturally, what was possible for a parent body was assumed to be possible for the earth, hence we postulated an iron–nickel core.

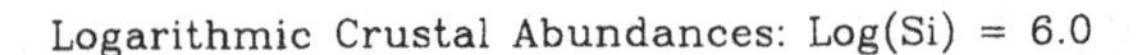

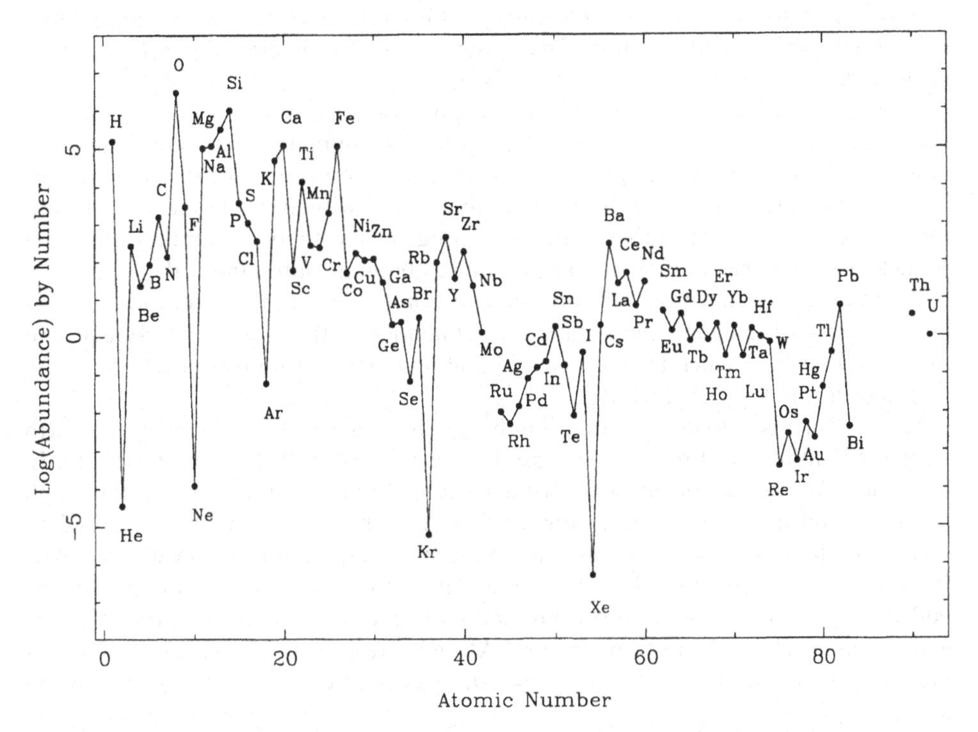

Figure 5.1. Abundances in the Earth's Crust (Carmichael 1982). Logarithms are base 10.

Goldschmidt noted that certain chemical elements were enriched in the metallic phases of meteorites. The same elements were depleted in the silicate phases. For other elements, the reverse was true, they were enriched in the silicates and depleted in the metallic phases. The abundances in these two phases showed complementary patterns.

The notion of depletion or enrichment of an element is usually made with respect to the abundances of the neighboring chemical elements in the Periodic Table. It is common knowledge that platinum and gold are rare in the earth's crust. Indeed, there is a little group of elements, rhenium, osmium, iridium, platinum, and gold, that may be seen in Figure 5.1 to be depleted by two or more orders of magnitude from a smooth extrapolation of the abundance curve, say from the even-Z elements gadolinium, dysprosium, and hafnium across to lead.

Goldschmidt called elements that were enriched in the iron–nickel phases of meteorites (and depleted in the earth's crust) siderophiles, meaning iron loving. Likewise, he called those elements that were enriched in the crust lithophiles, or rock loving.

One can see two main groups of (depleted) siderophiles in the earth's crust (Figure 5.1). The most obvious group has already been mentioned, rhenium through gold. A second group, ruthenium through silver, has an analogous atomic structure. Iron is less obviously depleted in the crust, first because of its high cosmic abundance (Figure 1.1), and second because of its presence in many minerals that are abundant in the crust.

Some lithophiles can be spotted by looking for departures from the *odd–even* alternation to be expected from the nuclear properties of matter. We shall discuss this important relation at several places in this discussion on cosmochemistry (Chapters 7 and 9). For the present, we simply state that nuclei with even numbers of protons (or neutrons) are more stable than their congeners with odd numbers of either particle. For this reason they are usually more abundant throughout the cosmos.

We have discussed earlier why the element potassium is enriched in the crust. As can be seen in the figure, the potassium abundance in the crust is nearly equal to that of the even-Z calcium. Compare the abundance distribution near potassium with that in the SAD (Figure 1.1).

Many elements with even Z are lithophiles. The sequence of lanthanide rare-earth elements (lanthanum through lutetium), both even- and odd-Z's, are lithophiles. In fact, the entire group of elements from barium through hafnium lies rather high in the crustal abundance curve (Figure 5.1), and this is reasonably attributed to their lithophile properties. Again, examine the corresponding region of the SAD. Barium ($Z = 56$) lies high, but the remainder of the elements, beginning with the odd–even pair lanthanum and cerium, and ending with the pair lutetium ($Z = 71$) and hafnium ($Z = 72$), fall low in the SAD with respect to the series of elements preceding them, starting with tin ($Z = 50$). Essentially, the reverse is true in the crust.

Goldschmidt was able to provide a sound thermodynamic basis for his geochemical classifications of siderophile and lithophile elements. In addition to these terms, he also introduced the words *atmophile*, and *chalcophile*. The former described elements such as nitrogen, carbon, oxygen, the noble gases, etc., which are found in the earth's atmosphere. The term chalcophile, which literally means "copper loving," is now used for elements with a positive affinity for sulfur. In meteorites, these elements associate with the mineral FeS (troilite).

The thermodynamic arguments given by Goldschmidt to support the chalcophile classification are complicated, and generally less satisfactory than those concerning the lithophiles and siderophiles. Consequently, we shall not reproduce them. However, there was an additional empirical basis for the notion of chalcophiles which Goldschmidt drew from metallurgy. When copper ore is smelted, a sulfur-rich phase forms which is known as matte. In this matte, the chalcophiles are, by definition, enriched.

Table 5.1 is adapted from Krauskopf's (1979) *Introduction to Geochemistry.* Entries above the line are in weight percent, those below the line are parts per million, by weight. One should note an (imperfect) trend for the same elements that are concentrated to the metallic phase of meteorites to be enriched in the metal phase of ore melts.

Goldschmidt's siderophile–lithophile classification has remained in active use in cosmochemistry. The chalcophile classification may become more useful, after we have had experience with planetary surfaces that are less depleted in sulfur than our

Table 5.1. *Chemical Analyses Illustrating Geochemical Classification. Adapted from Krauskopf (1979).*

	Meteorites			Metallurgical Products		
Element	metal	sulfide	silicate	metal	sulfide	silicate
Si	0.015	0	21.60	0.02	0.05	22.09
Al			1.83	0.05	0.0	9.11
Fe	88.60		13.25	73.58	22.92	3.0
Mg			16.63	0	0.05	7.46
Ca			2.07	0.003	~ 0.001	13.50
Na			0.82	~ 0.1	0.1	0.64
K			0.21		0.49	3.28
P	1 800	3 000	700	18 400	0	300
Cr	300	1 200	3 900	0	0	40
Ni	84 900	1 000	3 300	17 200	2 800	500
Co	5 700	100	400	24 000	2 500	40
V	6		50	800	~ 100	200
Ti	100	0	1 800	20	20	300
Zr	8	0	95			
Mn	300	460	2 050	0	6 400	2 000
Cu	200	500	2	64 400	462 000	2 340
Pb	56	20	2	20	2 200	200
Zn	115	1 530	76	8	16 800	3 700
Ag	5	19	0	150	2 520	0
Au	2	0.5	0	8	0	0
Pt	16	3	0	8	0	0
Sn	100	15	5	80	0	0
W	8	trace	18	0	0	30
Mo	17	11	3	66 400	0	20

own. One may see in Figure 5.1 that sulfur is depleted in the earth's crust relative to the even-Z oxygen, magnesium, silicon, and calcium. On the other hand, the Jovian moon Io has a surface chemistry that is dominated by sulfur, and it is probable that sulfur is enriched in the Martian soil. The problem of where the earth's sulfur complement went is unsolved. At one time it was speculated that the sulfur might be at the core–mantle boundary. A more recent idea is that it is in the outer core (Jeanloz 1990, see above). Yet another possibility is that it was never incorporated into the earth in the amounts we might expect because of its volatility.

In modern discussions of the chemistry of the planets, and also of grains in the interstellar medium, the notion of volatility plays a key role (cf. Richardson and McSween 1989). The opposite of the word volatile is refractory, a concept we have already encountered in our discussion of the refractory oxides in §2.5.1. It is simple enough to characterize a *substance* as volatile or refractory, depending upon the temperature at which it will condense from the gas phase. Materials that solidify from the gaseous state at high temperatures, or which, equivalently, vaporize at high temperatures, are called refractory. Those which solidify only at low temperatures, are called volatile. We will explore this in detail below.

A chemical element with a relatively low boiling temperature, such as sodium, may form compounds which will condense from the vapor phase at much higher temperatures than the pure element. By convention, the classification of an element depends upon the temperature at which it leaves the vapor phase, *whether as an element or as a compound.* In many cases, it is possible to make calculations that will show the condensation sequences of the chemical elements and compounds, thereby providing a sound basis for the volatile–refractory classification. However, the calculations depend upon a number of assumptions, and in many cases the assumptions are not straightforward. Consequently, since the time of Goldschmidt, cosmochemists have relied heavily on the analysis of cosmic materials (Table 5.1) to provide an empirical basis for the classification.

If one looks at the broad picture, there is no doubt that we have a sound basis for the classification of many elements, but there are still many details to be settled. Some of the problems will become clear after the discussions below.

5.2 The Siderophile–Lithophile Distinction

Consider the following reactions:

$$Mg(s) + \frac{1}{2}O_2(g) \rightarrow MgO(s), \ \Delta G^0 = -136.1; \tag{5.1}$$

$$MgO(c) + SiO_2(gl) \rightarrow MgSiO_3(c), \ \Delta G^0 = -10.0. \tag{5.2}$$

The ΔG's are in units of kilocalories ($4.18 \cdot 10^3$ joules), at 298 K. As in Chapter 4, we indicate the states of the substances: (c) means crystalline, (gl) glass, and (g) gas. What we can see is that the Gibbs energy change in forming the magnesium oxide represents the dominant part of the Gibbs energy change in the formation of the mineral enstatite. *These figures are representative of the formation of most oxides and silicates.* Therefore, in considering the thermodynamics of silicate formation, it should be sufficient, at least for purposes of orientation, to consider only the ΔG's of oxide formation.

Goldschmidt argued that since both reduced (metallic) iron and oxidized iron are present in the earth as well as in meteorites, iron would *compete* with the other elements for oxygen. He concluded that any element with an oxide for which ΔG was smaller (i.e., more negative) than the value for FeO would win over iron in the formation of oxides. Such elements would subsequently form silicates, and thus be lithophiles. Those elements for which the ΔG for an oxide formation was *less* negative than that of FeO would remain in the reduced phase, along with the metallic iron. Table 5.2, adapted from Krauskopf, gives the Gibbs energy of formation of a number of oxides at two temperatures.

The higher temperature is closer to the freezing temperatures of rocks, and is therefore probably more representative of the conditions under which the partitioning of the siderophiles and lithophiles took place. If we look, for example, at tin, its

Table 5.2. *Free Energies of Oxide Formation (kilojoules/mole)*

Oxide	298 K	1000 K
CaO	−603.5	−530.7
$(\frac{1}{2})ThO_2$	−584.4	−518.7
MgO	−568.9	−493.0
$(\frac{1}{3})Al_2O_3$	−527.4	−453.8
$(\frac{1}{2})ZrO_2$	−521.4	−455.1
$(\frac{1}{2})UO_2$	−515.9	−455.8
$(\frac{1}{2})TiO_2$	−444.7	−381.2
$(\frac{1}{2})SiO_2$	−428.2	−365.1
VO	−404.2	−343.8
MnO	−362.9	−311.8
$(\frac{1}{3})Cr_2O_3$	−351.0	−289.7
ZnO	−320.5	−248.6
$(\frac{1}{2})WO_2$	−266.9	−203.7
$(\frac{1}{2})MoO_2$	−266.5	−204.5
$(\frac{1}{2})SnO_2$	−260.0	−186.7
FeO	−251.4	−207.0
CoO	−214.0	−163.3
NiO	−211.6	−149.2
PbO	−189.3	−119.3
Cu_2O	−147.9	−95.5
HgO	−58.5	−48.0
Ag_2O	−11.2	

Sources: Robie, Hemingway, and Fisher (1978), *HCP*, and *JANAF*.

ΔG_f^0 at 298 K places it among the lithophiles. At 1000 K, it has a ΔG_f^0 that is greater (considering the sign) than that of iron which would indicate a siderophile. This is in accord with the concentration of tin to the metallic phases given in Table 5.1.

One can see from the table that nickel will lose in competition with iron for oxygen. Nickel plays an important role in the chemistry of the reduced phases of cosmic materials because of its relatively large abundance. One may see from Figure 1.1 that it is (cosmically) by far the most abundant of the siderophiles, apart from iron itself.

It is common to speak of "partition" of the elements into different minerals or mineral phases. If there was, at one time, a parent magma consisting of silicate minerals as well as iron, the siderophiles would be partitioned, preferentially, into the reduced phase as the melt froze. It would be desirable, especially for the trace elements, to know the quantitative ratios of this partitioning process, the so-called partition or distribution coefficients (see Carmichael, Turner, and Verhoogen 1974).

Goldschmidt's ideas readily account for certain regularities in the chemistry of

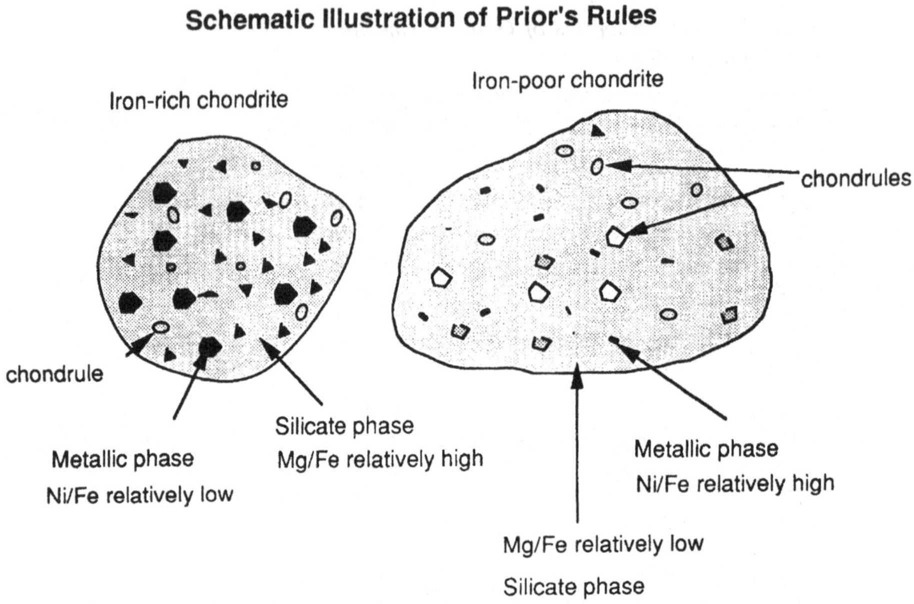

Schematic Illustration of Prior's Rules

Iron-rich chondrite
Iron-poor chondrite
chondrules
chondrule
Silicate phase
Mg/Fe relatively high
Metallic phase
Ni/Fe relatively low
Metallic phase
Ni/Fe relatively high
Mg/Fe relatively low
Silicate phase

Figure 5.2. Illustration of Prior's rules for Meteorites.

meteorites known as Prior's (1916) rules. These are illustrated, schematically, in Figure 5.2.

The stony meteorites known as chondrites typically contain small blebs of reduced metal (kamacite or taenite, the iron–nickel minerals). Prior noticed that the nickel-to-iron ratio in these blebs was variable, and that the most nickel-rich blebs were found in those meteorites having the smallest amounts of reduced metal. A second trend concerned the oxidized phase. The fraction of oxidized iron to oxidized iron *plus* oxidized magnesium decreases with the relative volume of reduced metal in the meteoritic samples.

Both of these trends are readily explained with the help of Table 5.2. It is only necessary to assume a competition for the reduced and silicate phases of magnesium and nickel with iron. Nickel will win over iron in the competition for the reduced phase, so as the relative volume of metal diminishes, the fraction of nickel in the reduced phase will increase. Similarly, as the reduced phase increases, competition for the silicate phase becomes keen. In this case the preference of magnesium over iron for the oxidized (silicate) phase becomes more apparent. Thus, the fraction of the silicate phase that contains magnesium (e.g., forsterite, Mg_2SiO_4) increases relative to the fraction that contains iron (e.g., fayalite, Fe_2SiO_4).

Prior argued that his rules implied a genetic relationship among the chondrites, as might happen if all were derived from some primitive material that had been subjected to varying amounts of oxidation. This is not widely held today. We shall return in a later chapter to the problem of genetic or evolutionary inferences made from correlated observational results.

Table 5.2 provides a good basis for the empirical results of Table 5.1. Those elements enriched in the silicate phases, for both the ore products and the meteorites, have more negative values of ΔG than iron. Those enriched in the metal phase have more positive values than iron. We do not show entries for oxides of gold or platinum. As is well known, these metals do not oxidize easily.

5.3 The Problem with the Chalcophiles

We could try to explain the chalcophile property on the basis of ΔG's for the formation of sulfides. As we have anticipated, this is not satisfactory. Let us attempt a simple comparison of the ΔG_f^0's for iron and copper, and try to draw an analogous conclusion to that made for nickel and iron with regard to the siderophile–lithophile classification. According to the data in the *Handbook of Chemistry and Physics* (61st Edition, Weast 1980), ΔG_{298}^0 is -23.3 kilocalories for FeS, while it is -11.7 for CuS. Consequently, iron would be expected to win over copper in a competition between the two reduced elements for a sulfide phase.

This is not compatible with the empirical classification of copper as a chalcophile, nor with the appearance of 46.2% copper (by weight) in the sulfide matte of Table 5.1 compared with 22.9% iron.

The resolution of this difficulty is not yet at hand. Almost certainly, the empirical classification results from processes that are more involved than for oxides and silicates. The analogy with lithophiles and the oxide formation is therefore unsatisfactory. In that case, we could say that once the oxide had formed, geochemical processes would lead naturally to the silicates. Note that in Table 5.1 we assumed a constant pressure for gaseous O_2. The processes that lead to association with the sulfide phases cannot be so simple. It would be necessary to know details of these processes to understand why an element would prefer the sulfide to the reduced (metallic) phase. We have noted that Goldschmidt does attempt to provide a thermodynamic basis for this classification, but the arguments will not be followed here.

5.4 Molecular Equilibria in the Gas Phase

Before we examine the condensation temperatures of various substances it is necessary for us to determine the partial pressures of expected molecules in the gas phases. This problem was attacked by H. N. Russell (1934) in a classic study of molecular formation in stellar atmospheres, where the actual problem of condensation was not considered. In modern work on this problem, condensation is thought to play a key role in the formation of interstellar solids.

The example given here will be limited to a discussion of abundant diatomic molecules and water, the latter as a token polyatomic species. The general principles will thus be illustrated, but we eschew a presentation of the tediously long equations necessary to represent a realistic condensation sequence at low temperatures. A few mineral condensations will be explicitly discussed in the following section.

For a diatomic molecule, of arbitrary atoms A and B, we have

$$AB \rightarrow A + B. \tag{5.3}$$

and by Equation 4.37,

$$\Delta G = \Delta G^0 + \mathscr{R}T \ln \left[\frac{P(A)P(B)}{P(AB)} \right]. \tag{5.4}$$

At equilibrium, we therefore have

$$\log \left[\frac{P(A)P(B)}{P(AB)} \right] = -\frac{\Delta G^0}{2.303\mathscr{R}T}. \tag{5.5}$$

We have replaced the activities by the corresponding partial pressures under the assumption that the gases are ideal.

The right-hand side of Equation 5.5 may be calculated with the help of thermochemical data, and cosmochemists have typically made use of such data. Astronomers, following Russell, have often made *ab initio* calculations of dissociation equilibrium constants for diatomic molecules, viz.

$$\frac{P(A)P(B)}{P(AB)} = kT \frac{Q_A Q_B}{Q_{AB}} \left(2\pi\mu kT/h^2\right)^{3/2} \exp(-D/kT). \tag{5.6}$$

The Q's are atomic and molecular partition functions, μ is the reduced mass of the molecule ($1/\mu = 1/m_A + 1/m_B$), D is the dissociation energy, and k is Boltzmann's constant ($k = \mathscr{R}/\mathscr{N}_a$). The derivation of Equation 5.6 is given in §4.12 and, for example, by Cowley (1970).

The right-hand side of 5.6 is often designated $K_P(T)$, the pressure-equilibrium constant for a given temperature. Here, we shall write it $K(AB)$, for convenience. The dimensions of $K(AB)$ are clearly those of pressure *for the case of a diatomic molecule.* In the astronomical literature, cgs units are commonly used, so published values of $K(AB)$ are typically (Aller 1963, Tsuji 1973) in dynes/cm^2. On the other hand, it would follow from the use of activities and the derivation of Equation 5.5 that the natural units of $K(AB)$ would be atmospheres.

The *JANAF*, and some other thermochemical tables, give $\log(K_P)$ for each temperature. Some arithmetic can be saved by using these values directly, but the user must be careful to note precisely how these K_P's are defined, and what the relevant standard states are. For example, for the element hydrogen, the standard state is defined to be gaseous H_2 for all temperatures. This is a departure from the usual situation, where the standard state is defined to be the most stable state for a given temperature. For sulfur, for example, the standard state is crystalline to 388.36 K, thence liquid to 717.75 K, and finally an ideal, diatomic gas. The phase changes are indicated in Figure 4.2.

For H_2, the *JANAF* tables give 0.00 for all temperatures for $\log(K_P)$, as well as for ΔG_f^0 and ΔH_f^0. This is because H_2 is defined as the standard state. The relevant value is to be found under H as a monatomic gas. *JANAF* gives data for the reaction

$$\frac{1}{2}H_2(g) \rightarrow H(g). \tag{5.7}$$

On the other hand, the $\log(K_P)$ given by Aller and Tsuji in the references cited is for the reaction

$$2H(g) \rightarrow H_2(g). \tag{5.8}$$

One may see from Equations 5.7 and 5.8, that the appropriate ΔG's will differ by a factor of 2 *and* a sign change in addition to a possible change in units! Consider the specific value of $\log(K_P)$ for 2000 K that may be calculated (see Problem 2 below) from the data tabulated by Tsuji: 0.441. *JANAF* gives -2.790 for H(g) at 2000 K. To get Tsuji's value, multiply this by 2, and add 6.0 to convert from *JANAF*'s atmospheres to dynes/cm^2. We get 0.420. The difference, about 5% in K_P, is reasonably attributed to the approximations used to evaluate the molecular partition functions.

Let us now consider the diatomic molecules that may be formed from the elements hydrogen, carbon, oxygen, and nitrogen. We shall also include water. We now define the *fictitious pressure* of any atomic species as the partial pressure that would hold if all of the molecules containing that species were dissociated. The fictitious pressure ratios are identical to the abundance ratios by numbers. We designate the appropriate fictitious pressures as, for example, $P_f(\mathrm{H})$, for hydrogen.

In any calculation of molecular equilibria, it is necessary to know (or assume) a value of the total gas pressure, the temperature, and the relative abundances of the elements. For the latter, it is common to use the symbol ϵ, and to choose numerical values relative to some arbitrarily chosen standard. In geochemistry, the common choice is $\epsilon_{\mathrm{Si}} = 10^6$, while in astronomy, one usually assumes $\epsilon_{\mathrm{H}} = 10^{12}$. Thus at the beginning of the calculation we know the ratios of the fictitious pressures, but their absolute values are not yet known. The solution begins by assuming (guessing!) a value for one of the fictitious pressures. If the temperature is high, it is reasonable to assume the partial pressure of monatomic hydrogen is close to the total gas pressure. Once this assumption is made, one may write a series of equations for each atomic species of the form

$$\left.\begin{array}{lcl} P_f(\mathrm{H}) & = & P(\mathrm{H}) + 2P(\mathrm{H_2}) + P(\mathrm{CH}) + P(\mathrm{NH}) + P(\mathrm{OH}) + 2P(\mathrm{H_2O}), \\ P_f(\mathrm{C}) & = & P(\mathrm{C}) + 2P(\mathrm{C_2}) + P(\mathrm{CH}) + \ldots, \\ P_f(\mathrm{N}) & = & P(\mathrm{N}) + 2P(\mathrm{N_2}) + P(\mathrm{NH}) + \ldots, \\ P_f(\mathrm{O}) & = & P(\mathrm{O}) + 2P(\mathrm{O_2}) + P(\mathrm{CO}) + \ldots. \end{array}\right\} \tag{5.9}$$

We can express the right-hand sides in terms of the four unknowns $P(\mathrm{H})$, $P(\mathrm{C})$,

Solar composition

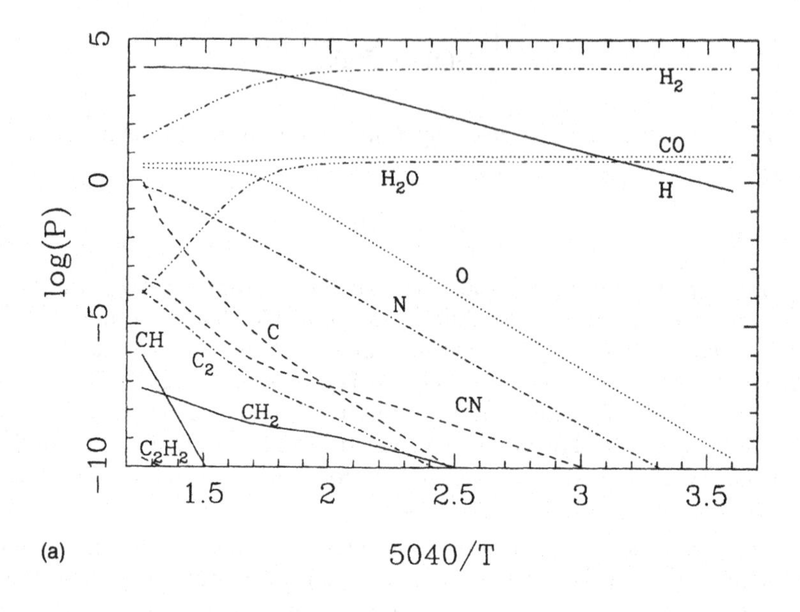

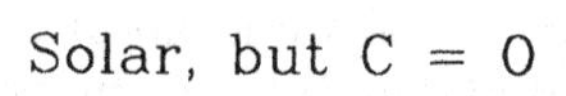

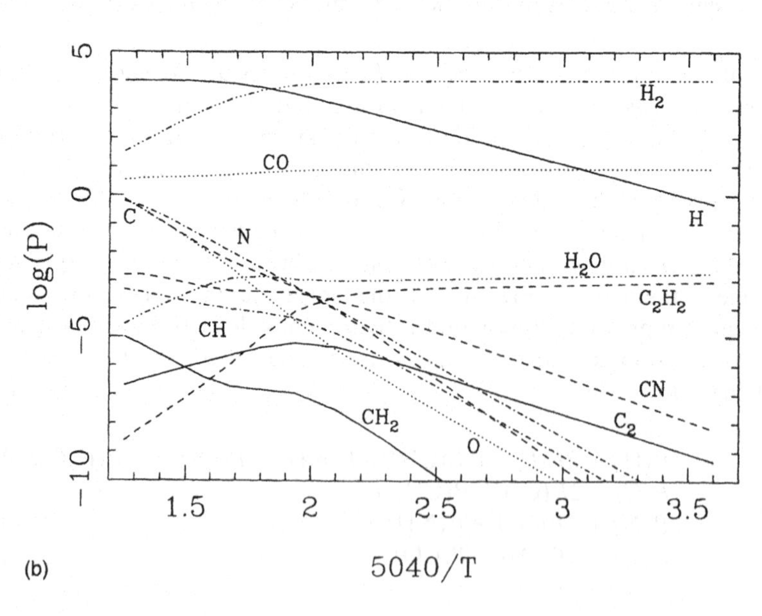

Figure 5.3. For legend see opposite page.

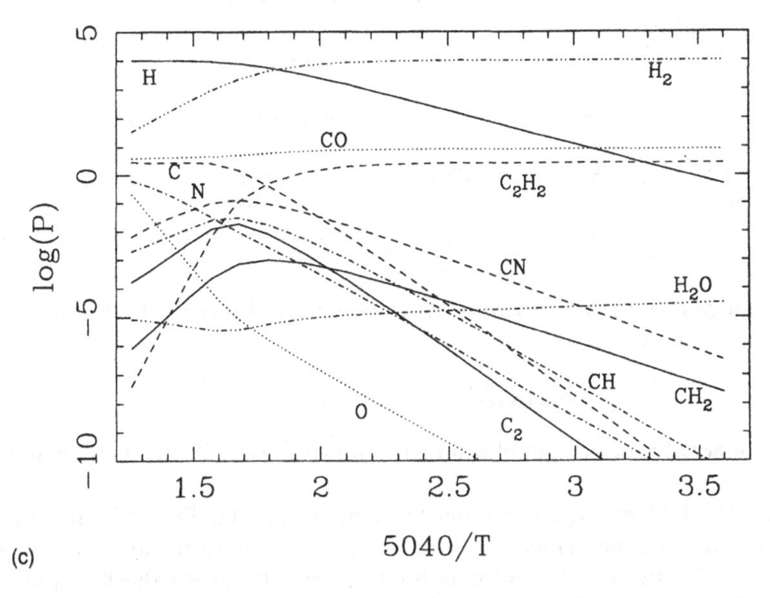

Figure 5.3. Molecular Equilibria at Moderately Low Temperatures. Courtesy of Carol Wissel

$P(\mathrm{N})$, and $P(\mathrm{O})$ with the help of the equilibrium constants. Thus

$$\left.\begin{aligned} P_f(\mathrm{H}) &= P(\mathrm{H}) + 2\frac{P^2(\mathrm{H})}{K(\mathrm{H_2})} + \frac{P(\mathrm{C})P(\mathrm{H})}{K(\mathrm{CH})} + \ldots, \\ P_f(\mathrm{C}) &= P(\mathrm{C}) + 2\frac{P^2(\mathrm{C})}{K(\mathrm{C_2})} + \frac{P(\mathrm{C})P(\mathrm{H})}{K(\mathrm{CH})} + \ldots, \\ P_f(\mathrm{N}) &= P(\mathrm{N}) + 2\frac{P^2(\mathrm{N})}{K(\mathrm{N_2})} + \frac{P(\mathrm{N})P(\mathrm{H})}{K(\mathrm{NH})} + \ldots, \\ P_f(\mathrm{O}) &= P(\mathrm{O}) + 2\frac{P^2(\mathrm{O})}{K(\mathrm{O_2})} + \frac{P(\mathrm{O})P(\mathrm{H})}{K(\mathrm{OH})} + \ldots. \end{aligned}\right\} \tag{5.10}$$

These relations are a coupled set of four quadratic algebraic equations. The solution is carried out by numerical methods. T. Tsuji (1973) employed a Newton–Raphson method, which gave "reasonably fast convergence" even when several hundred molecular species were treated. At the University of Michigan, we have used Newton–Raphson with good success. We recommend the use of symbolic manipulators for the partial differentiation that is required. These programs typically have the ability to write output in a form that can be used directly in FORTRAN or other numerical codes.

Tsuji's paper also gives a parameterization of the dissociation constants for a large number of polyatomic as well as diatomic molecules. For the former, the

Table 5.3. *Thermochemical Data for the Condensation of Solid Fe*

T (K)	$10^4/T$	ΔG^0 (cal)	$\log P_{\text{atm}}(\text{Fe})$
1500	6.67	46 998	−6.85
1700	5.88	40 607	−5.22
1900	5.26	34 338	−3.95

equilibrium constant may be defined with the help of Equation 4.37. For a diatomic molecule ($AB \rightarrow A + B$),

$$\log[K(\text{AB})] = -\Delta G^0/(2.303\mathcal{R}T). \tag{5.11}$$

We have noted that the equilibrium constants are given in the *JANAF* and similar tables.

The three plots in Figure 5.3 illustrate molecular equilibria in the gas phase for three values of the critical C/O ratio. The temperature range shown is 4000 to 1400 K. The sturdy CO molecule binds essentially all of the less abundant of the two elements, carbon or oxygen. In the case of the solar composition, where $\epsilon_C < \epsilon_O$, some oxygen is available for the formation of the strong TiO bands seen, for example, in the M stars. TiO is not shown here. When $\epsilon_C > \epsilon_O$ the oxygen partial pressure drops precipitously, and carbon species are evident: C_2, CN, and the intriguing C_2H_2, acetylene. These calculations do not include the formation of solid carbon. When $\epsilon_O > \epsilon_C$, the partial pressure of atomic carbon is always too low for solid carbon to form in this temperature range. When $\epsilon_C > \epsilon_O$, graphite is stable at the low-temperature end. However, its partial pressure is reduced, especially by C_2H_2. Under these circumstances, larger carbon molecules can form – perhaps even the *polycyclic aromatic hydrocarbons* or PAH's (see Chapter 14).

The calculations upon which the figures are based include a number of species not shown, although no molecules more complicated than C_2H_2.

5.5 The Condensation of Solids

The condensation of a pure elemental solid is simple to understand. Consider the condensation of metallic iron.

$$\text{Fe(g)} \rightarrow \text{Fe(c)}. \tag{5.12}$$

Following the developments in §4.6, we set the activity of the solid equal to unity, and assume the activity of the gas is equal to its pressure (in atmospheres). Then

$$\log[P(\text{Fe})] = \Delta G^0/(2.303\mathcal{R}T). \tag{5.13}$$

It is convenient to represent this equation in a plot of $\log[P(Fe)]$ vs. $10^4/T$, with T in degrees Kelvin. Thermodynamic data from the second edition of the *JANAF* tables are shown in Table 5.3, and plotted in Figure 5.4. The plot closely resembles

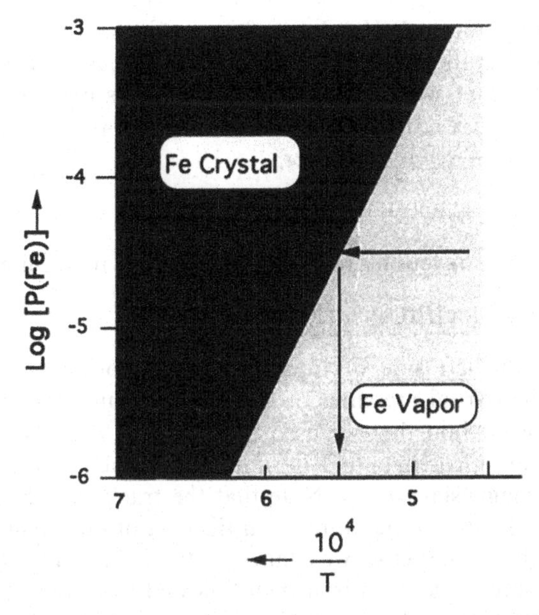

Figure 5.4. Equilibrium for Iron in Solid and Vapor Phases.

Figure 4.3; in the present situation, we deal with a phase change rather than a chemical reaction.

We can see from the figure that the temperature at which iron can be expected to condense from the gaseous phase is a function of the partial pressure of iron in the vapor phase. Figure 5.4, like Figure 4.3, may be divided into two zones. On the upper left, we have solid, metallic iron; on the lower right, iron vapor. The condensation temperature is fixed as soon as the partial pressure of iron is given.

The diagram indicates a "trajectory" for the vapor based on an assumed constant (partial) pressure of iron of $10^{-4.5}$ atm. The corresponding condensation temperature is about 1820 K. In reality, *all* of the vapor will not condense. There will always be some, finite vapor pressure in equilibrium with a solid for a given temperature, and workers in this field often speak of "50% condensation temperatures." These may be defined as the temperatures for which the equilibrium vapor pressure drops to 50% of its value at the point where the cooling trajectory initially crosses the equilibrium line. Details are left as a problem.

Iron can condense as a liquid at the high temperature–pressure domain of the current calculations. One needs to examine the competition between the liquid and solid phases to see which is relevant. This is also left as a problem.

When all molecular formation is negligible, the partial pressure of iron vapor is simply $\epsilon(\mathrm{Fe})/[\sum \epsilon]$ times the total gas pressure P_T. In general, this total pressure is fixed by a model of the solar nebula. The quantity $\sum \epsilon$ represents the sum of the relative numbers of atoms of all elements. For a fixed P_T, the partial pressure of iron will be sensibly constant with temperature in most cosmic mixtures, for wide ranges of temperature which permit gaseous Fe. This is because $P_T \approx P(\mathrm{H}) + P(\mathrm{He})$, or if the hydrogen is molecular, $P_T \approx P(H_2)+P(He)$. In the former case, we have $P(Fe) \approx P_T \times [\epsilon(Fe)/(\epsilon(H)+\epsilon(He))]$, and in the latter $P(Fe) \approx P_T \cdot [\epsilon(Fe)/(0.5\epsilon(H)+\epsilon(He))]$.

A number of complications enter when many solid compounds condense. We discuss them here, and again in §6.6. Grossman and Larimer (1974), in a detailed review, discuss the condensation of Al_2O_3. Consider

$$2Al(g) + 3O(g) \rightarrow Al_2O_3(c). \tag{5.14}$$

The equilibrium conditions for this reaction are readily written down. We find

$$2\log[P(Al)] + 3\log[P(O)] = (1/2.303\mathscr{R}T)\Delta G^0. \tag{5.15}$$

Grossman and Larimer plot the left side of Equation 5.15 as ordinate against $10^4/T$ (cf. also Figure 6.3). The partial pressures of gaseous aluminum and oxygen may be estimated by the same method that we used for iron, and the equilibrium intersect plotted. Figure 5.5, from Grossman and Larimer, shows such a plot, along with condensation tracks for magnesian spinel. Note that the tracks for the vapor pressure are not flat (zero slope) with temperature, as in the case of our assumption for the iron condensation. In the former case, we simply postulated a constant total gas pressure. The partial pressure of gaseous iron then followed from the assumed abundance (see Problem 1 below).

If we assume some detailed model of the solar nebula, including relative abundances, the temperature and pressure would in general be functions of some third state variable, for example, the density. We could not plot the curve giving the temperature–(partial) pressure evolution within the solar model unless the density were specified by some additional assumptions. In early work of this kind, an adiabatic equation of state was assumed for the solar nebula. In this case, one state variable would be sufficient to fix the others, and a cooling track could then be plotted, as in the upper part of Figure 5.5. This cooling track is the line coming down from the upper right-hand corner, to meet the heavy equilibrium line at the point marked $T_c = 1758$.

The lower part of the diagram, marked SPINEL SATURATION, contains the heavy equilibrium curve for the reaction

$$Mg(g) + 2Al(g) + 4O(g) \rightarrow MgAl_2O_4(s). \tag{5.16}$$

The solar nebula track representing the vapor phase comes down from the right as before, with the relevant ordinate being read from the scale on the right-hand side of the figure. The partial pressures are calculated as before from the assumed abundances and total pressure (given by T for an adiabatic gas). At $T = 1758$, however, the aluminum must all leave the gas phase, along with 3 oxygens for every 2 aluminum atoms. If the gas in the solar nebula were constrained within some volume, we would then expect a discontinuous drop in the curve representing the sum of the partial pressures. Precisely what happens depends on the assumptions of the model. In constructing this particular figure, it has been assumed that the nebula adjusts itself in such a way that this discontinuity in pressure has *not* taken place. It is reasonable to assume that when some solids condense, the nebula itself contracts in such a way as to prevent a discontinuity in the pressure on such curves.

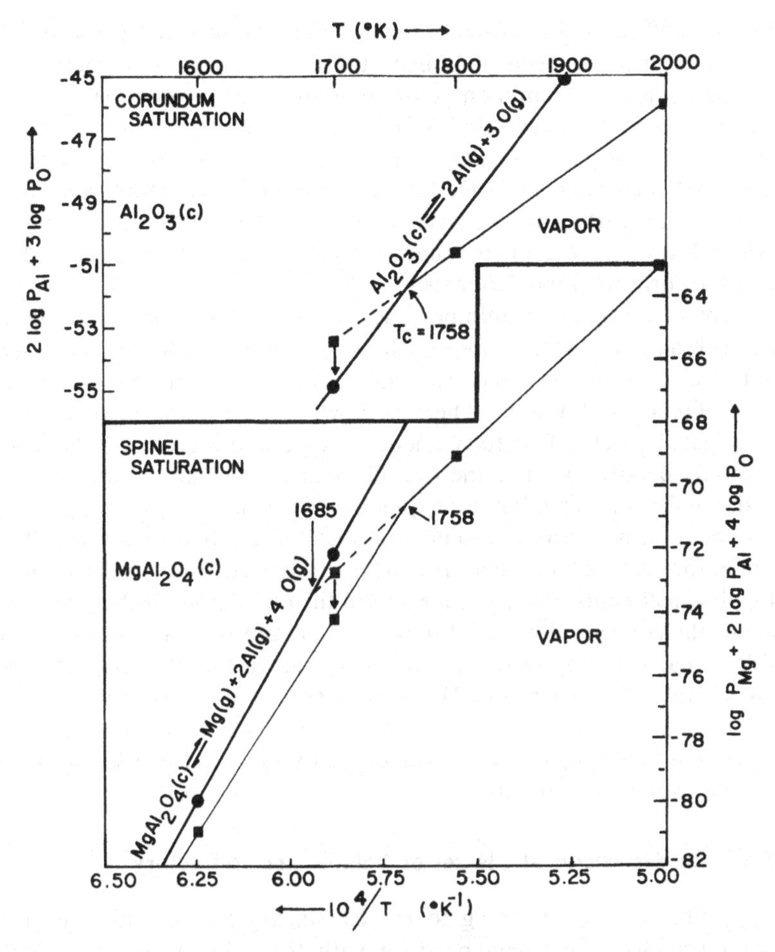

Figure 5.5. Condensation of Corundum and Spinel. From Grossman and Larimer, *Reviews of Geophysics and Space Physics*, **12**, 71, copyright American Geophysical Union.

Although the pressure does not drop discontinuously, it is clear that the slope of the partial pressure vs. $10^4/T$ curve must change, because the composition of the gas has changed. There is no longer *any* gaseous Al and the partial pressure of oxygen is proportionately lower. Because of this change in slope, the condensation of the magnesian spinel does not take place at $T = 1685$. In fact, solid $MgAl_2O_4$ would appear at a much lower temperature. We would have to consider the reaction

$$Mg(g) + O(g) + Al_2O_3(c) \rightarrow MgAl_2O_4(c) \tag{5.17}$$

and *its* equilibrium line to find that temperature. According to Grossman and Larimer, solid spinel appears at about $T = 1500$ K when solid Al_2O_3 reacts with gaseous magnesium and oxygen.

Our discussion of the condensation of corundum and spinel, complicated though it may seem, has nevertheless been simplified. It is almost certain that solid corundum does *not* condense directly from monatomic aluminum and oxygen. At the very least, we would expect the molecules O_2 and AlO to form in the gaseous phase. The *JANAF* tables contain data on gaseous AlO_2, Al_2O, as well as on positive and negative gaseous ion molecules such as AlO_2. These additional molecules will have little effect on the overall calculations provided the other molecules are "intermediate" species, which form over a limited range of temperatures and pressures, and they may all react to form the final products.

This concept of the relative unimportance of "transient" species is unnecessary for species such as gaseous AlO, for which we can find thermodynamic data. In a detailed set of calculations, we may account for all species, and put in the relevant chemistry insofar as we know it. There is, however, the question of unknown or poorly investigated species. Planetary scientists hope that a more complete analysis will change only details, and not the overall picture. We may, in any case, accept the above discussion of corundum and spinel as heuristic.

We must mention here an additional complication relative to condensation and equilibrium in the early solar system. If condensed material remains as a fine dust, it can react with constituents of the gas phase. On the other hand, if it coagulates, only the outside of the condensed material may so react, and the bulk of the condensate is incapable of participating in further gas-phase chemistry. We shall consider this problem in Chapter 6. Sharp and Huebner (1990), and Sharp (1988) discuss an alternate analytical procedure to the one given here. Their method minimizes the Gibbs energy of all relevant species. It has the advantage that it treats both gas and condensed phases in a similar way.

5.6 The Geochemical Classes of Volatile and Refractory

It is now possible to give meaning to the classification of chemical elements as volatiles or refractory. This may be done with the help of specific calculations such as the ones outlined in the preceding sections. *Chemical elements that are removed from the vapor phase at high temperatures are called refractory. Those which remain in the vapor phase until very low temperatures are called volatile.* If an element condenses out at a high temperature as an oxide, it is refractory, even though it might not condense as a pure element until a much lower temperature.

It should be noted that the condensation temperatures are dependent on the assumed model, and especially upon the assumed chemical composition. Perhaps the most sensitive factor is the carbon-to-oxygen ratio. If this is greater than unity, the condensation chemistry changes in a fundamental way. In particular, many of the high-temperature condensing oxides will not form; many of the chemical elements will condense as carbides or sulfides, at temperatures that depend on the thermochemistry of *these* molecules (Lattimer, Schramm, and Grossman 1978).

The word "refractory" is commonly used both in the sense of a geochemical classification and simply to mean a material that remains in the solid phase at high temperatures. Thus we may properly speak of aluminum as a refractory element, and of Al_2O_3 as a refractory oxide, with a subtle change in the meaning of the word.

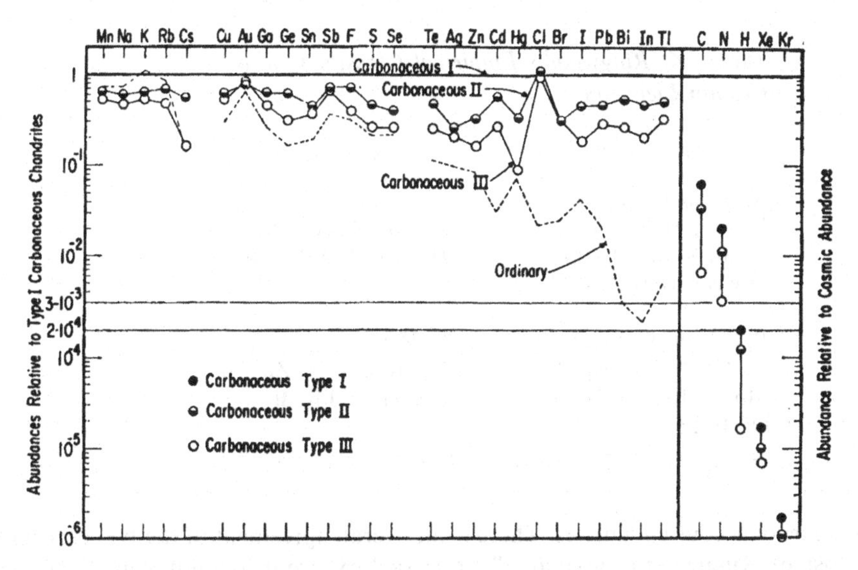

Figure 5.6. Meteorite Compositions as an Empirical Indication of Volatility. See text for description. Reproduced with permission from the *Annual Review of Astronomy and Astrophysics*, Vol. 9, 1971 by Annual Reviews Inc. Courtesy E. Anders.

Anders (1971) and Ringwood (1975) discuss the basis for an empirical classification with regard to volatility. Figure 5.6 is from the paper by Anders.

The abundances of various elements in three classes of chondritic meteorites are plotted relative to those in the so-called Type I carbonaceous chondrites (see Chapter 7). The latter are thought by many to be the least chemically altered solids from the primitive solar composition (§7.1). While the suitability of these meteorites as the "Holy Grail" may be subject to qualification, there is no doubt that they contain the most water of hydration of all the meteorites, and on this basis alone, we may proceed with a classification. Those elements that show the greatest depletions in meteorites with the *least* water of hydration would be considered to be the most volatile. Anders points out that the depletions of the elements in the Type II and Type III carbonaceous chondrites take place in approximately constant ratios, 0.3 and 0.6 relative to 1.0 for the Type I's. These depletions are independent of the element, as one might expect if the samples were part of a sequence, for example, if they represented stages in the heating of a primitive substance.

Ringwood's (1975) geochemical classification of selected elements is given in Table 5.4. This classification is based on geochemical experiments with silicate melts under high-temperature reducing conditions. He uses *involatile* rather than refractory; for the present purposes, we need not concern ourselves with the distinction. He also subdivides the major categories in an obvious way.

There is no reason to expect his classification to agree in detail with that of Anders, and, indeed, it does not. For example, the element chlorine, which Ringwood classes among the volatile nonmetals, shows little or no depletion in the Type II and Type

Table 5.4. *Ringwood's Empirical Classification of Certain Elements*

I. Involatile Group	II. Volatile Group
A. *Oxyphile subgroup* Li, Mg, Al, Si, Ca, Sc, Ti, Sr, Y, Zr, Nb, Ba, rare earths, Hf, Ta, Th, U	A. *Nonmetals* He, Ne, Ar, Kr, Xe, H, C, N, S, F, Cl, Br, I
B. *Siderophile Subgroup* Fe, Co, Ni, Cu, Ag, Au, Mo, W, Ru, Rh, Pd, Re, Os, Ir, Pt	B. *Metals* Na, K, Rb, Cs, Zn, Cd, Hg, Tl, In, Pb, Bi

III carbonaceous chondrites. There is an overall agreement in the two classifications. Most of Ringwood's "involatile" or refractory group do not appear on Anders's plot, whose purpose is to display the volatiles.

We can see that geochemical classifications are concepts that are subject to a great deal of qualification. The fact that the definitions are not precise is a difficulty, but it does not vitiate the value of these ideas. An enormous amount of cosmochemistry may be described with the help of such classifications.

5.7 The Geochemistry of Trace Elements: Ionic Substitution

Our discussion of condensation of elements from the gaseous phase makes it clear why some are classified as volatiles and others refractory. We now return to a discussion of the behavior of elements in the liquid and solid phase in order to clarify the lithophile–siderophile classification of a large number of elements whose cosmic abundance is low.

Eight elements dominate the chemistry of the earth's crust. Oxygen, silicon, aluminum, iron, calcium, sodium, potassium, and magnesium make up more than 99 percent of the mass of the crust. It is no accident that these eight elements are all found in the dominant mineral families, the olivines, pyroxenes, feldspars, and quartz. The fact that these elements are enriched in the crust is the result of their overall cosmic abundances and the properties of the minerals in which they are found. Potassium and sodium, for example, are enriched in the crust because of the properties of the ubiquitous feldspar minerals.

We have also mentioned that the crust is enriched in other elements, such as the lanthanide and two actinide rare earths (Th and U). A few of the trace elements are concentrated in special ores, but most of them are said to be dispersed, that is, found as impurities in the crystal structure of the major minerals. Due largely to the work of Victor Goldschmidt and his associates, the natural processes governing the distribution of trace species are now well understood.

Mason (1966) wrote

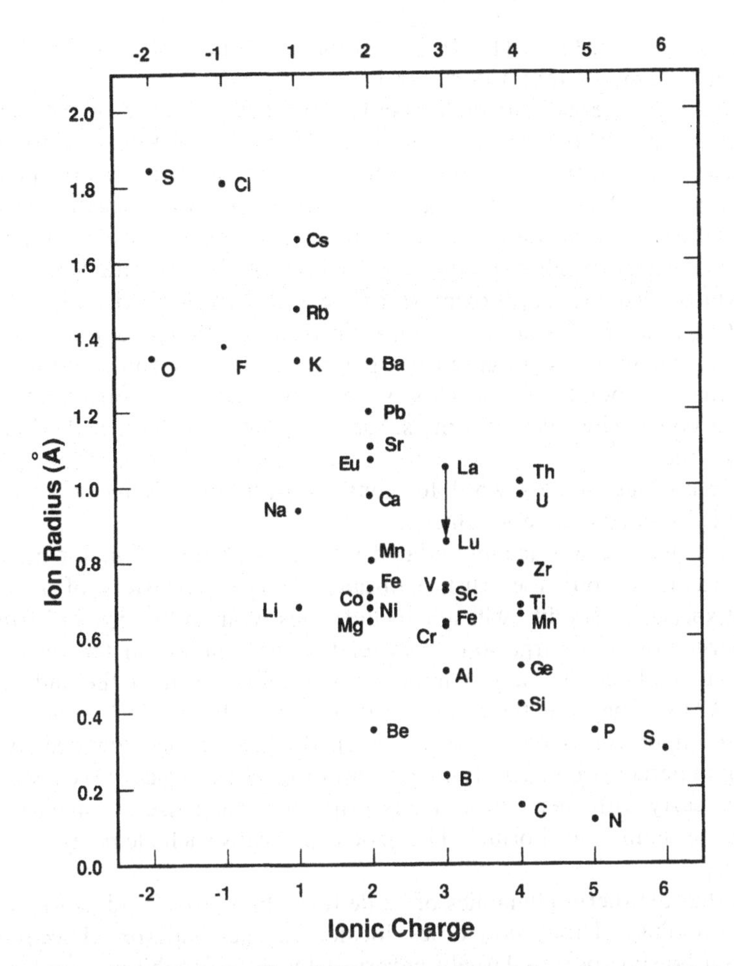

Figure 5.7. Ionic Radii and Valences. Adapted from Berry and Mason (1959).

It is a tribute to Goldschmidt's insight that he not only realized the significance of crystal structure determinations for geochemistry, but also devised a plan of research which led to a maximum of results in a minimum of time. Between 1922 and 1926 he and his associates in the University of Oslo worked out the structures of many compounds and thereby established the extensive basis on which to found general laws governing the distribution of elements in crystalline substances.

We may summarize his findings with the help of data on ionic radii and a few simple rules. It has been found empirically that extensive ion substitutions can occur in a crystal if the valence of the replacement ion is within one unit of that of the ion being replaced. Similarly, the radii of the ions should be within about 15 percent of one another for extensive substitution. Of two ions that can occupy the same site in a crystal, the ion with the smaller radius will form the stronger bond. Figure 5.7, adapted from Berry and Mason (1959), is a graphical display of crystal radii and valences for ions commonly found in minerals.

Goldschmidt's rules are easy to understand if one has a picture of the ions as little spheres that must fit together in a compact way. Such a picture is appropriate for what the chemist calls the ionic bond. The electron from the ion with positive valence (the cation) is detached, and transferred to the ion with negative valence (the anion). The result is that both ions are more or less spherical. We cannot prove it here, but it is a general result of quantum mechanics that states with zero angular momentum, which filled shells approximate, are spherically symmetrical.

In the *covalent*, or shared, bond, the valence electrons are shared, and a visualization would give something like a figure-eight pattern for the electron cloud between two atoms. Any real bond may be closely approximated by a superposition of these two bond types. The rules of ionic substitution work well when the bond is predominately ionic.

There is a marvelous, useless word for ionic substitution: *diadochy*. Another common term is isomorphous replacement.

The large ions that are not accommodated into the structure of major minerals tend to be concentrated into the crust. Consider the cooling history of a typical lava. The first species to solidify will not include ions with an "improper" size for their crystal structure. The large ions thus tend to be retained in the melt until the last material solidifies. These last minerals to solidify occur at the end of the Bowen series (§2.6). They are typically less dense than those which solidify first (anorthite is an important exception to this rule). Because of their low density, the late-solidifying materials are generally carried upward by geological processes, into the crust. They carry with them those misfits with the wrong size or valence, such as the actinides uranium and thorium. The geochemist calls such elements *large-ion lithophiles* (LIL).

It turns out that the thermodynamics of oxide formation *would* lead us to classify these LIL's as lithophiles. This is more or less incidental. Their widespread occurrence among common crustal rocks is largely governed by their (in)ability to substitute for abundant ions. If they could substitute in the olivines and pyroxenes they would be abundant in mantle rocks and less common in the crust.

We now turn to a discussion of the behavior of a group of elements of great cosmochemical interest, the rare earths. In most cases the radius of an ion increases, for a given charge, with atomic number, but there are important cases where the opposite happens. The common valence of the lanthanide rare earths is +3, and for these ions the radius decreases with nuclear charge, as shown in Figure 5.8.

Certain of the lanthanides show other valences than +3. Of these, perhaps the most important instance is that of the +2 state of europium.

The *lanthanide contraction* illustrated by Figure 5.8 arises because the successively heavier lanthanides are filling an inner electron shell, rather than adding outer electrons, as commonly happens for elements with increasing atomic number. The inner shell being filled is the $4f$ subshell, and by the rules of atomic structure, 14 electrons may be put into an f subshell. The wave functions for these $4f$ electrons are said to collapse, toward the nucleus of the lanthanide atoms, with the result that the overall size of the ions decreases rather than increases.

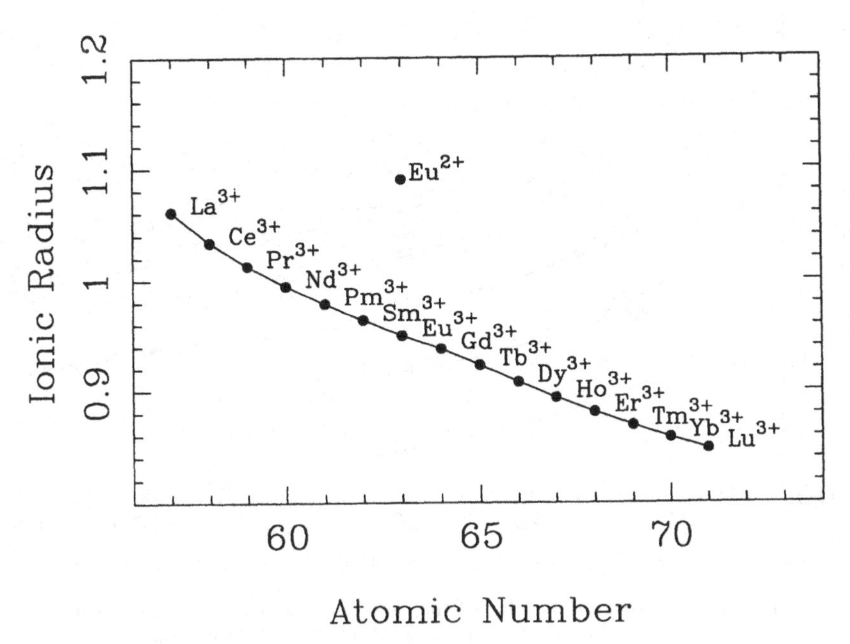

Figure 5.8. Ionic Radii for the Lanthanides.

Figure 5.9 from Cowan's (1981) *The Theory of Atomic Spectra and Structure* illustrates the "collapse" of the wave functions with high values of the angular momenta (d, f, g, ... electrons) with increasing atomic number Z. A contraction of the ionic radius with Z also occurs for the $3d$ (iron group) and $4d$ elements.

Both filled shells and filled subshells represent configurations of electrons with a special stability, and this property is true, to a lesser extent, of half-filled shells. When europium has the +2 rather than the +3 valence, its shell of f electrons is half full, f^7. The stability of the +2 valence state is therefore increased in the case of europium.

We can see from Figures 5.7 and 5.8 that when europium has the +2 valence state, its ionic radius is only slightly larger than that of the major element calcium. Consequently, we may expect that europium may be preferentially partitioned into calcium-rich rocks whenever the conditions are such that the +2 valence state is to be favored over that of the +3.

In many cosmic solids, a europium anomaly may be observed, on plots of lanthanide abundance as a function of atomic number Z. Figure 5.10 shows examples of positive and negative europium anomalies taken from lunar samples.

In plots of the lanthanide abundance distributions it is common to divide the abundances in the sample by the "chondritic" abundances in order to remove the odd–even effect. This is why the plots of Figure 5.10 are relatively smooth apart from the europium anomaly. Since the lanthanides are refractory elements, their

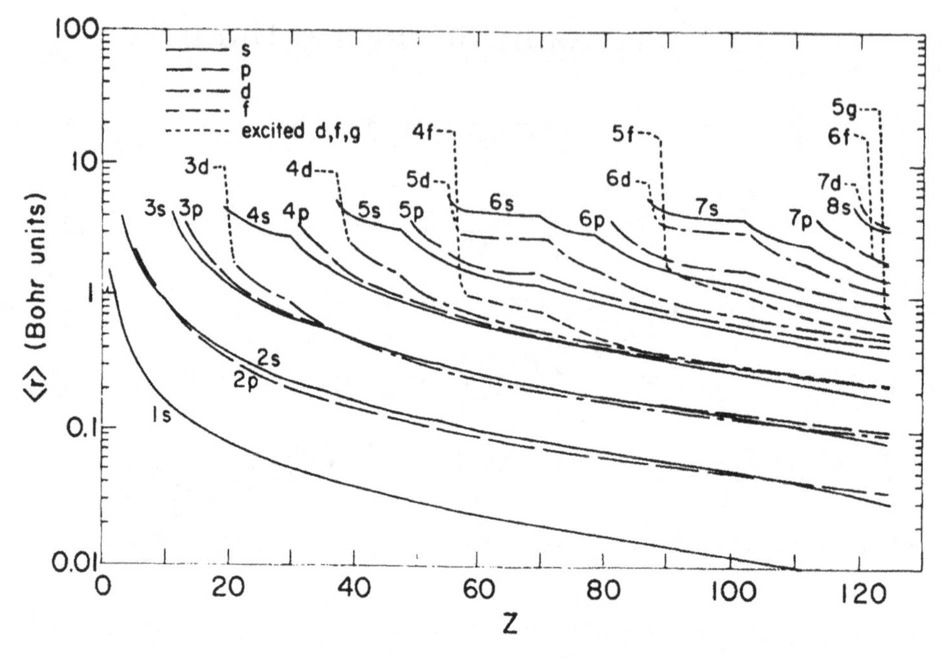

Figure 5.9. Expectation value of the radial coordinate r for various values of nl as a function of atomic number Z. Courtesy of the Regents of the University of California, Copyright 1981.

relative abundances in the chondrites are essentially the same as those in our SAD (Standard Abundance Distribution, Tables A4 and A5 of the appendix). When the sample abundances are the same as those in the SAD, the plots are horizontal lines with ordinates equal to unity.

The highland rocks in Figure 5.10 which are predominantly $CaAl_2Si_2O_8$ show a strong *positive* europium anomaly, as one would expect from the fact that the general level of oxidation on the Moon is low (see Chapter 6). The +2 valence (lower oxidation state) of europium is favored, so europium is partitioned into the anorthositic rocks. The rare earths as a group have abundances that are lower than the chondrites in the highland, anorthositic rocks. The ordinates for most of the lanthanides are less than unity. This is to be expected if the mineral anorthite crystallized early. The lanthanides, having the wrong ion size and valence, were, as a class, retained in the melt.

Note the slope of the distribution, for the highland rocks, which shows a relative enrichment of the lighter lanthanides. This is typical of the pattern of lanthanides in the terrestrial crust, where the usual explanation is that the larger the ion, the longer it is retained in the melt, and therefore carried up.

The lunar basaltic (mare) rocks show enhanced abundances with respect to the chondrites. This is in keeping with our view of this material as originating in partial melting hundreds of millions of years after the anorthositic crust was formed. The

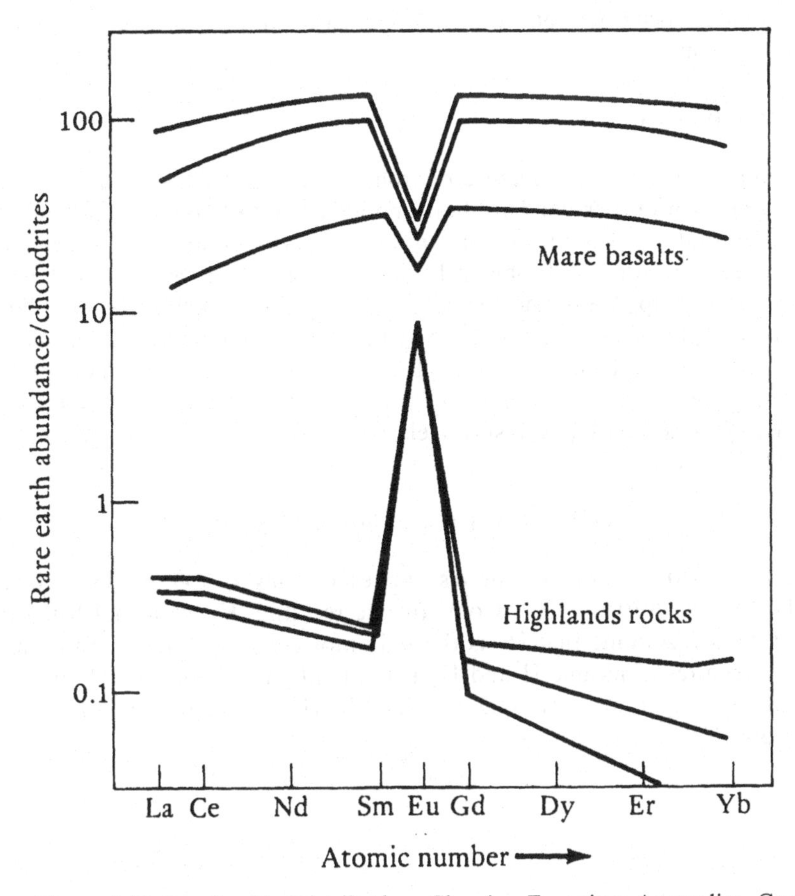

Figure 5.10. Lanthanide Distributions Showing Europium Anomalies. Courtesy of Harry McSween.

materials to be partially melted below the lunar impact basins which eventually became maria would be enriched in the LIL's including the lanthanides.

Many of the mare rocks show a negative europium anomaly, illustrating a concept of great importance in cosmochemistry. We say that the mare basalts of Figure 5.10 show the *complementary pattern* to that of the anorthosite. A genetic inference is possible, namely that the europium was unavailable to the source rocks of the mare basalts because of its earlier incorporation into the anorthosites. We may also use the notion of complementarity to account for the fact that the slopes of the lanthanide distribution are *not* downward, to the right, among the mare basalts. Wouldn't we expect the largest lanthanides to be concentrated in the newly melted source rocks of Figure 5.10? A possible explanation is again to be found in the notion of complementarity. If the source regions were already depleted in the lighter lanthanides because of the formation of the crust, the expected downward slope would not occur.

It is, of course, incumbent on the cosmochemist to show that these interpretations are the correct ones.

5.8 Problems

1. In Figure 5.4, the partial pressure of iron in the vapor phase is assumed to be $10^{-4.5}$ atm. If no other condensation or molecular formation is assumed, what would the total gas pressure be for the SAD (Tables A4, and A5) abundances?
2. For the gas pressure of Problem 1, find the difference between the (simple) condensation temperature and the 50% condensation temperature for solid iron.
3. Construct a plot similar to Figure 5.4, but for the condensation of the *liquid* rather than the solid phase of iron. From a comparison of the new plot with the earlier one, define the domains of pressure and temperature where iron will condense as solid and liquid respectively.
4. Tsuji (1973) gives

$$\log(K_P) = a_0 + \theta(a_1 + \theta(a_2 + \theta(a_3 + \theta a_4))) \quad (5.18)$$

where $\theta = 5040/T$. For H_2, the a's are respectively 12.739, -5.1172, 0.12572, -0.014149, and $6.3021 \cdot 10^{-4}$. Compare the resulting K_P with that listed in the *JANAF* tables. Assume that H_2 is the only molecule that forms, and obtain the partial pressures of atomic H and H_2 if the *total* gas pressure is 10^6 dynes/cm^2. Do this for temperatures of 2000 and 4500 K. (Hint: See the remarks following Equation 5.8.)

6
The Theory of the Bulk Composition of the Planets

6.1 Geophysical Constraints

The mean densities of planets constitute the most important constraint on their overall compositions. For example, it is inconceivable that Mars, with a mean density of 3.9, could be made primarily of troilite (FeS), whose density is 4.7.

For some of the planets, the mass is sufficient to cause a compression that significantly increases the density of their solid (and liquid) constituents. This compression is relatively unimportant for the Moon, but is of considerable importance for the earth and Venus. Naturally, if we are to make a comparison of the composition of planetary materials with the thermochemical calculations of Chapter 5, we must know the density of the planetary materials under zero (or low) pressures.

As will become clear, finding decompressed densities is like lifting oneself by the bootstraps! We cannot compute the decompressed density without a knowledge of the compressibility properties of the material, and we cannot know that with certainty without knowing the composition of the material itself – the very thing we are trying to find out. It turns out that we can make "plausible" guesses at the compressibility without knowing the precise composition of the material. Urey's (1952) study was again seminal.

Table 6.1 shows bulk and decompressed densities and moment of inertia factors (see below) for the Moon and planets. Distances from the sun are given in terms of the earth–sun distance. This *astronomical unit* (AU) is roughly 1.5×10^8 km. A precise value is given in Table A2 at the end of the book. The Moon is, on the average, 0.0026 AU from the earth.

The moment of inertia is a second bulk constraint that we have on the planets. Let us write for it,

$$I = \int (x^2 + y^2)\, \rho dxdydz = k\, Ma^2. \tag{6.1}$$

The integral is carried over the volume of the solid, whose rotational axis is taken to be the z-axis. This integral is often done in introductory courses in calculus for the case of a uniform density ρ. For a nonuniform density, it can be very tedious. On the right-hand side of 6.1, M is the mass of the body, and a is the perpendicular semi-axis; we have assumed rotational symmetry. The factor k is given in Table 6.1 for the planets. For uniform solids such as spheres and ellipsoids, a useful mnemonic for k is Routh's rule: k is the sum of the perpendicular semi-axes divided by 5. (This

Table 6.1. *Planetary Densities and Moment of Inertia Factors*

Planet	Distance from Sun(AU)	Mean Density (gm/cm^3)	Decompressed Density	Moment of Inertia Factor k
Mercury	0.39	5.43	5.3	?
Venus	0.72	5.24	3.9–4.7	?
Earth	1.00	5.52	4.0–4.5	0.331
Moon		3.34	3.3	0.392
Mars	1.52	3.94	3.7–3.8	0.36–0.37
Jupiter	5.20	1.33		0.25
Saturn	9.52	0.70		0.22
Uranus	19.2	1.30		0.23
Neptune	30.1	1.76		0.29
Pluto	39.7	1.1	?	?

Sources: Hagen and Boksenberg (1991), Lewis (1988), and Wood (1979).

works for spheres and spheroids. For a more general statement see Jeans 1935, p. 294.) Thus for an oblate spheroid with uniform density, we expect k to be $2/5$ or 0.4. The table shows that the moments of inertia factors are all less than 0.4, as we would expect.

Astronomers often use the moment of inertia factor as a measure of the *central concentration* of stars or planets. The greater the central concentration, the lower the value of k. A useful benchmark is the value $k = 0.27$ for a sphere with a density that decreases linearly from some central value to zero at the boundary. We can see from Table 6.1 that the terrestrial planets are all less centrally condensed than this linear model, but more than a uniform sphere. Of the Jovian planets, only Neptune appears to be less centrally condensed than the linear model.

For the earth only, the moment of inertia is known with precision. Let ΔI denote the difference in the moments of inertia about the z- and the x- or y-axes, and let I be the moment of inertia for the rotational, or z-axis. Observations of precession of the equinoxes yield $\Delta I/I$ very accurately. The interested student can find a useful discussion of the theory of this method in the old physics text by Leigh Page (1935), *Introduction to Theoretical Physics*. We shall refer to it again shortly. Observations of the regression of the line of nodes of orbits of earth satellites give ΔI. From these two results, the factor k may be specified to five or six figures, many more than can be accounted for with our present earth models.

For the Moon and the other planets, the moments of inertia are inferred from the observed flattening. The theoretical basis for this inference, which is only approximate, is remarkably tedious to derive, and will not be given here (see Kaula 1968).

The decompressed densities of planets are obtained by modeling their structure. Such models give the pressure and density (and possibly the temperature) as a function of depth. Of course, models cannot be constructed until some guess is made for the equation of state of the planetary material. Naturally, we must know something of the composition of the planets before we can specify the equation of state, and so the problem seems to be stuck in a vicious circle – we must bootstrap!

Some progress is possible if we do not aim at great precision. We can guess at a form for the equation of state for earth materials with the help of data from seismic waves, or compressibility experiments. The results of such information may be parameterized in the form of a relation such as

$$\rho = \rho_0 f(P), \tag{6.2}$$

where $f(P)$ is a function of the pressure and the assumed materials. The function $f(P)$ may be assumed to increase monotonically from unity at low pressures to a value close to 2 for the earth's core.

If an equation of state is specified in this form, $\rho = \rho(P)$, without involving the temperature, it is then straightforward to integrate the equation of hydrostatic equilibrium (see Equation 6.5) to obtain the pressure–density structure of a model planet. Given the density $\rho(r)$, the mass of the planet is clearly $M = \int \rho(r) \cdot 4\pi r^2 dr$. However, M will depend on ρ_0 ! This procedure may thus be considered an integral eigenvalue equation. Given M, only one value of ρ_0 will be satisfactory. This will be the accepted decompressed density.

6.2 Earth Models

Model planets should give proper values of the mass and moments of inertia, and conform to other constraints that we shall now discuss. It should come as no surprise that this is an ongoing enterprise, and only partial solutions may be articulated (see Hubbard 1984, Kaula 1986).

Extensive information is available on the interior of the earth from seismic data. We now have determinations of the velocities of pressure and shear waves as a function of depth in the earth. The geophysicists call these two kinds of waves primary (P) and secondary (S), because the faster P-waves arrive at a seismic station before the S. It is of mnemonic value that the P-waves happen to be (P)ressure (or longitudinal), and the S-waves (S)hear (or transverse). Figure 6.1 is based on the data given by Kennett and Engdahl (1991). The velocities of the P- and S-waves are often designated α and β rather than V_P and V_S.

From the theory of elastic solids, one may show that the velocities of pressure and shear waves are related to the bulk modulus K, and the density ρ. Discussions may be found in physics books that treat the theory of elasticity or geology texts that cover geophysics. The reader who wishes to see a full derivation in which the P and S velocities follow from a wave equation for an elastic solid may consult the discussion in Page's (1935) *Introduction to Theoretical Physics.* Page's treatment makes clear how the strange factor of 4/3 arises in Equation 6.3 below, and also clarifies the relationships among the elastic constants λ, μ, and the modulus of elasticity: $K = \lambda + \frac{2}{3}\mu$.

$$K = \rho\left(V_P^2 - \frac{4}{3}V_S^2\right). \tag{6.3}$$

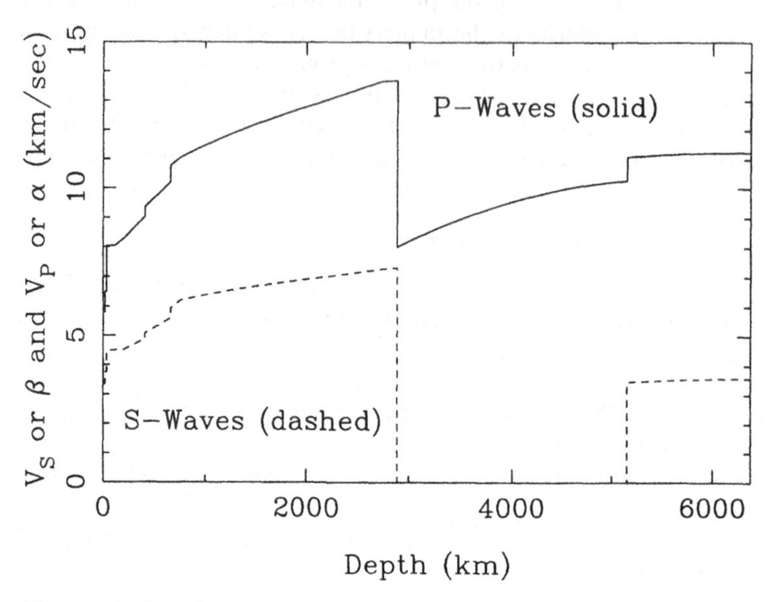

Figure 6.1. Seismic Wave Velocities in the Earth (Kennett and Engdahl 1991).

We shall make use of the bulk modulus for changes that take place at constant entropy, viz.

$$K = \rho \left(\frac{\partial P}{\partial \rho} \right)_S . \tag{6.4}$$

The relation between this K and that which emerges from the theory of deformable bodies, which is usually presented without any thermodynamic considerations, is discussed, for example, by Landau and Lifshitz (1970) in their *Theory of Elasticity*. In the following discussion we shall assume that seismic compressions take place on a time scale that is short compared with the time that heat generated by the compression could travel a distance of the order of one wavelength. A similar (adiabatic) assumption is made in the elementary treatment of sound waves in air.

The observed seismic wave velocities provide the ratio K/ρ from Equation 6.3, as a function of depth in the earth. For an adiabatic process, we may use Equation 6.4 and eliminate the pressure from the equation of hydrostatic equilibrium

$$dP = g\rho dr. \tag{6.5}$$

Thus if we designate the ratio K/ρ by ϕ, we have

$$\frac{d\rho}{\rho} = \frac{g}{\phi} dr. \tag{6.6}$$

The quantity ϕ may be found at each depth from 6.3 (see Figure 6.1).

The relation 6.6 is known as the Adams–Williamson equation. It is, in principle, integrable, from the surface of the earth inward, since we know g at the earth's surface. The density is then found numerically as a function of r. If one follows this procedure, it is found that no geologically reasonable starting density, ρ_0, will match the global constraints on an earth model, giving the proper mass and moment of inertia.

We shall not discuss the attempts to correct the adiabatic assumptions used in the derivation of Equation 6.6 (see Jeffreys 1976, p. 208). Instead, we show how the above relationships may be used as a guide to an empirical equation of state. Suppose we may expand K as a power series in the pressure, viz.

$$K = K_0 + bP + cP^2 + \ldots \tag{6.7}$$

(Murnaghan 1951, Meissner and Janle 1984).

If only the first two terms on the right are taken, we may write

$$\rho \frac{dP}{d\rho} = K_0 + bP. \tag{6.8}$$

This equation gives, on integration,

$$\rho = \rho_0 \left(1.0 + b\frac{P}{K_0}\right)^{1/b}. \tag{6.9}$$

Equation 6.9 is clearly an approximation. We have used it for heuristic purposes as an example of an equation of state. In this equation, ρ_0 is clearly the decompressed density. For planets with a core–mantle structure, it is reasonable to work with at least two decompressed densities. However, the condensation theory requires a single, appropriately averaged value.

Anderson (1988) discusses more general equations of state, among them the popular "Birch–Murnaghan" relation:

$$P = (3/2)K_0 \left[(\rho/\rho_0)^{7/3} - (\rho/\rho_0)^{5/3}\right] \left\{1 - \zeta \left[(\rho/\rho_0)^{2/3} + \ldots\right]\right\}. \tag{6.10}$$

In the applications to models of the earth, all workers have been faced with the problem of treating phase changes in materials whose compositions are not precisely known. The solid–liquid transformation at the core–mantle boundary is only the most obvious of these changes. Another well-explored phase change concerns the transition of olivine, which crystallizes in the orthorhombic system at low pressures, to the spinel structure, which is isometric. This transformation provides a natural explanation of the rapid rise in seismic velocities at depths near 400 km. An elaborate discussion of the experimental investigations of this phase change is given by Ringwood (1975) in his book *Composition and Petrology of the Earth's Mantle*. Important new results of high-pressure experiments are discussed by Hemley and Cohen (1992), specifically, the perovskite structure that can be assumed by silicates, such as $MgSiO_3$.

Current attempts to specify the earth's bulk composition may be divided into two classes, depending upon the degree to which meteorites are used as a guide. Mason and Moore (1982) get a reasonable approximation to the results of more elaborate studies by the following simple assumptions (but see Taylor 1988):

1. The iron alloy of the core has the composition of the average for nickel–iron in chondrites, and includes the average amount (5.3%) of FeS in these meteorites.
2. The composition of the mantle plus crust is the same as the oxidized materials of the average chondrites.

The introduction of some light material into the core is a subject of current interest. Mason and Moore's suggestion that it may be the sulfur from FeS is a natural one given their emphasis of meteoritic composition. O'Neill (1991a,b) has discussed the composition of the earth's core in connection with a general theory of the origin of the Moon. He concludes that the light-element component of the earth's core "remains enigmatic."

Ringwood has approached the problem of the current composition of the earth with a heavy reliance on seismic data, laboratory experiments, and the idea that a good indication of the composition of the earth's mantle may be obtained from studies of samples, peridotites and eclogites believed to have come from the upper mantle with only minor modifications from the materials through which they must have passed.

It is beyond the competence of the present writer to comment on the relative merits of these different approaches.

6.3 The Moon and Terrestrial Planets

Seismic data are available for the Moon, but based on measurements from four of the Apollo missions, and are by no means extensive enough to be applied as in the Adams–Williamson equation. At one time considerable significance was given to the failure of S-waves from a single seismic event on the far side of the Moon to reach one of the seismic stations. The inference was made that these waves were unable to penetrate a molten or partially molten lunar core. Unfortunately, there has been only one observation of this kind, so that more recent discussions of the lunar structure have not stressed the necessity for a molten core.

We know from the Moon's mean density that its bulk composition cannot be the same as that of the earth. More precisely, it must be depleted in iron.

The returned lunar samples certainly show a depletion of siderophiles, as might be expected in core formation. However, it is also possible that the Moon formed from materials that were once a part of the earth's mantle, after formation of the core. These materials would have already been depleted in siderophiles, as in one of theories of lunar origin offered by Ringwood (1979). More recently, much attention has been given to the possibility that the Moon may have been formed from debris blasted off the earth by a large, perhaps Mars-sized object. A lengthy discussion of these possibilities may be found in Hartmann, Phillips, and Taylor's (1986) *Origin of the Moon.*

There is still speculation that the Moon may have a small core with a radius of several hundred kilometers; both Fe and FeS have been suggested as major constituents.

In a zero-order approximation, one may say that the Moon's chemistry must resemble that of the earth's mantle. But there is little question that the chemistry of the Moon is distinct from the terrestrial crust and mantle, insofar as we may infer the latter. The Moon is a much dryer place than the earth. This is shown not only in the notable absence of surficial liquid water, but by the mineralogy of the returned solid samples. The Moon is essentially devoid of minerals that are "late" in the sense of the Bowen series. There are only traces of amphiboles or serpentines, and when these minerals are found, the common (OH) group is replaced by a halogen.

The state of oxidation of lunar materials is also considerably lower than terrestrial rocks. We have already mentioned (§5.7) the common +2 valence state of europium, and the ubiquitous positive europium anomaly in the highland rocks. More obvious than this is the common presence in lunar rocks of microscopic quantities of reduced iron. Some of this free iron may be accounted for in terms of meteoritic infall, but not all of it. Rough estimates of the relevant oxygen fugacities (pressures) are possible if one assumes an equilibrium between crystalline Fe and FeO (Wustite) with temperatures in the range 1270 K to 1500 K. The temperature range is chosen to match melting temperatures of lunar basalts. Oxygen fugacities in the range of 10^{-12}–10^{-15} atm are obtained. These are consistent with electrochemical measurements (Sato 1976), which yield 10^{-12}–10^{-16} atm. The measured fugacities are typically 0.5 to 1.0 orders of magnitude below this "iron–wustite buffer."

The present terrestrial oxygen fugacity is, of course, much higher (roughly 0.2 atm), and with a few bizarre exceptions, terrestrial rocks do not contain reduced iron. There is little doubt that there was less free oxygen in the early history of the earth.

The familiar minerals in which iron exhibits the +3 valence, hematite (Fe_2O_3) and magnetite (Fe_3O_4), are present in such small quantities in lunar samples that they could be explained entirely in terms of contamination by meteorites.

These considerations all lead to models of the composition of the Moon that are difficult to explain on the basis of a general hypothesis that will be discussed below in which the overall chemistry of a planet is determined by its location in a cooling solar nebula. Modern theories of the origin of the Moon account for oxidation and other chemical differences between lunar and terrestrial rocks.

In the case of Mars, our information is far less than for the Moon. Of the Viking experiments, the only marked failure was the seismometer on Lander I. Lander II returned weak signals that could not be distinguished with certainty from perturbations of the Lander by wind (Anderson *et al.* 1977). It seems clear that Mars is not active seismically like the earth, and is probably less active than the Moon. Most earthquakes may be attributed to plate tectonics. The much milder lunar quakes are caused by tidal perturbations from the earth. Mars, presumably, has no plate tectonic activity, and its satellites are too small to cause significant tidal stresses.

The Martian atmosphere has been rather thoroughly sampled, but the major mineralogy of the surface was not revealed directly by the X-ray fluorescence spectrometer that was used to analyze the soil samples. These instruments have little or no resolution among the lighter elements such as Na, Mg, Al, and Si. One can find estimates of the relative abundances of these elements. They are obtained by

Table 6.2. *Relative Masses of Core and Mantle–Crust*

	%			
	Mercury	Venus	Earth	Mars
Mantle & Crust	32–35	68–72	68.5	80–95
Core	65–68	28–32	31.5	5–20

comparing the X-rays observed from the Martian samples with "standard" terrestrial materials, whose compositions, of course, were known in complete detail.

It is even more unfortunate that no rocks were analyzed, and no direct mineralogical data are available. The soil samples that were studied, unlike the lunar soils, probably have undergone important chemical modifications, in view of the probable past history of liquid surface water. An unusual soil chemistry is required to explain the results of the Viking Lander life experiments.

We know respectively less about the bulk chemistry of Venus and Mercury. Table 6.2 gives estimates of the relative masses of the mantles and crusts, and the cores, of Mercury, Venus, and Mars. In the simplest approximation, these two divisions must be chemically similar to the analogous regions in the earth (Hubbard 1984, Zarkov 1986, §6.2). Entries in the table are based on the preceding references, and Taylor (1982). As explained in these sources, the figures vary from one model to another.

6.4 The Jovian Planets

Jupiter and Saturn have bulk compositions that may be crudely described as solar, with a small loss of volatiles. The masses of these planets are sufficient that escape, even of hydrogen and helium, from the atmospheres is small. Models of their structure have been modified in the last decade to be compatible with the observed energy losses from their surfaces, which exceeds by roughly a factor of 2 the energy gain from solar radiation. Current thought is that these planets are still emitting primordial heat. This is brought to the surface by convection currents in a liquid hydrogen mantle that encloses a small silicate core. The mantle has two zones; in the inner one, the liquid hydrogen has metallic properties.

Recent models of the formation of the giant planets have started with rocky cores. Pollack (1985, see his Figure 1) shows clearly why such cores are required for the outer Jovian planets Uranus and Neptune. By analogy, it may be assumed that Jupiter and Saturn also began with the accumulation of a rocky mass. Much additional information on the giant planets is given in the books *Protostars and Planets II* (Black and Matthews 1985) and *Protostars and Planets III* (Levy and Lunine 1993).

6.5 A Zero-Order Model for the Solar Nebula

In this section we will show how one may estimate the temperature as a function of radius in a model of the solar nebula. The actual model has *heuristic value only*; it illustrates many of the features (and uncertainties) of more realistic models.

Let us suppose that the primordial solar material collapsed isothermally from a spherical gas cloud. The reason for this collapse need not concern us here. We assume that the potential energy of the cloud becomes large with respect to its internal kinetic energy. This may be because the cloud was compressed by either a supernova blast wave, or a density wave, responsible for the spiral patterns in galaxies (Chapter 16).

The isothermal assumption may be justified roughly as follows. During the initial collapse, the particles within the cloud gain energy by collisions with one another. However, these collisions excite low-lying energy levels in the atoms and molecules, which may be radiated away so long as the gas cloud is opaque to the infrared radiation. This will be the case until polyatomic molecules form. We know that water, for example, has extensive rotation–vibration bands that prevent most of the infrared radiation from the sun and stars from penetrating the earth's atmosphere.

Let the initial temperature of the cloud be T_i. A rough assumption that will be sufficient for our purposes is that the cloud becomes opaque at some radius R_a, and from this point, the collapse will be assumed to be adiabatic. This means that the equation of state may be written

$$P = K\rho^{\gamma}, \tag{6.11}$$

where γ is the ratio of the specific heats. It is unrealistic to assume a *constant* γ, but the error involved is not significant in view of other uncertainties.

We may use Equation 6.11 to eliminate the pressure from the ideal gas law, thus obtaining a relation of the form $T = T(\rho)$. A relation frequently used in modeling the solar nebula has been the so-called Larson (1969) adiabat:

$$T = 5 \cdot 10^8 \rho^{2/3}. \tag{6.12}$$

This, or a similar relation, may be derived as soon as one chooses an initial temperature and volume $(4\pi/3)R_a^3$ at any point in time during the adiabatic collapse. The temperature at the start will be that of the primordial cloud, since the first collapse is assumed to be isothermal.

Eventually, the residual angular momentum of the cloud will prevent collapse perpendicular to the rotational axis. Collapse will proceed in the direction of the axis of rotation, and the cloud will flatten. We shall assume the cloud may eventually be approximated by a cylindrical disk of thickness t with surface density σ (gm/cm^2). Let us try to find an expression for σ as a function of radius.

We therefore assume a uniform spherical mass of radius R_a, which collapses directly into a disk. In making this assumption, we are combining two stages into one. After the adiabatic phase, the cloud could continue to collapse in a roughly spherical form, and then, after reaching some radius smaller than R_a, collapse into a disk. There is no need for us to introduce the additional notation to describe this.

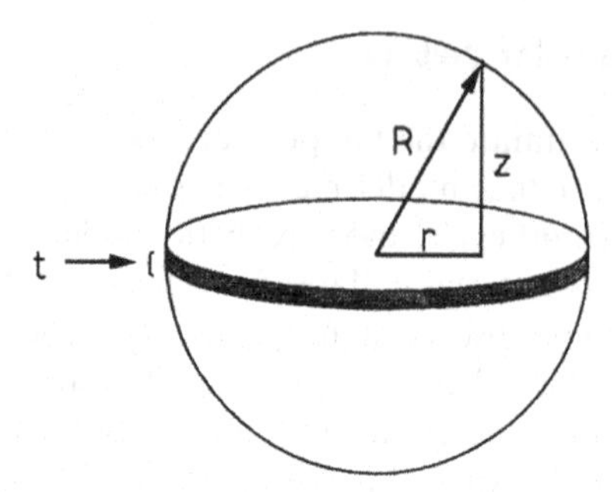

Figure 6.2. A Heuristic Model of the Solar Nebula.

Obviously, it would be more realistic to assume more central concentration than that of a uniform-density cloud. For example, we could assume a mass with an equation of state given by

$$P = K\rho^{1+1/n}, \tag{6.13}$$

where n is a free parameter. An object that obeys this law is called a polytrope. This particular assumption leads to analytical complications that are unnecessary for our present task.

The surface density, σ, after collapse of a uniform sphere is given by

$$\sigma = \int_{+z}^{-z} \rho_a dz. \tag{6.14}$$

The density ρ_a is the density of the cloud when adiabatic collapse begins, and the limits on the integral are given by $z = \pm\sqrt{R_a^2 - r^2}$, where r is a radial coordinate in the disk as shown in Figure 6.2.

If we let the thickness t be another of the model parameters, the density in gm/cm^3 at any point within the disk is given by

$$\rho = (2\rho_a/t)\sqrt{(R_a^2 - r^2)}. \tag{6.15}$$

Cameron and his coworkers (see Cameron 1985 in *Protostars and Planets II*) have made very elaborate models of the solar nebula, including vertical structure, and convection zones.

We note that Equation 6.15 gives us the ability to specify the density at any point within the model, and by using Equation 6.14 and the ideal gas law, we have the temperatures and total gas pressures as well. Assumptions about the chemical composition and the molecular equilibrium relations then lead to the partial pressures necessary for the calculation of condensation sequences.

An adiabatic equation of state will now allow us to specify the temperature-pressure evolution of the solar nebula. This allowed us to plot the track, or "cooling trajectory," for the nebula in Figure 5.5. We must also assume that the composition of the gas is uniform, an assumption whose validity is coming more and more into question.

6.6 Equilibrium and Disequilibrium Condensation

The theory that was developed in §5.4 is sufficient in principle to allow calculation of partial pressures of all species in the gas phase. As soon as the first solids begin to condense, we must make certain additional assumptions about the possible interactions of the gaseous materials with the condensed phases. We may imagine two extreme situations. In the first, suppose the condensed material remains in the form of very small dust particles, which are at all times in contact with the gas phase. Thus for example, iron might condense first as a crystalline solid, but it would then be allowed to react with oxygen to form wustite (FeO) or with sulfur to form troilite (FeS).

Lewis (see below) and his coworkers have called this kind of interaction *equilibrium condensation*. Whatever chemistry is thermodynamically possible is imagined to take place. The planets are then formed from the solid materials that are in equilibrium with the gas phase at a specific temperature. Such a process could be imagined provided the solids remained as a fine dust as long as chemical reactions with the gas phase were possible.

An opposite extreme is their disequilibrium model. In this model, one removes any condensed material from the system of equilibrium relations. This situation could be realized physically if the solids condensed sufficiently rapidly into large particles that the bulk of their mass was shielded from the ambient gas. The gas would then be excluded from contact with the bulk of the solid, whose mass depends on the cube of the radius. The surface would remain in contact with the gas, but we neglect this relatively small amount ($\sim r^2$-dependence) of material.

An important question concerns the formation of the earth's core. It is quite clear that the metallic core was never in complete chemical equilibrium with the material of the mantle. If this had been the case, the mantle materials would not contain the Fe^{3+} that may be inferred from observations of fresh ocean basalts and alpine peridotites. It is generally thought that the core formed very quickly after the bulk of the earth had accumulated, and that the rapidity of this process precluded a complete equilibrium. However, it may not be totally excluded that the first solids to accumulate into the earth were primarily solid Fe with a possible admixture of FeS. If an iron-rich core were present at the beginning, then reactions of the iron with the residual gas phase would not be possible. This would be an important case of disequilibrium condensation.

The equations of molecular equilibrium that were presented in Chapter 4 describe gas-phase chemistry only. If solid phases are also to be considered, certain generalizations must be made. Consider the condensation of the hypothetical crystalline molecule AB from the gas phase, where we shall assume the abundance of B is much larger than A.

The partial pressure of A will then drop rapidly to zero as soon as the path of the $\log[P(\mathrm{A})]$ vs. $1/T$ curve intersects with the equilibrium line for solid and vapor as shown in Figure 6.3 (see also the figures in §5.5). The heavy line is determined entirely from thermodynamic data for the species AB, and has nothing to do with the model solar nebula. As we discussed in §5.5, the ordinate for the thin line (solar nebula) is $\log(P_A) + \log(P_B)$. If AB is the only molecule involving A, then P_{A} is

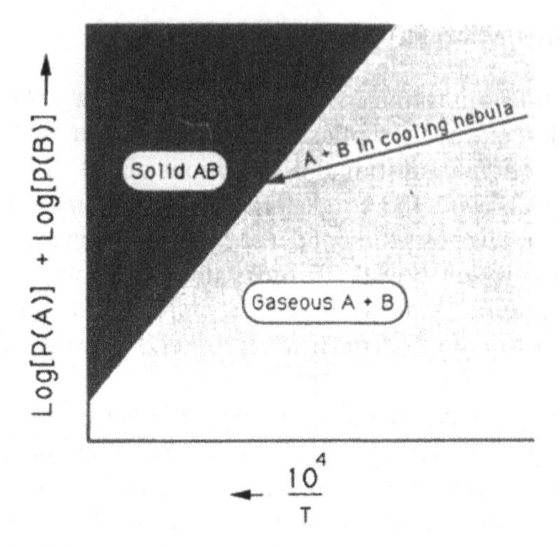

Figure 6.3. Condensation of the Hypothetical Molecule AB.

proportional to $\epsilon_A \cdot P_{Total}$, where P_{Total} is the gas pressure of the model nebula. A similar relation holds for P_B. There is a locus of points for the nebula because we assume P_{Total} is a function of the temperature.

For disequilibrium condensation, we would remove the term corresponding to AB from the relations of the form 5.10 for all temperatures and pressures to the left of the condensation line, and the fictitious pressure of the more abundant constituent B must be suitably reduced: $\epsilon_B > \epsilon_A$. Assume also that A and B are elements that have not participated in prior molecule formation or condensation. Then after condensation, we set $P_f(B) \rightarrow P_f(B) - P_f(A)$, and *then* $P_f(A) = 0$; there is no more 'A' in the gas phase.

The molecule AB would thenceforth be removed from consideration, and the calculations would proceed with the adjusted abundances or fictitious pressures. This was the cause of the slope change in Figure 5.5 for the track of the nebular pressures.

If the condensate remains in equilibrium with the gas, the program must look for further reactions of AB with the gas phase. Consider for example, reactions of the form

$$C(g) + AB(c) \rightarrow ABC(c). \qquad (6.16)$$

At the point of the solar nebula's cooling trajectory for which this reaction proceeds from left to right, one would remove the appropriate amount of the constituent C from the fictitious pressures.

The "appropriate amount" of C depends on how much AB there is. In our example, we would reduce $P_f(C)$ to zero, if $\epsilon_A > \epsilon_C$, or to $P_f(C) - P_f(A)$ if $\epsilon_A < \epsilon_C$.

An equilibrium condensation program must test each species for possible removal in this way, and adjust the abundances in the gas phase accordingly. This is an enormously complicated process, and it is acknowledged that the calculations inevitably skip over chemistry that would occur if true equilibrium were maintained.

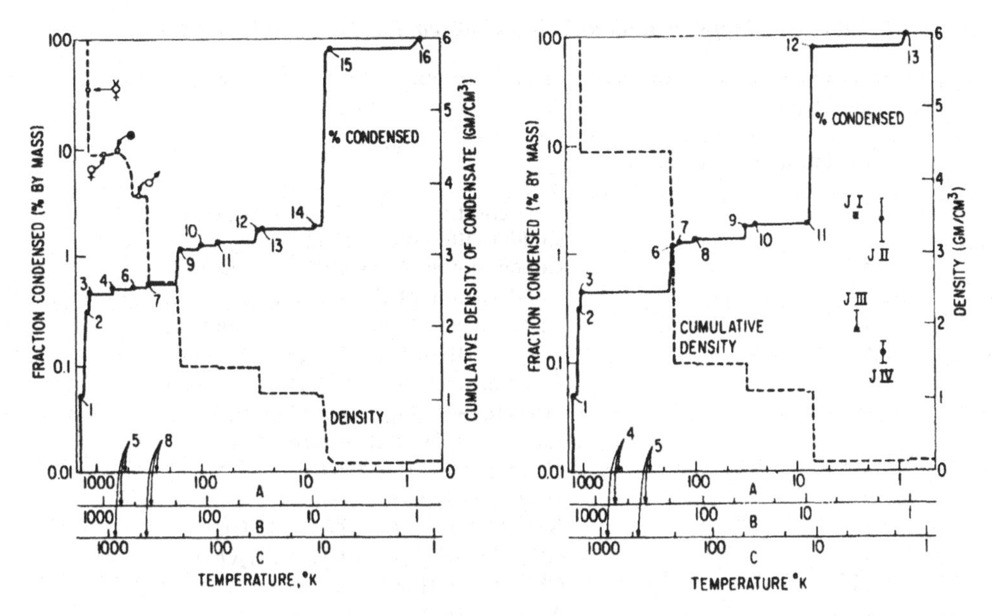

Figure 6.4. Condensation Sequences in a Model Solar Nebula. Reproduced with permission from the *Annual Review of Astronomy and Astrophysics*, Vol. 14, ©1976 by Annual Reviews Inc.

As we discussed in §5.5, modifications to the total gas pressure resulting from a condensation depend on the detailed assumptions of the model. If the total gas pressure is assumed to be constant during the condensation of various phases, then it must be assumed that the nebula undergoes some contraction to compensate for the loss of the relevant partial pressures. It must be remembered that as far as the terrestrial planets are concerned, the bulk of the gas pressure will be contributed by molecular hydrogen at the relevant condensation temperatures. The condensing materials would therefore give rise to only small perturbations of the total pressure.

6.7 The Condensation Sequences of Lewis and Coworkers

Figure 6.4 shows two plots from the review paper of Barshay and Lewis (1976).

Two curves are plotted in these diagrams. One curve gives the percentage of material in the solar nebula that has condensed at a given temperature, the other is the density of that condensed matter. Three temperature scales are given, corresponding to three assumed total pressures in the nebula: 10^{-6} (A), 10^{-4} (B), and 10^{-2} (C) atmospheres. The fact that it is possible to present the calculations in this way shows that the sequences of condensations are relatively insensitive to the total pressure. One can see from Equation 5.13 that the equilibrium relation between solid and vapor depends logarithmically on the pressure (activity, or fugacity for ideal gases) and linearly on the temperature. There *should* be some small total

Table 6.3. *Steps in Equilibrium and Disequilibrium Condensation*

Step Number	Reaction
Equilibrium Condensation	
1	Condensation of Ca, Al, and Ti oxides
2	Condensation of Fe-Ni alloy
3	Condensation of enstatite ($MgSiO_3$)
	Condensation of (Na,K)$AlSi_3O_8$
4	Fe-Ni alloy + $H_2S(g) \rightarrow FeS(s)$ and NiS(s)
5	$CO(g) + 3H_2(g) \rightarrow CH_4(g) + H_2O(g)$
	Calcium silicates + $H_2O(g) \rightarrow$ tremolite
6	Fe metal + $H_2O(g) \rightarrow FeO(s) + H_2(g)$
7	Enstatite + $H_2O(g) \rightarrow$ serpentine
8	$N_2(g) + 3H_2(g) \rightarrow 2NH_3(g)$
9	$H_2O(g) \rightarrow H_2O(s)$
10	$NH_3(g) + H_2O(s) \rightarrow NH_3 \cdot H_2O(s)$
11	$CH_4(g) + 8H_2O(s) \rightarrow CH_4 \cdot 8H_2O(s)$
12	$CH_4(g) \rightarrow CH_4(s)$
13	$Ar(g) \rightarrow Ar(s)$
14	$Ne(g) \rightarrow Ne(s)$
15	$H_2(g) \rightarrow H_2(s)$
16	$He(g) \rightarrow He(s)$
Disequilibrium Condensation	
1	Condensation of Ca, Al, and Ti oxides
2	Condensation of Fe-Ni alloy
3	Condensation of enstatite ($MgSiO_3$)
4	$CO(g) + 3H_2(g) \rightarrow CH_4(g) + H_2O(g)$
5	$N_2(g) + 3H_2(g) \rightarrow 2NH_3(g)$
6	$H_2O(g) \rightarrow H_2O(s)$
7	$NH_3(g) + H_2S(g) \rightarrow NH_4SH(s)$
8	$NH_3(g) \rightarrow NH_3(s)$
9	$CH_4(g) \rightarrow CH_4(s)$
10	$Ar(g) \rightarrow Ar(s)$
11	$Ne(g) \rightarrow Ne(s)$
12	$H_2(g) \rightarrow H_2(s)$
13	$He(g) \rightarrow He(s)$

pressure dependencies in the condensation sequences, but they must be too small to require individual curves for each assumed total pressure.

The solid curve marking the percentage of condensed material begins near the lower left in both equilibrium and disequilibrium plots and moves upward and to the right in a series of steps indicating the condensation of major solid materials. Typical reactions corresponding to these steps are given in Table 6.3. The changes take place in vertical steps because it is assumed that the solids begin and end condensation at a fixed (condensation) temperature, T_c. At the highest temperatures, very little of the material is condensed, while at the lowest, which go all the way to 1 K (!), all of the material is solidified.

The density of the condensed material is closely coupled with the condensation temperatures. A simple heuristic argument may be given to explain this correlation. Denser materials have more closely packed atoms. If we visualize ionic bonding in crystals, the bonds will be stronger, simply because of the inverse-square electrostatic forces, when the ions are closer together. This means that the denser materials should be able to survive as solids at higher temperatures than the less dense ones. By the same reasoning, we expect them to condense first out of the gas phase.

A similar relation holds, incidentally, for the melting point, and explains why the dense, mafic minerals typically melt after the felsic ones. However, we know that there is the important exception to this generalization, that the feldspar anorthite ($CaAl_2Si_2O_8$) melts at relatively high temperature. Thus it is not density *alone* that governs the melting or sublimation (solid $\rightarrow$ gas) of materials. Nevertheless, this single property is an important guide and mnemonic.

One may place the planets on a diagram of this kind. The terrestrial planets are plotted. The Jovian planets are not plotted because it is thought their composition depends not only on condensation sequences of this kind, but also on the planets' gravitational fields. We shall return to them shortly. Major differences between the equilibrium and disequilibrium sequences involve the transformation of reduced iron to troilite (FeS) and to wustite (FeO). In the disequilibrium sequences, these materials cannot form because, as we have already discussed, after solid Fe condenses, it is removed from the Russell equations. Thus the equilibrium sequence allows some discrimination among the predicted densities of the terrestrial planets. On the disequilibrium sequence, we would expect Mercury to have a high density, while Venus, earth, and perhaps Mars would all be expected to have decompressed densities of the order of 4.5.

The Jovian planets could be placed on the decompressed density curve too, if we knew what these values were. Surely they are less than unity! However, it was once assumed that the Jovian planets retained nearly a full complement of hydrogen and helium. This was as much because of the gravitational attraction of their large masses as because of the temperature in the solar nebula where they formed. We must invoke an important additional consideration to explain the composition of the Jovian planets.

A theory of the formation of the terrestrial planets also requires assumptions beyond the discussion of condensation sequences. If we grant that solid materials with the proper density are available in the solar nebula there is still the problem of coagulation of the solids, and, more significantly, dispersal of the remaining gas. One can see from Figure 6.4 that the terrestrial planets remove some 30 percent or less (3% for Mercury) of the putative nebular gas with solar composition. The standard assumption is that the extra matter was swept away by a youthful, vigorous (T Tauri) solar-type wind.

There is a certain amount of astronomical evidence to support such an assumption. The present solar wind is driven energetically by the same sources that heat the chromosphere and corona. While the details of the heating mechanism are still not well understood, their manifestations, in the form of chromospheric activity, have been well observed. From stellar observations, we know that such chromospheric activity decreases with the age of stars. In some of the youngest stars, named after the prototype T Tauri, one observes indications of violent chromospheric activity (Bertout 1989). The youth of these stars is indicated by their location, in association with aggregations of dust and gas, and in clusters (§13.1) with hot, massive stars whose energy output is so prodigious that their hydrogen would be exhausted in a few tens of millions of years.

6.8 Difficulties with the Condensation Sequences

The simple picture of planetary composition based on condensation must be modified in fundamental ways. Before we present some of the difficulties, let us note that the basic thermodynamic concepts have not been discarded. Condensation itself will remain of key importance in cosmochemistry. Indeed the fundamental geochemical classifications, volatile and refractory, are based on condensation. It will be well for readers to keep this in mind as they read the remainder of this section.

Alternative suggestions to condensation for the high decompressed density of Mercury are now available. It is now widely thought (see Wetherill 1988) that severe bombardment of the planet may have removed a silicate mantle in much the way that lunar material was blasted from the early earth (§6.3). This would leave a dominant, metallic core, and account for the high bulk density. Wetherill's calculations also show extensive mixing of planetesimals from regions of the solar nebula with very different heliocentric radii. This mixing would smear out any unique density–radius that would arise in our heuristic model nebula.

Lewis (1988) attempted to account for the density of Mercury using condensation models which included materials from a plausible range of heliocentric radii. The resulting densities were too low – 4.2 gm/cm^3, rather than Mercury's decompressed density of 5.3. Mercury is surely the keystone of the condensation model for bulk planetary compositions, and if its density cannot be explained, we must surely question the entire theoretical structure.

In addition to the bulk densities of the planets, some information on volatility in solids as a function of radial distance from the sun is now available from meteorites. Palme and Boynton (1993) point out that the relative abundances of moderately volatile elements is greater in the earth than in certain meteorites known as eucrites (Chapter 7). These basaltic meteorites probably originated in a parent body that formed somewhere in the asteroid belt, beyond the orbit of Mars. Our simple model of equilibrium condensation would therefore lead us to expect the eucrites to be enhanced in these moderately volatile elements. Taylor (1992) points out many shortcomings of the theory and summarizes newer ideas.

Thus far, we have not considered the question of whether there is sufficient *time* for a given phase to condense, before the temperature of the nebula drops and still new solids form. Exploration of kinetic (time-dependent) aspects of the chemistry of the early solar system is clearly an important topic for future work. Prinn (1993) discusses *kinetic inhibition* of equilibrium as a result of mixing of materials on a short time scale.

We have mentioned that the earth's core could not have been in chemical equilibrium with the mantle. This is revealed not only by the state of oxidation of the mantle, but also by the content of siderophiles, which are depleted, but not enough for equilibrium to have taken place. This fact alone has been sufficient for some authors to discard the notions of equilibrium condensation entirely. An alternative approach has been to attempt to construct planetary models with the help of meteorites. The general procedure is very closely related to the work discussed in §6.2, where the earth's bulk chemistry is obtained from meteorites. Taylor (1982, Section 9.8), for example, pictured the formation of the earth as beginning with the accretion of the core from nickel–iron-rich materials. This was followed by bombardment with meteoroids with different compositions which would supply the trace quantities of crust–mantle siderophiles as well as the earth's complement of volatiles.

Equilibrium condensation certainly does not account for the earth's volatiles, nor can it be expected to. While the earth's oceans account for only some 0.02% of the bulk mass, even that small amount is not allowed in the theoretical calculations that we have presented.

Table 6.4 gives an overview of the compositions of the atmospheres of Venus, earth, and Mars. Note that the water vapor on Mars is highly variable, both in time and with position on the planet and with altitude in the atmosphere. The indicated range also neglects real variations with altitude. The entry for water vapor on Venus is for altitudes less than 55 km; von Zahn *et al.* (1983) give 1 to 40 ppm, at cloud tops. Carr (1981) gives the fraction by volume as 3×10^{-4}. Observers quote ranges of *precipitable microns* from essentially zero up to about 100. Lewis and Prinn (1984) give the following useful conversions: one precipitable micron (pr μm) of water is 10^{-4} gm/cm^2, which is equivalent to 0.1 cm·amagat. In order to convert these numbers to fractions by volume, we have assumed the Martian atmosphere contains 5.9×10^{17} moles of CO_2.

In 1978, the Pioneer Venus mission struck a strong blow against equilibrium condensation. The amount of ^{36}Ar in the atmosphere of Venus was much too high.

The bottom part of Table 6.4 gives the masses of several volatile elements relative to the total planetary mass (gm/gm). Consider ^{36}Ar. Venus has relatively more than the earth, which has more than Mars. Argon-36 is not the major isotope in the atmosphere of the earth or Mars, ^{40}Ar is. However, the geochemistry of ^{40}Ar is strongly influenced by that of potassium, because ^{40}Ar is the daughter of ^{40}K. In the SAD, there is some three times more ^{40}Ar than ^{40}K, yet in the earth, the overwhelming percentage of argon almost certainly derived from ^{40}K. The reason is that *all* of the noble gases are severely depleted in the earth (see Figure 5.1). This fact cannot be explained in terms of loss from the atmosphere, because these noble gases are too heavy to escape. The canonical explanation is that they were never in the earth to begin with. Some trace amounts are present, of course, but they are insignificant relative to expectations based on the SAD. The earth formed from the accretion of solids that were already depleted in the noble gases and other volatiles.

This much is generally consistent with the equilibrium condensation model. But with this general picture, we would not expect to find more ^{36}Ar on Venus than on

Table 6.4. *Atmospheric Compositions of Terrestrial Planets*

	Venus	Earth	Mars
Fraction of Lower Atmosphere (by Volume)			
N_2	4. (−2)[a]	7.81 (−1)	2.7 (−2)
O_2	2–4. (−5)[b]	2.09 (−1)	1.3 (−3)
Ar	5–12. (−5)	9.34 (−3)	1.6 (−2)
H_2O	0.1–1. (−4)	≤4. (−2)	0.001–1. (−3)
CO_2	9.6 (−1)	2–4. (−4)	9.5 (−1)
Ne	5–13. (−6)	1.82 (−5)	2.5 (−6)
He	1. (−5)	5.24 (−6)	
CH_4		1–2. (−6)	
Kr	5. (−8)	1.14 (−6)	3. (−7)
H_2		4–10. (−7)	
N_2O		3.0 (−7)	
CO	2–3. (−5)	0.1–2. (−7)	8. (−4)
Xe		8.7 (−8)	8. (−8)
SO_2	1.5 (−4)	≤2. (−8)	≤ 2. (−8)
HCl	<1. (−5)	≤1.5 (−9)	≤ 1.5 (−9)
Fraction of Planetary Mass (gm/gm)			
^{12}C	2.6 (−05)	1.5 (−05)	1.1 (−08)
^{14}N	2.2 (−06)	1.4 (−06)	1.9 (−10)
^{20}Ne	2.9 (−10)	1.0 (−11)	4.4 (−14)
^{36}Ar	2.5 (−09)	3.4 (−11)	2.2 (−13)
^{84}Kr	4.7 (−12)	1.7 (−12)	1.8 (−14)
^{130}Xe	1.6 (−13)[c]	1.4 (−14)	2.1 (−16)

Sources: Lewis and Prinn (1984), Pepin (1989), Donahue (1991).
[a]Parenthesized numbers indicate powers of ten. [b] Single experiment.
[c]Using data from Zahnle (1993).

the earth. The argument runs as follows: argon is a volatile; the percentage of most volatiles (e.g. water) increases from the innermost toward the outermost planets; so we expect more argon on the earth than on Venus. We cannot count ^{40}Ar because that is related to ^{40}K, but there should *not* be more ^{36}Ar on Venus than on the earth.

The ^{36}Ar observations are now just one of a series that are difficult to understand in terms of equilibrium condensation. Current ideas involve *ad hoc* mechanisms. One theory is that the noble gases currently observed in the terrestrial planets were implanted in the accreting planetesimals by the solar wind. On this model, we would have expected the planetesimals that were closest to the sun to have picked up more of these noble gases.

Another reasonable suggestion is that the noble gases were adsorbed on the accreting solids and then carried into the inner planets. If this were true, the relative abundance of these gases would depend upon the properties of the carrier. If this

Table 6.5. *Non-SAD Abundances in the Giant Planets*

Property	Planet			
	Jupiter	Saturn	Uranus	Neptune
$<\rho>$(gm/cm^3)	1.33	0.70	1.30	1.76
E(out)/E(in)	1.7	1.8	1.1	2.6
[He/H]	−0.15	−0.70	0.0	?
[C/H]	0.37	0.3 to 0.8	1.3	1.4 to 1.9
[N/H]	0.3	0.3 to 0.6	$\ll$0?	$\ll$0?

material were prevalent at the higher temperatures, it could account for the ^{36}Ar anomaly.

Table 6.4 shows that the ratio of ^{36}Ar to ^{20}Ne does not vary appreciably among the atmospheres of Venus, earth, and Mars. Indeed, this is generally true for the noble gases; their relative abundances are similar even though their percentage abundance is highly variable among these three planets. Whatever mechanism is responsible for the noble gases in the terrestrial planets did not severely fractionate them. Pepin (1992) and Zahnle (1993) review the general topic of noble gases in terrestrial planets. Additional material and references may be found in §8.7.

Good examples of trace species as clues to important epochs in the history of planets come from additional isotopic anomalies among the atmospheres of the terrestrial planets. For example, Hunten *et al.* (1989) make a good case for the loss of large amounts of volatiles from the atmosphere of Venus and Mars. Such losses, of course, would be modifications of any original composition that might be accounted for by condensation.

The giant planets have generally been thought to have compositional fractionations that do not fit into the condensation picture as presented in §6.7. Some of the relevant data are summarized in Table 6.5. The abundances are logarithms of ratios in the planetary atmospheres relative to similar ratios in the sun, and have been taken from Gautier and Owen (1989). The *bracket* notation is one often used in stellar astronomy (see Equation 13.4). Atreya (1986) discussed the compositions of the atmospheres of the giant planets. He also treated the observed departures from solar ratios of certain elements. It is very important for theories of the formation of these planets to know if these departures are also true for the current bulk compositions.

The row of Table 6.5 labeled E(out)/E(in) gives the ratio of the energy radiated by these planets to that absorbed from the sun. The figures are from Atreya's Table 2.1, with an update for Neptune following Voyager's observations. Quite possibly the internal heat sources of these planets stir the material sufficiently that one may obtain an indication of the bulk compositions from a careful study of the atmospheres.

6.9 Problems

1. Refer to Equation 6.13, the P–ρ relation for polytropes. Show that for a perfect gas undergoing an adiabatic change the polytropic index is 3. Comment on the thermodynamic implications for material that obeys 6.13 with some index other than 3. First consider only perfect gases, and then enlarge your discussion to materials that obey more general equations of state.
2. A very rough estimate of the partial pressure of oxygen on the Moon at the time of solidification may be obtained by assuming the reaction $Fe(c) + \frac{1}{2}O_2 \rightarrow FeO(c)$ was in equilibrium (see §6.3). Assume the range of temperatures from 1270 to 1500 K, and verify the order of magnitude of the pressures (fugacities) discussed in that section.
3. Calculate the condensation temperature of graphite for a solar nebula model with a total gas pressure of 0.01 atm. *Ignore molecule formation*, and use the abundances from Table A4. If graphite actually condensed at the temperature you obtain, the element carbon would be considered refractory. In fact, it is usually considered a volatile (or atmophile) element. Explain.
4. A 2 solar mass protostellar cloud collapses isothermally to a sphere with a temperature T_0. From this point, further collapse follows a Larson adiabat 6.12. Calculate the density ρ_0 and radius R_0 for this sphere assuming $T_0 = 10$ K, and $T_0 = 500$ K.
5. For the case of $T_0 = 10$ K, calculate ρ, T, and P after collapse to a disk of thickness of 10 solar radii (see Equation 6.14). Assume a uniform density with height in the slab. In order to get the pressure, you will have to assume a value for the molecular weight. Comment on an appropriate value.

7

Meteorites and the Standard Abundance Distribution (SAD)

7.1 An Overview

We may divide the meteorites into three broad categories, the stones, the stony-irons, and the irons. Meteorite samples are described as "falls" or "finds." If a meteorite is observed to fall, and brought to a museum curator, it is called a fall. Finds are meteorites that have not been seen to fall, or at least not by the person who discovers them.

It is now reasonable to speak of two major divisions of meteorites – those discovered in the last several decades in Antarctica, and all of the rest. The Antarctic meteorites have roughly doubled the available samples of solid cosmic debris. It is difficult to know precisely how many independent falls are represented, since all of these samples are finds. However, we must leave differences among the Antarctic and non-Antarctic samples to the references (see Koeberl and Cassidy 1991).

The non-Antarctic meteorites are named by the location of the fall or find. The names are often exotic. For the Antarctic meteorites, locations are also used for the names, but these are supplemented by alphanumerical codes. Most of the world's classified (non-Antarctic) meteorites are listed in the *Catalogue of Meteorites* (Graham, Bevan, and Hutchison 1985). They include listings for selected Antarctic meteorites.

Most of the meteorites in museums are irons, while the opposite is true of falls – most of the latter are stones. This distinction is important to keep in mind, because there is a strong selection effect favoring finds of irons. Stony meteorites, especially some of the chondrites, quickly weather or crumble until they are difficult to distinguish from terrestrial rocks. The iron meteorites, on the other hand, have a distinctive appearance, and they certainly do not crumble into dust.

Most theories of the origin of meteorites invoke the notion of parent bodies, as we discussed in §5.1 (cf. McSween 1987). The chemistry of the meteorites differs sufficiently that it is common to think of several of these planets which differentiated into mantle and core, and were later broken up by an interplanetary collision. It can by no means be ruled out that certain of the meteorites originated in the nuclei of comets. It is well established that certain of the meteor (shooting star) *showers* originate from material found in the orbits of old, "burned-out" comets. Unfortunately, the debris from these showers has not reached the ground in

Table 7.1. *Broad Categories of Meteorites: Percent of Falls*

stones	chondrites	carbonaceous chondrites	4.9
		ordinary chondrites	80.2
	achondrites		8.8
stony irons			1.3
irons			4.8

recoverable amounts, although some fragments may have been recovered by special experiments involving high-flying air- and spacecraft (Brownlee 1985).

In order to have an idea of the relative frequency of the bodies that strike the earth, it is necessary to consider only the falls. Table 7.1 gives an outline of meteorite classification along with the percentage of falls in each category. The figures are based on Table 7.2 (below).

The name chondrite derives from the word chondrule, a name given to spherical inclusions in these rocks, whose origin is a question of keen interest (King 1983). Figure 7.1 is an illustration from McSween (1987) of the cut face of a chondrite showing rounded chondrules, as well as white, calcium–aluminum-rich inclusions.

The chondrules come in a variety of sizes, of the order of a millimeter. Not all are well rounded. Sometimes they are easily distinguished from a background matrix. In other chondrites, the rock texture can be chaotic, and it is difficult for the non-expert to distinguish chondrules from other, sometimes amorphous, inclusions.

The major minerals of most stony meteorites are ones that we have already encountered, the olivines, pyroxenes, plagioclase, and nickel–iron alloys. Table 7.2 is a classification according to Mason (1979), which lists the major mineral phases. The parenthesized figures are, with one exception, the number of observed falls in each class.

Workers now classify the carbonaceous chondrites into the classes (or groups) CI, CM, CO, CV, and recently, CK (Kallemeyn, Rubin, and Wasson 1991). Each group has a prototype meteorite, which can be found, for example, in the *Catalogue of Meteorites.* Thus, I stands for *Ivuna,* M for *Murchison,* O for *Ornans,* V for *Vigarano,* and finally K for *Karoonda.* The groups show distinct chemical differences (Wasson 1985, Fig II-3). Another notation is very common: C1, C2, and C3. These designations are virtually synonymous with CI, CO and CM, and CV (Wasson 1974, pp. 20ff).

We must refer to Wasson (1974) or Mason (1979) for a discussion of the classification of chondrites by petrological types (see also van Schmus and Wood 1967).

We have already discussed the use of meteorites as clues to the primitive composition of the solar system, as well as indicators of the bulk chemistry of planets. The carbonaceous chondrites show the least evidence of chemical fractionation. Among these, the CI's are widely recognized as the best indicators of the primitive composition of the solar system (Goles's Holy Grail).

Wasson (1974) used Ivuna for the CI prototype because of its mnemonic value:

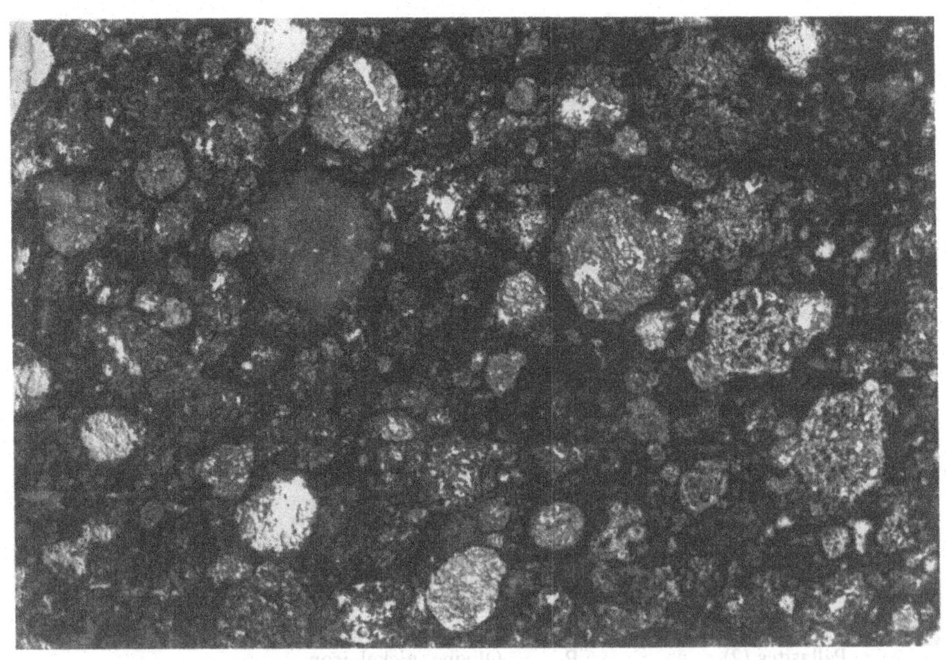

Figure 7.1. Chondrules. A thin section of the meteorite Vigarano showing rounded chondrules and white, calcium–aluminum inclusions. Vigarano is a carbonaceous chondrite of type CIII, or CV3 (see text). Courtesy of Harry McSween.

CI suggests the old Wijk type C I as well as C1. However, the most notorious of the CI's is known as Orgueil. These stones were seen to fall in France in 1864. Figure 7.2 is a microscopic view of this remarkable meteorite. It shows the notorious lack of chondrules in the carbonaceous chondrites, along with myriad veins of carbonate and sulfate minerals against a dark background of phyllosilicates. The reader will note that none of these minerals are "early" in the sense of being near the top of the Bowen series (§2.6).

It has become recognized that certain rare meteorites may well be fragments from the Moon and Mars. These matters are discussed by numerous authors (Wasson 1985, Dodd 1986, McSween 1987).

The cosmochemistry of Moon rocks and meteorites may be investigated in exquisite detail because they may be taken into terrestrial laboratories. There is nevertheless a keen interest in and huge literature on the chemistry of comets and asteroids. Time and production constraints prevent more than a few literature citations on the ingenious techniques that have been used to study the compositions of these bodies. Comets were reviewed by Arpigny (1991), and Newburn *et al.* (1991). For asteroids, we recommend the review by McSween (1989), and the references of Zappala (1991).

Table 7.2. *Mason's Meteorite Classification*

Class	Symbol	Principal Minerals
Chondrites		
Enstatite (11)	E	Enstatite, nickel–iron
Bronzite (224)	H	Olivine, bronzite, nickel–iron
Hypersthene (256)	L	Olivine, hypersthene, nickel–iron
Amphoterite (49)	LL	Olivine, hypersthene, nickel–iron
Carbonaceous (33)	C	Serpentine, olivine
Achondrites		
Aubrites (8)	Ae	Enstatite
Diogenites (8)	Ah	Hypersthene
Chassignite (1)	Ac	Olivine
Ureilites(3)	Au	Olivine, clinobronzite, nickel–iron
Angrite (1)	Aa	Augite
Nakhlite (1)	An	Diopside, olivine
Howardites (17)	Aho	Hypersthene, plagioclase
Eucrites (20)	Aeu	Pigeonite, plagioclase
Stony-irons		
Pallasites (2)	P	Olivine, nickel–iron
Siderophyre (1)(find)	S	Orthopyroxene, nickel–iron
Lodranite (1)	Lo	Orthopyroxene, olivine, nickel–iron
Mesosiderites (6)	M	Pyroxene, plagioclase, nickel–iron
Irons		
Hexahedrites (4)	Hx	Kamacite
Octahedrites (27)	O	Kamacite, taenite
Ataxites (1)	D	Taenite

7.2 Meteorite Ages

Meteorites may be dated by the same radioactive methods that are used to assign ages to terrestrial rocks. A variety of techniques are currently in use: Rb–Sr, Nd–Sm, K–Ar, ^{39}Ar–^{40}Ar, etc. Some of these methods will be discussed in detail in Chapter 8. For the present, we note that all of these techniques measure slightly different things. If one wants a precise definition of the age measured, one must consider the specific experiment. However, roughly speaking, most rock-dating methods are designed to measure the time since the materials solidified sufficiently to hold the radioactive daughter atoms near the site of their creation.

Some of the time intervals that are relevant to the cosmochemistry of meteorites are illustrated in Figure 7.3. From [1] to [2] general stellar element production enriches the interstellar gas. The corresponding time interval must be of the order of the age of the Galaxy less that of the sun: in round numbers, perhaps 8 to 10

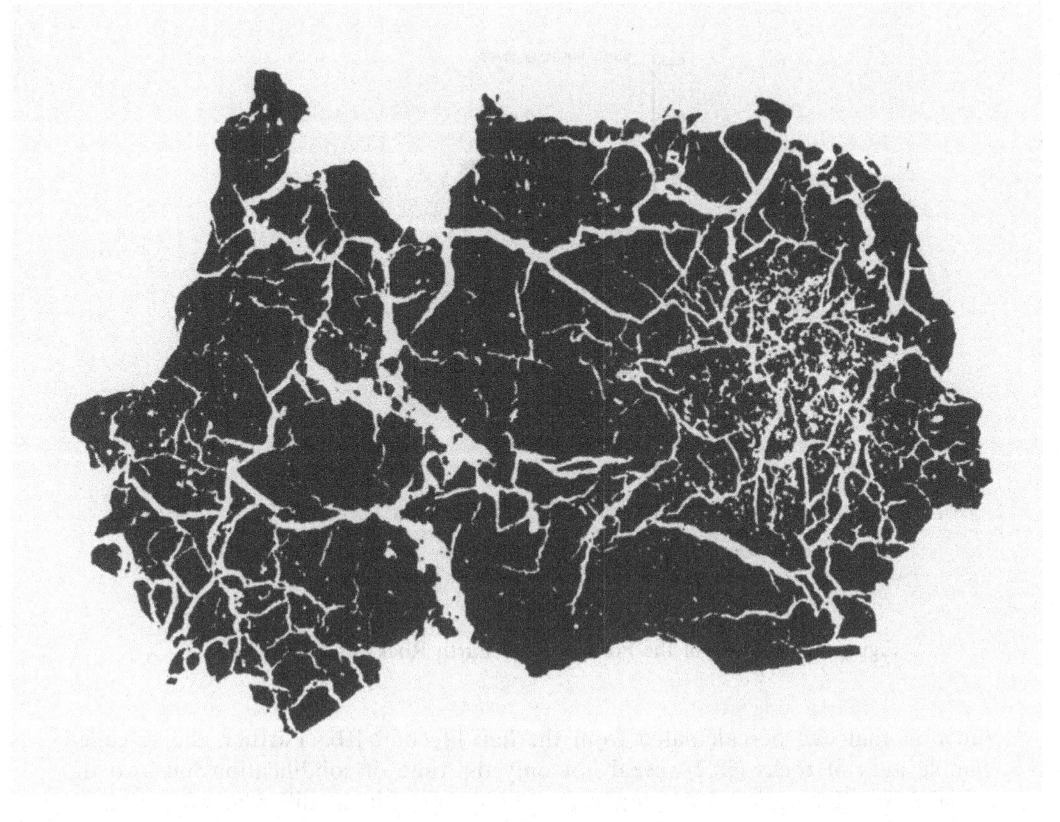

Figure 7.2. The Orgueil Meteorite (Courtesy Harry McSween)

billion years. Until recently, it was thought that any solid material formed during this time was subsequently vaporized. We now know this cannot be entirely true.

At [3], the solar nebula collapses so that its composition is no longer influenced by general nucleosynthesis. Apart from minor influence by cosmic rays, the unstable nuclei such as ^{40}K or ^{87}Rb that were made in stars decay with no source of replenishment. At [5] solid particles condense from the nebula, and this condensation continues over some interval of time, up to the point [6]. The small bodies that formed during the interval [5] $\rightarrow$ [6] probably remain cold. However, with the formation of planetary bodies [6], there is the possibility of remelting and resetting the radioactive clocks. The sample ages (§8.2) of surface rocks measure the interval of time since the rocks were *last* melted, roughly any interval of time within [9] $\rightarrow$ [10].

We can estimate the interval ([5] $\rightarrow$ [6]) of time over which the earliest meteorites solidified. The picture is as follows. The primordial solar system material was *almost* entirely gaseous. A radioactive parent, say ^{87}Rb, begins to decay to ^{87}Sr while the solar materials are all in the gas phase. The first solids to solidify from this gas should have relatively less ^{87}Sr (daughter) and more ^{87}Rb (parent) than solids formed later. The increase in the daughter-parent ratio is a simple time-dependent

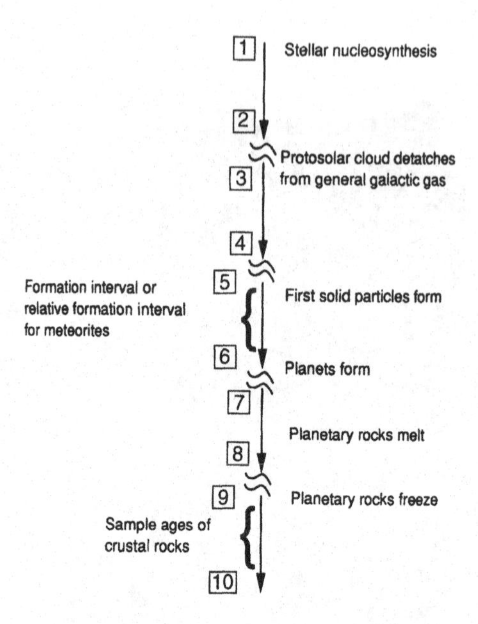

Figure 7.3. History of the Formation of Earth Rocks and Meteorites.

function that can be calculated from the half-life of ^{87}Rb. Further, the so-called sample ages of rocks (§8.2) reveal not only the time of solidification but also the initial ^{87}Sr/^{87}Rb ratio. By comparing this initial ratio in different meteorites, it has been possible to delineate intervals of time of the order of tens of millions of years. During these intervals some of the oldest materials in the solar system solidified.

Figure 7.4, taken from Faure (1986), illustrates the evolution of the initial ^{87}Sr/^{87}Sr ratio in certain old meteorite samples.

With some very important qualifications that have already been noted, the majority of the matter (perhaps!) was homogenized. In order for the inferred formation interval to be valid, we must assume secondary nuclear processes affecting parent and daughter materials used to date the meteorites were minimal. By this we mean that nuclear reactions other than the decay may be neglected. It is entirely possible, for example, that cosmic rays could induce reactions that would affect the populations of parent and/or daughter nuclides. §8.3 gives an analytical discussion of the evolution of the initial ratios.

The *intervals* of time determined from the initial ^{87}Sr/^{86}Sr ratios are conceptually well-defined quantities. Interpretations are not so straightforward. We might assume that the Allende (ALL) meteorite solidified prior to Angra Dos Reis (ADOR), which solidified before the basaltic achondritic samples (BABI). We pretend here, for simplicity, that BABI represents a single rock rather than the mean for several. BABI actually means basaltic achondrite, best initial (value) (Papanastassiou and Wasserburg 1969). It is equally valid to assume that all three sets of materials solidified at the same time, but that the clocks for ADOR and "BABI achondrites"

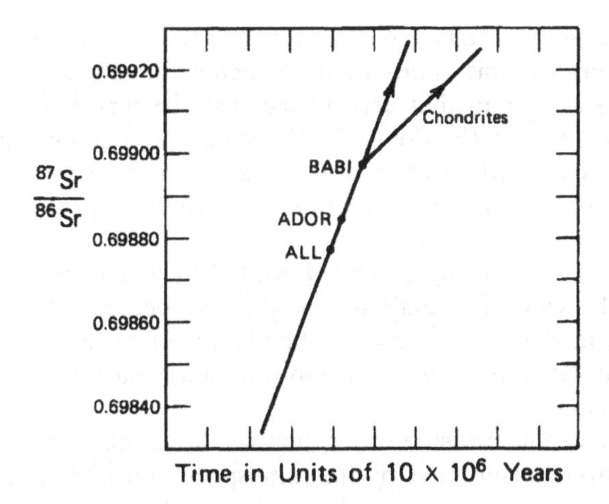

Figure 7.4. Evolution of the $^{87}Sr/^{87}Sr$ ratio in the early solar system. From G. Faure, *Principles of Isotope Geology, 2nd ed.*, reprinted by permission of John Wiley & Sons, Inc., ©1986.

were reset by metamorphic events a few tens of millions of years after the initial solidification.

Information on a closely related time interval is available from the presence of what are called extinct radionuclides in meteorites. Podosek and Swindle (1988) define these as radioactive isotopes "with lifetimes of the order of 10^6 to 10^8 years, long enough to survive the interval between nucleosynthesis and the formation of the solar system but short enough so that they are essentially fully decayed and now extinct in the solar system." Clayton (1968, p. 593) uses the term extinct radioactivities to describe essentially the same phenomena. One also sees extinct isotopes or extinct nuclides. The case of ^{26}Al is an example (§8.8).

It is generally assumed that the origin of most radioactive (and other) nuclides is stellar nucleosynthesis. These topics will be taken up in Chapter 10 and again in Chapter 16. Since the birth of the Galaxy, some 10 to 20 billion (10^9) years ago, stars have been producing heavy elements. The rate of production of these elements depends on factors that lie within the domains of the astronomer (star formation and death rate) and the nuclear astrophysicist (details of nucleosynthesis). Stellar nucleosynthesis gradually enriched the interstellar gas from which our solar system formed (see Chapter 16). Once the sun and solar nebula separated from the general background gas the abundances were no longer subject to modification by the general stellar synthesis of nuclides within the Galaxy. Until fairly recently, it was widely thought that supernova explosions stirred the galactic gas thoroughly, so that separation from it really had connotations of cutting an umbilical cord.

Certain models of galactic nucleosynthesis specify the rates of production of the heavy elements, including the radioactive ones. From these models, one can specify the relative amounts of parents and daughters in the Galaxy as a function of time. In this way it is possible to get some insight into the overall history of nucleosynthesis in our Galaxy by examining isotopic ratios of other parent–daughter pairs with long half-lives. The idea is similar to that used to measure the time interval over which the oldest meteorites formed (Figure 7.4). The situation is complicated by the fact that nucleosynthesis stopped abruptly in the case of the detachment of the solar nebula, and it continued at some uncertain rate in the case of the Galaxy as a whole.

We have postulated that nucleosynthesis ended with detachment of the solar nebula. It is certain that this can only be approximately true, because we know of secondary processes that can change the relative abundances of nuclides after the umbilical cord has been cut. Nuclear reactions due to solar and galactic cosmic rays are examples of such processes.

Cosmic-ray exposure ages can be measured in meteorites. Materials at the surface of the earth are shielded from cosmic rays by the atmosphere, but the surfaces of meteoroids and presumably meteorite parent bodies are unshielded. They are subject to constant bombardment by particles energetic enough to cause nuclear reactions. The cosmic-ray exposure ages are simply a measure of the amount of damage done by these energetic particles.

Shielding from cosmic rays is provided by roughly a meter of rock. Consequently, the exposure ages are a measure of the time such shielding was absent. If we consider the breakup of a body larger than some tens of meters in radius, then the bulk of the material will have been shielded. Thus an exposure age of some sample may be roughly interpreted to mean the time interval since the breakup of a parent body until the object fell on the earth.

It is possible to make calculations of the general properties of the orbits of such fragmentary bodies in the solar system. Such calculations lead to reasonably well-defined "lifetimes" in space, before the smaller bodies are swept up by one of the major planets. The picture that one has in mind, let us say for a body originating in the asteroid belt, is as follows. Perturbations, usually from Jupiter, but possibly involving Mars, eventually change the orbital eccentricity of the body until it is cast into a Mars- or earth-crossing orbit. Once this is accomplished, there is a finite probability that the body will be accreted by the planet. Hartmann (1993) gives ~ 500 million years as a rough estimate of the life in free space of a planetesimal.

Cosmic-ray exposure ages for the iron meteorites show a distribution, as would be expected from the stochastic model sketched above. Typical ages are in the ranges of $(0.2\text{–}1.0) \times 10^9$ years. Exposure ages for chondrites tend to be less, by factors of 20–100, than those of irons (Wasson 1974), a fact that may be related to the relative ease of breakup of iron and stone planetesimals. If the stones fragment more frequently than the irons, they will constantly be presenting more fresh material to the cosmic rays for bombardment than the irons (Dohnanyi 1970).

7.3 Meteorites and the SAD

It is generally accepted that the abundances in the Type I (CI) carbonaceous chondrites provide the best guide to abundances of nonvolatile elements in the envelope and photosphere of the sun. Various compilations (see Cameron 1982, Anders and

Grevesse 1989) traditionally used by astronomers and nuclear astrophysicists make this assumption.

A nonnegligible number of elements are sufficiently volatile that abundances must be taken from other sources than the CI's. For example, solar spectroscopic abundances have been used for hydrogen, carbon, nitrogen, and oxygen, while solar wind measurements provide data for the noble gases. "Adjustments" are made in a few cases, based on both theoretical preconceptions (from nucleosynthesis) and eclectic selection of data sources (see Anders and Ebihara's 1982 discussion of Br). Tables A4 and A5 at the end of this book contain the SAD according to Anders and Grevesse (1989).

Compilations of the SAD give not only the elemental abundance determinations but also isotopic abundances, and an attempt is made to extrapolate these back to the time of formation of the solar system, some 4.55 billion years ago. While isotopic abundances play a major role in both the adjustments and the interpretation of the SAD, most relative abundances of the isotopes are based on terrestrial rather than meteoritic measurements.

The *International Union of Pure and Applied Chemistry* (IUPAC 1989) tabulates results of analyses of standard samples of what are called "normal materials." Normal materials are defined (!) to be free from isotopic variations. The samples reside in national or international laboratories, and are made available to researchers. The water sample known as SMOW (§8.5) is an example. However, the very notion of normal materials predates the accuracy achievable by ultraclean laboratory techniques (Wasserburg *et al.* 1969), and considerable attention is now being paid by IUPAC to isotopic variations in both geological and extraterrestrial samples (see IUPAC 1992).

For some elements, terrestrial standards are clearly inappropriate for (primordial) SAD isotopic abundances, especially for those species affected by radioactive decay. For example, terrestrial argon is dominated by ^{40}Ar which derives from radioactive ^{40}K. In the SAD, the most abundant Ar isotope is ^{36}Ar. We must refer to Anders and Grevesse (1989) for further details.

Until the early 1970s it was thought that there was little reason to consider isotopic fractionations of either terrestrial or extraterrestrial materials with the exception of a few geochemically well-understood cases. Terrestrial fractionations of hydrogen and deuterium and of ^{16}O, ^{17}O, and ^{18}O are well studied, and generally understood. Isotopic variations in carbon, nitrogen, and sulfur are also known (Hoefs 1987). The current situation (see IUPAC 1989) is quite different, especially for extraterrestrial materials, and one may certainly ask if the notion of an SAD is still viable.

The current postulate, by no means universally shared, is that while interesting inhomogeneities definitely exist in solar system material, they are due to an admixture of a relatively small mass fraction, 0.01%, and therefore need not perturb the overall picture. In the case of the oxygen isotopic anomalies, which are more general, and typically larger than some of the anomalies studied in trace elements, the percentage of "unmixed" material that must be postulated to account for the observations is considerably larger. We shall give some quantitative figures in Chapter 8.

The motivation for finding the SAD or Holy Grail is to enable us to determine the nuclear history of matter – the origin and history of the nuclides independent

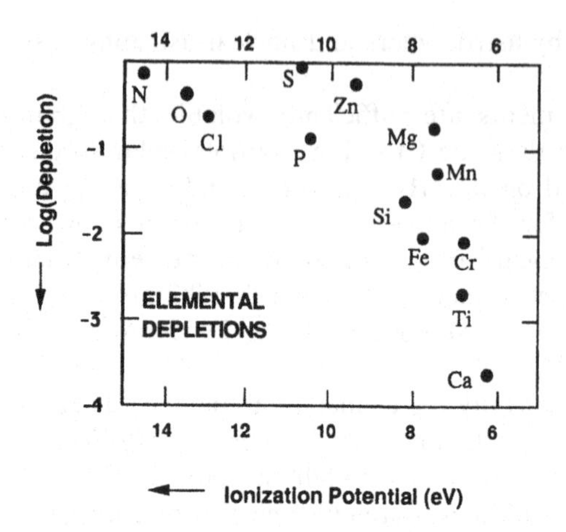

Figure 7.5. Depletions of Heavy Elements in the Interstellar Gas.

of chemistry. This has been the traditional task of the nuclear astrophysicist. The chemical history has been relevant only insofar as it allowed this task to be carried out.

Now it is quite clear that the process of selecting the SAD involves explicit assumptions about the influence of nuclear properties (see Chapter 9), and sometimes specific reactions of nucleosynthesis, on the SAD. The entire process flirts with circularity. While the overall procedure seems to be logically sound, one may surely inquire about the extent of the influence of this circularity on details of the problem of the history of matter. For example, it seems to be a postulate of nuclear astrophysics that abundances in the sun are determined exclusively by nuclear and not by chemical processes. The possibility that this may not be entirely true does not seem to have been seriously considered.

We now think that the interstellar medium consists of a solid and a gas phase in which severe chemical fractionations have taken place. The interstellar gas has been depleted in elements heavier than helium. Figure 7.5 shows "typical" depletions according to Jenkins (1987). We have used his Equation (6) with $\log n(H_{tot}) = 0.5$. In addition, we have estimated "typical" positions for the elements S, Ni, and Ca with the help of depletion diagrams in the article cited. Additional elements are discussed by Hobbs *et al.* (1993).

Workers often use condensation temperatures as the abscissae of these plots, rather than first ionization energies as we have done. Both plots are rather similar. We have chosen ionization potentials for two reasons. First, the available condensation temperatures have mostly been calculated for conditions chosen to resemble those in the solar nebula. They may be very different from those representative of the conditions under which the interstellar grains formed (see Lattimer, Schramm, and Grossman 1978). Second, cosmic-ray abundances show correlations with first ionization potential (Meyer 1985). A mechanism discussed by Alfvén and Arrhenius (1975) called the critical velocity could account for some fractionation of abundances that would depend on ionization potential.

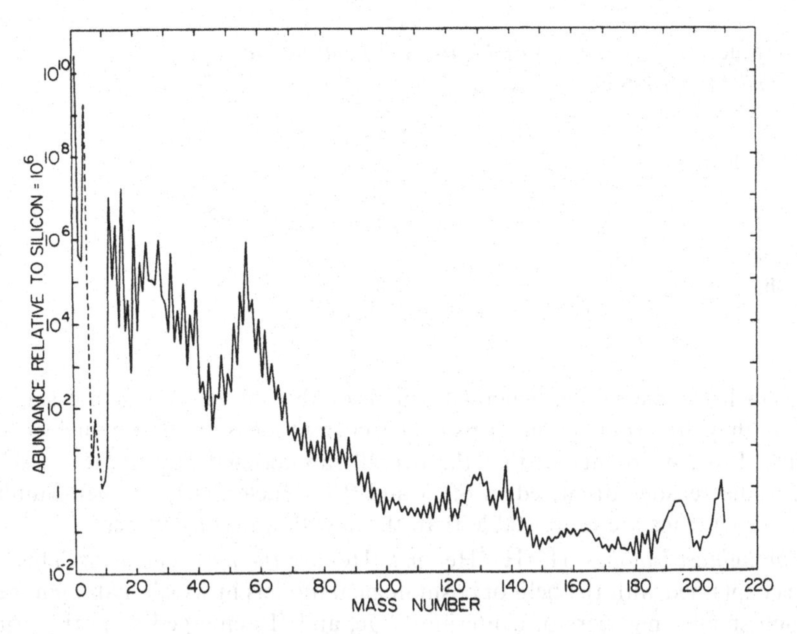

Figure 7.6. The SAD: Abundances vs. Mass Number (Cameron 1982, with permission).

Fractionations of this nature left their imprint on planetary abundances. Solid and gaseous phases became physically separated, and chemically differentiated abundances resulted. We discussed this at some length in Chapter 6. In the case of the terrestrial and Jovian planets, the observations clearly demanded a separation mechanism, and one was found: the protosolar, T Tauri wind.

In the case of stellar abundances, separation of gas and solid phases has been discussed only very recently (see Holweger and Stürenburg 1993). It was widely assumed by astronomers that stars were unaffected by such processes. Nevertheless, when two phases coexist in space, the potential for a chemical separation is always present. The main question is whether separation occurs on a scale that would influence stellar abundances. Surely some stars must have formed with a complement of solid materials (grains) that was not exactly the same as the rest of the interstellar medium. Venn and Lambert (1990) invoked a process of this kind to explain the chemical peculiarities of certain main-sequence stars named after the prototype, λ Boo.

In the sections that follow, we shall ignore questions of this kind, and assume the SAD is both representative of the sun and due totally to nuclear processes.

7.4 The SAD and Nuclear Processes: Overview

The influences of nuclear processes are illustrated somewhat more clearly on a plot of abundances vs. mass number A than they are on plots using atomic number Z (see Figure 7.6).

Table 7.3. *Abundances of Li, Be, and B in the Sun and Meteorites.*

Element	log(ϵ) SAD, Anders and Grevesse (1989)	
	Meteorites	Sun
Li	3.3	1.2
Be	1.4	1.2
B	2.9	2.6

We shall list some of the general features of this curve, and at the same time show how they are explained in terms of our current ideas about the nuclear history of matter. First, we point out that the overall saw-toothed appearance of the plot arises for the reasons discussed in §§1.3 and 9.3. Nuclei with an even number of protons or neutrons are more stable than those with an odd number.

(1) *The lightest isotopes,* H, ^{2}H, ^{3}He, *and* ^{4}He, *are the most abundant.* This observation is explained with the help of cosmological models in which hydrogen, helium, and much smaller numbers of deuterium, ^{3}He, and ^{7}Li emerged from the primeval fireball known as the big bang (§10.7).

(2) *Isotopes of lithium, beryllium, and boron fall in a trough.* It has been realized since the early work on abundances of the elements that some singular processes must be invoked to explain the marked deficiency of isotopes of Li, Be, and B. Suess and Urey (1956) in their seminal paper on cosmic abundances state that "The low abundances of these elements can easily be understood as a consequence of their instability at high stellar temperatures and their possible thermonuclear reactions with protons." The idea goes back much further. Goldschmidt suggested in 1923 that the fragility of these nuclei was responsible for their low cosmic abundance (*Geochemistry* 1954, see pp. 73 and 206). Current ideas are that destruction of these three elements takes place in stellar *envelopes.* The terms in the theory of stellar structure that correspond to core, mantle, and crust of planetary astronomy are core, envelope, and photosphere. The temperatures needed for lithium, beryllium, and boron to be destroyed by proton bombardment are roughly 2.5, 3.5, and 5 million kelvins. In the sun, these temperatures are reached roughly half-way to the center (Table 9.3(*c*)), which is deeper than it is currently thought complete mixing currents reach. This means either that our calculations of the depth of mixing are wrong, or that the solar lithium has been depleted during some pre-main-sequence phase.

Table 7.3 gives a comparison of abundances determined from the solar photosphere with those recently adopted by Anders and Grevesse (1989). In evaluating this material one must keep in mind that the solar spectroscopic determinations are unusually difficult (boron could only be determined from the far ultraviolet), and the meteoritic determinations for Be and B are problematical. It seems clear that in the case of lithium, the abundance in the solar photosphere does not reflect that of the solar nebula.

There is no doubt that current observations of lithium, beryllium, and boron in

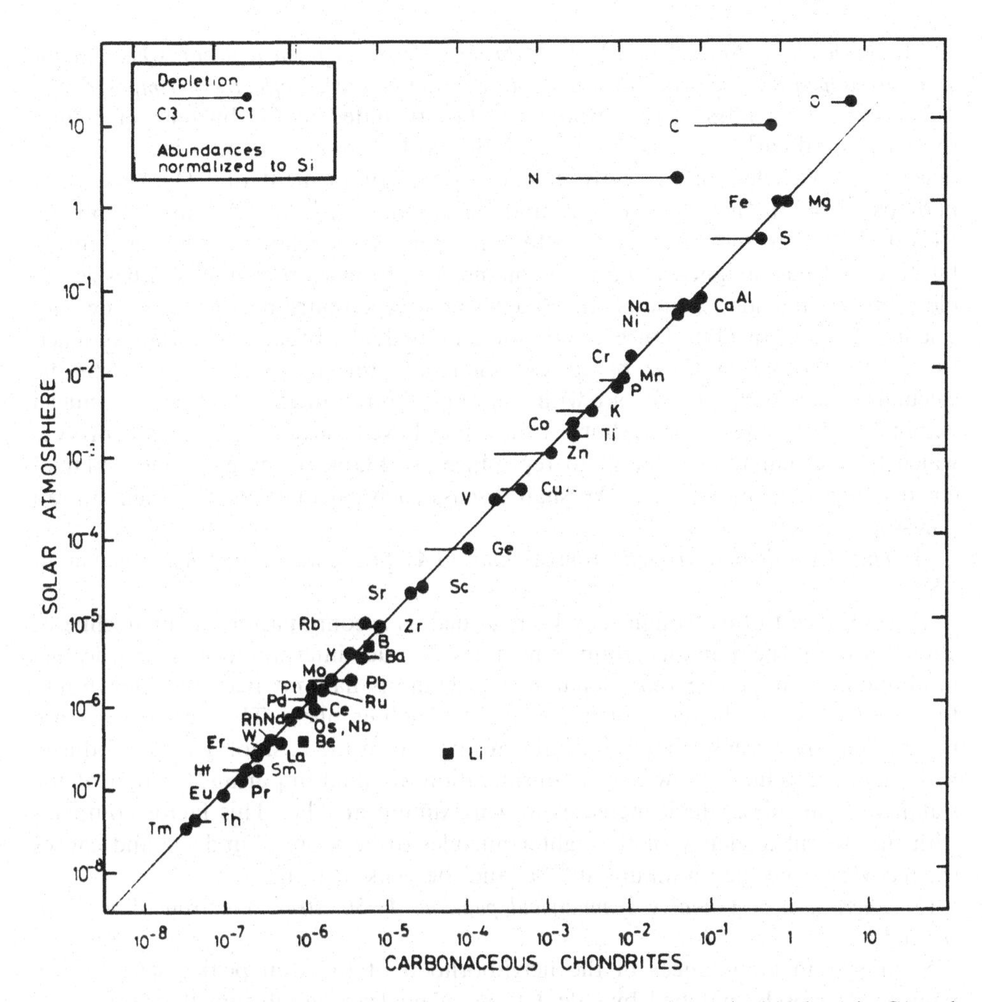

Figure 7.7. Abundances in the Solar Atmosphere Compared with those in C1 and C3 Carbonaceous Chondrites. Courtesy H. Holweger and International Astronomical Union.

stars pose severe constraints on various scenarios of stellar evolution as well as calculations of stellar structure. In the sun, for example, we find severe depletions of lithium, but smaller depletions of boron and beryllium compared with the meteoritic values. Mixing in the sun must be finely tuned in such a way as to account for this observation, which is illustrated in Figure 7.7.

Among late-type stars, the presence of a strong line of Li I at 6708 Å has been recognized since the work of Herbig (1965) as an indication of stellar youth. The Spites (1982) showed that Li/H ratios of about 10^{-10} could be observed in very old halo and disk stars, in apparent contradiction to an overall trend which seemed to be well established. This question was discussed by Reeves (1991). We shall return to the light elements in §§10.5-10.7.

(3) *The nuclides from* $A = 12$ *(C) through* 40 *(Ca) form a group whose major structure appears to be a downward slope with the odd–even effect superimposed.*

The even-A nuclides in this group may be attributed to the burning of helium to carbon, and carbon to heavier species, primarily through the addition of helium nuclei or α-particles. B^2FH attributed the solar complement of ^{16}O, ^{20}Ne, "and perhaps ^{24}Mg" to "helium burning," and ^{24}Mg, ^{28}Si, ^{32}S, ^{36}Ar, ^{40}Ca, and "probably ^{44}Ca and ^{48}Ti" to the "α-process." Modern terminology retains the concept of He burning, but has assigned a variety of names to processes that may synthesize the old α-process nuclides. Some of these names are oxygen burning, neon burning, and silicon burning. In abundance tabulations, individual isotopes are often assigned to one or another of these processes. Generally this means that some specific calculation has been carried out with the aim of reproducing the SAD for nuclei within a certain range. The authors of the paper have referred to the cosmic process which their calculations attempt to reproduce as "blank burning," where "blank" could be carbon, oxygen, etc. We shall discuss such calculations and their overall philosophy below.

(4) *There is a definite trough, from A equal to* 41 *to* 50, *with a decided minimum at* ^{45}Sc.

The entrance to this trough may be regarded as a continuation of the decline of abundances for the previous group of nuclides. A crude interpretation of this decline in abundances might be that stellar nucleosynthesis has not had the opportunity to manufacture the heavier nuclides from the lighter ones. The lighter ones are more abundant *because* they were built up first, from the cosmologically produced hydrogen and helium. However, a consideration of equal importance is that of the stability of nuclides, which increases to a maximum at ^{56}Fe. This factor competes with the overall tendency of the lighter nuclides to be more abundant, and causes the recovery from the minimum at ^{45}Sc, and the peak at iron.

(5) *There is a very nearly symmetrical peak at* $A=56$, *stretching from* $A \approx 45$ *to* $A \approx 67$.

Solar system abundances in the neighborhood of the iron peak itself may be remarkably closely matched by calculations of nuclear equilibrium that are of the same nature as the chemical equilibrium considered in §4.8. B^2FH therefore referred to the nuclear processing that created these abundances as the *e*-process (§10.4). More recent developments have introduced both refinements and a proliferation of terminology. Bodansky, Clayton, and Fowler (1968) showed that one could reproduce the observed abundances from ^{28}Si through $A \approx 62$ by assuming a "quasi-equilibrium." The ratios of nuclei to α-particles, protons, and neutrons were taken from equilibrium relations equivalent to our Equation 4.77, except for the proton-to-neutron ratio. This was assumed to be fixed. Bodansky *et al.* called this process *silicon burning*. The process is still of heuristic value, but it has been superseded by detailed calculations. The new methods replace the equilibrium or quasi-equilibrium assumptions by a network of rate equations. Workers who use these methods try to include all relevant reactions in a *reaction network*. The calculations based on such networks have been called by a variety of names, such as "neutron-rich silicon burning" (Cameron 1979), or "explosive nucleosynthesis."

(6) *There is an abrupt change of slope in the decline of abundances from the iron*

peak that is apparent near $A \approx 70$. The abundance decrease very nearly halts. It continues from $A \approx 90$ to ≈ 110, after which there is a shelf (110–140) upon which considerable structure may be seen. There is another decline ($A \approx 145$–150) followed by a similar shelf with structure. From $A \approx 180$ to 209, the abundances actually increase, on the average, with A.

B^2FH invoked neutron addition processes to explain these observations. Beyond the iron peak, the synthesis of the elements could no longer come as a by-product of stellar energy generation, since the binding energy per nucleon reaches a maximum at ^{56}Fe. Two distinct addition processes were envisioned, a slow and a rapid one, called the *s*- and the *r*-processes. Slow or rapid meant with respect to the β-decay rates of stable nuclei. If neutrons were added so slowly that all possible β-decays had time to occur prior to the addition of the next neutron, nucleosynthesis would follow a (nearly) unique path up through the valley of β-stability. Rapid neutron addition would, at least for a time, populate a portion of the nuclide chart to the right and below the valley of β-stability. Eventually, these neutron-rich nuclides would β-decay to stable forms, leaving a characteristic isotopic signature.

The characteristic neutron addition times of the *r*- and *s*-processes are vastly different. In the *r*-process, neutrons must be added within fractions of a second, corresponding to neutron densities of 10^{23} cm^{-3} or so. By contrast, if the necessary β-decays are to occur, there must be time intervals of the order of 10^4 years, with corresponding neutron densities of only 10^5 cm^{-3}.

It is inconceivable that these two processes could occur in the same star at one time. The traditional site of the *s*-process is the interior of a red giant. The site of the *r*-process has always been more problematical, although the supernova is usually invoked as the site of both *r*-processing and equilibrium or quasi-equilibrium nuclear burning.

The general characteristics of the abundance curve from $A \approx 70$ through 209 (^{209}Bi) may be accounted for in terms of these two processes and the theory of nuclear structure. In particular the various bumps are generally well explained in terms of the so-called neutron "magic" numbers, which will be discussed below.

(7) *Numerous proton-rich nuclides cannot be accounted for in terms of any of the processes enumerated thus far. B^2FH assigned them to a p-process whose site was, and remains, uncertain.* Lambert (1992) reviews some possible origins.

We shall return in Chapter 10 to the subject of the processes of nucleosynthesis.

7.5 Problems

1. Among the minerals listed in Mason's classification (Table 7.2), which indicate rapid cooling? (Hint: cf. §2.5.3.)
2. Which of the minerals listed in Table 7.2 are "late" in the sense of being relatively low in the Bowen series (Figure 2.6)?
3. In Chapter 8 we shall show how it is possible to get an estimate of the ratio of ^{26}Al to the stable ^{27}Al at the time of solidification of the host mineral (anorthite). Assume these isotopes were present in equal amounts at the time of detachment of the solar nebula (point [3] of Figure 7.3). With this assumption, calculate the $^{26}Al/^{27}Al$ ratio 10 million years later.

4. The cosmic abundance distribution contains the fingerprints of various processes of nucleosynthesis. In particular, our current understanding of neutron addition processes is based on the detailed shape of this curve. Examine the features of the curve (Figure 7.6) beyond the iron peak, and pick out features that you think any satisfactory theory should be able to explain.

8

An Introduction to Isotope Geology with an Emphasis on Meteorites

8.1 Introduction

The isotopic abundances of cosmic materials may change for a number of reasons. If a substance contains radioactive nuclei, there will be a continual decrease in the parent and a buildup of the daughter isotopes. Bombardment of materials by cosmic rays or other high-energy particles can also alter the isotopic complement of a sample. During radioactive decays or nuclear fission, particles are emitted which can affect the surrounding nuclei. Fission fragments remain in the neighborhood of the parent nuclei. A third possibility is fractionation, by either diffusion or small mass-dependent effects in chemical reactions. All three of these contingencies have been mentioned or intimated previously. We shall now take up certain aspects of these processes in detail.

It will not be possible for us to discuss most of the dating techniques. The interested reader may consult the textbooks of Faure (1986) or Durrance (1986). Richardson and McSween (1989) have an excellent chapter on radioactive dating.

8.2 Rubidium–Strontium Dating; Sample and Model Ages

One of the most straightforward methods of age determination makes use of the decay of ^{87}Rb to ^{87}Sr. We shall discuss this particular method here in detail, because of its pedagogical advantages. We shall have time to mention only briefly other methods, some of which are now more actively pursued than rubidium–strontium.

Both rubidium and strontium are geochemically dispersed, that is they occur primarily as impurities in major minerals. Strontium has four stable isotopes, ^{84}Sr, ^{86}Sr, ^{87}Sr, and ^{88}Sr, whose respective abundances are given by Cameron (1982) as 0.5, 9.9, 7.0, and 82.6%. Rubidium has one stable isotope ^{85}Rb, and one long-lived radioactive one, ^{87}Rb, with relative abundances 72.1 and 27.9%.

^{87}Rb β-decays to ^{87}Sr with a half-life of $4.9 \cdot 10^{10}$ years. The value is uncertain by perhaps as much as several percent, so many geologists have adopted a half-life of $5.0 \cdot 10^{10}$ years as a "standard" to facilitate the intercomparison of dates.

Let us consider a magma containing, for example, molten minerals of the kinds already familiar to us – olivines, pyroxenes, feldspars, and oxides. While liquid, the melt will contain ions of rubidium and strontium that may migrate freely, and which we may therefore assume to be homogenized. At the moment of freezing, however, the rubidium and strontium will be squeezed into the major minerals. This will happen in a way that will depend upon the sizes of the cations rubidium and strontium, and those in the major minerals for which they must substitute.

Rubidium and strontium will partition very differently as a result of their differing ion sizes and valences. A glance at Figure 5.7 will show that Rb may be expected to be enriched in orthoclase ($KAlSi_3O_8$), where the rubidium ions can occupy the sites normally filled by K^+. Strontium will be enriched in calcium-bearing minerals such as anorthite ($CaAl_2Si_2O_8$), with the Sr^{2+} ions substituting for Ca^{2+}. It is important to realize that some substitution will occur in every mineral, and that these amounts may be measured by mass spectrometry with accuracies that seem incredibly good to this author who was trained in analytical stellar spectroscopy.

Consider now the abundance ratios of the radioactive parent and daughter (^{87}Rb and ^{87}Sr) to the stable isotope ^{86}Sr. During partition (freezing) the behavior of the two isotopes of Sr cannot be distinguished. The atomic properties, ion size and valence are identical. The small isotopic effects that occur in principle, have not been detected in atoms as massive as those of strontium. In the second part of this chapter, we will discuss the cases of lighter elements where the mass differences do cause some fractionations.

It is common in work of this nature to introduce a stable nuclide such as ^{86}Sr that has no *radiogenic* contribution. There are several reasons for doing this. First, the chemical properties of ^{87}Sr and ^{86}Sr are essentially identical. In addition, the mass spectrographic techniques now commonly in use (see Wasserburg *et al.* 1969) measure abundance ratios more easily than absolute values.

Figure 8.1 is a schematic plot of the ratio $^{87}Sr/^{86}Sr$ vs. $^{87}Rb/^{86}Sr$ for various minerals of a rock, denoted M_1 through M_4. Consider first the situation in a freshly frozen rock sample. The ratio $^{87}Sr/^{86}Sr$ will be identical in all of the minerals, while $^{87}Rb/^{86}Sr$ will be quite different. The reason for this is that the ion sizes of ^{86}Sr and ^{87}Sr are virtually identical, while those of ^{86}Sr and ^{87}Rb are quite different (see Figure 5.7). Both ions are typically *trace* species, accommodated within the host mineral as discussed in §5.7. Their relative abundances are strongly dependent on the sizes of the ions.

The points corresponding to the various minerals will thus be strung out in a line whose ordinate is the initial $^{87}Sr/^{86}Sr$ ratio, that is, the value at the time of freezing ($t = 0$). In time, the ^{87}Sr will increase and the ^{87}Rb decrease according to the law of radioactive decay. In the equations that follow, we will use the symbols of the isotopes, viz. ^{86}Sr, ^{87}Sr, and ^{87}Rb to stand for the numbers of these species in the minerals. Since we shall employ ratios, it is immaterial whether we consider these as absolute numbers per mineral fragment, or number densities (assuming, of course, uniform number densities within the fragments!).

The $^{87}Rb/^{86}Sr$ ratio will obey the law

$$(^{87}Rb/^{86}Sr)_t = (^{87}Rb/^{86}Sr)_0 \cdot \exp(-\lambda t), \qquad (8.1)$$

where the decay constant λ is related to the half-life of rubidium by the relation

$$\tau_{1/2} = 0.6931/\lambda. \qquad (8.2)$$

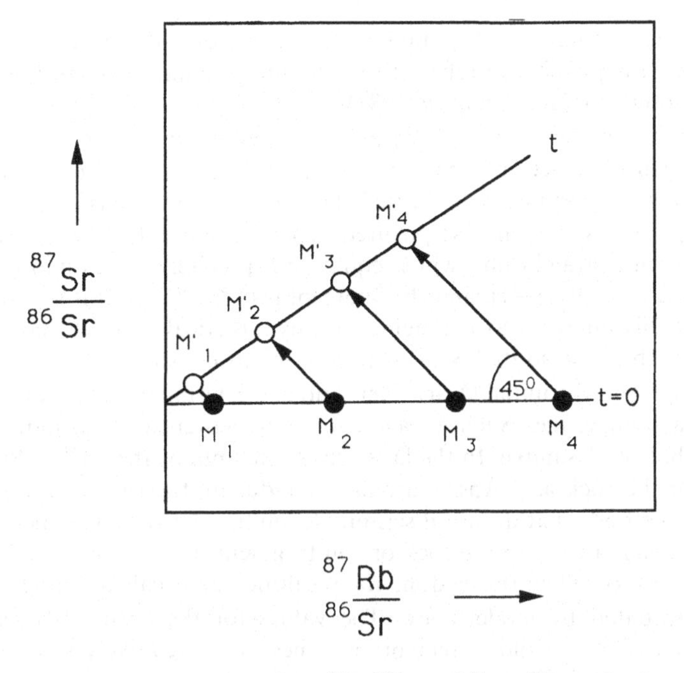

Figure 8.1. Evolution of the $^{87}Sr/^{86}Sr$ and $^{87}Rb/^{86}Sr$ ratios in a hypothetical rock.

The subscript '0' refers to the time of freezing, and 't' to the time *interval* since freezing. The numerical constant 0.6931 is the natural logarithm of 2. Equation 8.2 follows immediately from 8.1 by putting $(^{87}Rb/^{86}Sr)_t = (1/2)(^{87}Rb/^{86}Sr)_0$.

At time t, the ratio $^{87}Sr/^{86}Sr$ present in any mineral is the sum of the initial ratio, plus a contribution from the decay of ^{87}Rb. The *radiogenic* contribution is often designated by an asterisk written as a superscript, thus

$$(^{87}Sr)_t = (^{87}Sr)_0 + (^{87}Sr)^*. \tag{8.3}$$

But

$$(^{87}Sr)^* = (^{87}Rb)_0 - (^{87}Rb)_t = (^{87}Rb)_t \cdot [\exp(\lambda t) - 1], \tag{8.4}$$

where the last step follows readily from Equation 8.1.

If we now divide Equation 8.4 through by $(^{86}Sr)_t$, we obtain the relation for obtaining the *sample age* of a rock by the rubidium–strontium technique.

$$(^{87}Sr/^{86}Sr)_t = (^{87}Sr/^{86}Sr)_0 + (^{87}Rb/^{86}Sr)_t \cdot [\exp(\lambda t) - 1]. \tag{8.5}$$

After analysis of a single mineral, there remain two unknowns in Equation 8.5, t and $(^{87}Sr/^{86}Sr)_0$. In order to determine a sample age, one needs to determine $(^{87}Sr/^{86}Sr)_t$ and $(^{87}Rb/^{86}Sr)_t$ in at least two minerals. It is then possible to solve simultaneously for both the time since freezing, and the initial $^{87}Sr/^{86}Sr$ ratio. In practice, one measures these ratios in a number of minerals, and finds a solution by the method of least squares.

The traditional least squares method minimizes vertical or y-deviations from an adopted mean line. It is possible to refine it to take into account errors in both the ordinate and the abscissa (Cameron *et al.* 1981).

Let us consider the time evolution of the individual points in Figure 8.1. As the ^{87}Rb decays, an equal increment of ^{87}Sr appears, so the points all move up, and to the left, following a 45 degree line, as indicated. The more ^{87}Rb there is in a sample, the bigger the steps. For the ideal case pictured here, the points for the individual minerals all remain on a straight line, whose equation is precisely our Equation 8.4 . The slope of the line is $[\exp(\lambda t) - 1]$, and the intercept is $(^{87}\mathrm{Sr}/^{86}\mathrm{Sr})_0$. It is sometimes helpful to think of this intercept as belonging to a hypothetical mineral that could accommodate *no* ^{87}Rb at all, so its $^{87}\mathrm{Sr}/^{86}\mathrm{Sr}$ ratio remains constant.

When the rock age and the initial $(^{87}\mathrm{Sr}/^{86}\mathrm{Sr})$ ratio are determined simultaneously, the age is called a sample age. Such an age cannot be assigned to an individual mineral, or to a whole *rock* sample. In the latter case, one obtains the $^{87}\mathrm{Rb}/^{86}\mathrm{Sr}$ and $^{87}\mathrm{Sr}/^{86}\mathrm{Sr}$ ratios for the rock as a whole, usually by reducing the rock to a powder before analysis by methods that do not discriminate among the various minerals. It is still possible to assign an age to the rock or soil fragment by assuming a value of the initial $^{87}\mathrm{Sr}/^{86}\mathrm{Sr}$ ratio. When this is done the resulting age is called a model age.

Lunar soils were dated by model ages. The values for the $(^{87}\mathrm{Sr}/^{86}\mathrm{Sr})_0$ ratios assumed were taken from the oldest meteorites. These were relatively low values, compared to those found from model ages for lunar rocks, so the corresponding slopes, in the sense of Equation 8.4, were larger. This gave rise to the apparent paradox that the lunar soils were older than the rocks from which they were derived, by the long-term pulverization of meteoroid impact. We shall discuss this question again with the help of concepts to be raised in the following section.

8.3 Evolution of the Initial Ratio: $(^{87}\mathrm{Sr}/^{86}\mathrm{Sr})_0$

Consider a well-mixed gas such as the primordial solar nebula, which slowly cools. Let us call t_0 the time at which the first solid materials appear (point [5] of Figure 7.3). If this solid consists of several minerals, that is, it is a rock rather than a mineral grain, then the memory of the $(^{87}\mathrm{Sr}/^{86}\mathrm{Sr})$ ratio at the instant of solidification is frozen into the solid. It may be determined by the methods of the preceding section.

Suppose another solid forms at a somewhat later time, t_1. If this solid is a rock, the memory of the $(^{87}\mathrm{Sr}/^{86}\mathrm{Sr})$ ratio at time t_1 will be frozen into it as well, but since $t_1 > t_0$, $(^{87}\mathrm{Sr}/^{86}\mathrm{Sr})_{t_1} > (^{87}\mathrm{Sr}/^{86}\mathrm{Sr})_{t_0}$ because more ^{87}Rb has had time to decay.

Several billions of years later, when we analyze these rocks, we find different values of the $(^{87}\mathrm{Sr}/^{86}\mathrm{Sr})_0$ ratios, and we would like to infer from this something about the time interval $t_1 - t_0$ (within [5] to [6] of Figure 7.3). For simplicity in printing, let us put $t_0 = 0$, and $t_1 = t$. Then when the second sample solidified, we have

$$(^{87}\mathrm{Sr}/^{86}\mathrm{Sr})_t = (^{87}\mathrm{Sr}/^{86}\mathrm{Sr})_0 + [(^{87}\mathrm{Rb}/^{86}\mathrm{Sr})_0 - (^{87}\mathrm{Rb}/^{86}\mathrm{Sr})_t], \tag{8.6}$$

where it is important to note the subscripts '0' now refer to the time of freezing of the first solid (at t_0), and not to the frozen-in initial ratio of the second sample. This notational change can be confusing if it is not thoroughly understood.

We must also be clear about the physical basis for Equation 8.6. It is valid, of course, for any closed system as a whole. The solar nebula now contains, by our hypothesis, a gas and a solid phase. These two systems, gas and solid, are not closed, because we imagine that the gas is continually condensing to form more solid. We can be sure that the ratio of Rb/Sr will be very different in the gas and solid phases, because of the differences in the way the two ions substitute in the major mineral phases. Surely, the solid will be relatively impoverished in the large ion Rb^+, which has a proclivity for substitution in minerals with low melting and condensation temperatures. This means that the $^{87}Sr/^{86}Sr$ ratio should increase faster in the residual gas than in the solids. The point is that the $^{87}Rb/^{87}Sr$ ratio will change in the solid and gas phases for *chemical* reasons as well as radioactive decay. Let us take for the moment Equation 8.6 to refer to the nebula as a whole – solid plus gas phase.

If we let the time t in Equation 8.6 be a variable, we may find the increment

$$\delta(^{87}Sr/^{86}Sr) = (^{87}Rb/^{86}Sr)_0 \cdot [\exp(-\lambda t)]\lambda\delta t, \tag{8.7}$$

and we may solve for the time increment δt corresponding to the *observed* increment $\delta(^{87}Sr/^{86}Sr)$ in the *initial ratios* of the two rocks. In doing this, we must be able to ignore possible fractionations between the gas and solid phases mentioned above if we wish to assume the time increment thus found is characteristic of the gas plus solid. This should be a valid assumption as long as the fractional amount of rubidium and strontium that have left the gas phase is small. Alternatively, we may simply regard the initial ratios in the two rock samples as characteristic of the gas from which they solidified. We have

$$\delta t = \frac{\delta(^{87}Sr/^{86}Sr)}{(^{87}Rb/^{86}Sr)_0\lambda \cdot [\exp(-\lambda t)]}. \tag{8.8}$$

Now the denominator of 8.8 contains the time t, which by assumption is the same as the interval δt, if we adopt the time of solidification of the first rock as $t = 0$. We have, then,

$$(^{87}Rb/^{86}Sr)_0 \cdot \exp(-\lambda t) = (^{87}Rb/^{86}Sr)_t, \tag{8.9}$$

where t is the time of solidification of the second rock. For cases where sample ages are available, $\delta(^{87}Sr/^{86}Sr)$ represents the difference in the initial ratios. The denominator of 8.8 or the right-hand side of 8.9 contains the unmeasured $^{87}Rb/^{86}Sr$ ratio at the time the second rock froze. However, the decay constant is a small number, and for a time of the order of $5 \cdot 10^9$ years, the $\exp(-\lambda t)$ is only 0.93. The percentage error made by using the present $(^{87}Rb/^{86}Sr)$ ratio (instead of the value when the second rock froze) is not large insofar as the determination of δt is concerned.

A traditional ploy that avoids the problem of different partitioning of ^{87}Rb and ^{87}Sr into gas and solid has been to use the abundance of Rb and Sr in the sun. It is necessary to suppose that the isotope ratios found in various compilations may be used to obtain the desired $^{87}Rb/^{86}Sr$ for the sun.

One may also use ^{87}Rb/^{86}Sr ratios in a relatively unfractionated meteorite. In practice, this means taking the value directly from some tabulation such as that of Anders and Grevesse (1989) or Cameron (1982) which may be considered representative, according to the compilers, of such meteorites. Note that Cameron's values "correct" for radioactive decay over the age of the solar system. Differences between sun and meteorites are not large, but they are not trivial either. Holweger (1979) gave Rb/Sr = 0.22 (by number) for the sun, while Anders and Ebihara (1982) report 0.27 for the same ratio. Anders and Grevesse give 0.30.

As an exercise, let us work out the time interval for

$$\delta(^{87}\mathrm{Sr}/^{86}\mathrm{Sr})_0 = 0.698\,98 - 0.698\,77, \tag{8.10}$$

corresponding to the measurements reported as BABI and ALL in our Figure 7.4 (Faure 1986). We need the value in the solar system as a whole of (^{87}Rb/^{86}Sr) at the time of the BABI solidification. For simplicity, we again speak of BABI as though it were a single sample (see §7.2). Using $\lambda = 1.37 \cdot 10^{-11}$ yr^{-1}, and (^{87}Rb/^{86}Sr) = 1.97/1.76 = 1.12, directly from Anders and Ebihara, we find $\delta t = 2.1 \cdot 10^{-4}/(1.37 \cdot 10^{-11} \times 1.12) = 1.4 \cdot 10^7$ years.

This interval is small with respect to either the BABI sample age or that of Allende. We may apply a small correction based on the BABI age of $4.7 \cdot 10^9$ years. We have $\exp(1.37 \cdot 10^{-11} \cdot 4.7 \cdot 10^9) = 1.07$, so the interval just deduced should be decreased by some 7%.

It is remarkable that it has been possible to resolve an interval of time of the order of 14 *million* years, from the present distance in time of some 4.5 *billion* years.

8.4 Further Remarks on Radioactive Dating

It is by no means assured that the plot of (^{87}Sr/^{86}Sr) vs. (^{87}Rb/^{86}Sr) for a rock sample will yield a straight line. If the points do not fall along a line, the time determination is said to be discordant. A standard explanation of discordant times is partial redistribution of the ^{87}Sr or ^{87}Rb, the latter is more typically mobile because of its larger ion radius. Reheating of the rock can cause such effects. If the reheating is severe, the radioactive clock can be *reset.* The effect is roughly the same as if the rock refroze from a melt. All of the (^{87}Sr/^{86}Sr) ratios in the minerals are once again the same, and the enrichment of ^{87}Sr from ^{87}Rb, greater in the rubidium-rich minerals, begins again.

We cannot give the details of other methods of isotopic dating here. Among the invaluable methods used today are those based on the decay of ^{40}K to ^{40}Ar, and the neodymium–samarium techniques (O'Nions *et al.* 1979, Faure 1986). Each of these methods has its own techniques, and interpretations. For example, the potassium–argon methods are related to the time when a sample has hardened to the point where it can retain the daughter, ^{40}Ar. This time is similar, but not identical, to the time when the rubidium and strontium partitioned, permanently, into crystals. In the reference books cited above, details are given of these and other dating methods.

With our current background, we are in a position to make some sense out of the model age for a soil fragment, or a whole-rock sample. For the lunar soils, for example, a common choice for the initial ratio was that of BABI. Since this initial ratio is smaller than those found by the method of sample ages for any lunar rock, it is easy to understand why the model ages of the soils are higher than those of the "parent" rocks. The model age is a rough measure of the time since the BABI samples froze. It is not precisely that, since rubidium and strontium would have partitioned differently in the BABI samples and the Moon.

8.5 Stable Isotope Geology

Isotopic variations up to several percent are common in the earth for a number of the lighter elements. The best-studied cases are those of hydrogen, carbon, oxygen, and sulfur, but variations in a number of other elements are known. The terrestrial nitrogen isotopes do not show large fractionations, $\leq 1\%$ are typical, but the Viking experiment measured $^{14}N/^{15}N = 165$ on Mars, which should be compared with the terrestrial value of 277 (see Carr 1981, and references therein). This represents an enormous enrichment of ^{15}N in the atmosphere of Mars relative to that of the earth.

According to a current interpretation outlined by Carr, enough ^{14}N was once present to bring the ratio in line with the terrestrial value. The Martian atmosphere was thus once much richer in N_2. Photodissociation and preferential loss of the lighter isotope from the atmosphere account for the current observation. An unavoidable consequence of this hypothesis is that a large portion of the original volatile complement was lost from the planet. We shall have occasion to discuss similar, very large isotopic fractionations in later chapters – in the interstellar medium, and in stellar atmospheres. For the most part, we shall deal with much smaller separations.

Because isotopic variations are typically small, it is customary to discuss the fractional changes in units of 0.001 (*per mil*) rather than 0.01 (*percent*). The symbol ‰ is used analogously to the %-sign. Standard isotope ratios are adopted. For hydrogen (deuterium and hydrogen) and oxygen (^{16}O, ^{17}O, and ^{18}O) these ratios come from a sample designated SMOW, for standard mean ocean water. The US National Institute of Standards and Technology (NIST, formerly NBS) has a program for the determination and dissemination of such reference materials. Some of the original standard samples are no longer available, and work must be done with respect to secondary standards, and transformed to the original system. A general discussion and references may be found in Mason and Moore (1982).

The symbol δ is used to denote the variation per mil from the standard. Thus if R_0 is the ratio of ^{18}O to ^{16}O in SMOW, and R is the same ratio in some sample under study, one would determine

$$\delta^{18}O = [(R - R_0)/R_0] \cdot 1000.0‰ \,. \tag{8.11}$$

It is extremely useful in isotope geology to employ the three-isotope correlation diagram (cf. Podosek 1978), in which values of δ for one isotope pair are plotted against those for another. Figure 8.2 is a schematic plot of δ for $^{17}O/^{16}O$ vs. δ for $^{18}O/^{16}O$. The figure is from Podosek (1978), and is based on the work of R. N. Clayton, and his collaborators at the University of Chicago, who made the

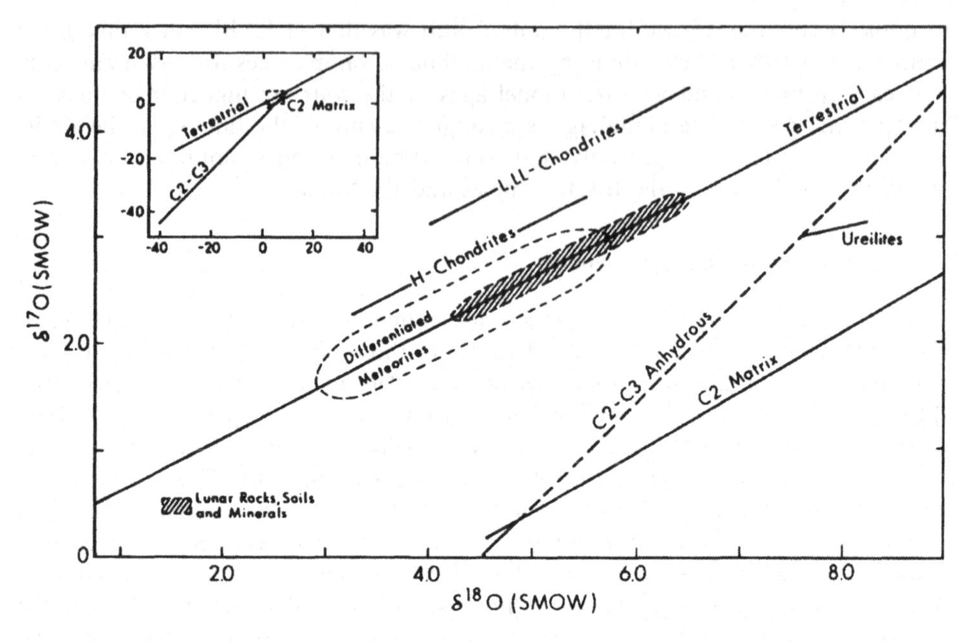

Figure 8.2. Oxygen Isotopic Correlation Diagram. Reproduced from *Annual Review of Astronomy and Apstrophysics* **16**, 293, ©Ann. Rev. Inc. 1978. Courtesy F. Podosek.

pioneering study of isotopic anomalies in meteorites (see Clayton, Grossman, and Mayeda 1973). The various terrestrial samples fall along a single line with slope 1/2. Certain of the chondrites and *differentiated meteorites* also plot along this line, but other meteorite samples do not. Let us consider the various processes that cause fractionation, and attempt to account for the general distribution of points in such a diagram. We shall discuss some of the problems raised for individual meteorite classes in the sections that follow.

There are two basic kinds of processes that lead to geochemical fractionations of isotopes. The first process involves the breaking of weak intermolecular forces, as in the evaporation of a liquid. This process accounts for much of the fractionation of the hydrogen and oxygen isotopes in H_2O on the earth. There is obviously a preference of the vapor for the lighter molecules. If one boils a pot of water, one can be sure that the heavier water molecules, for example, D_2O, $H_2{}^{17}O$, $H_2{}^{18}O$, will be enriched in the residual liquid, while the vapor will be richer in the lightest water molecule, $H_2{}^{16}O$. A similar process happens when ocean waters evaporate.

This isotopic fractionation is in some ways similar to the chemical differentiation that results from the partial melting of a miscible, solid solution. One could draw phase diagrams similar to Figure 2.7 for melting of the albite–anorthite system. In the case of water, the two phases could be ice and liquid water, or they could be liquid and vapor. In either instance, the higher-temperature phase would be enriched in the lighter molecules.

A second kind of fractionation of isotopes depends on the difference in the chemical bonding of the isotopes in molecules. A useful working rule is that the heavy isotopes are preferentially retained in the sites where the binding is strongest.

A heuristic explanation of this effect can be given for binding that results in oscillatory motion about some equilibrium point, as in a diatomic molecule or a crystal. The energy of the oscillator is quantized, so that $E = h(n\nu_0 + 1/2)$, where n is a quantum number equal to $1, 2, 3, \cdots$, and $1/2$ gives the quantum mechanical zero-point energy (see §11.4). For an ordinary harmonic oscillator, the vibrational frequency ν_0 is $(1/2\pi)\sqrt{K/m}$, where K is the force constant. The binding force (K) of atoms can be expected to depend on the electronic structure, and therefore to be very nearly independent of the mass. Consequently, the frequency ν_0 will decrease as the inverse square root of the mass. Thus, heavier isotopes will lie deeper in an oscillator potential well than lighter ones. This is illustrated in Figure 8.3.

The figure shows potential energy curves for H_2 and D_2 (diatomic deuterium). They were constructed by fitting a Lennard–Jones potential to the constants for H_2 and D_2 given by Huber and Herzberg (1979). This analytical approximation to molecular potential curves is discussed by many authors (e.g., Castellan 1983, p. 673). The full line is for H_2, while the points are for D_2. The curves are indistinguishable at the scale of the figure. The inset shows the region of the minima of the curves, with the ground vibrational levels for H_2 (solid) and D_2 (dashed).

The detailed calculation of the equilibrium distribution of an isotope involves the consideration of rotational and translational degrees of freedom, when they are applicable, and can become rather involved. Broecker and Oversby's (1971) excellent text gives some examples.

Fortunately, detailed calculations are unnecessary to understand most of the fractionations illustrated in Figure 8.2. We only need to know that the difference in the mass of isotopes is responsible. This leads us to assume that any process that leads to a given change in δ_{17} will produce *twice* that change in δ_{18}, because the mass difference is twice as large. This reasoning may be shown to be valid for *small* fractionations. Podosek (1978) discusses a more general relation.

We can therefore understand a fractionation line for oxygen isotopes that has a slope of 1/2. For many years, such fractions were the only ones that were observed. However, beginning in the late 1960s, analyses revealed that many meteoritic samples plot off the line occupied by so many terrestrial and lunar materials (see the inset of Figure 8.2).

In order to understand a distribution of points over an *area* of a three isotope correlation diagram let us consider the results of a mixture of two substances which fall along any line, for example, AB in Figure 8.4.

Any mixture of A and B will fall along the line AB, and the relative distances along the line will be given exactly by the relative amounts of substances with compositions A and B. The principle invoked is nothing more complicated than that of forming a mathematical mean. The mean must fall within the range of the values upon which it depends. Similarly, if we have a third composition, say C, any mixture of the three will plot within the field bordered by the triangle ABC. Two possible mechanisms can lead to the occupation of an area on one of these

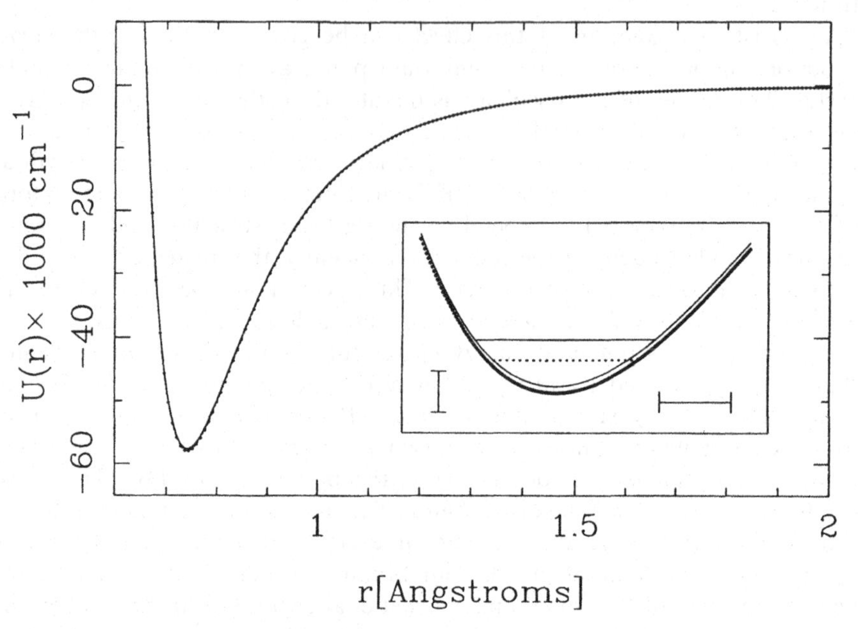

Figure 8.3. Binding of H_2 and D_2. The region of the minima of both curves is shown in the inset. The vertical bar shows 2000 cm^{-1}, or 0.25 eV. The horizontal bar is 0.02 Å. The ground vibrational level of D_2 is 630 cm^{-1} or 0.078 eV *below* that for H_2.

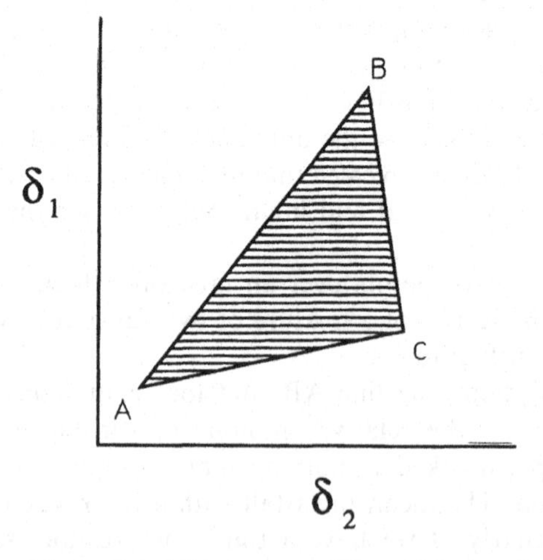

Figure 8.4. Mixtures on a Three–Isotope Correlation Diagram.

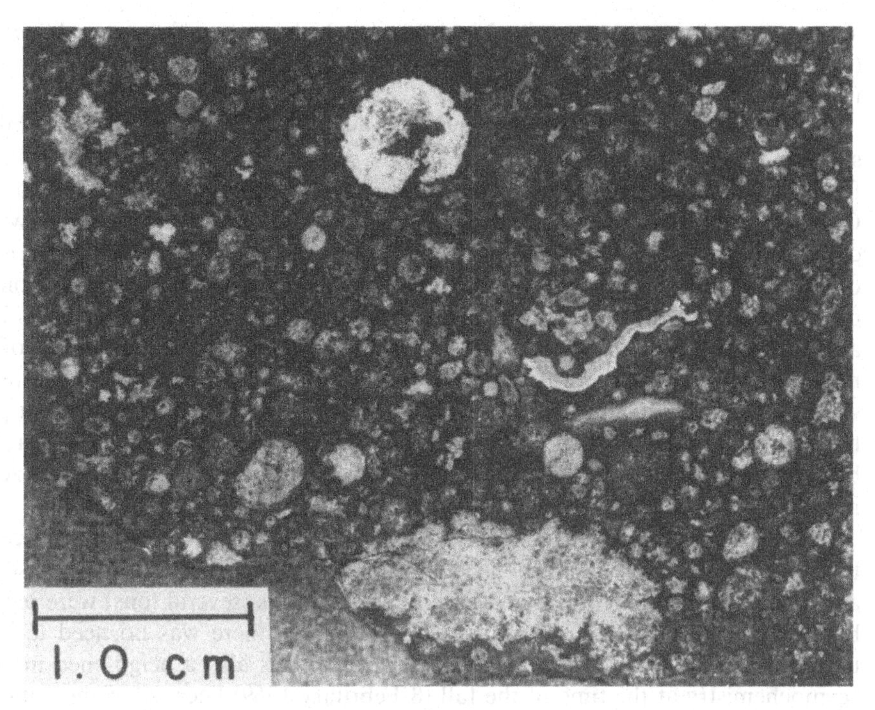

Figure 8.5. A Slab Surface of the Allende Meteorite. Reproduced with permission from the *Annual Review of Earth and Planetary Science*, Vol. 8, p. 559, ©1980 by Annual Reviews Inc. Courtesy L. Grossman.

diagrams: mixing of materials with three (or more!) compositions, or the mixing of two components, followed by isotopic fractionation.

8.6 Oxygen Anomalies and the Refractory Inclusions of Meteorites

Some of the carbonaceous chondrites are quite unlike the terrestrial and lunar igneous rocks that we discussed in Chapter 3. These solids were certainly not heated, as a whole, to anything like the temperatures necessary to melt the entire assemblages, although this does appear to be true for the chondrules. In general, we may consider these rocks to be assemblages of materials, some of which have had distinct histories. In other words, we cannot explain the whole rock in terms of solidification from a uniform-composition melt. Figure 8.5 shows a slab surface from the Allende meteorite, which has invoked so much cosmochemical interest. Our discussion follows Grossman (1980).

One can see the rounded chondrules of various sizes, a large "ameboid olivine aggregate," and filaments of various shapes. Many of these materials are not in chemical equilibrium – given time, or elevated temperatures to speed up reactions that are thermodynamically possible, the mineralogy of this rock would be quite different. Particularly bizarre are some of the rims, or zones (usually ≤ 60 microns) about some of the coarse-grained inclusions, which contain the minerals nepheline ($NaAlSiO_4$) and sodalite ($3(NaAlSiO_4) \cdot NaCl$). Petrologists call these minerals unsaturated.

Unsaturated minerals are not found in association with any of the SiO_2 minerals such as quartz or the high-pressure forms tridymite and cristobalite (sometimes found in shocked materials). They are thermodynamically "hungry" for SiO_2. Consequently, they could not have condensed in equilibrium with a gas with the composition of the SAD. The unsaturated minerals are not at the outermost rims of the inclusions, which are chemically and mineralogically *zoned.* A typical sequence might be Fe-bearing spinel, perovskite, forsterite, nepheline, hedenbergite and diopside (see Grossman 1980). It is therefore plausible that many of these materials have had quite different histories. Indeed, it is frequently suggested that some of the calcium–aluminum rich inclusions may have been solids from a part of the interstellar medium that was chemically or isotopically distinct from that of the bulk of the solar-system material. It is likely that some were never vaporized and mixed in with the majority of the solar nebula.

Cosmochemists have frequently made the point that the great prominence given to the Allende (CV or C3) carbonaceous chondrite has arisen largely because of the availability of the material for analysis. Large amounts (several tons) were recovered by cosmochemists or their surrogates and for once, there was no need to depend upon the largess of museums or collectors. There was also a heightened interest in cosmochemistry at the time of the fall (8 February 1969) because of the anticipated Apollo lunar samples.

Figure 8.6 shows some of the oxygen anomalies in the "coarse-grained inclusions" of Allende. The departures from the SMOW standard are quite large, ≥ 40 per mil. While terrestrial anomalies of this size occur in ^{18}O, those of ^{17}O are much smaller, and consistent with isotopic fractionation based on mass.

The line DA in Figure 8.6 has a slope very close to unity, and if extrapolated toward the largest negative values of δ, for both ^{17}O and ^{18}O, it would pass near the point $(-1000, -1000)$, which corresponds to pure ^{16}O. Consequently, the inclusions are often interpreted as a mixture of material with "normal" oxygen isotopes plus an admixture of nearly pure ^{16}O. It has become common to refer to first (normal) composition as solar or uncontaminated and to the second as exotic or E. A small number of samples plot off the line AD. These may be interpreted as either fractionations from the line AD, or indications of the presence of additional components, possibly with compositions B or C. The line ABC has the proper slope for fractionations.

8.7 Isotopic Anomalies in the Noble Gases

The noble gases are strongly depleted, even in the volatile-rich CI carbonaceous chondrites. As a result, their abundances are highly subject to perturbations by any mechanism responsible for the inclusion or removal of a noble gas from a cosmic solid.

It makes no sense to attempt to derive noble gas abundances (i.e., element-to-silicon ratios), for tabulation in the SAD, from meteorites. Unfortunately, solar

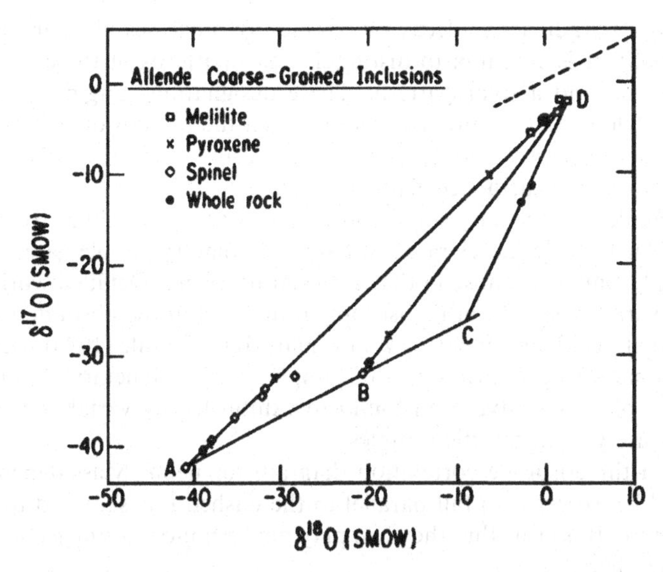

Figure 8.6. Oxygen Isotopes In Allende. From Clayton and Mayeda (1977) with permission. Courtesy of American Geophysical Union.

spectroscopic analyses are especially difficult for these elements. The first excited states lie far above the tightly bound ground state, which means that the strongest lines fall in the solar ultraviolet, an especially difficult region for quantitative analysis. Transitions from excited states do fall in the more tractable regions of the spectrum accessible from the ground, but the levels are highly excited. In practice, this means either that the lines are too faint to be seen, or (like the He I D_3) that they arise in the solar chromosphere. In the latter case quantitative analysis is most difficult.

The most abundant of the inert gases is helium. We cannot call it rare! It has lines at 5876 Å (D_3) and 10 830 Å that are studied with ground-based instruments. However, these lines arise from some 20 or more electron volts of excitation, and are found in "active" solar regions (Foukal 1990).

Compilers of cosmical abundances have turned to a variety of sources such as the photospheric analyses of hot stars and interstellar emission regions, measurements of the gases of the solar wind, and for the isotopic ratios, the analysis of meteorites. All of these sources are checked for consistency. In the case of ^{36}Ar, Cameron (1973, 1982) and Anders and Ebihara (1982) *interpolated the abundance between* ^{28}Si *and* ^{40}Ca *with the aid of calculations of nucleosynthesis!* This procedure is only reasonable because variations from the SAD both in the solar neighborhood and in the cosmos in general are not large. Nevertheless one must think of the amount of "bootstrapping" that goes on in cosmochemistry when one sees the results of theoretical abundance calculations compared with the SAD. The isotope ^{36}Ar fits well.

Isotopic abundances of noble gases have been extensively studied in meteorites, and a confusing terminology has arisen. If the noble gases do not owe their origin to nuclear reactions such as radioactive decay or cosmic-ray processes, they are called trapped (Swindle 1988). It is common to distinguish two patterns of these trapped noble gases, a planetary and a solar pattern. These designations originate with a paper by Signer and Suess (1963) written at a time when the pattern of noble gases in the sun was even more poorly known than it is now, and when the only planet available for analysis was the earth (see Zahnle 1993).

According to Swindle (1988), the "solar" noble gases probably come from the solar wind or are solar particles accelerated in flares. "Planetary" noble gases show isotopic ratios roughly similar to those in the earth's atmosphere. Detailed work has revealed important variations, both in the isotopic ratios and in the absolute values of noble gases in the atmospheres of earth, Venus, and Mars. Zahnle (1993) uses the abundances in carbonaceous chondrites as an example of the "planetary" pattern.

Oxygen and the noble gases have been found to exhibit highly variable isotopic patterns in a wide variety of meteoritic samples.

Figure 8.7 shows a three-isotope correlation diagram for neon. Mass-dependent fractionations would be expected to fall parallel to the dashed line sketched toward the bottom of the figure. It is clear that they can play only a minor role in accounting for the data presented here.

The major features of this diagram may be accounted for by mixing of materials with the "solar" and "planetary" patterns, along with "spallogenic" fractionation, that is, cosmic-ray interactions which produce roughly equal amounts of the three isotopes.

Planetary and solar neon are also called neon-A and neon-B, respectively. A third component has been recognized, which plots below the triangular area shown in Figure 8.7 It is called neon-E, exotic neon; it may result from the same kinds of processes that create the exotic oxygen mentioned above. Neon-E is nearly pure ^{22}Ne, and may be the frozen-in product of nucleosynthesis (Anders 1988).

In those meteorites showing "planetary" noble gas patterns, the ^{3}He/^{4}He ratio is about $1.5 \cdot 10^{-4}$, while in those showing "solar" patterns the value is a factor of 2.6 higher. Much of the ^{4}He in meteorites derives from the α-decay of uranium and thorium. The ^{3}He is of "spallogenic" origin, that is, derived by spallation from cosmic rays. In spallation processes a light nuclear fragment is split from a larger nucleus by a suitable projectile. The helium isotopic ratios will be relatively unperturbed by these sources only in those meteorites that have a rich complement of noble gases. The meteoritic ^{3}He/^{4}He ratios can be compared with the values from the solar wind. These values are variable, with means in the range from 4 to 5 times 10^{-4}. The apparent excess of ^{3}He on the sun is often attributed to the burning of deuterium which can occur in the convective envelope.

Argon, like neon, has three stable isotopes, but unlike neon, there is always a dominant radiogenic source for one of them; ^{40}Ar comes primarily from the decay of ^{40}K. Mason (1979) tabulates some noble gas abundances in six meteorite samples, five chondritic, and one iron. In these the ^{36}Ar/^{38}Ar ratios varied from 0.63 in the iron to 4.9 in dark inclusions of the chondrite Fayetteville. There is little to be made of this information at this time.

The krypton isotopes generally show a subdued variation that has not stimulated the kind of interest that has been focused on xenon.

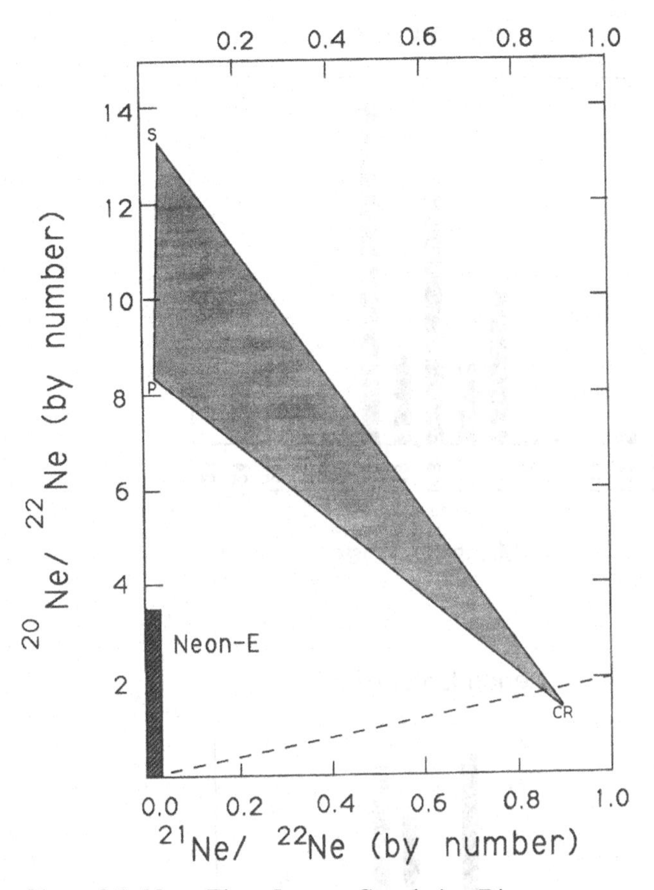

Figure 8.7. Neon Three Isotope Correlation Diagram.

In the case of xenon, the noble gas cosmochemistry is highly developed. The reader may find it helpful to refer to Figure 8.11 below in the discussion that follows. It is a region of the nuclide chart near xenon, and shows the relevant xenon and iodine isotopes.

It was suggested by Aten (1948) that atmospheric ^{129}Xe is more abundant than would be expected from the abundances of its neighbors, and that there was a contribution from the now extinct radionuclide ^{129}I. It is not clear how good a case one can make for this from the atmospheric abundance of xenon. Figure 8.8 is a plot of isotopic abundances of Xe and Sn from the tabulation of Anders and Ebihara (1982). It is not immediately clear that the large value of ^{129}Xe is unusual until one reflects that the isotope has an odd number of neutrons. We know that nucleon *pairs* lead to stability (§§1.3, 5.1, 9.2, and 9.3). Why should ^{129}Xe be more abundant than its even-N congeners?

Certain meteorites have an unusually large ^{129}Xe abundance, and in those meteorites, the case for ^{129}I is even stronger. Figure 8.9 shows xenon isotopic abundances,

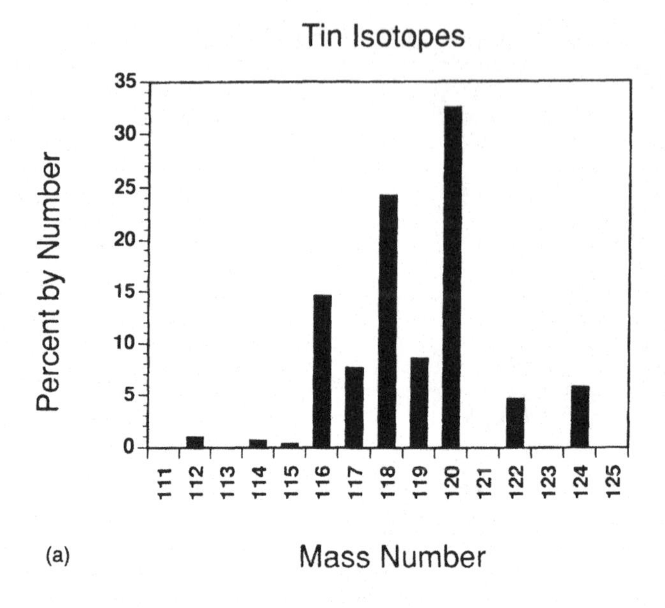

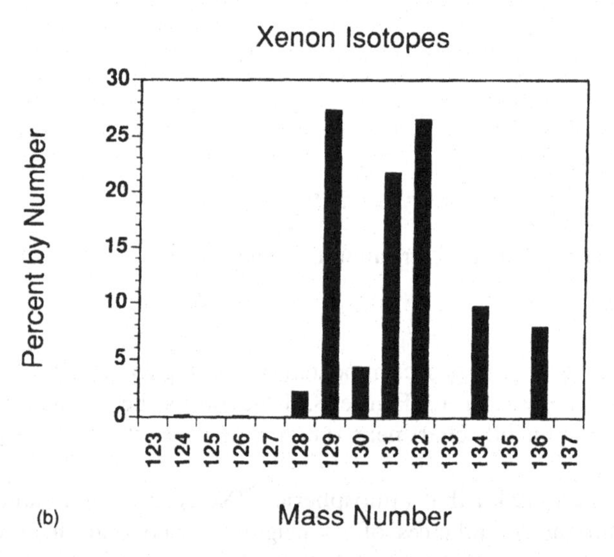

Figure 8.8. SAD Isotopic Abundances for Tin and Xenon

normalized to $^{132}Xe = 100$, in three meteorites. Saint-Severin clearly has an unusually large amount of ^{129}Xe, and it is not an extreme case.

It is not enough merely to point out the spike at ^{129}Xe to complete the case that the origin of this spike is ^{129}I! By the rules of cosmochemistry, one must demonstrate that the ^{129}Xe anomaly is *correlated* with the iodine content of the sample. This has not been particularly easy because of the volatile nature of iodine

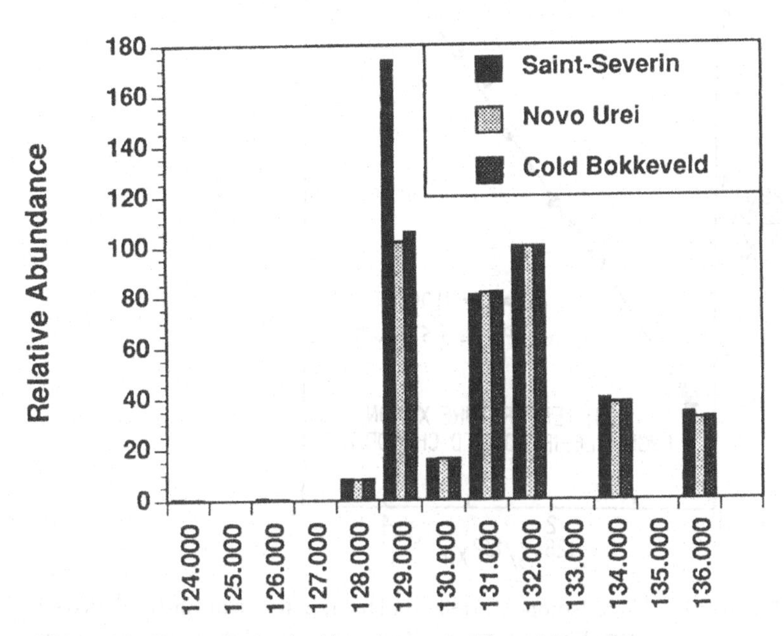

Figure 8.9. Xenon Isotopic Abundances in Three Meteorites.

and its complicated geochemistry. Goldschmidt (1954) called iodine an atmophile, and suggested that it was "without doubt concentrated in the primordial atmosphere of the earth." Since the large ^{129}Xe excesses were found in meteorites, it is not clear how to use this atmophile property for the present purposes.

Jeffery and Reynolds (see Wasserburg and Papanastassiou 1982) demonstrated the iodine association. They created ^{128}Xe *from* ^{127}I by neutron irradiation in the following steps:

$$n + {}^{127}\mathrm{I} = {}^{128}\mathrm{I} + \gamma; \tag{8.12}$$

$$^{128}\mathrm{I} = {}^{128}\mathrm{Xe} + \beta^- + \bar{\nu}. \tag{8.13}$$

The newly created ^{128}Xe could be released from meteoritic samples by stepwise heating. A very similar procedure is used in the ^{39}Ar–^{40}Ar method of dating. The temperature of the sample is raised by incremental steps, with analysis of the residual gas after each step. The last steps drive the noble gas from the most tightly bound locations of the rock. These are usually at the center of the sample, but if the neutron-irradiation-derived gas is found in more than one mineral, there will be a difference in the abilities to retain the total gas, and also in the gas contents per gram. If the ^{129}Xe is derived from ^{129}I, there should then be a correlation between the ^{129}Xe and the ^{128}Xe recently created from ^{127}I (see Figure 8.9). The original ^{128}Xe in the meteorites was small, so the ^{128}Xe that was driven off from heating may be thought of as entirely derived from the ^{127}I by the neutron bombardment.

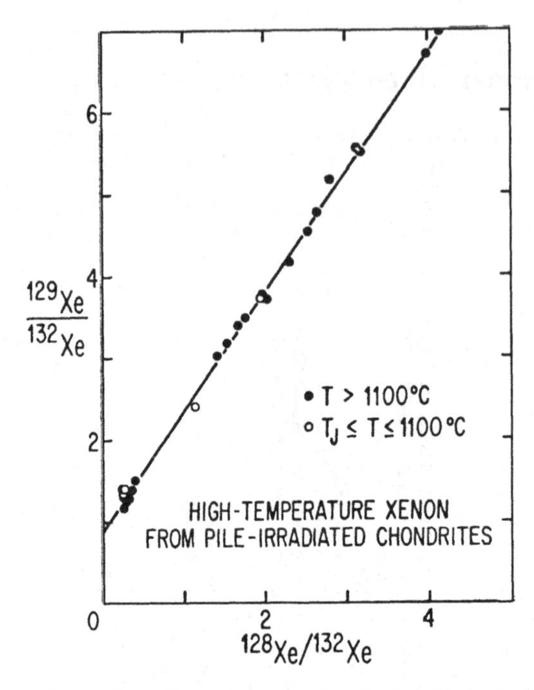

Figure 8.10. Three-Isotope Correlation Diagram for Neutron-Irradiated Chondrites. From Wasserburg and Papanastassiou (1982) with permission.

Figure 8.10, taken from Wasserburg and Papanastassiou (1982), shows that the ^{129}Xe is highly correlated with the ^{128}Xe, thus demonstrating convincingly the presence of ^{129}I in these meteorites at the time of their solidification. The half-life of ^{129}I is $1.7 \cdot 10^7$ years, which is certainly long enough for the isotope to become incorporated into solid material *as iodine* prior to radioactive decay. We shall show below that there is good evidence that the even shorter-lived ^{26}Al entered into some of the oldest meteoritic material.

The lighter isotopes of xenon are shielded (see §10.8) from the neutron-rich side of the valley of β-stability. It is therefore impossible for them to contain fission fragments of heavy nuclei. However, this is not true for the heavier isotopes, and it has become generally accepted that they contain a "fissiogenic" component.

Figure 8.11 is a portion of the nuclide chart showing the xenon isotopes and those of a few of the lighter elements. The black triangles in the lower right corners of the nuclide boxes indicate fission fragments of ^{235}U. The heavier isotopes of xenon, ^{129}Xe–^{136}Xe, are thus fission fragments of ^{235}U. They are also fission fragments of most of the other actinides, and in particular, it has been suggested that ^{244}Pu is largely responsible for the excesses of the heavier Xe isotopes that are found in some meteorites.

Figure 8.12 is a plot of the relative (logarithmic) abundances of xenon in several samples. The ordinates are $\log(^A\mathrm{Xe}/^{130}\mathrm{Xe}) - \log(^A\mathrm{Xe}/^{130}\mathrm{Xe})_{\mathrm{SAD}}$, where A stands for the various mass numbers of other xenon isotopes. We use a *bracket* notation for this

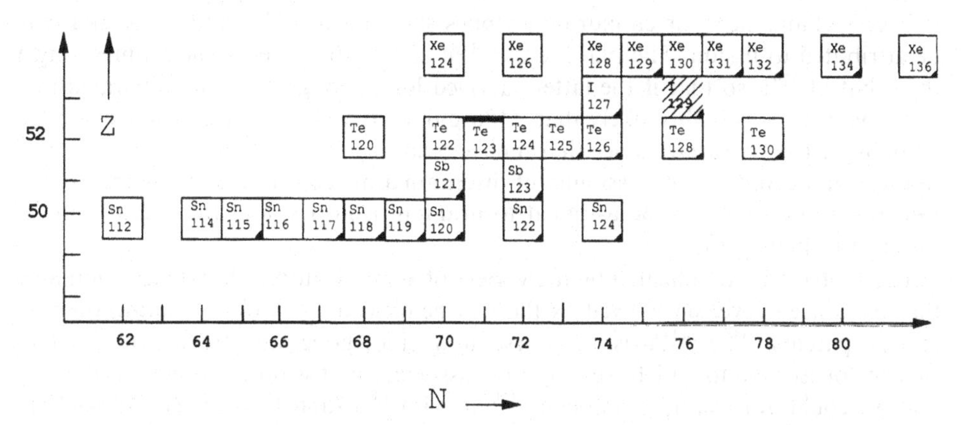

Figure 8.11. Nuclide Chart Showing Isotopes of Xenon and Lighter Elements.

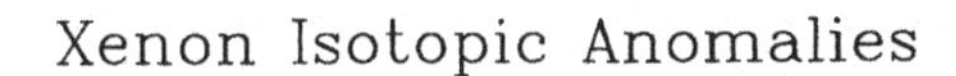

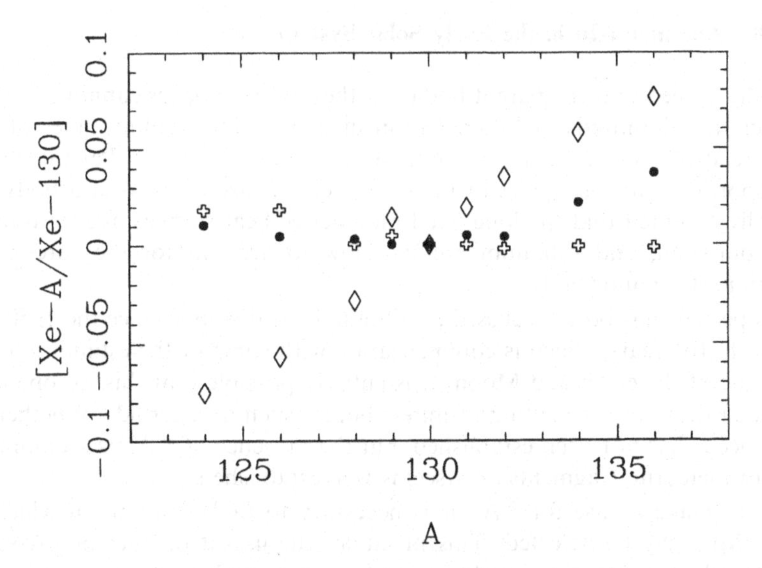

Figure 8.12. Isotopic Anomalies of Xenon in Several Sources. The Data are from a table of Swindle (1988). For discussion, see text.

logarithmic ratio, often seen in stellar abundance work (§13.4.1). A flat, horizontal line means the Xe isotopes have the SAD pattern. The filled circles represent average carbonaceous chondritic abundances. The lighter isotopes of these meteorites and the lunar rocks show small deviations from the SAD pattern. The open diamonds are values for the terrestrial atmosphere. The nearly linear distribution is reasonably attributed to mass fractionation. The discordant point at ^{129}Xe may arise from the decay of ^{129}I.

Excess abundances for chondritic isotopes show up in ^{134}Xe and ^{136}Xe and may be attributed to fission. The half-life of ^{244}Pu is $8.3 \cdot 10^7$ years, some 5 times longer than that of ^{129}I, so that if the latter survived long enough to be incorporated into solid material, it is highly likely that ^{244}Pu did so as well. Kuroda (1960) suggested ^{244}Pu in a letter to *Nature*. Since that time a great deal of attention has been devoted to this isotope and to the possibility of discriminating, among the anomalous xenon isotopes, those that may be assigned to uranium isotopes, and those that may be assigned to plutonium.

The problem is complicated by the variety of xenon isotopic abundance anomalies that have been revealed as well as the geochemical difficulties in relating them to specific parents. The ^{244}Pu is, of course, long since gone, and there is no obvious element or isotope to which we may now associate its fission fragments in the way that we could, for example, associate ^{129}Xe with ^{127}I (and thus to the extinct ^{129}I). Anders and his associates have argued that some of the anomalous xenon may originate from fission of the putative superheavy elements that may exist with Z of 110 or 114.

8.8 Aluminum-26 in the Early Solar System

It is generally agreed that the parent bodies of the iron meteorites cannot have been much larger than hundreds of kilometers in diameter. The cooling times of such bodies are relatively short. A simple dimensional argument shows that the cooling time is proportional to the square of the characteristic size (radius) of a body, and it is generally accepted that the long-lived radioactive heat sources due to uranium, thorium, potassium, and rubidium are too slow to account for the core–mantle differentiation of a minor body.

The isotope ^{26}Al has been discussed as a possible heat source since the 1950's. Its half-life is $7.1 \cdot 10^5$ years, which is commensurate with some of the estimates for the accretion time of the earth and Moon. It is entirely possible that this isotope could have provided the heat for melting of minor bodies such as asteroids. Whether this did in fact occur has not been established, but the presence of ^{26}Al as aluminum in a number of meteorite fragments now seems well established.

In order to make a case for ^{26}Al, it is necessary to find a sample in which the daughter, ^{26}Mg, may be detected. This precludes the use of olivines or pyroxenes, for example, which contain so much magnesium that small portions of radiogenic ^{26}Mg would be hidden. Good candidates for radiogenic ^{26}Mg are the aluminum-rich minerals for which magnesium is a trace element. Anorthite ($CaAl_2Si_2O_8$) is an ideal candidate. The rare mineral hibonite, approximately $CaO \cdot 6(Al_2O_3)$, has provided good evidence for ^{26}Al.

It is also necessary to show that the anomalous ^{26}Mg *correlates* with the amount of aluminum in the sample. This may be done with a plot of $\delta(^{26}Mg/^{24}Mg)$ vs. $^{27}Al/^{24}Mg$, as shown in Figure 8.13, taken from Wasserburg and Papanastassiou (1982). A plot of this kind may be extrapolated to zero aluminum content, to reveal the "original" $^{26}Mg/^{24}Mg$ ratio 0.14, which is the terrestrial ratio.

If we assume that all of the excess ^{26}Mg has come from ^{26}Al, we may write an

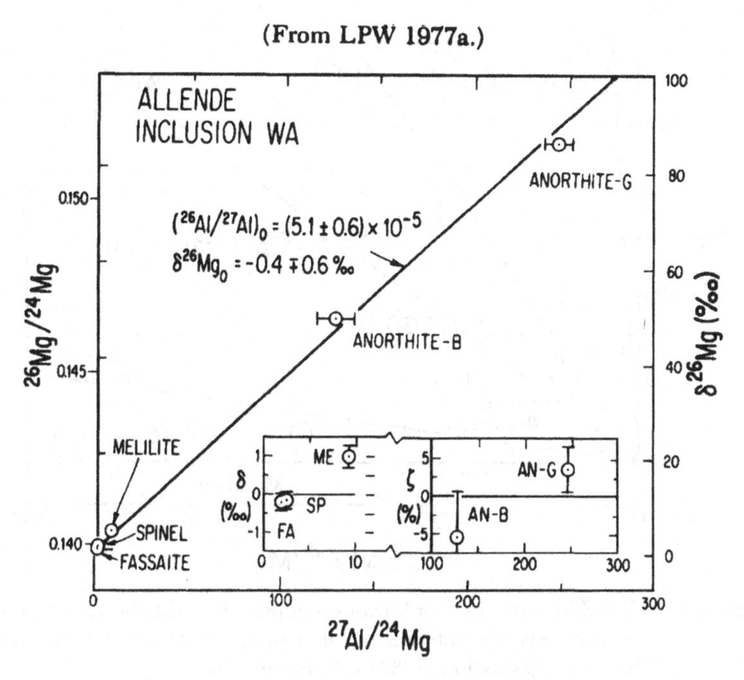

Figure 8.13. ^{26}Al in Allende. Deviations of the data from the line fit are shown in the inset. From Wasserburg and Papanastassiou (1982) with permission.

equation very similar in nature to the relations developed in §8.2. Thus

$$^{26}\mathrm{Mg} = (^{26}\mathrm{Mg})_0 + (^{26}\mathrm{Mg})^* = (^{26}\mathrm{Mg})_0 + (^{26}\mathrm{Al}), \tag{8.14}$$

where ^{26}Al refers to the material present at the time the sample formed. Normalize by ^{24}Mg, and divide and multiply on the right by the observed ^{27}Al. Then

$$\left(\frac{^{26}\mathrm{Mg}}{^{24}\mathrm{Mg}}\right) = \left(\frac{^{26}\mathrm{Mg}}{^{24}\mathrm{Mg}}\right)_0 + \left(\frac{^{26}\mathrm{Al}}{^{27}\mathrm{Al}}\right)\left(\frac{^{27}\mathrm{Al}}{^{24}\mathrm{Mg}}\right). \tag{8.15}$$

Equation 8.15 is a straight line (Figure 8.13) with a slope equal to the ^{26}Al/^{27}Al ratio. From Equation 8.14, we see that the proper interpretation of this ratio is the ^{26}Al/^{27}Al ratio *at the time of formation of the rock*, since this equation only makes sense (for a given rock) in that context. It can be seen that the fraction of ^{26}Al in the sample of Figure 8.13 was quite small, $\approx 5 \cdot 10^{-5}$. This is a much lower ratio than we would expect for freshly synthesized materials, judging on the basis of relative abundances of *stable* nuclides in this portion of the nuclide chart. Compare, for example, the relative abundances of ^{35}Cl and ^{37}Cl or ^{39}K and ^{41}K, which are respectively 0.32 and 0.07. From very simple grounds, with no theoretical calculations at all, we would find it difficult to account for an ^{26}Al/^{27}Al ratio lower than, say, 0.001, without invoking decay *since* synthesis. The observed "initial" ^{26}Al/^{27}Al ratio is also smaller than theoretical calculations would predict (see Ward and Fowler 1980).

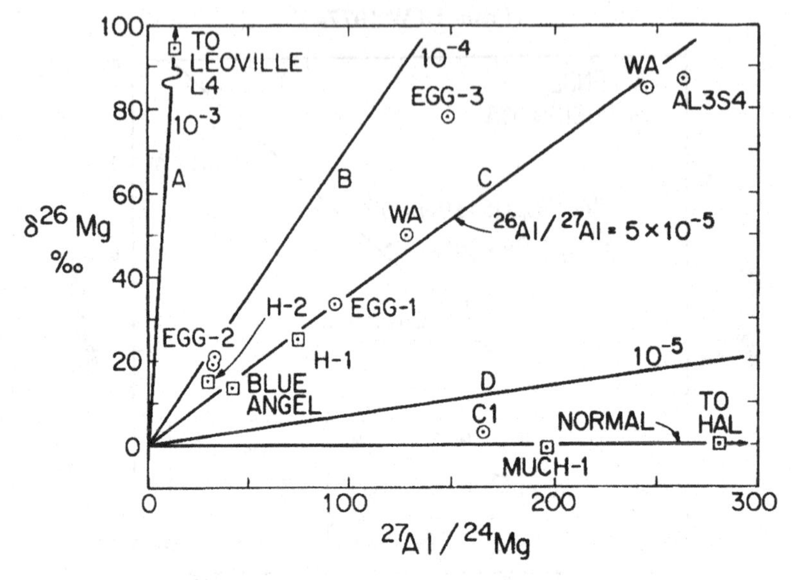

Figure 8.14. The ^{26}Mg Anomaly in Various Samples. The distribution of points may be explained by different histories of the materials. From Wasserburg and Papanastassiou (1982) with permission.

One can think of two plausible explanations. First, some of the ^{26}Al had time to decay before it became incorporated (*as aluminum*) in the solid. Second, freshly synthesized aluminum was *mixed* with material that had solar isotopic ratios, and then incorporated into the solid before very much of the ^{26}Al had a chance to decay. Obviously a combination of these possibilities is likely. However, if we take some of the estimates of the ratio of ^{26}Al to ^{27}Al likely to emerge from nucleosynthesis, we may give a quantitative value for the mixing ratio of freshly synthesized to solar material in order to explain the observations. While values of ^{26}Al/^{27}Al ranging from $4 \cdot 10^{-4}$ to 1 have been suggested (see Wasserburg and Papanastassiou), we shall use the figure 0.1 for purposes of illustration here.

Let x be the fraction of freshly synthesized material mixed with a fraction $1-x$ of solar matter, to give $5 \cdot 10^{-5}$ for the combined ^{26}Al/^{27}Al ratio. We assume 0.1 for the freshly synthesized ^{26}Al/^{27}Al, and 0.0 for the ratio in solar matter. Then $x(0.1) + (1-x)(0.0) = 5 \cdot 10^{-5}$, whence $x = 5 \cdot 10^{-4}$.

Wasserburg and Papanastassiou plot (see Figure 8.14) excesses of ^{26}Mg in a variety of samples from Allende as well as other meteorites vs. ^{27}Al/^{24}Mg. The points are *not* concordant, that is, they do not define a unique slope. The straight lines are drawn for several ^{26}Al/^{27}Al "initial" ratios. We may easily account for the *distribution* of points by invoking a mixing-decay history that is different for the various samples. Of particular interest is the Leoville sample, with the large value of ^{26}Al/^{27}Al, which, according to Wasserburg and Papanastassiou, requires confirmation.

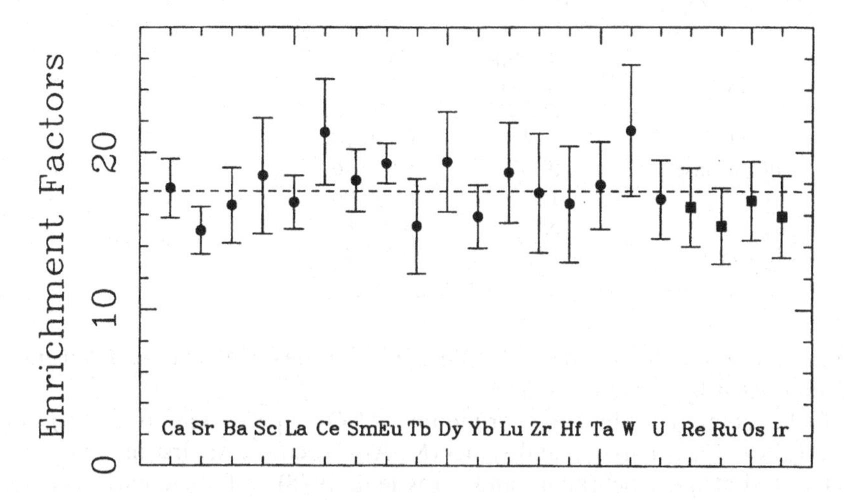

Figure 8.15. Enrichment Factors in Allende Inclusions. After Grossman, Ganapathy, and Davis (1977).

8.9 The Nature of the Allende Inclusions

For most of the Allende inclusions, isotopic anomalies are found for only a few elements, oxygen, magnesium, and some of the noble gases. However, a few of them have anomalies in virtually every element for which studies have been made. These rare samples have been labeled FUN inclusions by Wasserburg and his colleagues, to imply effects of fractionation (F) and unidentified nuclear (UN) influences. Undoubtedly, a literal interpretation is also implied. Faure (1986) remarks that these workers have given us FUN as well as the BABI we ADOR.

Grossman, Ganapathy, and Davis (1977) assembled abundance results for 21 elements in 10 samples of coarse-grained inclusions from Allende. Enrichment factors from their Table 5 are plotted in our Figure 8.15, along with the corresponding errors. These enrichments are defined by dividing the mean abundance results for the inclusions by the corresponding results for C1 chondrites.

The figure shows that all of these species, and calcium, are enhanced by the same factor. Such a uniform enhancement is not to be expected because of the different chemical properties of these elements. Abundances range from a few hundred parts per 10^9 for Re and U to 17% for Ca. We have plotted the lithophiles as circles and the siderophiles as squares.

It is clear that these materials were *never in equilibrium with reduced iron.* This, of course, is also true of the C1 chondrites. The common denominator of all of the elements in Figure 8.15 is a high condensation temperature. A plausible scenario for the history of the materials that were analyzed is that they condensed at high temperatures and that the solids never equilibrated with the gas after it had cooled,

Table 8.1. *Rubidium–Strontium Data for Lunar Rock 15016,46*

Sample	$^{87}Rb/^{86}Sr$ ($\times 10^2$)	$^{87}Sr/^{86}Sr$
Whole rock	2.14	0.700 16
Plagioclase	0.55	0.699 36
s1	1.62	0.699 94
s2	5.82	0.701 86
Ilmenite	13.5	0.705 44

or with the solids that condensed from the cooler gas. The early-condensing solids are dominated by refractory oxides.

Of the common minerals, corundum (Al_2O_3), perovskite ($CaTiO_3$), melilite ($Ca_2Al_2SiO_7$–$Ca_2MgSi_2O_7$), and spinel ($MgAl_2O_3$) condense first in a gas for which $C/O < 1$ (Lattimer, Schramm, and Grossman 1978). Of these early condensates, only the melilites are silicates. Consequently we may expect enhancements of calcium and aluminum relative to silicon. The precise sequences leading to the enrichments of all of the elements shown in the figure are not known. Some could have condensed as pure metals (W, Re, Ir), others as oxides (most lanthanides), but presumably all did so at relatively high temperatures, before the bulk of the silicon condensed, and they were somehow isolated, chemically, from the bulk of the remaining gas and solid materials.

Isotopic anomalies in the Allende inclusions were an early indication of what are now called *presolar* condensates. Although these materials are found in primitive meteorites, they do not (by definition) carry the SAD pattern, and are now regarded as remnants of matter with a distinct nuclear history from that of the solar system. Presolar matter is found among carbonaceous phases of primitive meteorites as well as the Ca–Al inclusions. We must leave a discussion of *microdiamonds*, graphitic carbon, and silicon carbide to the references (Anders 1988, Ott 1993). Isotopic anomalies have also been found in the interplanetary dust particles, or IDP's. Sadly, we must again refer to the literature for additional information on this rapidly developing field (see Kerridge and Matthews 1988, Gruen 1991).

8.10 Problems

Table 8.1 shows Rb–Sr data for a rock from Apollo 15 taken from Evensen, Murthy, and Coscio (1973). The samples s1 and s2 are not mineralogically well defined, but are probably mostly pyroxene.

1. Use the BABI initial $^{87}Sr/^{86}Sr$ value of 0.698 98 to determine a model age for the whole-rock fragment of Table 8.1.
2. Determine a sample age for the rock of Table 8.1.
3. Your model age should be somewhat greater than your sample age. How do you account for the difference?

4. In Chapter 10, $^{26}Al/^{27}Al$ ratios ranging from 0.13 to 0.30 are presented in Table 10.2. Assume a fraction x of freshly synthesized aluminum is mixed with material having the SAD composition to produce the Leoville sample of Figure 8.14. What must the mixing fraction x be for a $^{26}Al/^{27}Al$ ratio of 0.2?

9
Some Concepts from Nuclear Physics

9.1 Introductory Remarks

In this chapter we shall examine the atomic nucleus from the point of view of cosmochemistry. A detailed treatment cannot be given, but we try to make a number of important results plausible. Rigorous treatments leading to these results are often both tedious and quite sophisticated. We must leave them to the references.

We shall not be concerned here with matters of subnuclear, or high-energy physics. This rapidly developing domain has made an impact primarily on the cosmochemistry of the primordial fireball (big bang). We cannot entirely avoid the early universe. It synthesizes hydrogen, helium, and a few other light nuclei, after the decay of exotic denizens of the domain of high-energy physics (§10.7). Our current ideas about the synthesis of the heavier nuclei may be generally understood without appeal to elementary particle physics.

Some of the long-lived radioactive isotopes (see Chapter 8) preserve a memory of the earliest times of nucleosynthesis, and therefore give constraints on the age of the universe itself. This is one way cosmochemistry is connected to cosmology, and ultimately, to high-energy physics. Another is the question of the cosmological abundance of deuterium, which can set important constraints on cosmological models. Neither cosmology nor cosmochemistry is a mature topic in the sense of certain domains of classical physics or chemistry. Future work may show that cosmochemistry is more closely related to particle physics than appears at the present time.

9.2 The Semi-empirical Mass Formula

Figure 9.1 is a Chart of the Nuclides, showing the stable and longer-lived isotopes on a plot of the number of protons Z vs. the number of neutrons N. The portion of N–Z-space occupied by these nuclei is known as the *valley of β-stability*. This is because the masses of the nuclei are lower than others with the same number of nucleons. Unstable, or radioactive isotopes that lie to the right, or neutron-rich side, decay back to it by β-emission. For example, we are familiar with

$$^{87}\mathrm{Rb} \rightarrow {}^{87}\mathrm{Sr} + e^{-} + \bar{\nu}_e, \qquad (9.1)$$

where $\bar{\nu}_e$ is an electron antineutrino. The emitted electron is known as a β-particle for historical reasons: the three kinds of emission from radioactive elements were originally known as α-, β-, and γ-rays (see below). Unstable isotopes that lie to the left, or neutron-poor, side of the valley of β-stability, decay toward it by emitting a positive electron and an electron's neutrino.

Positron emission, or β^+-decay, takes place with the emission of a positive electron (Problem 1 below). It is typical of nuclides with proton excesses. Occasionally such nuclei will interact with one of the electrons of the inner atomic shell, often called the K-shell. An example of cosmochemical significance occurs for the isotope ^{40}K,

$$^{40}\mathrm{K} + e^- \rightarrow {}^{40}\mathrm{Ar} + \nu. \tag{9.2}$$

This process is known as *electron capture* (EC). It is interesting that the ^{40}K nucleus does not always decay in this fashion. Indeed, nearly 90% of the time, it does an ordinary β-decay to ^{40}Ca. The nuclear decay is said to *branch*, and the ratio of the β^--decays to all decays in a sample is called the *branching ratio* for that decay mechanism. We will encounter this phenomenon again.

Quite a number of nuclear properties are evident from a simple inspection of Figure 9.1. First, there are no known stable nuclei with charges (atomic number, or Z) greater than 83 (bismuth). The naturally occurring trans-bismuth elements, such as uranium and thorium, have no stable isotopes.

It is reasonable to attribute the maximum charge that a stable nucleus may have to the competition between the strong forces that bind the nucleus, and the Coulomb repulsion that would, in the absence of the nuclear forces, force such a configuration of charges apart. It is obvious that the nuclear binding force must depend on the presence of neutrons. Indeed, apart from the hydrogen nucleus, all nuclei are made up of both protons and neutrons. As the proton charge builds up in the heavier nuclei, a relatively larger proportion of neutrons to protons is required for a stable configuration. We can see the latter from the fact that the slope (dZ/dN) of the valley of β-stability decreases with increasing Z (or N).

It is extremely instructive to attempt to write the mass of a nucleus with a given N and Z as a function of N and Z. To a zeroth approximation, the nuclear mass is simply N times the neutron mass plus Z times the proton mass. However, the nuclei are less massive than one would calculate in this way, and the difference between the sum of the individual nucleon masses and the mass of the composite nucleus is a measure of the nuclear binding energy. Einstein's relation $E = mc^2$ gives this energy.

Nuclear masses of stable nuclei are measured with mass spectrographs similar to those used to determine isotopic abundances for radioactive dating. Consequently, relative masses may be determined more accurately than absolute ones. Following a decision of the International Union of Physics and Chemistry in 1961, the nuclear mass unit has been defined as one twelfth of the ^{12}C atom (including electrons). This mass is given in Table A1 of the Appendix and is $1.660\,54 \cdot 10^{-27}$ kg, or 931.494 MeV.

The masses of many nuclei are determined with the help of *Q-values*, which may be defined as follows. For generality, suppose two nuclei with rest masses m_1 and

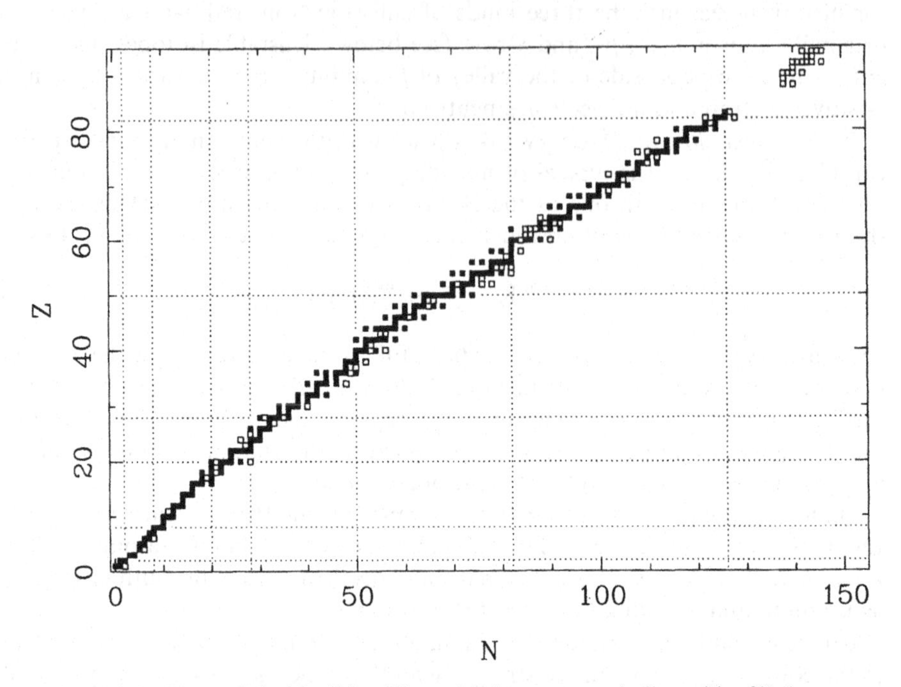

Figure 9.1. Chart of the Nuclides. The stable isotopes are indicated by filled squares. Magic numbers are indicated by dotted lines at N and Z = 2, 8, 20, 28, 50, 82, and $N = 126$. Some of the longer-lived radioactive nuclides are plotted as open squares. Some of these have half-lives that are sufficiently long that they are generally considered *stable*, even though they may be found in tables of radioactive nuclei (^{40}K, ^{87}Rb, etc.).

m_2 react to form two different nuclei with masses m_3 and m_4. Then by definition

$$Q = (m_1 + m_2 - m_3 - m_4)c^2. \tag{9.3}$$

The lower-case m's are masses of the nuclei without the electrons. Traditionally, the nuclear mass (upper-case M) includes the mass of the atomic electrons, and the relatively small binding energy of the electrons is rarely taken into account explicitly. We ignore it here.

If the number of electrons in the products and reactants of Equation 9.3 is the same, then the masses $M_1, M_2, \ldots$ including electron masses may be used interchangeably with the m's. However, if electrons are created, as in β-decay, the new electron masses must enter explicitly (see Problem 1 below). The *binding energy*, BE, of a nucleus is defined by

$$M(A,Z) = (A - Z)M_n + ZM_{\mathrm{H}} - \mathrm{BE}(A,Z). \tag{9.4}$$

Here, M_n is the neutron mass, M_{H} is the mass of a hydrogen atom, and the mass number $A = Z + N$. Because of the sign choice in 9.4, the binding energy is *positive*.

There is no rigorous way to derive the function BE(A, Z). Finding an optimum form may be regarded as an ongoing problem in nuclear physics.

We shall present an equation that has primarily heuristic value at the present time. This simplified version is known as the Weizsäcker (1935) mass formula. The overall procedure is to use physical insight to write the functional form for BE. In general, the relation will involve several terms, whose coefficients we may adjust by the method of least squares. If we knew no physics, we might be inclined to try a power series in the variables Z and N. Since the first work on the mass formula was done by physicists, they chose their function BE on the basis of a physical model. The nucleus was modeled as analogous to a drop of liquid. These efforts produced the *liquid drop model* of the atomic nucleus.

The semiempirical mass formula is usually written in terms of the mass number A and Z rather than N and Z. The volume of a drop of nucleons should be proportional to A times some mean volume per nucleon, so if we picture a drop of "nuclear fluid," its radius should be proportional to $A^{1/3}$, and its surface area to $A^{2/3}$. It is interesting to note that these relations apply to an *incompressible* liquid, and not at all to the model that successfully describes atoms. The "radius" of an atom is *not* proportional to the one-third power of the number of orbital electrons!

It is useful to think of the mass formula as the equation of a surface involving the three variables M, Z, and either N or A. We must keep in mind that while the mathematical surface can be defined for *continuous* values of N, A, or Z, the physical parameters have only discrete, integer values. When we think of neutron addition processes, the Z vs. N representation is convenient, since Z remains fixed (Figure 9.1). For β^+- or β^--decay, the sum of Z and N is constant, so it is useful to think of M as a function of Z and A.

Let us now write out the *Weizsäcker mass formula* and then comment on the meaning of the various terms. The fitting coefficients $a_1, a_2, \ldots$ can be found by the method of least squares.

$$\begin{aligned} M(A,Z) &= (A-Z)M_n + Z\left(M_p + M_e\right) - a_1 A + a_2 A^{2/3} \\ &\quad + a_3 \frac{\left[(A/2)-Z\right]^2}{A} + a_4 \frac{Z^2}{A^{1/3}} + \delta. \end{aligned} \tag{9.5}$$

The terms are explained as follows. The first are simply the neutron, proton, and electron masses as before. The remaining terms express the binding energy or $-$BE of Equation 9.4. The term with a_1 simply states that the binding energy should depend linearly on the number of nucleons. The term proportional to $A^{2/3}$ expresses a dependence on the surface area of the nuclear drop. One may think of it as a surface energy. The reader unfamiliar with the notion of surface energy may find it useful to examine §14.6, and especially Figure 14.18, where surface energies are discussed within the context of an ordinary fluid.

The numerator of the term with $[(A/2)-Z]^2 = (N-Z)^2/4$ expresses a parabolic dependence of the mass upon the departure from equality of protons and neutrons. It represents a proclivity of nuclei to have equal numbers of protons and neutrons. A heuristic justification for this goes as follows. The nucleons are closely packed, so that the Pauli exclusion principle is important. Nature can achieve the densest

Table 9.1. *Fitting Coefficients for the Weizsäcker Formula (MeV)*

A range	$a_1 = a_{vol}$	$a_2 = a_{surf}$	$a_3 = a_{sym}$	$a_4 = a_{coul}$	a_{pair}
20–200	15.77	17.81	91.44	0.717	11.52
20–80	15.62	17.37	97.08	0.707	12.53
50–150	15.00	15.68	82.57	0.652	10.35

packing by keeping the numbers of protons and neutrons equal, because the Pauli principle applies to them *individually* – 10 protons *and* 10 neutrons can be packed into a smaller volume than 20 protons *or* 20 neutrons.

The nuclide chart shows a tendency for neutron excesses in larger nuclei, so there is an A in the denominator of the term with a_3. This will make the term relatively less important for massive nuclei. The packing efficiency is more critical for the smaller nuclear volumes.

The term involving $Z^2/A^{1/3}$ is analogous to the potential energy of an aggregate of charge. Astronomers are familiar with a similar expression for the gravitational binding energy of a spherical mass (the gravitational analogue of charge), which depends on the square of the mass divided by the radius.

The δ-term is subtle. In our fits, we have written it

$$\delta = \pm a_{\text{pair}} A^{-1/2}. \tag{9.6}$$

The plus sign is used with odd–odd nuclides, and the minus sign with even–even. The δ-term expresses the particular "pairing" stability of nuclei for which the number of protons or the number of neutrons is even. The reason for this stability will be explained in the next section, dealing with the nuclear shell model. The odd–even abundance effect (§§1.3, 5.1 and 7.1) is a consequence of this pairing. For the present, we simply appeal to the data.

For *even–even nuclei* (A and Z both even), an amount is subtracted from the mass, commensurate with greater stability. For *odd–odd nuclei*, an amount is added. The δ-term is taken to be zero if there is only one unpaired neutron or one unpaired proton. This δ-term changes the mass formula from an equation for a single surface in M, Z, and A or N space to one for *three* surfaces, displaced slightly in the vertical or M direction. This is necessary to explain how one can get several stable *isobars*, that is, nuclei with the same value of A.

The value of the fitting coefficients varies from study to study, depending on the nuclear masses chosen for the fits. We have made several fits for the present text using material from the National Nuclear Data Center at Brookhaven. Three results are displayed in Table 9.1. They should be used with the form 9.6 and give $M(A, Z)$ in MeV. The nuclides used either are stable, or have half-lives equal to or greater than 1000 years.

The average square deviations for the fits of Table 9.1 were 3.3, 2.4, and 2.9 MeV^2 respectively.

We may imagine a surface above the nuclide chart, with a minimum or *valley* in the region of the stable nuclei. It may be useful to think of three closely parallel surfaces, because of a factor δ in the mass equation, 9.6. We can hold one of the variables fixed, and examine the behavior of the mass as the remaining variable changes. "Cuts" at constant A are shown in Figure 9.2. In the neighborhood of the minimum, these "cuts" resemble parabolas, with the stable nuclei near the minima. We note again that only *integral* A, Z, and N are physically meaningful.

There are many reasons why the systematics of nuclear masses are important for cosmochemistry. In most cases, the laboratory physicist may turn to a tabulation of measurements, such as the tables of Lederer and Shirley (1978) or Wapstra, Audi, and Hoekstra (1988) for masses. These authors tabulate the mass excess

$$\Delta \equiv M - A. \tag{9.7}$$

Here, M is the mass of the *neutral atom*, on a scale with $\Delta(^{12}C)$ defined to be zero.

Figure 9.2 illustrates two cuts through the valley of β-stability at *constant* A. For odd A, the δ-term of Equation 9.5 is zero, since if A is odd, either the number of protons *or* the number of neutrons must be odd. The plot for $A = 101$ was made using the first set of coefficients in Table 9.1, while that for $A = 130$ used the third set. There is only a single intersection of the surface $M = M(A, Z)$ with the plane $A = 101$. For $A = 130$, the numbers of neutrons and protons must be either both even or both odd. In the former case the mass formula contains the term $-|\delta|$, while for the latter a term $|\delta|$ is added.

For $A = 101$, the mass formula shows why there is only one stable isobar, since any nuclide may β^-- or β^+-decay to a more tightly bound isobar below it. The stable nuclide with $A = 101$ is ^{101}Ru at $Z = 44$. The situation for the even-A isobars is more interesting. As Z increases by one unit, one jumps from the odd–odd (upper) curve to the even–even (lower), back to the even–even curve, and so on. We would expect two β-stable isobars at $A = 130$, since the lower curve for both $Z = 54$ (^{130}Xe) and $Z = 56$ (^{130}Ba) is below the upper curve at $Z = 55$ (^{130}Cs). Interestingly, ^{130}Te, $Z = 52$, is also virtually stable. According to the Brookhaven Data, the half-life is $1.25 \cdot 10^{21}$ years!

Additional properties of these mass relationships are taken up in the problems at the end of the chapter.

Our Equation 9.6 involves only a few fitting parameters to describe the masses of the several hundred nuclides. Far more elaborate mass formulae are possible. In particular, modern work on the mass formula specifically incorporates the shell effects that we shall discuss in the next section. We must refer to the book by Myers (1977) or the papers edited by Haustein (1988) for details.

In nuclear cosmochemistry, it is necessary to know the masses of species that have never been measured in the laboratory. Such information is traditionally obtained by extrapolation of a semiempirical mass formula into the domains often far removed from those in which it was established.

An extremely important application of this procedure comes in the calculation of the buildup by very rapid neutron addition (r-process) of the heavy elements (Mathews and Ward 1985, Mathews and Cowan 1990). Under the conditions in

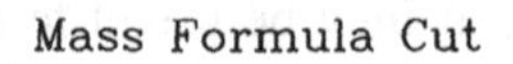

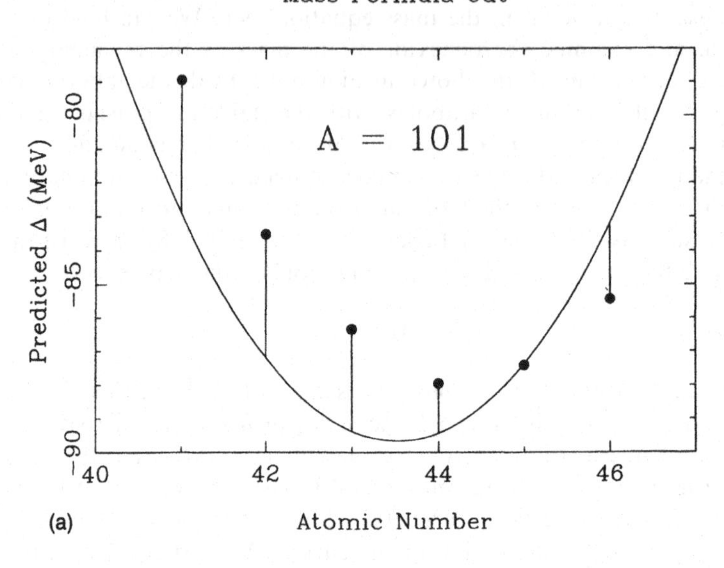

Mass Formula Cut

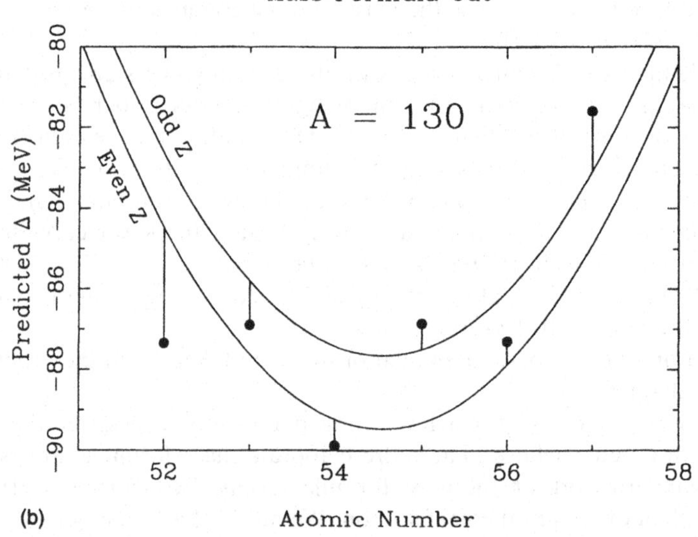

Figure 9.2. Cuts Through the Valley of β-Stability at Odd and Even A. The upper curve for $A = 130$ was made using $+a_{\text{pair}}$ in 9.5 and 9.6, while the lower curve was made with $-a_{\text{pair}}$. An intermediate curve corresponding to $a_{\text{pair}} = 0$ is mathematically possible, but not physical because A, Z, and N must be integers. Actual mass excesses from Lederer and Shirley (1978) are shown by the filled circles with lines connecting them to the predictions of the mass formula. The largest errors are only about 5%.

which this happens, a portion of the nuclide chart well to the neutron-rich side is populated. The relevant nuclides are mostly too short-lived to be measured in the laboratory. Consequently, their β-decay rates must be calculated with the help of a mass formula. The r-process is discussed in more detail in §10.9.

9.3 The One-Particle or Shell Model of the Nucleus

Much of the background for the nuclear history of matter was developed in the years immediately following the Second World War. One particularly interesting symbiosis was that of the cosmic abundances and the so-called shell model of the nucleus developed independently by Maria Mayer and Hans Jensen (see Mayer and Jensen 1955). A primary goal of the early work on nuclear structure was to understand the particular stability exhibited by nuclei with special numbers of protons or neutrons: 2, 8, 10, 20, 28, 50, 82, 126 (see Figure 9.1).

The stability is manifested in a variety of ways. The binding energies per nucleon are generally higher for magic-number nuclei than for those of their neighbors, and their cross sections for destruction, for example, by neutron capture are lower than those of their neighbors. This is shown in Figure 9.3. Interestingly enough, the *abundances* of these nuclei are also relatively high, telling us immediately that the properties of the atomic nuclei manifest themselves in the cosmic abundances.

Maria Mayer's interest in nuclear structure was piqued by research she was doing with Edward Teller on the origin of the chemical elements. In preparation for this work, she had been compiling information on the abundances of the isotopes, and her attention was drawn to the magic numbers which stood out both in the properties of the atomic nuclei and in their abundances.

It is almost certainly no accident that Jensen's first paper on the shell model was written in collaboration with Hans Suess, who, with Harold Urey, was soon to produce the abundance table that stimulated the modern theory of stellar nucleosynthesis. Suess (1987) covers many of the topics discussed in the present volume. In his book one may find a detailed discussion of nuclear systematics in the context of the SAD, or as it was once called, the "cosmic" abundance distribution.

These special numbers of nucleons found by Meyer, protons or neutrons, were called magic by no less of a virtuosity than the Nobel Laureate Eugene Wigner, because they seemed to defy understanding.

As soon as it was realized that the nucleus consisted of a system of particles, it was natural to attempt the same kind of analysis that had been so successful with the electronic structure of atoms. Could quantum mechanics also describe the structure of the atomic nucleus? Atomic structure, we might say, also manifests certain "magic" numbers. The atomic magic numbers, 2, 10, 18, etc., give the noble gas structures.

At the time the shell structure of the nucleus was being worked out the theory of atomic structure was already highly developed. It is questionable whether one could say even now that nuclear physics has achieved a level of success equal to that reflected in Condon and Shortley's (1935) classic *Theory of Atomic Spectra.* Though the modern analytical tools of nuclear physics are vastly more sophisticated than

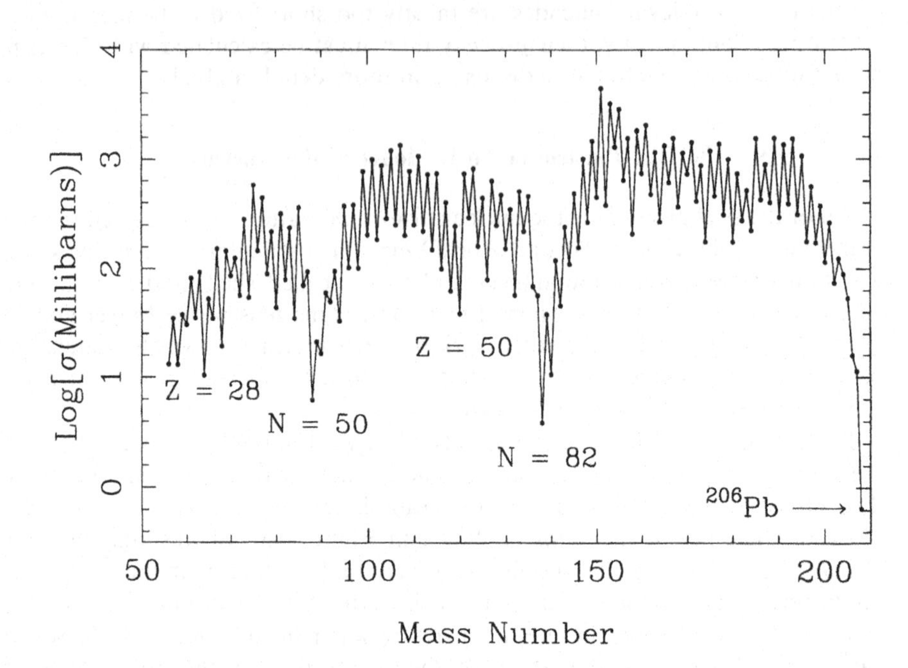

Figure 9.3. Neutron Capture Cross Sections Reflect the Neutron Magic Numbers. These nuclides are on the *s*-process path (§10.8).

those discussed by Condon and Shortley, the problems are intrinsically much more difficult, and progress has been correspondingly slower.

The developments made by Jensen and Mayer brought nuclear physics to about the level of atomic physics following the solution of the hydrogen atom using wave mechanics rather than Bohr's postulates. Let us therefore look briefly at the aspects of atomic structure that were illuminated by this work (see also §11.2). We will then be in a position to understand the insights into nuclear structure provided by the single-particle shell model.

Schrödinger's wave mechanics introduced three quantum numbers, now called n, l, and m. They describe the energy states of hydrogen and ionized helium rather closely. With the postulate of an additional degree of freedom, the electron spin, it became possible to enumerate the number of possible electron states that could belong to the quantum numbers n and l. The quantum number l arises in both atomic and nuclear physics. We note that it is a measure of the angular momentum, and that states with $l = 0, 1, 2, 3, \ldots$ are called s, p, d, f, ... If intrinsic spin is included, there can be at most $2(2l+1)$ electrons with a given value of n and l. This means, for example, there can be two s electrons, six p electrons, ten d electrons, etc.

We need to know one other rule in order to be able to understand the structure of the periodic table. The allowed values of l range from zero to $n-1$. This means

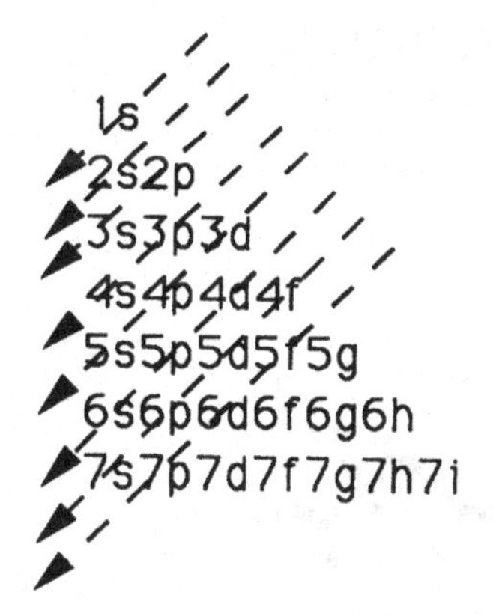

Figure 9.4. Mnemonic for Filling of Atomic Subshells.

there can only be s states for $n = 1$ (lowest Bohr orbit), s and p for $n = 2$, s, p, and d for $n = 3$, and so on.

The first period of the periodic table is thus completed at helium, when there are two s electrons. No more electrons are allowed for $n = 1$, and the next electrons must go in the second shell with $n = 2$. There, we may have two s electrons and six p electrons, and no more. This gives eight electrons in the $n = 2$ shell, and upon adding the two electrons from the $n = 1$ shell, we get a total of 10 electrons in the next noble gas, neon. Argon has only eight additional electrons; its atomic number is 18. Consequently, in the $n = 3$ shell, only states with $l = 0$ and 1 are filled, even though d electrons are possible. To understand the periodic table from this point on, we must recognize that the shells (with a given n) can be partially filled. In the fourth period, for example, the filling of the $3d$ shell is completed with the addition of eight $4s$ and $4p$ electrons for a total of 18 at krypton, with atomic number 36. The $4d$ and $4f$ shells remain unfilled until the fifth and sixth periods.

We present a simple, mnemonic scheme for remembering the order of shell and subshell filling in the periodic table (Figure 9.4). The arrow shows the order in which the subshells are filled, and makes it easy to remember, e.g., that the $3d$ shell is not filled until the fourth period, and the $4f$ is filled first in the sixth period. The scheme is not perfect (see Problem 5 below).

How is it that the solution of a one-electron problem allows this kind of insight into the many-electron problem? The answer is that many-electron states can be built up from these single electron solutions with the help of the *Pauli exclusion principle* (§4.9): no two electrons can have the same four quantum numbers. This principle applies to all particles with half-integral spin and therefore to nucleons as well as electrons.

It is possible to use these single-particle solutions because, to a rather good approximation, we can assume that each electron moves in the smeared-out, *spherically symmetrical potential* caused by the nucleus and the other electrons.

In atomic physics, the assumption of spherical symmetry is called the central field approximation. From it, the (θ,ϕ)-solutions of the wave equation follow directly.

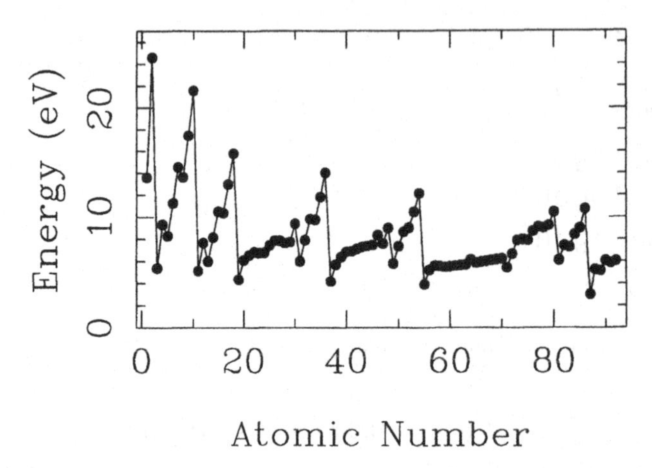

Figure 9.5. Atomic Ionization Potentials vs. Z. Atomic "magic" numbers show up in the jumps in ionization potentials when the atomic shells are filled.

These solutions describe the *s*-, *p*-, etc. electrons. The assumption of spherical symmetry in the nucleus will bring about analogous states. It will be important for us to realize here that this spherical symmetry can manifest itself both for atoms and for nuclei. We sometimes see cosmochemical abundance patterns for which we cannot distinguish between nuclear and atomic effects.

The single-particle wave functions for hydrogen do not enable one to predict the energy levels of multi-electron atoms. But they do afford some insight into their structure. What aspects of the structure of complex atoms may be understood with a single-particle picture plus notions of the Pauli principle?

Figure 9.5 is a plot of the ionization potential or ionization energy for atoms. One can see that this ionization energy rises to a maximum at the location of the closed shells, the noble gases, and declines precipitously thereafter. In addition to the high ionization energies of the noble gases, there is the closely related property of the energy levels. The first excited states lie relatively far above the ground state so that it takes more energy to excite an electron from a closed shell than from an open shell.

Let us now outline the basis for the nuclear shell model, and see how some nuclear properties analogous to those illustrated in Figure 9.5 can be understood in terms of it.

The Schrödinger equation for a single particle moving in a potential $V(r)$ can be immediately written down. For certain simple forms of $V(r)$, it is straightforward to carry out a solution. Two such forms for $V(r)$ are the so-called square well and the harmonic oscillator potentials shown in Figure 9.6. Figure 9.6 also shows a more realistic approximation (Woods and Saxon 1954) to the potential. Early workers simply interpolated between the square well and the harmonic oscillator.

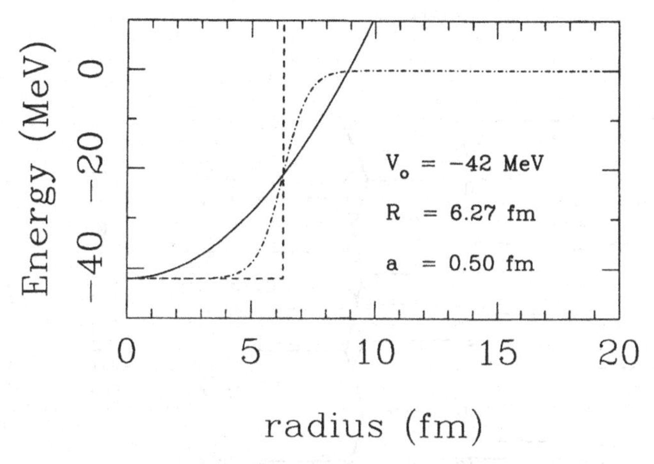

Figure 9.6. Approximate Potentials for the Nuclear Shell Model. The solid line shows the potential for the isotropic (three-dimensional) harmonic oscillator. The dashed lines show the square well, and the dot–dashed line a Woods–Saxon potential with the parameters shown. The units of the abscissa are fermis or femtometers (fm, 10^{-13} cm).

One more ingredient is necessary to predict the magic numbers, and the insight required to find this was the special contribution for which Mayer and Jensen were awarded the Nobel Prize. In nuclei, interaction between the spins of the particles and their orbital motion is generally a much more important effect than in atomic structure. In the atom, analogous *spin–orbit interactions* give rise to what is called *fine structure.* The effects are small with respect to the level separations that can be accounted for with the help of the quantum numbers n and l. In the nucleus, the spin of the nucleons interacts strongly with the orbital motion, causing a splitting of the same order of magnitude as that due to differences in the orbital angular momentum, with quantum number l.

Figure 9.7 shows the energy levels that can arise for a single particle in a harmonic oscillator potential (left), and a square well (second from left) in the absence of spin-orbit interactions. The third series of levels are interpolated linearly between these two. The principal quantum numbers and angular momentum quantum numbers are written beside each level. The former has an analogous (but not identical) meaning to that for the atom, while the latter is the same. The fourth series of levels includes the effects of particle spins. In addition to the orbital angular momenta indicated by s, p, d, etc., a total angular momentum quantum number j is assigned.

Each state is split into two levels, with j having the values $l + 1/2$ for the lower energy and $l - 1/2$ for the upper. This sequence may be reversed. Atomic states may be similarly labeled, but one usually chooses a different set of labels.

For a single particle, one may use either the quantum numbers n, l, m_l, and m_s

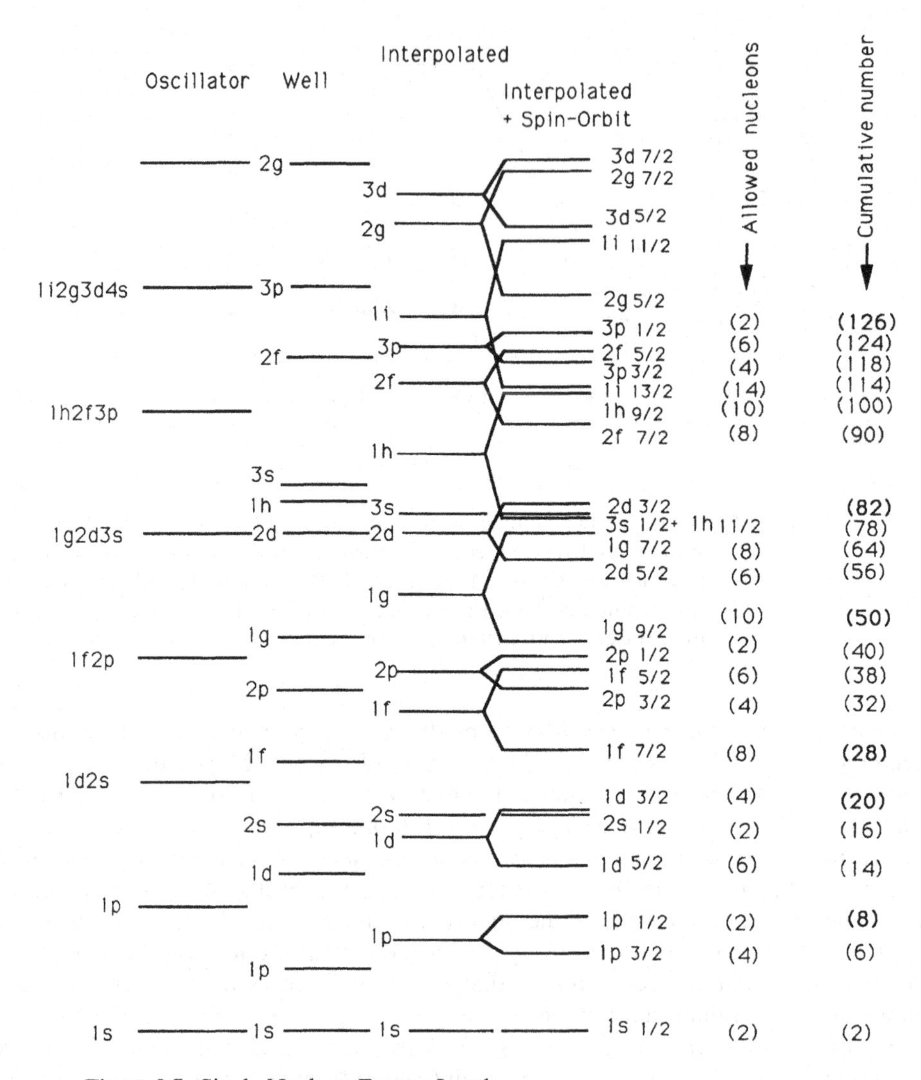

Figure 9.7. Single-Nucleon Energy Levels.

to label single particle states, or n, l, j, and m_j. The same number of states for a given n and l result. This is shown in Table 9.2, which divides logically into three sections. The first two rows specify the quantum numbers n and l, and the following three rows show possible values of m_l and m_s, with a row summing the numbers of allowed particles with the quantum numbers specified in the columns. The last three rows show the possible values of the quantum numbers j, and m_j, and the sums of the allowed numbers of particles. The sums are the same, for a given n and l.

Whenever spin–orbit interaction is important, the n, l, j, m_j scheme is the better

Table 9.2. *Single-Particle Quantum Numbers*

$n =$	1	2				3								
$l =$	0	0	1			0	1			2				
$m_l =$	0	0	-1	0	$+1$	0	-1	0	$+1$	-2	-1	0	$+1$	$+2$
$m_s =$	$\pm\frac{1}{2}$	$\pm\frac{1}{2}$	$\pm\frac{1}{2}$	$\pm\frac{1}{2}$	$\pm\frac{1}{2}$	$\pm\frac{1}{2}$	$\pm\frac{1}{2}$	$\pm\frac{1}{2}$	$\pm\frac{1}{2}$	$\pm\frac{1}{2}$	$\pm\frac{1}{2}$	$\pm\frac{1}{2}$	$\pm\frac{1}{2}$	$\pm\frac{1}{2}$
Sum:	2	8				18								
$j =$	$\frac{1}{2}$	$\frac{1}{2}$	$\frac{1}{2}$	$\frac{3}{2}$		$\frac{1}{2}$	$\frac{1}{2}$	$\frac{3}{2}$		$\frac{3}{2}$		$\frac{5}{2}$		
$m_j =$	$\pm\frac{1}{2}$	$\pm\frac{1}{2}$	$\pm\frac{1}{2}$	$\pm\frac{1}{2}$	$\pm\frac{3}{2}$	$\pm\frac{1}{2}$	$\pm\frac{1}{2}$	$\pm\frac{1}{2}$	$\pm\frac{3}{2}$	$\pm\frac{1}{2}$	$\pm\frac{3}{2}$	$\pm\frac{1}{2}$	$\pm\frac{3}{2}$	$\pm\frac{5}{2}$
Sum:	2	8				18								

of the two labeling schemes. In nuclei, the individual spins and orbital angular momenta interact much more strongly than in atoms.

The essential point of Figure 9.7 is that $2j+1$ independent functions belong to a given value of j, and therefore $2j+1$ particles can belong to a given j without violating the Pauli exclusion principle. This also applies in the nuclear case. The second to last column of Figure 9.7 gives $2j+1$ for each level. The last gives the cumulative number of nucleons. The magic numbers correspond to the big gaps. The gaps at 2, 8, 20, 28, 50, and 82 are reasonably well illustrated. The figure was constructed with a trial value of the spin–orbit splitting as a part of a class exercise. Its heuristic value is apparently exhausted by the time the higher levels are reached.

The energy levels for the three-dimensional harmonic oscillator potential may be written

$$E_{n,l} = \hbar\omega\left(2n + l - \frac{1}{2}\right). \tag{9.8}$$

We may think of $\hbar\omega$ as simply a proportionality factor. The quantum number n takes on the values $1, 2, 3, \ldots$, and l has the same meaning of angular momentum that it does in atomic spectroscopy. The lowest level with zero angular momentum is called the $1s$, the next lowest $2s$, and so on. Likewise, the lowest level with 2 units of angular momentum is called the $1d$, etc. This notation is unlike that of the atom, where there is no $1p$ or $1d$, etc. In the pure $1/r$ potential, the energies are independent of l, but this is not the case in general. Another amusing difference between atomic and nuclear spectroscopy is that atomic spectroscopists omit j, they use $s, p, d, f, g, h, i, k, l, \ldots$. In nuclear spectroscopy the j-label is retained to mean 7 units of angular momentum.

The degeneracies shown on the left of Figure 9.7 are accounted for by the functional form of 9.8.

For the (infinite) potential well, the radial wave functions are spherical Bessel functions (j_l) whose indices correspond to the orbital angular momenta. Again, $l = 0, 1, 2, \ldots$ correspond to nucleons in *s*, *p*, *d*, ... orbitals. In this case, the different n-values arise from the *nodes* or zeros of the Bessel functions: $n = 1$ corresponds to the first node, $n = 2$ to the second, and so on. Clayton (1968) gives an approximate formula for the energy of the spherical well (we omit his zero point):

$$E_{n,l} \simeq \pi^2 \left(n + \frac{1}{2}\right) - l(l+1). \tag{9.9}$$

The energy here is in the units $\hbar^2/(2MR^2)$, where M is the mass of the particle, and R the radius of the well.

Again, the lowest s level is called $1s$, the lowest p level the $1p$, and so on. The spin-orbit interaction leads to a total angular momentum $j = l \pm 1/2$, with corresponding level displacements proportional to $-l/2$ and $+(l+1)/2$. In the terminology of atomic spectroscopy (Chapter 11) the nuclear doublets are inverted, that is, the level with the higher j-value lies below that with the lower j. Tabulations of nuclear energy levels (e.g., Lederer and Shirley 1978) give the energies in MeV and the *spin* and *parity* of each level. Within the context of the shell model, the spin is the vector sum of the j's of the nucleons, while the parity is even or odd as the sum of the l's is even or odd, as in atomic physics (see Chapter 11).

Figure 9.8 is the analogue for neutrons of Figure 9.5, which illustrated atomic shell structure. It shows the difference in the binding energies BE, for the Nth and $(N-1)$th neutrons, for constant $N - Z = 20$ and 22 respectively. The jump at the magic number 82 is not impressive, relative to the atomic case, but it is clearly present. The plot shows the alternation of the binding energies between neutron pairs – the ubiquitous odd–even effect. When the number of neutrons is even, the energy necessary to remove one is greater than that required to remove an unpaired neutron. The pairing energy appears in Equation 9.6 as the δ-term. This effect is of great importance in cosmochemistry, and is mentioned in a number of places in this book.

Wave functions for two-particle states can be made from products of single-particle wave functions. Consider spatial wave functions for a pair of neutrons (or a pair of protons) that have their spins opposed. It is possible to show that on the average, the neutrons are closer together in space than if the spins are parallel. Since the nuclear forces are attractive, the pairing of nucleons, with spins opposed, leads to strong binding. Exactly the opposite effect occurs for atomic electrons, because the force between them is repulsive, The tightly bound states are the ones where the electron spins are parallel (see Problem 4 below).

There are a variety of ways of illustrating the effects on the nucleus of the shell structure. One way is to compare the predictions of a mass formula, such as Equation 9.5 which does not allow for shell effects, with experimental data. A number of such illustrations may be found in books on nuclear physics (e.g., Preston 1962, section 6-3). Modern work on the mass formula takes shell effects explicitly into account.

Departures from spherical symmetry are much more important for nuclei than for atomic structure. It follows that the nuclear shell effects, though very significant, need

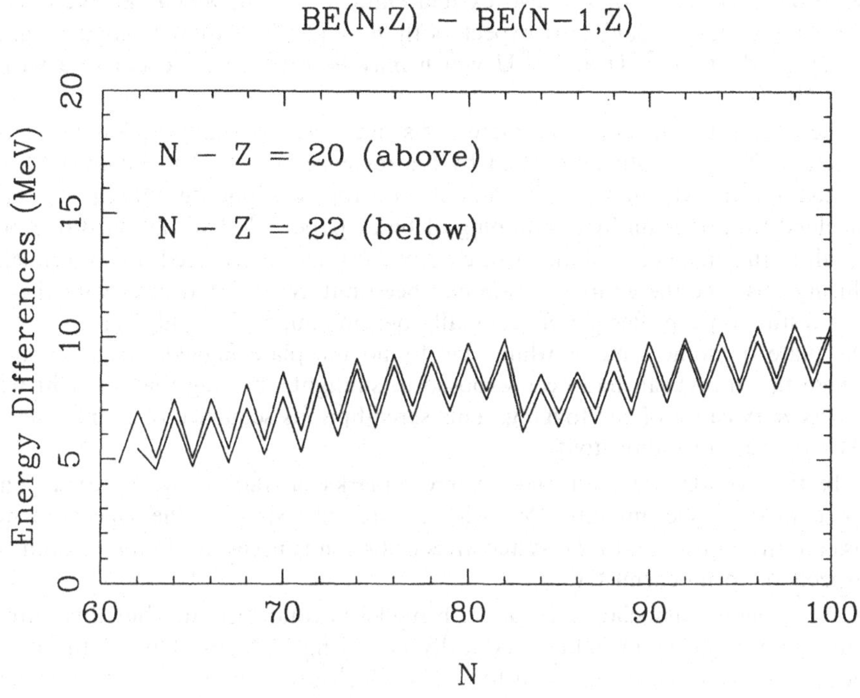

Figure 9.8. Neutron Binding Energy Differences. The graph gives BE(N, Z) minus BE$(N-1, Z)$ as a function of N. The jump is due to the magic number at $N = 82$. Data courtesy of National Nuclear Data Center, Brookhaven National Laboratory.

to be supplemented by additional considerations. Some effects may be described as distortions of the overall figure of the nucleus, giving rise to energy of vibration as well as rotation. These are often called *collective effects*, and they become important for nuclides away from the closed shells.

In a classical study of the abundances of heavy nuclides, Seeger, Fowler, and Clayton (1965) were able to explain a "hump" in the abundance distribution of the lanthanide isotopes. They did this by including departures from spherical symmetry, collective effects, or deformation energies in their mass formula that was used for species far on the neutron-rich side of the valley of β-stability. It is presently uncertain whether their solution is in fact the explanation of the peculiar "little maximum" that can be seen in the abundances near $A = 165$ in our Figure 7.6.

9.4 The Classical Forms of Radioactivity, and Fission

Near the beginning of the present century, α-, β-, and γ-"rays" were known to be emitted from certain heavy elements. We may recall that the discovery of radioactivity by Becquerel in 1896 resulted from experiments involving the compound potassium uranyl sulfate. The ^{238}U in the compound can α-decay to the ground state of ^{234}Th, or to one of two excited states of ^{234}Th, which can then decay by γ-emission. One would also expect both β- and γ-radiation from the numerous decay products of ^{235}U and ^{238}U which may be expected to occur in a variety of excitations.

According to the story, Becquerel's discovery was partially accidental. His idea was that X-rays might be created by fluorescence, and he had planned to put his potassium uranyl sulfate in the sunlight, on top of a photographic plate that was shielded from the sun by a wrapping of black paper. He had obtained one positive result in this manner. A number of cloudy days had interrupted the experiment and during this time the wrapped plate had been put away in a drawer with the uranyl salt resting on top. Becquerel eventually became impatient, and decided to develop the plate on some kind of whim. He found the plate fogged, just as in the case where the uranyl salt had been exposed to sunlight, showing that sunlight was not a necessary cause of the fogging. This serendipitous sequence of events constitutes the discovery of radioactivity.

In this section, we shall make a few remarks on radioactive processes and the calculation of relevant rates. We will also describe some of the aspects of nuclear fission, since insight into the systematics of fission is necessary to understand certain aspects of cosmochemistry.

We generally associate α-decay with nuclei beyond bismuth, but there are some important α-decays in lighter, naturally occurring isotopes. One of the more important is the α-decay of ^{147}Sm to ^{143}Nd which provides one of the more recently exploited radioactive dating methods (Faure 1986, chaps. 12–14). A number of the proton-rich lanthanides and some heavier species (^{190}Pt, ^{204}Pb) decay by α-emission. These decay times are usually quite long. Beyond ^{209}Bi ($Z = 83$), all of the nuclides are unstable and any isotope with a charge greater than that of bismuth must either α-decay, β-decay to a nuclide that can α-decay, or fission.

The *unstable* ^{8}Be decays into two α-particles in 10^{-16} sec! This short half-life is the reason why the relatively dense conditions of stellar interiors are required for helium burning, as will be discussed in the next chapter.

Most ground-state α-decay rates of relevance to cosmochemistry have been measured. The determinations are not easy in the case of the longer-lived nuclides, but theoretical calculations are available as a last resort. In principle, α-decay is straightforward. The α-particle is constrained by the potential barrier of the nucleus. At every reflection from the well, it has a probability of tunneling through. This probability is the square of the amplitude of the α-particle's wave function for the region beyond the potential (outside the nucleus) divided by the square of the amplitude of the wave function corresponding to the incident wave (the α-particle inside the nucleus).

Many β-decay rates are observed, but in the theory of nucleosynthesis it is necessary to make calculations for those nuclei that cannot be produced, and observed to β-decay in the laboratory.

A free neutron undergoes β-decay with a half-life of 618 sec. The uncertainty is no longer an important factor in cosmological (early universe) production of deuterium and helium. This matter will be discussed in §10.7. The neutron decay may be written

$$n \rightarrow p + e^- + \overline{\nu_e}. \quad (9.10)$$

It is often useful to think of nuclear β-decay as though one of the neutrons of the nucleus underwent "free" decay. Even though the other nucleons prevent this decay from being free, and the time scale is different, the net result is that an electron and a neutrino are emitted, and there is one more proton and one less neutron (in the nucleus).

Inverse β-decay occurs, for example, when a proton and an antineutrino produce a neutron and a positron. This may be derived from Equation 9.10 by applying one of the rules of particle physics: one may move a particle from one side of an equation such as 9.10 and place its antiparticle on the other side, provided energy and momentum conservation requirements are met. Thus a valid relation would be

$$\nu_e + n \rightarrow p + e^-. \quad (9.11)$$

The relation 9.11 is essentially the basis for the famous *solar neutrino experiments* (Bahcall 1989).

If we replace the e^- in 9.11 by an e^+ on the left, and then turn the arrow around, we would have an apparent proton decay. Energy conservation forbids the decay of the free proton to give a neutron, a positron, and a neutrino. However, such a reaction can take place in a nucleus because the masses of neutron-poor nuclei are greater than those on the valley of β-stability as is shown in Figure 9.2. Thus we may have, for example, the β^+-decay of ^{26}Al:

$$^{26}\mathrm{Al} \rightarrow ^{26}\mathrm{Mg} + e^+ + \nu_e. \quad (9.12)$$

When β-decay rates cannot be observed in the laboratory, they must be estimated from theoretical considerations. The decay rate of a nucleus is governed by two factors, one involving purely nuclear properties, and another related to the number of energy states available to the emerging electrons and neutrinos. The first factor is usually estimated from nuclear theory and semiempirical methods. The second may be calculated with the help of nuclear mass formulae, since the energy states available to the electron and neutrino are directly related by the difference in the masses of the initial and final nuclei. Very crudely, we may say that the β-decay rate will be faster if there are more energy states available to the emitted electron and neutrino. The *number* of these states depends on the difference between the masses of the initial and final particles, because this energy determines the volume of momentum space available to the decay particles.

Since many important β-decays involve species far from the valley of β-stability where mass measurements do not exist, it is necessary to make use of a semiempirical mass formula in the calculation of the rates.

Classical γ-emission is the nuclear analogue of the spontaneous emission of a photon from an atom. The difference is that nucleons rather than electrons are excited, and consequently the photons released are much more energetic than those emitted from atoms.

Typical nuclear energy levels are of the order of 10^6 eV, which corresponds roughly to a temperature $(E \propto kT) \approx 10^{10}$ K. Consequently, one can see that on the earth's

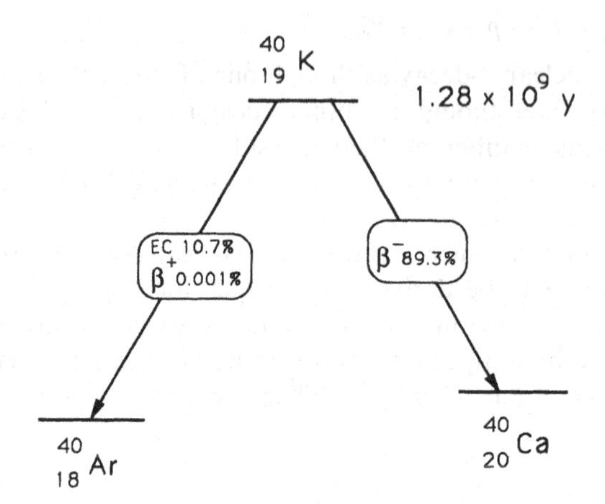

Figure 9.9. Radioactive Decay Modes of ^{40}K (schematic).

surface, and even in the interiors of most stars, the overwhelming majority of nuclei will be in their ground states. However, γ-emission was observed in the laboratory long before accelerators were available to provide energies of the necessary order to create excited nuclear states. The γ-rays were observed as a result of radioactive decays, in which an intermediate state was an excited daughter nucleus. Such nuclei might lose their energy by one of several processes, one of which is γ-emission. Figure 9.9 illustrates the decay routes open to ^{40}K.

The ground state of ^{40}K is some 1.5 MeV above the ground state of ^{40}Ar, as is shown in the figure. This energy is directly related to the masses of the nuclei. $M(^{40}\text{K})c^2 - M(^{40}\text{Ar})c^2 = 1.5$ MeV. Similarly, the mass of ^{40}Ca is slightly less than 0.2 MeV above ^{40}Ar. Since ^{40}Ca is doubly magic ($N = 20, Z = 20$), its first excited state lies rather high, at 3.35 MeV, well above the location of ^{40}K in its ground state. Hence there is only one possible state of ^{40}Ca to which ^{40}K may decay, the ground state, and this is done by β^--emission. This particular decay mode is the favored one for ^{40}K, occurring 89.3% of the time. The remaining decay modes may take place either to the ground state of ^{40}Ar, or to an excited state (not shown) lying 1.46 MeV higher. Both of these transitions take place by electron capture, that is

$$^{40}\text{K} + e^-(\text{K-shell}) = {}^{40}\text{Ar} + \nu_e, \tag{9.13}$$

but the transition to the excited state of ^{40}Ar is vastly more likely than to the ground state. As a result of this, most decays of ^{40}K to ^{40}Ar would be followed by the emission of a γ-ray with energy $h\nu = 1.46$ MeV.

Let us now turn briefly to a discussion of nuclear fission. Two classes of fission may be distinguished, spontaneous and induced. In the former case, the ground state of the nucleus may be pictured as a drop of (nuclear) fluid in a state of vibration. A zero-order model of fission would consist of two nuclear fragments in oscillation with respect to one another, rather like a diatomic molecule. On each oscillation,

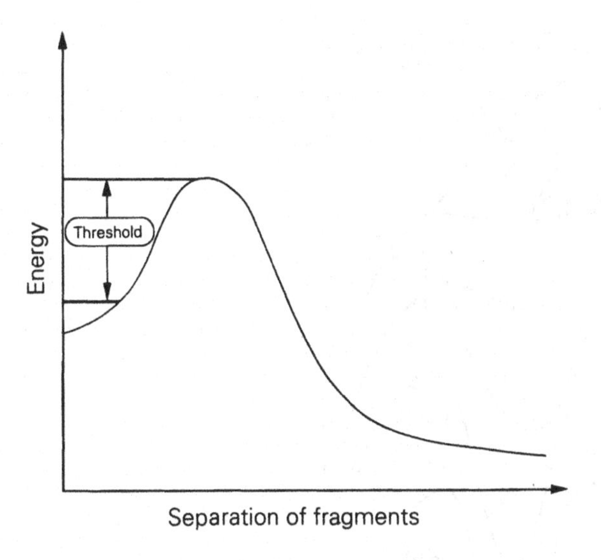

Figure 9.10. The Potential Barrier for Fission.

there is a finite probability that the potential barrier will be penetrated, and the separation will become permanent. This model is quite similar to that for α-decay, but the two nuclear fragments are much larger than α-particles.

Spontaneous fission is negligible for the long-lived isotopes of uranium and thorium, but it is the principal decay mode for ^{254}Cf and a number of the heavier nuclides listed by Lederer and Shirley (1978). Fission is much more likely from excited states of heavy nuclei. The famous fission of ^{235}U responsible for the first atomic bomb is induced by neutron bombardment. As a result of absorption of the neutron, an excited ^{236}U nucleus is formed. We may calculate the approximate energy level of the ^{236}U nucleus by adding the masses of the free neutron and ^{235}U and subtracting the mass of ^{236}U. The result is about 6.5 MeV, which is roughly the same as the energy of the fission barrier (threshold), illustrated in Figure 9.10. As a result decay of this excited nucleus by fission becomes competitive with other decay modes.

Induced fission by particles with low energies typically results in a spectrum of nuclei with peaks in two mass ranges, as illustrated in Figure 9.11. In fission work, one speaks of independent yields at each A. Ideally, the independent yields represent the fission fragments prior to subsequent reactions. Most of the fragments β-decay because they are neutron rich. *Cumulative* yields are approximately the sums at constant A of the independent yields. The independent yields do not sum exactly to the cumulative yields because neutron emission and absorption can change the A-values of the fragments.

The mass distribution of the fission fragments can be highly variable. Both the targets themselves, and the means by which fission is induced, can be important. This is illustrated in Figure 9.11 taken from Halpern (1959).

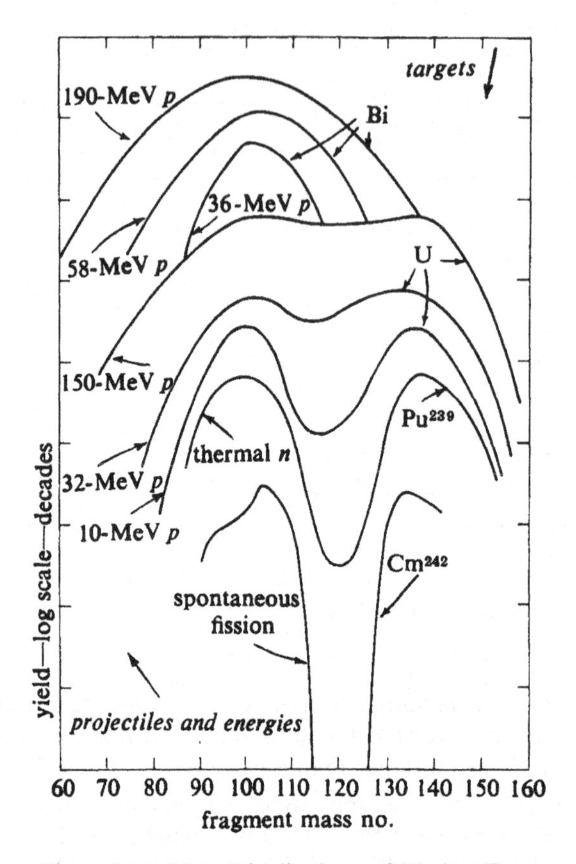

Figure 9.11. Mass Distributions of Fission Fragments From Several Targets and Projectiles. From *Annual Review of Nuclear Science*, Vol. 9, p. 245. ©Annual Reviews Inc., 1959.

The bimodal nature of the low-energy fission yields is related to the neutron magic numbers, but this cannot be the entire story.

In Chapters 11 and 12, we shall discuss the limited information available to the stellar spectroscopist on isotopic abundances. In the region of the fission fragments, little useful information has been obtained to date. It is therefore useful to present the fission products as a function of atomic number Z rather than A. This is done in Figure 9.12 for plutonium fission products. It is interesting to note the persistence of the odd–even effect, especially between xenon and neodymium. On the "slopes" of the yield curves, the odd–even effect is obscured.

At the present time, nuclear fission is thought to play an indirect role in the synthesis of the elements. Only the r-process can build the heavier nuclei belonging to the actinide rare earths, including the common long-lived uranium and thorium. The heavier actinide plutonium we now think was present in the early solar system on the basis of isotopic anomalies in xenon discussed in §8.7. The actinide series ends with lawrencium, atomic number 103. Other heavy species may be created in

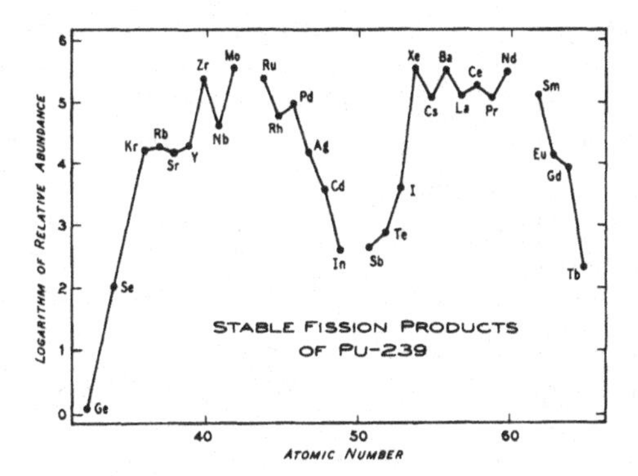

Figure 9.12. Distribution of Stable Elements from ^{239}Pu.

supernovae, including putative "super-heavy" elements with mass numbers near 298 (e.g., $Z = 114$, $N = 184$). A great deal of effort has been devoted to this region of the nuclide chart. There is reason to think that it may be an "island of stability" in which nuclides beyond bismuth might be stable, or at least relatively long-lived.

If the r-process neutron addition lasts long enough, the heaviest nuclei will fission, and their products will become seeds for further neutron addition. A steady state could eventually establish itself in which the nuclei cycle from fission products to massive species and back to fission products. The relative numbers of the various nuclei then remain constant, while their absolute values increase by a factor of 2 for each cycle, as the free neutrons become incorporated.

In the classic r-process calculations of Seeger, Fowler, and Clayton (1965) fission occurred for nuclei with $Z = 94$, $N \approx 85 - 90$. The fission products peaked near A-values of 95 and 127. This so-called long-time r-process solution did not give significant abundances at the $A = 80$ peak, and it can therefore not be primarily responsible for synthesis in this region. The actual r-process must be some kind of combination of long-time and short-time neutron additions in which the specific role of fission in the production of features on the abundance curve is unclear.

Fission processes have been invoked to explain the "hump" in the abundances that occurs in the region of the lanthanide rare earths in the SAD (Schramm and Fowler 1971). The fissioning nuclides would be associated with a putative shell closing at $N = 184$. Fission has also been suggested as a possible explanation of the very unusual distributions of lanthanides in certain chemically peculiar stars (see Steinberg and Wilkins 1978).

9.5 Nuclear Reactions and Their Rates

Most nuclear reactions may be pictured as a two-body collision process. The rates are described with the help of appropriate cross sections, obtained from either measurements in the laboratory or theoretical calculations.

According to standard terminology, a bombarding particle called a projectile collides with a target that may be considered stationary in an appropriate frame.

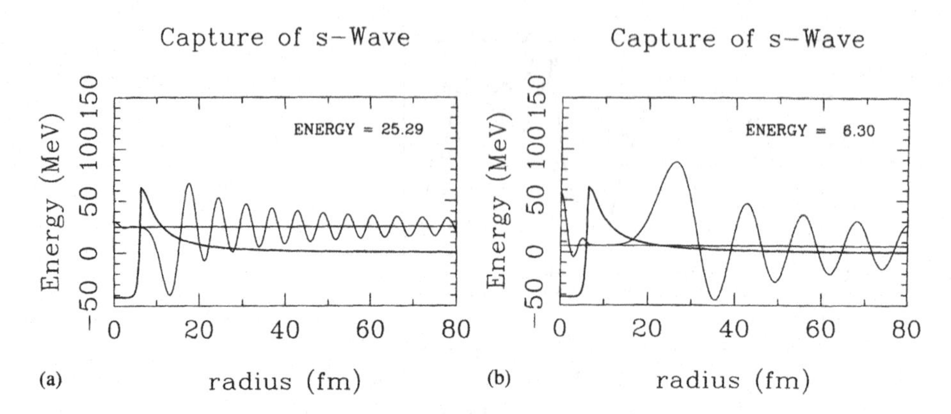

Figure 9.13. Nonresonant (a) and Resonant (b) Proton Capture.

If there are N_T targets per cm^3 with interaction cross section σ (cm^2), a projectile with relative velocity v will experience $N_T v\sigma$ collisions per second. This follows immediately from the fact that $v\sigma$ is the volume swept out in one second by the surface area σ moving with the relative velocity v. The number of targets within this volume is $N_T \sigma v \times$ (1 sec). Since there are N_P projectiles per cm^3, there will be $N_T N_P v\sigma$ collisions per cm^3 per second. If the targets and projectiles are the *same* particles, as in hydrogen burning by the *p–p* reaction, the appropriate rate is then $N^2 v\sigma/2$.

A very useful model for the calculations of cross sections in nuclear astrophysics involves the formation of a compound nucleus in which the nucleons of target and projectile are momentarily combined. This compound nucleus then "decays" by one of a number of possible channels. Reactions involving compound nuclei are usually treated using a method known as the Hauser–Feshbach formalism. We refer to Thielemann, Arnould, and Truran (1986), and Cowan, Thielemann, and Truran (1991) for modern applications of this technique.

Figure 9.13 illustrates *resonant* and *nonresonant* proton capture by an atomic nucleus. The heavier line represents the potential, which is essentially a Woods–Saxon potential as in Figure 9.6. A Coulomb potential is added for large values of r. The oscillating function is the radial part of the wave function, whose square is proportional to the probability density.

In Figure 9.13(a), a proton coming in from the right has an energy that is incommensurate with the nuclear energy levels. The small amplitude within the nuclear potential indicates a relatively small capture probability. In Figure 9.13(b), the energy of the incoming particle is lower, but it is commensurate with a natural resonance of the *compound nucleus.* The level is called *quasi-bound.* In such cases, the probability of capture is much larger, as shown by the relatively larger amplitude within the nuclear radius.

Reactions between charged particles involve the inverse process to the tunneling out of an α-particle in α-decay. Of course, no tunneling is necessary if the energy of the incoming particle is sufficient. However, under astrophysical conditions, the

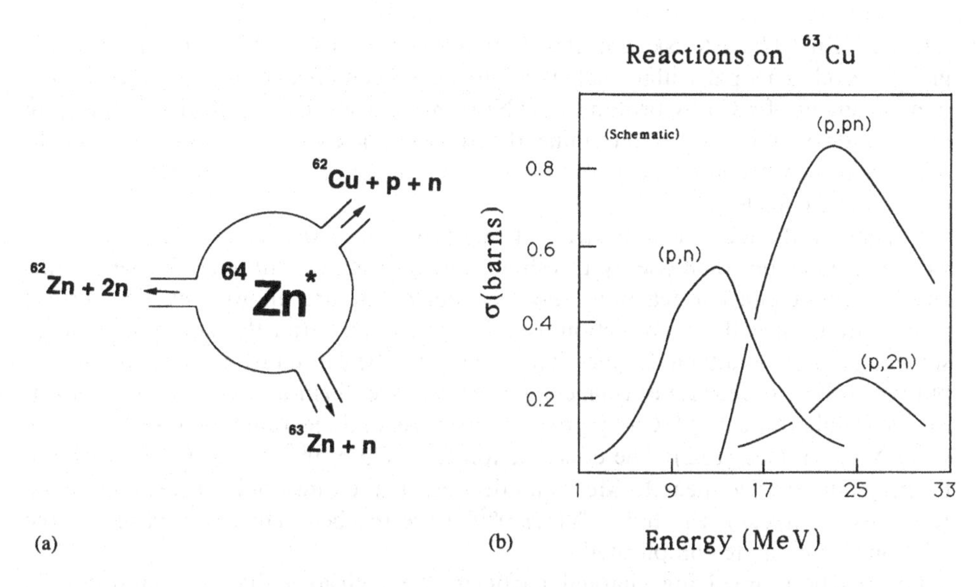

Figure 9.14. Proton Bombardment of ^{63}Cu. (a) Decay Channels of Compound Nucleus, (b) cross sections. Adapted from Ghoshal (1950).

temperature usually rises slowly, and the first reactions to occur always do so by tunneling, when two charged species are involved.

Electrostatic repulsion plays an important role in governing nuclear reactions in stars. If we ignore the quantum mechanical penetration of a nucleus' *Coulomb barrier*, we may estimate the temperature necessary for a nuclear reaction by setting the potential energy of two nuclei that are "just" in contact to the thermal energy $(3/2)kT$. The temperature obtained in this way is too high, because we have ignored tunneling through the Coulomb barrier by particles in the Maxwell–Boltzmann high-energy tail. The net result is that the required temperature for nuclear reactions to begin is lowered. With this in mind, we see that the first kinds of nuclear reactions to be expected will involve particles with the lowest charges – protons, deuterons, and helium nuclei. Reactions involving higher Coulomb barriers will require appropriately higher temperatures.

Astronomers had begun to speculate on the fusion of hydrogen into helium as the source of solar energy some years before the phenomena of barrier penetration had been articulated by Gamow and others in the late 1920's. However, Eddington's work on stellar structure had provided the first rough estimates of the central temperatures of stars, and without tunneling, they seemed much too low. Apparently, Eddington took a fair amount of criticism from his colleagues at the Cavendish Laboratory of Cambridge University for suggesting fusion as the source of solar energy; there did not appear to be enough energy to bring the nuclei together. We quote here Eddington's (1927) famous reply: "The critics lay themselves open to an obvious retort; we tell them to find *a hotter place*."

In Figure 9.14(a), we illustrate three decay modes for the specific compound nucleus $^{64}Zn^*$. The asterisk indicates the nucleus is *excited*, that is, not in its stable ground level. This particular nucleus might have been created, for example, by the bombardment of ^{63}Cu by protons, or ^{60}Ni by α-particles. It is up to the theoretician or laboratory physicist to determine the relative yields of the possible channels, which will then become a part of the cross section for an overall reaction involving one of the channels.

In general, the cross sections are a strong function of the relative velocity. There will, of course, be some energy threshold, and projectiles with energies below this energy cannot effect a reaction. For reactions that are allowed by the various conservation rules, the cross section usually rises rapidly after the threshold, reaches some maximum, and then declines. The decline can be caused by a variety of factors, but typically, it is because of *competing reactions*. We illustrate this in Figure 9.14(b) for the bombardment of ^{63}Cu by protons. For a general reaction $a + X \rightarrow Y + b$, we write $X(a,b)Y$ (see §10.2). The cross section for $^{63}Cu(p,n)^{63}Zn$ reaches a maximum for projectile energies near 13 MeV, and declines as the thresholds for the competing reactions $^{63}Cu(p,2n)^{62}Zn$ and $^{63}Cu(p,pn)^{62}Cu$ are reached. Thus, $\sigma = \sigma(v)$, and the relation can be quite complicated.

For reactions involving charged particles, the collisions that are important in astrophysical situations involve particles in the high-energy tail of the Maxwell–Boltzmann velocity distribution. The average collision frequency or reaction rate involves an integral over the velocity distribution and the velocity-dependent cross section. These two functions are shown schematically in Figure 9.15, where it has been assumed that the functional dependence of the cross section is $\propto \exp(-\sqrt{E_G/E})$, during the rise from threshold. The quantity E_G is a constant. The figure is from Rolfs and Rodney (1988).

The number of particles (say, projectiles) with energies in the range of the Gamow peak depends exponentially on the temperature. Indeed, if we call the normalized Maxwell–Boltzmann velocity distribution $\phi(v)$, and let μ be the reduced mass of the projectile and target,

$$\phi(v) = 4\pi \left(\frac{\mu}{2\pi kT}\right)^{3/2} v^2 \exp\left(-\mu v^2/2kT\right). \tag{9.14}$$

This may be written as a function of the energy $E = \mu v^2/2$ of relative motion by setting $\phi(v)dv = \phi(E)dE$. Then

$$\phi(E) = 4\pi\sqrt{2E}\left(\frac{1}{2\pi kT}\right)^{3/2} \exp\left(-E/kT\right). \tag{9.15}$$

The Gamow peak always occurs at $E/kT > 1$, where the exponential dominates the value of the function $\phi(E)$. As a result of this, the rates of nuclear reactions involving Coulomb barriers and Gamow peaks are also very strong functions of the temperature.

The cross sections shown in Figure 9.14 are called *nonresonant*. When the relative energy of the incoming particle, designated E on Figure 9.13(b), coincides with a level in the compound nucleus, the reaction cross sections can be enormously enhanced. Cross sections of the order of thousands of millibarns are then possible at

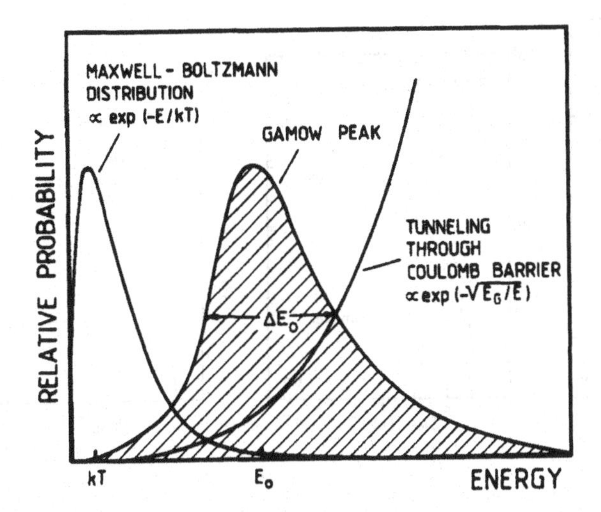

Figure 9.15. The Gamow peak. The Maxwell–Boltzmann function declines rapidly with increasing energy. The cross section increases rapidly with energy. The nuclear reactions take place at the energy domain where these two functions overlap, a region called the "Gamow peak." This peak has been amplified here for purposes of illustration. Courtesy of C. E. Rolfs and W. S. Rodney. From *Cauldrons in the Cosmos*, ©1988 The University of Chicago Press.

optimum projectile energies. (One barn is 10^{-24} cm^2.) The overall shape of the cross section σ as a function of energy may be influenced by these resonances, and can be very complicated, with numerous, sometimes overlapping, peaks. We must refer to textbooks on nuclear physics for more details. For our purposes, the complexities of nuclear structure are hidden in the symbol $\sigma(v)$ or $\sigma(E)$.

Nobel laureate W. A. Fowler has spent much of his career in the shepherding of nuclear reaction rates of relevance to astrophysics. Together with his colleagues, he has combined experimental and theoretical work on reaction rates, and presented the results in terms of analytical formulae which are easy to adapt to machine computation. We cite the work by Harris *et al.* (1983) as well as the book by Rolfs and Rodney (1988). The references given there will lead to the many papers of this monumental work.

Many of the reactions of importance in stars take place at energies much lower than those typically encountered in modern accelerators, which have been designed to reach increasingly higher energies. Much work of importance in nuclear astrophysics has therefore been done with (low energy) Van de Graaff generators. The first working version of these machines was built in 1929. Replicas may be seen in the laboratories of Hollywood's mad scientists. Even with modernized versions of Van de Graaff's machine, it has not been possible to probe all of the astrophysically relevant energy domains for some important reactions, such as $^{12}C(p, \gamma)^{13}N$. The most relevant astrophysical energies are of the order of a few tens of keV, short of the lowest measured energies.

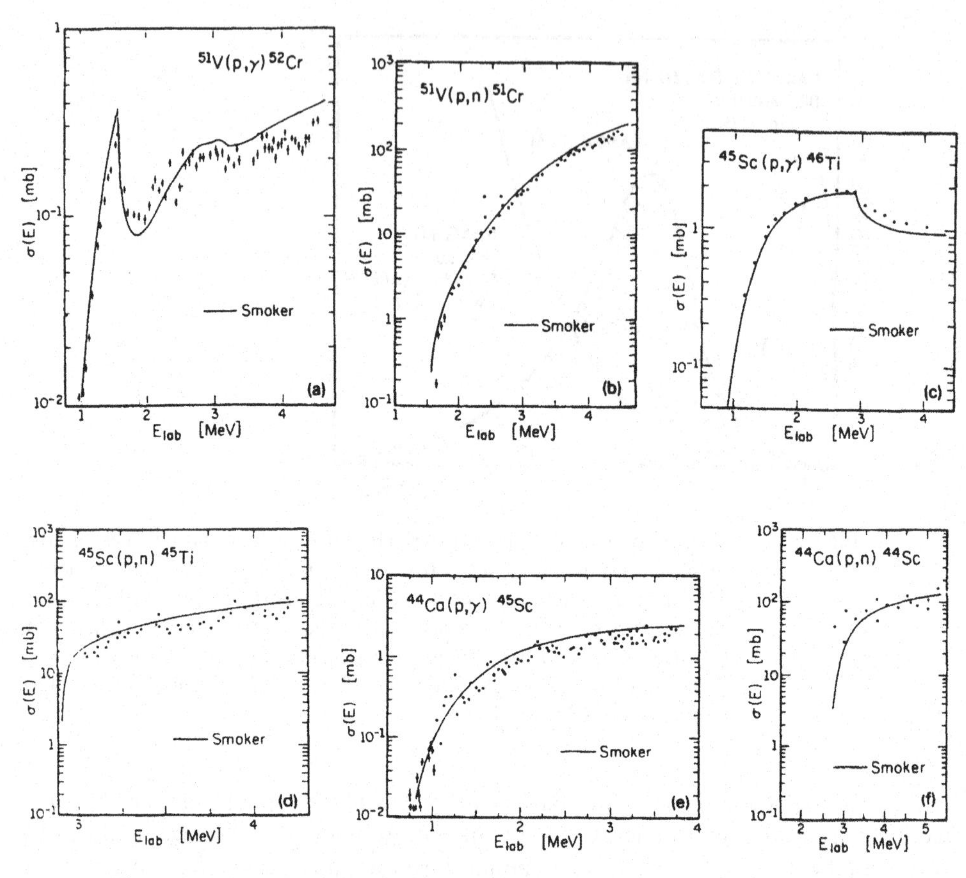

Figure 9.16. Experimental and Theoretical Nuclear Cross Sections. From *Physics Reports*, **208**, p. 267, 1991.

It is nevertheless important to emphasize that the empirical basis for nuclear astrophysics is well established. This is not to say that all cross sections are well known or that all rates are accurately calculated. If one considers the broad domain of nuclear astrophysics, there are surely many regions where liberal extrapolations are justified by physical insight combined with optimism and faith. Nevertheless, there *is* a broad empirical foundation to the entire structure that results from endless hours of laboratory work.

Figure 9.16 is taken from a review of Cowan, Thielemann, and Truran (1991) of the *r*-process and related topics. Many of the reaction rates used in nucleosynthesis come from theoretical calculations. These authors compare the results of the theoretical cross sections with experiment. The solid line gives the cross sections from the program developed by Thielemann, Arnould, and Truran (1986), called "Smoker." The code is based on Hauser–Feshbach or the compound nucleus formalism. The agreement with the experimental points, while not perfect, is most encouraging.

We wish to call attention primarily to the number of experimental points on these graphs. The atomic nucleus, while far from being completely understood, has been extensively investigated by laboratory studies.

Given the nuclear reaction rates, it is straightforward to obtain *energy generation rates* ϵ(erg/gm/sec) from them. Clearly

$$\epsilon = (\text{reactions/gm/sec})(\text{ergs released/reaction}). \quad (9.16)$$

In most cases, the energy that goes into neutrino production must be subtracted from the energy releases per reaction, since the neutrinos simply escape from the stars.

It is common to divide the stellar main sequence into an *upper* and a *lower* portion, depending upon whether the main source of stellar energy is derived from the proton–proton (p–p) chains or carbon cycles. These processes will be discussed in more detail in §10.2.

Figure 9.17 shows energy generation rates for hydrogen-burning. The ordinate of Figure 9.17 gives the energy generation rate ϵ in erg/gm/sec, divided by the mass density ρ and the fractional abundance of hydrogen (X_{H}) squared. It is common in calculations dealing with stellar structure and evolution to express abundances in terms of mass fractions (X) rather than relative numbers of atoms. For the SAD X_{H} is about 0.77, while the corresponding fraction by number is $N_{\mathrm{H}}/\sum N \approx 0.94$. For reference, we mention that the mass fraction for helium, commonly known as Y in the astronomical literature, is 0.208. The remaining mass fraction, Z, is 0.0187 (See Table A4).

We have used Equation (10.11) of Schwarzschild (1958) for the p–p chain. For the carbon cycle, we have used only the original Bethe (1939) cycle (Cycle I of Figure 10.1), with the critical $^{14}\mathrm{N}(p,\gamma)^{15}\mathrm{O}$ rate from Caughlan and Fowler (1988). The rates are plotted vs. the temperature in units of 10^6 K. The density and abundance factors are divided out. We have assumed $X_{\mathrm{N}} = 0.0933$ (SAD) in order to put both curves on the same diagram. Spectral types (§13.2) are placed very approximately, according to the *central* temperatures of main-sequence stars according to the models of Table 9.3 below as well as those given by Böhm-Vitense (1992). These are essentially *zero-age models*. The central temperature of the sun today is several million degrees hotter than that of a zero-age, one-solar-mass star ($1\ M_{\odot} = 1.99 \times 10^{33}$ gm) .

The changeover from dominance by the p–p chain to the carbon cycle takes place in the late A–F stars, a domain where many interesting spectroscopic phenomena are observed.

It is immediately apparent that the temperature dependence of the carbon cycle is much higher than that of the proton–proton chain. Large Coulomb barriers for CNO nuclei cause the steeper temperature dependencies illustrated. This provides the physical basis for the division of the main sequence into its upper and lower branches.

In the upper main-sequence stars, the steep temperature dependence of the carbon cycle results in a relatively large concentration of the energy generation toward the center of the star. This energy is created so copiously that radiative transport (by photons) is unable to remove it as efficiently as convection, that is, the mass transport

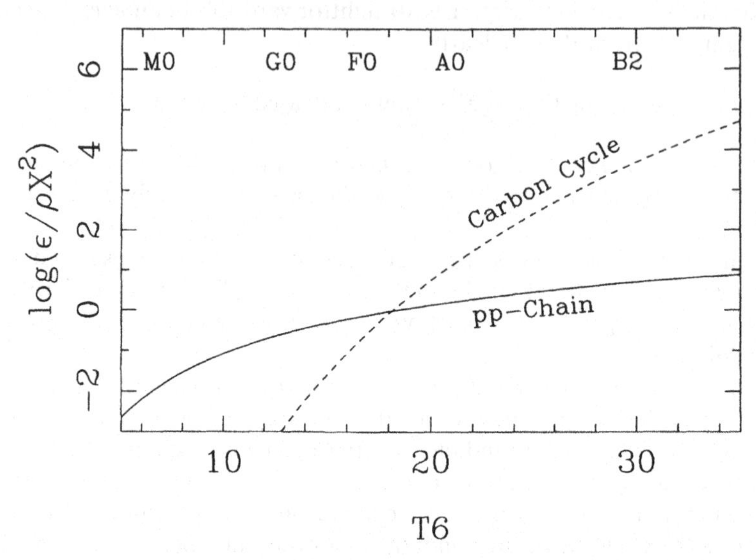

Figure 9.17. Energy Generation by the *p–p* Chain and Carbon Cycle. The abscissa is the temperature in units of 10^6 K.

of hot material by fluid motions. The net result is that upper main-sequence stars have convective cores and radiative envelopes. Conversely, lower main-sequence stars have radiative cores and convective envelopes.

The term envelope in stellar structure is roughly equivalent to mantle in planetary structure. The reason why the envelopes of lower and upper main-sequence stars are in convective and radiative equilibrium respectively is related to the thermodynamic properties of the gases, and is only indirectly attributable to nuclear energy generation. It is useful to remember that the modes of energy transport in stellar cores and envelopes are complementary with respect to convection and radiation. The student might use the temperature dependence of energy rates to infer the situation in stellar cores.

Tables 9.3(a–d) were kindly provided by P. Demarque, D. Guenther, and J. Howard. They give an overview of the internal structure of four representative stars. The central temperature of the 10-solar-mass (B2) star is only about a factor of two greater than that of the young sun. Nevertheless, the steep temperature dependencies of the nuclear reaction rates cause the core-envelope solutions to be totally different – the B2 star has a convective core and a radiative envelope, while the sun has a radiative core and a convective envelope.

The cosmochemical consequences of the convection in stellar envelopes are considerable, and the full implications have not yet been explored. We have already mentioned (§7.4) how convective currents in late-type stars can carry the light nuclei of Li, Be, and B to their destruction. If the envelopes of stars are stable against

Table 9.3(*a*). *Model of a 10 Solar Mass Main-Sequence Star*

r/R^a	M_r/M^b	L_r/L^c	$\log(\rho)$	$\log(P)$	$\log[T(\mathrm{K})]$	κ^d
0.03	0.001	0.029	0.949	16.598	7.495	0.37*
0.04	0.002	0.062	0.945	16.592	7.493	0.37*
0.06	0.008	0.183	0.934	16.574	7.486	0.37*
0.08	0.018	0.356	0.919	16.550	7.478	0.37*
0.10	0.034	0.545	0.899	16.518	7.467	0.38*
0.12	0.057	0.709	0.875	16.480	7.453	0.38*
0.14	0.087	0.832	0.846	16.434	7.437	0.38*
0.16	0.124	0.909	0.813	16.381	7.419	0.39*
0.18	0.168	0.950	0.776	16.322	7.398	0.40*
0.20	0.219	0.970	0.734	16.254	7.373	0.41*
0.22	0.275	0.977	0.687	16.178	7.346	0.41*
0.24	0.335	0.982	0.635	16.095	7.316	0.42
0.26	0.398	0.997	0.577	16.004	7.285	0.44
0.28	0.462	1.000	0.511	15.904	7.252	0.44
0.30	0.525	1.000	0.439	15.798	7.218	0.45
0.32	0.582	1.000	0.365	15.691	7.187	0.45
0.34	0.638	1.000	0.281	15.574	7.153	0.45
0.36	0.691	1.000	0.188	15.447	7.119	0.45
0.38	0.741	1.000	0.086	15.311	7.085	0.46
0.40	0.785	1.000	−0.022	15.169	7.052	0.47
0.43	0.822	1.000	−0.131	15.028	7.019	0.49
0.44	0.851	1.000	−0.231	14.899	6.990	0.52
0.47	0.879	1.000	−0.347	14.750	6.956	0.56
0.48	0.895	1.000	−0.424	14.650	6.933	0.58
0.50	0.915	1.000	−0.541	14.500	6.899	0.62
0.52	0.932	1.000	−0.657	14.350	6.865	0.65
0.54	0.946	1.000	−0.772	14.200	6.830	0.68
0.56	0.957	1.000	−0.888	14.050	6.795	0.71
0.58	0.966	1.000	−1.003	13.900	6.761	0.73
0.60	0.973	1.000	−1.119	13.750	6.726	0.75
0.62	0.979	1.000	−1.236	13.600	6.692	0.76
0.64	0.984	1.000	−1.353	13.450	6.659	0.78
0.66	0.988	1.000	−1.510	13.250	6.615	0.84
0.68	0.991	1.000	−1.627	13.100	6.581	0.89
0.71	0.994	1.000	−1.782	12.900	6.536	0.94
0.72	0.995	1.000	−1.898	12.750	6.501	0.98
0.74	0.997	1.000	−2.053	12.550	6.455	1.02
0.76	0.998	1.000	−2.206	12.350	6.407	1.07
0.78	0.998	1.000	−2.357	12.150	6.359	1.11
0.81	0.999	1.000	−2.558	11.884	6.293	1.14
0.82	0.999	1.000	−2.708	11.684	6.242	1.15
0.84	1.000	1.000	−2.933	11.384	6.168	1.12
0.86	1.000	1.000	−3.162	11.084	6.096	1.10
0.89	1.000	1.000	−3.473	10.684	6.005	1.15
0.90	1.000	1.000	−3.785	10.284	5.915	1.27

Table 9.3(*a*) (*continued*).

r/R^a	M_r/M^b	L_r/L^c	$\log(\rho)$	$\log(P)$	$\log[T(K)]$	κ^d
0.92	1.000	1.000	−4.098	9.884	5.825	1.39
0.94	1.000	1.000	−4.566	9.284	5.690	1.54
0.96	1.000	1.000	−5.209	8.484	5.521	2.01
0.98	1.000	1.000	−6.145	7.284	5.245	2.85
1.00	1.000	1.000	−8.810	3.784	4.405	2.90

Notes:
^{a} $R = 2.68 \cdot 10^{11}$ cm
^{b} $M = 10.0 \cdot 1.99 \cdot 10^{33}$ gm
^{c} $L = 2.14 \cdot 10^{37}$ ergs/sec
P and ρ in cgs units.
^{d} Opacity in cm^2/gm. The asterisk means the stellar temperature gradient is adiabatic.
Age = 0.000 17 Gyr, $X = 0.702$, $Z = 0.019$.

Table 9.3(*b*). *Model of a 2.5 Solar Mass Main-Sequence Star*

r/R^a	M_r/M^b	L_r/L^c	$\log(\rho)$	$\log(P)$	$\log[T(K)]$	κ^d
0.02	0.001	0.032	1.682	17.163	7.352	0.61*
0.04	0.005	0.128	1.674	17.149	7.346	0.62*
0.06	0.015	0.333	1.657	17.121	7.335	0.63*
0.08	0.034	0.559	1.633	17.081	7.319	0.65*
0.10	0.064	0.732	1.602	17.030	7.299	0.67*
0.12	0.105	0.836	1.564	16.968	7.274	0.71*
0.14	0.157	0.887	1.519	16.893	7.245	0.75*
0.16	0.219	0.920	1.464	16.805	7.211	0.80
0.18	0.288	0.958	1.399	16.705	7.176	0.84
0.20	0.363	0.984	1.323	16.593	7.140	0.89
0.22	0.437	0.994	1.240	16.474	7.105	0.93
0.24	0.513	0.998	1.141	16.338	7.067	0.97
0.26	0.582	0.999	1.038	16.199	7.032	1.02
0.28	0.646	1.000	0.928	16.054	6.997	1.08
0.30	0.700	1.000	0.817	15.911	6.964	1.15
0.32	0.750	1.000	0.697	15.757	6.931	1.24
0.34	0.795	1.000	0.570	15.596	6.897	1.35
0.36	0.832	1.000	0.441	15.434	6.864	1.48
0.38	0.861	1.000	0.324	15.287	6.834	1.62
0.40	0.885	1.000	0.204	15.137	6.804	1.78
0.42	0.906	1.000	0.085	14.987	6.773	1.97
0.44	0.923	1.000	−0.034	14.837	6.743	2.16
0.46	0.937	1.000	−0.154	14.687	6.711	2.34
0.48	0.949	1.000	−0.273	14.537	6.681	2.49

Table 9.3(*b*) (*continued*).

r/R^a	M_r/M^b	L_r/L^c	$\log(\rho)$	$\log(P)$	$\log[T(\text{K})]$	κ^d
0.50	0.959	1.000	−0.392	14.387	6.650	2.66
0.52	0.967	1.000	−0.512	14.237	6.619	2.86
0.54	0.974	1.000	−0.631	14.087	6.589	3.09
0.56	0.979	1.000	−0.750	13.937	6.558	3.33
0.58	0.983	1.000	−0.869	13.787	6.527	3.60
0.61	0.988	1.000	−1.027	13.587	6.484	4.01
0.62	0.991	1.000	−1.144	13.437	6.451	4.28
0.64	0.993	1.000	−1.260	13.287	6.417	4.53
0.66	0.994	1.000	−1.376	13.137	6.383	4.76
0.69	0.996	1.000	−1.530	12.937	6.337	5.02
0.70	0.997	1.000	−1.644	12.787	6.301	5.17
0.72	0.998	1.000	−1.796	12.587	6.253	5.34
0.74	0.999	1.000	−1.948	12.387	6.205	5.46
0.76	0.999	1.000	−2.082	12.211	6.162	5.57
0.78	0.999	1.000	−2.233	12.011	6.113	5.64
0.81	1.000	1.000	−2.461	11.711	6.041	5.61
0.82	1.000	1.000	−2.615	11.511	5.995	5.61
0.84	1.000	1.000	−2.848	11.211	5.928	5.78
0.86	1.000	1.000	−3.083	10.911	5.863	6.26
0.89	1.000	1.000	−3.396	10.511	5.777	7.20
0.90	1.000	1.000	−3.630	10.211	5.711	7.89
0.92	1.000	1.000	−4.023	9.711	5.603	8.80
0.94	1.000	1.000	−4.499	9.111	5.478	11.09
0.96	1.000	1.000	−5.050	8.411	5.327	14.84
0.98	1.000	1.000	−5.965	7.211	5.041	17.21
1.00	1.000	1.000	−8.990	3.111	4.041	24.19

Notes:
[a] $R = 1.20 \cdot 10^{11}$ cm
[b] $M = 2.5 \cdot 1.99 \cdot 10^{33}$ gm
[c] $L = 1.51 \cdot 10^{35}$ ergs/sec
P and ρ in cgs units.
[d] Opacity in cm^2/gm. The asterisk means the stellar temperature gradient is adiabatic.
Age = 0.01 Gyr, $X = 0.702$, $Z = 0.019$.

the mixing currents that arise as a part of convection, it is possible for a chemical differentiation to take place. Gravitation, which causes heavier atoms to diffuse downward, competes with radiation. Those species capable of absorbing the momentum from the net outward flux of the star are pushed up. It turns out that a sizable portion of upper main-sequence stars exhibit chemical peculiarities that are difficult to explain without the invocation of chemical differentiation *in situ*. We shall return to this matter in later chapters.

Table 9.3(c). *Model of a 1 Solar Mass Main-Sequence Star*

r/R^a	M_r/M^b	L_r/L^c	$\log(\rho)$	$\log(P)$	$\log[T(\mathrm{K})]$	κ^d
0.03	0.001	0.021	1.891	17.158	7.133	1.46*
0.04	0.002	0.046	1.888	17.151	7.131	1.47*
0.06	0.008	0.135	1.876	17.132	7.123	1.51*
0.08	0.018	0.257	1.860	17.106	7.113	1.55*
0.10	0.035	0.390	1.839	17.071	7.099	1.61*
0.12	0.058	0.510	1.814	17.029	7.082	1.69*
0.14	0.088	0.642	1.783	16.978	7.062	1.79
0.16	0.126	0.769	1.746	16.919	7.041	1.90
0.18	0.170	0.868	1.702	16.853	7.018	2.03
0.20	0.219	0.931	1.653	16.780	6.993	2.17
0.22	0.272	0.967	1.597	16.698	6.968	2.33
0.24	0.327	0.987	1.536	16.612	6.943	2.50
0.26	0.385	0.999	1.468	16.518	6.918	2.68
0.28	0.442	1.005	1.395	16.419	6.892	2.88
0.30	0.496	1.008	1.319	16.317	6.868	3.10
0.32	0.550	1.008	1.234	16.207	6.842	3.33
0.34	0.596	1.008	1.155	16.104	6.819	3.57
0.36	0.646	1.007	1.059	15.981	6.793	3.86
0.38	0.684	1.006	0.976	15.877	6.771	4.14
0.40	0.725	1.004	0.878	15.753	6.746	4.53
0.42	0.759	1.003	0.784	15.637	6.724	4.93
0.44	0.785	1.002	0.703	15.536	6.704	5.31
0.46	0.813	1.001	0.608	15.418	6.682	5.80
0.48	0.832	1.001	0.536	15.329	6.665	6.20
0.51	0.860	1.000	0.415	15.180	6.637	7.04
0.53	0.876	1.000	0.335	15.081	6.618	7.65
0.54	0.891	1.000	0.254	14.981	6.599	8.30
0.56	0.904	0.999	0.174	14.881	6.579	9.00
0.58	0.916	0.999	0.094	14.781	6.559	9.75
0.60	0.927	0.999	0.014	14.681	6.539	10.56
0.63	0.940	0.999	−0.104	14.531	6.508	11.72
0.65	0.948	0.999	−0.182	14.432	6.486	12.58
0.67	0.955	0.999	−0.259	14.332	6.464	13.58
0.69	0.962	0.999	−0.336	14.232	6.440	14.81
0.71	0.967	0.999	−0.409	14.133	6.415	16.44
0.72	0.972	0.999	−0.479	14.033	6.385	18.85
0.74	0.976	0.999	−0.539	13.939	6.351	22.62*
0.77	0.981	1.000	−0.629	13.790	6.291	31.05*
0.78	0.984	1.000	−0.688	13.690	6.251	37.75*
0.80	0.988	1.000	−0.778	13.540	6.191	47.30*
0.83	0.991	1.000	−0.868	13.390	6.132	58.07*
0.84	0.993	1.000	−0.958	13.241	6.072	70.94*
0.86	0.995	1.000	−1.048	13.091	6.012	85.54*
0.88	0.997	1.000	−1.168	12.891	5.932	110.36*
0.90	0.998	1.000	−1.316	12.641	5.832	158.13*

Table 9.3(*c*) (*continued*).

r/R^a	M_r/M^b	L_r/L^c	$\log(\rho)$	$\log(P)$	$\log[T(\mathrm{K})]$	κ^d
0.92	0.999	1.000	−1.463	12.392	5.733	241.88*
0.95	1.000	1.000	−1.820	11.785	5.488	965.14*
0.97	1.000	1.000	−2.123	11.285	5.297	
0.98	1.000	1.000	−2.753	10.285	4.947	
1.00	1.000	1.000	−6.273	5.285	3.754	0.33

Notes:
[a] $R = 6.07 \cdot 10^{10}$ cm
[b] $M = 1.0 \cdot 1.99 \cdot 10^{33}$ gm
[c] $L = 2.72 \cdot 10^{33}$ ergs/sec
P and ρ in cgs units.
[d] Opacity in $\mathrm{cm}^2/\mathrm{gm}$. The asterisk means the stellar temperature gradient is adiabatic.
Age = 0.042 Gyr, $X = 0.702, Z = 0.019$.

Table 9.3(*d*). *Model of a 0.65 Solar Mass Main-Sequence Star*

r/R^a	M_r/M^b	L_r/L^c	$\log(\rho)$	$\log(P)$	$\log[T(\mathrm{K})]$	κ^d
0.04	0.001	0.015	1.888	17.027	7.003	2.61*
0.06	0.003	0.045	1.882	17.017	6.999	2.65*
0.08	0.008	0.097	1.873	17.001	6.993	2.71
0.10	0.015	0.167	1.861	16.982	6.985	2.78
0.12	0.026	0.259	1.846	16.958	6.976	2.87
0.14	0.040	0.368	1.828	16.929	6.965	2.98
0.16	0.058	0.482	1.808	16.897	6.953	3.11
0.18	0.080	0.598	1.784	16.860	6.940	3.26
0.20	0.107	0.706	1.757	16.817	6.925	3.45
0.22	0.136	0.789	1.727	16.772	6.909	3.65
0.24	0.170	0.849	1.694	16.722	6.893	3.87
0.26	0.207	0.892	1.656	16.668	6.876	4.12
0.28	0.246	0.923	1.616	16.610	6.859	4.39
0.30	0.285	0.946	1.573	16.550	6.842	4.67
0.32	0.327	0.963	1.526	16.485	6.824	4.98
0.34	0.372	0.976	1.474	16.414	6.806	5.32
0.36	0.412	0.984	1.424	16.347	6.790	5.66
0.38	0.457	0.991	1.365	16.269	6.771	6.06
0.40	0.496	0.994	1.311	16.199	6.755	6.44
0.42	0.537	0.997	1.250	16.119	6.736	6.88
0.44	0.576	0.998	1.189	16.040	6.719	7.34
0.46	0.610	0.999	1.131	15.966	6.703	7.79
0.48	0.646	1.000	1.065	15.882	6.686	8.33
0.50	0.676	1.000	1.005	15.806	6.670	8.84

Table 9.3(*d*) (*continued*).

r/R^a	M_r/M^b	L_r/L^c	$\log(\rho)$	$\log(P)$	$\log[T(\mathrm{K})]$	κ^d
0.52	0.708	1.000	0.937	15.720	6.652	9.73
0.54	0.733	1.000	0.881	15.648	6.637	10.60
0.56	0.759	1.000	0.819	15.569	6.620	11.69
0.58	0.785	1.000	0.750	15.480	6.600	13.12
0.61	0.813	1.000	0.674	15.378	6.574	15.21
0.63	0.832	1.000	0.620	15.302	6.553	17.34
0.65	0.851	1.000	0.563	15.217	6.525	20.91
0.67	0.871	1.000	0.505	15.121	6.487	27.06*
0.69	0.890	1.000	0.445	15.021	6.447	35.18*
0.70	0.898	1.000	0.415	14.971	6.427	39.93*
0.72	0.913	1.000	0.355	14.871	6.387	50.91*
0.74	0.926	1.000	0.295	14.771	6.347	63.94*
0.77	0.943	1.000	0.205	14.621	6.287	87.26*
0.78	0.952	1.000	0.145	14.521	6.247	104.94*
0.81	0.963	1.000	0.055	14.371	6.187	127.67*
0.82	0.970	1.000	−0.005	14.271	6.147	144.02*
0.84	0.977	1.000	−0.095	14.121	6.087	171.70*
0.86	0.984	1.000	−0.215	13.921	6.007	217.92*
0.88	0.988	1.000	−0.305	13.771	5.947	262.51*
0.90	0.993	1.000	−0.449	13.521	5.847	368.58*
0.92	0.996	1.000	−0.587	13.271	5.748	544.47*
0.94	0.998	1.000	−0.782	12.921	5.611	
0.96	1.000	1.000	−1.160	12.295	5.376	
0.98	1.000	1.000	−1.782	11.295	5.025	
1.00	1.000	1.000	−5.642	5.795	3.633	0.16

Notes:
[a] $R = 3.97 \cdot 10^{10}$ cm
[b] $M = 0.65 \cdot 1.99 \cdot 10^{33}$ gm
[c] $L = 3.84 \cdot 10^{32}$ ergs/sec
P and ρ in cgs units.
[d] Opacity in $\mathrm{cm}^2/\mathrm{gm}$. The asterisk means the stellar temperature gradient is adiabatic.
Age = 0.247 Gyr, $X = 0.702$, $Z = 0.019$.

9.6 Problems

1. Positron emission or β^+-decay may be written symbolically $(A,Z) \rightarrow (A,Z-1) + e^+ + \nu$. If we use the nuclear masses M *including the electron masses* show that

$$M(A,Z) = M(A,Z-1) + 2m_e + Q \tag{9.17}$$

for β^+-decay, while

$$M(A,Z) = M(A,Z+1) + Q \tag{9.18}$$

for β^--decay.

2. The mass formula has a single minimum for odd-A isobars, so that one would expect only one stable odd-A isobar. This is generally so, but there are a few places on the nuclide chart where the minimum of the mass equation falls very nearly at half-integral Z, and for such points one might expect to find two isotopes that either are stable or have relatively long half-lives. One case where this is so is for $A = 87$. The two nuclides are ^{87}Rb and ^{87}Sr, used in radioactive dating. If one takes a half-life of 10^5 years as "nearly" stable, then there are a dozen pairs of such odd-A isobars. Find some of them.
3. The capture cross sections for "room-temperature" neutrons (0.025 eV) show the influence of shell closings, but much less dramatically than the 30 keV neutrons of Figure 9.3. You may demonstrate this for yourself if you have digitized data, or are willing to plot by hand. Why should the higher-energy neutron cross sections be more effective in displaying the magic numbers?
4. Real atoms depart occasionally from the scheme of Figure 9.4 in filling their shells. For example, the lowest configuration of neutral chromium is $3d^5 4s$. One might have expected $3d^4 4s^2$. This irregularity may be accounted for in terms of the special stability of a half-filled shell: explain. Find some additional examples of the departures from sequential filling of the d and f shells.
5. Consider two fermions for which the spatial wave functions are respectively symmetric and antisymmetric for exchange of the particles. If you fix the coordinates for one of the particles, you can then plot the probability distribution for the other. Compare the resulting probability distributions for the case of symmetric and antisymmetric spatial functions, and demonstrate that the overlap is greater (the particles are more likely to be closer together) for the symmetric functions. This means the spin functions must be antisymmetric or opposed (why?). This exercise illustrates the physical basis for the odd–even effect of elemental abundances and nuclear stability. It also explains the Hund rule of atomic spectroscopy, and why the ground states of most diatomic molecules are singlet (Σ's, see Chapter 11). (Hint: Look for a node in the combined probability function for the antisymmetric functions.)

10

Energy Generation in Stars and Nucleosynthesis

10.1 Introduction

Most stars are converting hydrogen to helium by either the proton–proton chain, or the CNO cycle. We will discuss these processes in §10.2. Nuclear energy is available by the combination or fusion of nuclei less massive than $A = 56$ (Fe). This nucleus has the maximum binding energy per nucleon. By far the largest fraction of energy in this process is released in the first step, the formation of helium from hydrogen. A rough figure for the ratio of the time that a star spends during hydrogen burning (its *main-sequence lifetime*) to all other phases of its lifetime is ten to one. The stellar lifetimes are terminated by supernovae explosions, or by quiescent deaths that may involve mass loss followed by the formation of a white dwarf or neutron star.

While the fusion of hydrogen into helium is undoubtedly one phase of nucleosynthesis, it is difficult to know the extent to which the observable helium throughout the universe was made in stars. It is relatively straightforward to demonstrate that the present luminosity of the galaxy, if constant over a lifetime of some 10 billion years, is insufficient to account for the present hydrogen-to-helium ratio. We shall return to this question in §10.7.

Helium, as we shall see shortly, is burned to carbon and oxygen, which can be returned to the interstellar medium in either quiescent mass loss or supernovae explosions. While the helium abundance in the Galaxy is arguably constant, this is surely *not* the case for carbon. Thus, the cosmological synthesis that is attractive for helium and a few additional fragments, is not invoked for carbon or heavier nuclides.

The picture of the origin of the chemical elements primarily through stellar nucleosynthesis was sketched out in the 1950's by B^2FH and A. G. W. Cameron (see §1.1). The book *Cauldrons in the Cosmos* by Rolfs and Rodney (1988) provides an excellent overall introduction. Clayton's (1968) *Principles of Stellar Evolution and Nucleosynthesis* was reprinted with updated references in 1983, and is still of great value. Comprehensive, well-researched reviews of nucleosynthesis and related phenomena in supernovae have been written by Trimble (see Trimble 1991).

10.2 The Burning of Hydrogen and Helium

The reactions of nuclear astrophysics may generally be described by time-dependent differential equations of the form

$$\frac{dN_i}{dt} = (\text{creation rate}) - (\text{destruction rate}), \tag{10.1}$$

where i runs from 1 to the number of species involved in the reactions. In modern work, the rates are calculated for all relevant nuclei. In the case of light species, the relevant rates have been fitted to convenient interpolation formulae which we discussed in §9.5.

We shall give a very brief overview of the burning of the lighter elements. Clayton's (1968) more detailed description is available for those who wish to pursue this topic more closely. From the point of view of cosmochemistry, it is necessary to integrate the rate equations with those of stellar structure – including the explosive, late stages of stellar structure – before one can obtain insight into such questions as the carbon-to-oxygen or oxygen-to-neon ratios. We will mention these calculations below, but it is difficult to provide heuristic insight into the results of such complicated calculations. On the other hand, there are analytic approaches to the synthesis of the elements heavier than, say, silicon, which retain great heuristic value. For this reason, our discussion will become more detailed when we discuss silicon burning in §10.4.

Hydrogen burning by the proton–proton chain can involve three sequences of reactions. Following Clayton (1968) we designate them PP I, PP II, and PP III. We use a common notation for nuclear reactions. Let a generic reaction be $a + X \rightarrow Y + b$. This is often written $X(a, b)Y$, where a may usually be considered the "projectile" and X the "target," as these terms were used in §9.5. For the p–p chains, then:

$$\left.\begin{array}{l} {}^1\mathrm{H}(p, e^+\nu)^2\mathrm{D}, \\ {}^2\mathrm{D}(p, \gamma)^3\mathrm{He}, \\ {}^3\mathrm{He}({}^3\mathrm{He}, pp)^4\mathrm{He}, \end{array}\right\} \text{PP I}$$

(or)

$$\left.\begin{array}{l} {}^3\mathrm{He}(\alpha, \gamma)^7\mathrm{Be}, \\ {}^7\mathrm{Be}(e^-, \nu)^7\mathrm{Li}, \\ {}^7\mathrm{Li}(p, \alpha)^4\mathrm{He}, \end{array}\right\} \text{PP II}$$

(or)

$$\left.\begin{array}{l} {}^7\mathrm{Be}(p, \gamma)^8\mathrm{B}, \\ {}^8\mathrm{B}(\ , e^+\nu)^8\mathrm{Be}, \\ {}^8\mathrm{Be}(\ , \alpha)^4\mathrm{He}. \end{array}\right\} \text{PP III}$$

The extent to which the alternative reactions are followed depends on the temperature of the hydrogen burning. Note that the Coulomb barriers are different for the reactions listed above, where the nuclear charges are higher than +1. The branching

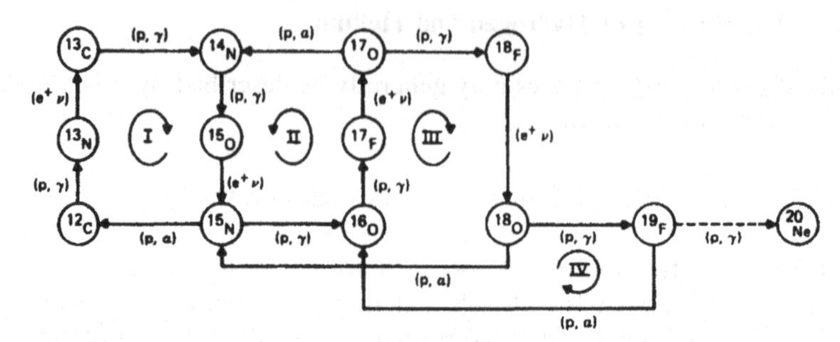

Figure 10.1. The CNO tri-cycle. Courtesy C. E. Rolfs and W. S. Rodney from *Cauldrons in the Cosmos*, University of Chicago Press, ©1988.

ratio for ^{7}Be between decays by the weak interaction (e^-, ν) and a reaction with a proton to form ^{8}B depends on the temperature and density.

Notice that all of the intermediate species are ultimately destroyed if the reactions go to completion. In low-mass stars, there is the possibility that the Coulomb barrier for $^3\text{He} +^3$ He is important. At the end of hydrogen burning, in these stars, there will by a sizable portion of ^{3}He present. However, such stars do not explode, or lose mass. Moreover, their hydrogen is burned very slowly, and it may take billions of years for the ^{3}He to be built up.

So it is difficult for the excess ^{3}He to manifest itself in an observable portion of the universe (but see Rood, Bania, and Wilson 1992). This means that we must attribute the highly unusual ^{3}He abundances observed in solar cosmic rays (Simpson 1983) and on the surfaces of certain chemically peculiar stars (Hartoog and Cowley 1979) to mechanisms that are unrelated to the proton–proton chain.

In galactic cosmic rays the abundance of ^{3}He is unusually high, but this is quite well accounted for in terms of spallation (§8.7) from the more massive cosmic-ray nucleons. Similar processes are thought to produce the Li, Be, and B that are observed in the sun and stars.

The net effect of the *p–p* chain is simply to convert hydrogen into helium in stars whose masses are in the range where the entire star may be considered a mass sink – the processed material has little chance of being returned to the interstellar medium. The situation is rather different for those stars that are operating on the CNO cycle.

Figure 10.1 illustrates what is now referred to as the CNO tri-cycle, with an additional branch to ^{19}F. The diagram is from Rolfs and Rodney (1988). The old Bethe–Weizsäcker CN cycle is called Cycle I in the diagram. Of the reactions in this cycle, the slowest by far under the conditions obtaining during CNO burning is $^{14}\text{N}(p,\gamma)^{15}\text{O}$. Thus it is easy to see that there should be a pileup at ^{14}N.

In Cycle II, the $^{16}\text{O}(p,\gamma)^{17}\text{F}$ reaction is actually slower by nearly two orders of magnitude than the $^{14}\text{N}(p,\gamma)^{15}\text{O}$. But the $^{15}\text{N}(p,\gamma)^{16}\text{O}$ that leads into Cycle II is nearly five orders of magnitude less probable than $^{15}\text{N}(p,\alpha)^{12}\text{C}$. Thus the cosmically abundant ^{16}O is fed by Cycle II into Cycle I (and destroyed) faster than it is replaced by (p,γ) on ^{15}N. If the CNO cycle operates for a sufficient time virtually all of the CNO isotopes will be turned into ^{14}N. If the hydrogen is very rapidly consumed, the

^{16}O may not be significantly depleted, but ^{12}C and ^{13}C decrease rapidly on typical time scales of less than a million years.

In addition to the excess of ^{14}N an important characteristic of CNO-processed material is the ratio of the isotopes of ^{12}C to ^{13}C. This ratio should be about 4 for material that has been through the CNO cycle. The SAD ratio is about 90. The isotope shifts of the atomic lines of carbon are too small to be useful, but molecular isotopic shifts are readily measurable. Modern detectors have made it possible to study isotopic shifts in an increasing number of molecules (Gustafsson 1989, Smith and Lambert 1990), including the ubiquitous CO.

Extensive investigations are now being made of isotopic abundances in giant molecular clouds with the help of millimeter-wavelength observations of a variety of molecular species detected in the last several decades (Wannier 1980, Irvine 1991, 1992).

There are a few stars whose atmospheres are unusually rich in helium, or alternatively, hydrogen poor. Two well-known examples are the peculiar binary stars β Lyrae (Mazzali 1987) and υ Sagittarii (Jaschek, Andrillat, and Jaschek 1990). The traditional interpretation of such objects is that the hydrogen-rich envelope either has been mixed with the helium core, or has been stripped off altogether. The fact that both of these stars are in binary systems makes mass loss plausible. Additionally, there are the fascinating R CrB stars whose atmospheres are hydrogen poor and carbon rich (Whitney, Soker, and Clayton 1991).

One of the greatest triumphs in the theory of the history of the chemical elements was the discovery that the nuclide gaps at masses 5 and 8 could be bridged by the stellar triple-α process. This reaction was suggested by Salpeter (1952). The discovery that the $^{8}Be(\alpha,\gamma)^{12}C$ rate was resonant was made in W. A. Fowler's laboratory as a result of a celebrated prediction by Fred Hoyle of the existence of an energy level in ^{12}C at just the proper position. Hoyle's (1982) own version of this story may be found in *Essays in Nuclear Astrophysics.*

The 3α process may be written

$$^{4}\mathrm{He}(\alpha,)^{8}\mathrm{Be}(\alpha,)^{12}\mathrm{C}^{*}(\ ,\gamma\gamma)^{12}\mathrm{C}, \tag{10.2}$$

where the '*' indicates an excited state of the ^{12}C nucleus – the state predicted by Hoyle. Whether a stable ^{12}C nucleus is formed or not depends on the behavior of the excited state or compound nucleus. There are two important branches to consider: reemission of an α-particle, and emission of a γ. Actually, two γ-emissions are necessary for the excited ^{12}C nucleus to reach the ground level, hence the $\gamma\gamma$ in Equation 10.2. The excited ^{12}C has an additional stabilizing decay channel involving the emission of an electron–positron pair. This is about sixty times less probable than the γ-emission.

The third α must be added on a time scale comparable to (or shorter than) the lifetime of the ^{8}Be that is made from the fusion of the first two α's. Since the half-life of ^{8}Be is about $7 \cdot 10^{-16}$ seconds, it is clear that an equilibrium abundance of $^{12}C^{*}$ must be established very rapidly. It is therefore possible to estimate the rate of the 3α process by multiplying the equilibrium number density of $^{12}C^{*}$ nuclei by the rate of γ-emissions from the excited state. The equilibrium number density of

^{12}C is found from considerations that are essentially the same as for the chemical reactions of Chapter 4 (see Problem 1 below).

The conditions under which helium burning begins and ends in stars is critically dependent on the stellar mass. We shall not give a detailed description, but refer instead to §3-7 of Mihalas and Binney (1981).

Helium burning can lead to the α-rich nuclides ^{16}O, ^{20}Ne, and ^{24}Mg, by (α, γ) reactions. In addition a number of interesting reactions involving α-addition can take place starting with the abundant ^{14}N seed. Consider the sequence $^{14}\mathrm{N}(\alpha,\gamma)^{18}\mathrm{F}(\ ,e^{+}\nu)^{18}\mathrm{O}(\alpha,\gamma)^{22}\mathrm{Ne}$. The end member could possibly be relevant to the so-called Neon-E, or exotic neon, observed in certain meteorites, which we discussed in §8.7 (see Figure 8.7).

The β^{+}-decay of ^{18}F is a very significant reaction in the sequence leading from ^{14}N to ^{22}Ne, because it leads to an overall increase in the nuclear neutron-proton ratio. Production of the α-rich species does not do this. Such β^{+}-decays are required to account for the observed intermediate species with odd Z, such as Na, Al, P, and Sc, for which there are no stable isotopes with $Z = N$. Some of these species may be produced in processes other than those closely related to helium burning, but the need for the β^{+}-decays remains.

A class of cool carbon-rich giant stars are known whose envelopes possess deep convection zones. It is generally thought that the excess of surficial carbon may be attributable to mixing of material from a carbon core to the surface layers. This scheme is unlikely as an explanation for the so-called WC or carbon-rich Wolf–Rayet stars. These objects are highly luminous, and are certainly young. Their unusual spectra can now be understood in terms of stellar evolution with mass loss (Maeder 1991, van der Hucht and Hidayat 1991). Hydrogen-rich layers are lost, and products of nuclear burning are revealed.

We have already mentioned the R CrB variables. The spectra of these stars at maximum light are those of supergiant F's (§13.2), apart from the unusual strength of features due to carbon – both molecular and atomic. Mass loss rather than mixing is the favored scenario for an explanation of these as well as the Wolf–Rayet stars.

10.3 Carbon, Oxygen, and Neon Burning

Significant divisions in stellar evolution come when $^{12}\mathrm{C} + ^{12}\mathrm{C}$ reactions begin. The most massive stars enter this burning phase quiescently, while stars of intermediate mass may be entirely disrupted by carbon ignition. Trimble (1982) gives the range 4–8$M_\odot$ as the "traditional risk zone" for this catastrophe. Stars (somewhat) less massive than this will fail to burn carbon at all, and will probably become white dwarfs after losing a sufficient amount of mass during their giant phases.

The favored reaction channels for the $^{24}\mathrm{Mg}^{*}$ compound nucleus formed from ^{12}C + ^{12}C are ^{20}Ne + ^{4}He, ^{23}Na + ^{1}H, and ^{23}Mg + n.

After carbon burning, and before consumption of ^{16}O, there is an episode in which ^{20}Ne is decomposed by (γ, α) reactions. The α's are consumed by the nuclei present, including ^{20}Ne, in this process, which is called neon burning. The next major reactions consume ^{16}O. The major product of oxygen burning is ^{28}Si. The original B^2FH paper discussed an *α-process* (cf. §7.4). The nuclides ^{24}Mg, ^{28}Si, ^{32}S, ^{36}Ar, ^{40}Ca, ^{44}Ca and ^{48}Ti were considered to have been formed, at least in part, by that process. This process has been subsumed in the current nomenclature, but one occasionally sees references to "α-elements" (§13.4.1). Usually, this means an element assigned by B^2FH *primarily* to their α-process.

In the following section, we shall discuss how certain models of supernovae are capable of yielding abundances that match those in the SAD with astonishing accuracy, including elements from carbon through those of the iron peak. Fowler's (1984) Nobel lecture gives a superb introduction to this topic along with many references (see also Trimble 1991). Our overall understanding of supernovae increased enormously as a result of observations and theoretical work associated with the outburst known as Supernova 1987A (§10.11).

10.4 Silicon Burning, Equilibrium, and Quasi-Equilibrium Processes

Silicon is consumed in a series of reactions that lead to ^{56}Ni, which is doubly magic, but β^+-unstable. It is usually assumed that these reactions occur in the course of a stellar explosion. If this explosion takes place on a time scale that is short with respect to β^+-decays, the initial proton-to-neutron ratio will remain nearly frozen into the stellar material. We may then expect a deficit of neutron-rich species with respect to strict equilibrium calculations, which assume that the proton-to-neutron ratio will be determined only by the temperature and density. In the former case, we speak of quasi-equilibrium burning of silicon, as opposed to the latter case of pure equilibrium.

In strict thermal equilibrium, the relative number densities of nuclear species are determined by the relations derived in §4.12 (see Equations 4.76 and 4.77). Let us consider, for example, the cosmochemically significant ratio of the radioactive ^{26}Al to the stable isotope ^{27}Al. If we use n to represent a free neutron, we have

$$^{27}\mathrm{Al} \rightarrow {}^{26}\mathrm{Al} + n. \tag{10.3}$$

This has the form of a simple dissociation, and obeys the relation 4.77 as long as we may ignore degeneracy (§§4.9 and 4.10) and assume the particle energies are non-relativistic. Then

$$\frac{N(^{26}\mathrm{Al})N(n)}{N(^{27}\mathrm{Al})} = \frac{Q_{\mathrm{int}}(^{26}\mathrm{Al})Q_{\mathrm{int}}(n)}{Q_{\mathrm{int}}(^{27}\mathrm{Al})} \cdot (2\pi\mu kT/h^2)^{3/2} \cdot \exp(-\mathrm{BE}/kT), \tag{10.4}$$

where BE stands for the binding energy of the neutron in ^{27}Al. It is given by

$$\mathrm{BE} = M(^{26}\mathrm{Al}) + M(n) - M(^{27}\mathrm{Al}). \tag{10.5}$$

The nuclear partition functions may be calculated by summing over the known energy levels so long as the thermal energy kT is small with respect to the binding energies of the nucleons, which is, roughly speaking, the nuclear analogue of the ionization energy for atoms. For higher thermal energies, there is a problem with convergence of the partition function that need not concern us here (see Fowler,

Table 10.1. *Lowest Spins and Energies of ^{26}Al and ^{27}Al*

^{26}Al		^{27}Al	
J	Energy(MeV)	J	Energy(MeV)
5	0.0	2.5	0.0
0	0.228 2	0.5	0.843 76
3	0.416 9	1.5	1.014 46
1	1.057 8	3.5	2.210
2	1.759	2.5	2.734
1	1.851		
4	2.068 7		
2	2.069 5		
1	2.072		

Engelbrecht, and Woosley 1978). The (internal) partition function for the neutron is 2, as it is for all particles with spin 1/2: $g = 2J + 1 = 2S + 1 = 2$. The addition of angular momenta, orbital and spin, is discussed further in Chapter 11 in connection with atomic spectra.

Table 10.1 gives the spins and energies of the lowest levels in ^{26}Al and ^{27}Al from Lederer and Shirley (1978) used in the calculation of the nuclear partition functions of these two nuclides.

At very low temperatures, the partition functions are well approximated by the statistical weights of the ground states, which are $2 \cdot 5 + 1 = 11$ for ^{26}Al and 6 for ^{27}Al. Workers in stellar interiors often use the symbol Tx to mean the temperature in units of 10^x K. Then, for ^{26}Al and ^{27}Al, at $T9 = 4.0$ the corresponding values for the partition functions are 13.8 and 6.4. At $T9 = 4.8$ they have increased to 14.7 and 6.7 respectively.

It is convenient to write 10.4 in logarithmic form:

$$\log\left[\frac{N\left(^{26}\text{Al}\right)}{N\left(^{27}\text{Al}\right)}\right] = -\log[N(n)] + 33.76 + 1.5 \cdot \log(T9) + \log\left(\frac{2Q_{\text{int}}\left(^{26}\text{Al}\right)}{Q_{\text{int}}\left(^{27}\text{Al}\right)}\right) - (0.4343 \cdot \text{BE})/kT9, \qquad (10.6)$$

where 0.4343 is $\log_{10}(e)$. If BE is in MeV, the appropriate value of Boltzmann's constant is 0.086 173 MeV per 10^9 K. The student should verify the numerical value of the constant on the right.

We have now developed the tools with which we may determine the equilibrium (or quasi-equilibrium) value of the ratio of ^{26}Al to ^{27}Al. The ratio depends on the value of two free parameters, $T9$ and $N(n)$. How may they be realistically fixed?

One approach might be to assume that the ^{26}Al is manufactured along with the rest of the elements between silicon and the iron peak in a nuclear explosion, the results of which produce the abundances of the nuclides that make up the SAD.

Table 10.2. *Some Best Match Quasi-equilibrium Conditions*

T_9	$\log[N(p)]$	$\log[N(n)]$	$\log(\rho_{cgs})$	$^{26}Al/^{27}Al$
4.0	27.89	19.72	7.5	0.13
4.2	28.29	20.43	8.0	0.17
4.4	28.66	21.09	8.5	0.21
4.6	29.00	21.68	9.0	0.25
4.8	29.31	22.23	9.0	0.30

We then ask for the values of $T9$ and $N(n)$ resulting in the most satisfactory fit, and use them in Equation 10.6. Table 10.2 is taken from Bodansky, Clayton, and Fowler (1968). They give a number of values of the free parameters of the quasi-equilibrium calculation leading to best fits with what were at the time thought to be optimum SAD abundances. *We caution the reader that the tabulated values, and indeed the quasi-equilibrium method itself, must be regarded as primarily of historical and heuristic interest.*

The final column lists the $^{26}Al/^{27}Al$ ratio resulting when the data from the first columns are used in Equation 10.6. These figures, in spite of our caveat, are entirely reasonable from the point of view of the discussion in §8.8, where we assumed 0.1 as a working value. Active researchers in this field have obtained a very wide variety of values for this ratio (see Figure 8.14). Zinner *et al.* (1991) report a pristine, or quasi-equilibrium value of 0.2 (cf. Table 10.2), in a SiC grain from the Murchison carbonaceous meteorite. We refer to the paper by Clayton and Leising (1987) for a review of the ^{26}Al problem.

The general quasi-equilibrium process results may be readily written down. We shall write abundances in terms of that for ^{28}Si, although any other nuclide could serve equally well. The relevant reaction involving ^{28}Si and $^{A}_{Z}El$ is

$$^{A}_{Z}\mathrm{El} = ^{28}\mathrm{Si} + \delta_p p + \delta_n n, \tag{10.7}$$

where $\delta_p = Z - 14$, the number of *additional* protons in $^{A}_{Z}El$, and δ_n is the number of additional neutrons. We regard this simply as a reaction of the general form 4.36, for which the equilibrium relations 4.76 apply. Recall that the total partition functions Q contain the volumes, and that the dimensionless, script $\mathcal{N}$'s become particle densities N (cm^{-3}) on division by these volumes. We have, then,

$$\frac{N\left(^{28}\mathrm{Si}\right) N_n^{\delta} N_p^{\delta}}{N\left(^{A}_{Z}El\right)} =$$

$$Q_{\mathrm{int,Si}} \cdot \left(2\pi M_{\mathrm{Si}} kT/h^2\right)^{3/2} \cdot \left(2^{2/3} \cdot 2\pi M_n kT/h^2\right)^{3\delta_n/2}$$

$$\times \left(2^{2/3} \cdot 2\pi M_p kT/h^2\right)^{3\delta_p/2}$$

$$\times Q_{\mathrm{int,El}}^{-1} \cdot \left(2\pi M_{\mathrm{El}} kT/h^2\right)^{-3/2} \cdot \exp(-B/kT), \tag{10.8}$$

where B is the difference in the masses of the quantities on the right- and left-hand sides of Equation 10.7. Note that Bodansky *et al.* included α-particles in their dissociation relation, but their results are the same as ours, so long as the α's are in equilibrium with their constituent protons and neutrons, as the authors assumed.

We can obtain the strict *equilibrium* process results, the e-process of B^2FH, from Equation 10.8 by requiring that the neutrons and protons are in equilibrium with one another:

$$n \rightarrow p + e^- + \bar{\nu}_e. \tag{10.9}$$

The relevant equilibrium relation is straightforward so long as classical statistics may be assumed for the electrons:

$$\frac{N(p)N(e)}{N(n)} = \frac{2 \cdot 2}{2} \cdot [2\pi\mu kT/h^2]^{3/2} \cdot \exp(-B/kT). \tag{10.10}$$

The quantity μ is now the reduced mass of the electron.

Realistically, classical statistics cannot be assumed for the electrons when the mass density has values of the order of those given in Table 10.2. Under these conditions, the electrons are said to be degenerate, a situation first encountered by astronomers in connection with the structure of white dwarf stars. In a qualitative way, one can say that electron degeneracy will drive the equilibrium relations toward those for quasi-equilibrium. The electrons are so densely packed that there is no room in phase space for the electron that appears during a reaction such as 10.9. The proton–neutron ratio is therefore essentially frozen, which is the same situation as in quasi-equilibrium.

We must now turn to a discussion of what might be termed the state of the art calculations that have superseded equilibrium and quasi-equilibrium methods. The modern methods all have in common explicit time dependencies of both the overall temperature and density and the species involved in the reaction networks. The general principles of such reaction networks are essentially the same as those already discussed in terms of the burning of lighter nuclear fuels. The main difference is that it is necessary to keep track of more species, and the time scales are of the order of seconds rather than eons, as in the case of the burning of hydrogen and helium.

Pioneering work by Truran, Cameron, and Gilbert (1966) with such networks was followed by studies of Arnett, Woosley, and others (see the references in Rolfs and Rodney 1988 and Trimble 1991).

There have been two approaches to the overall time dependencies of temperature and density. In one, a supernova explosion has been simulated. For example, Woosley, Arnett, and Clayton (1973), in calculations that superseded those of Bodansky *et al.* (1968), assumed that the density and temperature were given by

$$\rho(t) = \rho_i \cdot \exp(-t/\tau_{\mathrm{HD}}), \quad T9 = T9_i \cdot (\rho/\rho_i)^{\gamma-1}. \tag{10.11}$$

The *hydrodynamical* time scale τ_{HD} was taken to be $446\chi\rho_i^{1/2}$ in seconds, with ρ in gm/cm^3. The free parameters were an initial density, ρ_i, an initial temperature, $T9_i$, and an "arbitrary scaling parameter," χ. The redundant numerical factor, 446, appears for historical reasons. They were to be fit by comparisons with the SAD. The authors fixed γ at 4/3, and made a careful study of the temperature–density domains in which quasi-equilibrium was valid.

An important property of these calculations that is entirely missing in quasi-equilibrium is the question of *freeze-out*. As time increases in Equation 10.11, the physical conditions pass through domains in which quasi-equilibrium will be invalid. The results of the network calculation may then be significantly different from any results that might be simulated with the older technique. The problem is most serious for the trace nuclear species, as we might have anticipated from the fact that quasi-equilibrium fits the more abundant isotopes reasonably well.

Studies that use parameterizations such as Equation 10.11 have the disadvantage that there is only the most rudimentary input from the current theory of stellar structure and evolution. An early, very important attempt to correct this deficiency was the work of Arnett (1969), who used the equations of stellar structure to follow, with some simplifications, the detonation of a carbon–oxygen core of 1.37 solar masses in the simulation of a supernova event arising from the death of a star in the mass range from 4 to 9 solar masses. The nuclear reactions were followed by a time-dependent network integration.

The stellar core was processed essentially entirely to iron and neighboring iron-peak nuclei, and disrupted. There was no remnant core, which might represent a neutron star or pulsar. The conversion of the nuclear fuels to the iron group is explicable in terms of our quasi-equilibrium calculations. More specifically, we might have expected either all iron, or all silicon if the mean conditions were near those that we have investigated for silicon burning. A simple calculation shows that the smallest departures from the conditions shown in Table 10.2, which lead to approximately SAD abundances, will give either all silicon or all iron.

Let us apply a logarithmic form of Equation 10.8 to the case where ${}^{A}_{Z}El$ is ^{56}Ni, which decays into ^{56}Fe. Then $\delta_n = \delta_p = 14$, and if we assume the logarithmic partition function ratios are approximately constant, we have

$$\log\left(\frac{N[^{28}\mathrm{Si}]}{N[^{56}\mathrm{Ni}]}\right) = -14\cdot\log(N_p) - 14\cdot\log(N_n) + 42\cdot\log(T9) - \frac{0.4343\cdot B}{0.086\,17\cdot T9} + \text{constant}. \quad (10.12)$$

Clearly the ratio of ^{56}Ni to ^{28}Si depends on the 14th power of the proton and neutron densities. The sensitivity to the temperature is not well described by a simple power because of a competition between the temperature that appears in the exponential (Equation 10.8) and those that occur in the translational partition functions. Woosley *et al.* wrote a relation showing an *inverse 79.3th power dependence* of the remaining mass fraction of silicon on $T9$ (see their Equation (60)).

In Chapter 16 we shall take up the question of the heavy element abundances in galaxies in relation to their sources, insofar as we understand them. For the present, we remark that a galaxy of some 10^{11} solar masses, each with the SAD abundances, would contain about $1.4\cdot10^{8}$ solar masses of iron, since the mass fraction of iron in the SAD is about $1.4\cdot10^{-3}$. If the Galaxy were 14 billion years old, its iron content could be accounted for with 1 event per century that produced 1 solar mass of iron. This rate – one per century – is only "somewhat" lower than current estimates for the supernovae in our own Galaxy thought to produce iron.

In the most recent work on the synthesis of elements from silicon through the iron peak there has been a return to the approach taken by Arnett (1969) in which reaction networks have been closely incorporated with stellar structure models. The sustained efforts of S. E. Woosley and his collaborators have been joined by numerous workers of many nationalities. We note particularly the work of K. Nomoto and his coworkers. The literature on this topic is immense, and we mention only a few sources. We cite the review by Woosley and Weaver (1986), the article by Nomoto, Shigeyama, and Tsujimoto (1991) as well as the books edited by Petschek (1990) and Woosley (1991). Additional references are given by Trimble (1991), and in the *Transactions of the International Astronomical Union*, or IAU (e.g., McNally 1991). Articles on supernovae and nucleosynthesis will be cited in the reports of various IAU commissions, primarily 35 (Stellar Constitution), but also 27 (Variable Stars), 47 (Cosmology), and 48 (High Energy Astrophysics).

Two very different models have been used, corresponding to what the astronomer broadly calls Type I and Type II supernovae. Historically, these objects were differentiated by their *light curves* – that is, the plots of brightness as a function of time – and their locations within galaxies. Type I's occur in both elliptical and spiral galaxies, while Type II's are found only in the disks of spirals.

The modern classification scheme is based primarily on spectra (§10.11). Supernovae showing hydrogen lines in their spectra are by definition Type II. Others are Type I. We discuss the II's first.

The progenitors of Type II's are relatively massive stars, the favorite ranges being 8 or 9 solar masses and above. In such stars, the nuclear fuels all burn non-explosively, so that at the end of the life of the star there is an onion-like series of composition layers, each containing "later" nuclear fuels. The outermost shells are thus hydrogen with "cosmic" helium, followed by helium, carbon and oxygen, and silicon layers located successively deeper. The idea of stars with this kind of structure is rather old. Figure 10.2 shows two models taken from Hoyle's (1955) *Frontiers of Astronomy*. B^2FH, and Rolfs and Rodney (1988) give more technical descriptions of the expected behavior of such stars.

The production of nuclear energy by fusion ends with iron, which has the isotope with the maximum binding energy per nucleon. In heavier nuclei, the Coulomb repulsion and surface energies begin to play an increasingly important role relative to the strong nuclear binding forces. At the end of the valley of β-stability natural radioactivity occurs for all isotopes (trans-bismuth elements). These trans-iron nuclei are increasingly less tightly bound than iron.

The further contraction of a star with an iron core can thus produce no more (central) nuclear energy. Stars in this configuration, however, will continue to contract, and the core temperatures will rise. The energy source is now gravitation. Eventually, temperatures will be reached at which the iron nuclei will dissociate. This dissociation is a sink for energy, so the situation in which energy is provided by nuclear fusion reactions is reversed. Kinetic energy is now consumed by the disintegrating nuclear fragments. This prevents the central temperature of the star

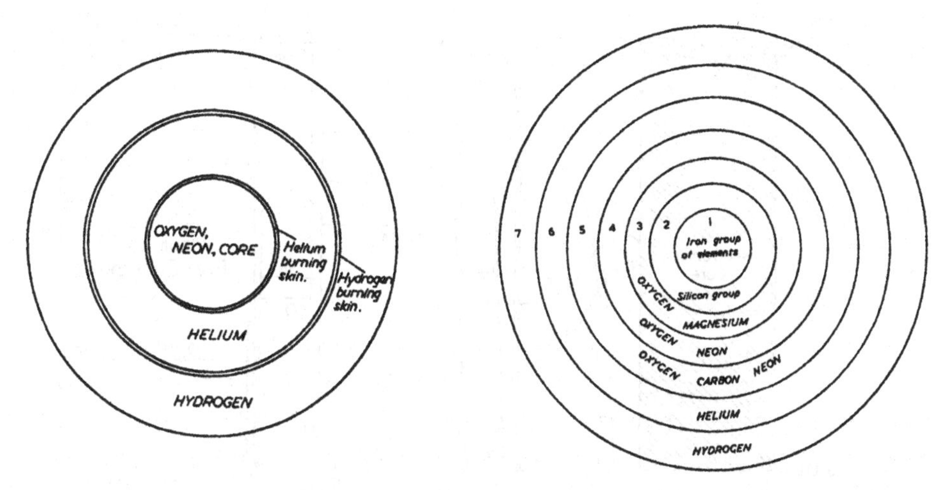

Figure 10.2. Early Onion-Shell Stellar Models.

from rising, and removes pressure support. The result is an implosion, which all logic argues must somehow reverse and cause the explosion seen as a supernova (of Type II).

We must leave to the references a discussion of the technical difficulties encountered by theoreticians who have attempted to reproduce the anticipated "bounce" of such models. In the calculations illustrated in Figure 10.3, the bounce was artificially induced, and, as nearly as this writer can discern, this treatment of the bounce is about the only "free" parameter of the calculations. As we know from our experience with the equilibrium process, the relative portion of the core that is blown off must be critical in determining the relative amounts of silicon and iron. If one is attempting to fit the SAD pattern, for example, one can vary this fraction expelled in such a way as to optimize the fit.

In Figure 10.3(a,b), we show mass fractions of a 25-solar-mass star, just prior to the induced explosion, and the resulting ejected masses (Woosley and Weaver 1982).

After the induced bounce, the inner stellar zones are processed by high temperatures and densities in much the same way as in the parameterized calculations discussed earlier. However, the outer layers, containing the lighter elements, are not significantly processed. It is clearly one of the more remarkable achievements of the theory of stellar structure and nucleosynthesis that the resulting blown-off material matches the SAD pattern rather well. It might appear that another kind of Holy Grail had been found, namely, an object that would account for the SAD.

As of this writing, it is not clear whether the remarkable coincidence shown in Figure 10.4 from the 1982 Woosley–Weaver paper may be attributed to the discovery of a unique model, or an indication that the SAD pattern represents some kind of most probable distribution, capable of being achieved in a variety of ways. Woosley and Weaver (1985) discuss similar fits using newer calculations for a 25-solar-mass model which is better than that shown in Figures 10.3 and 10.4. In the most recent

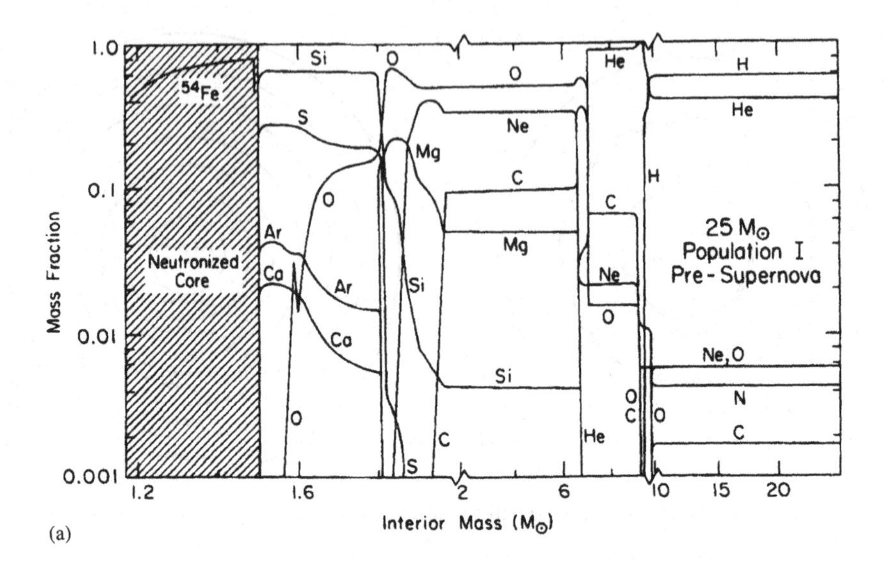

(a)

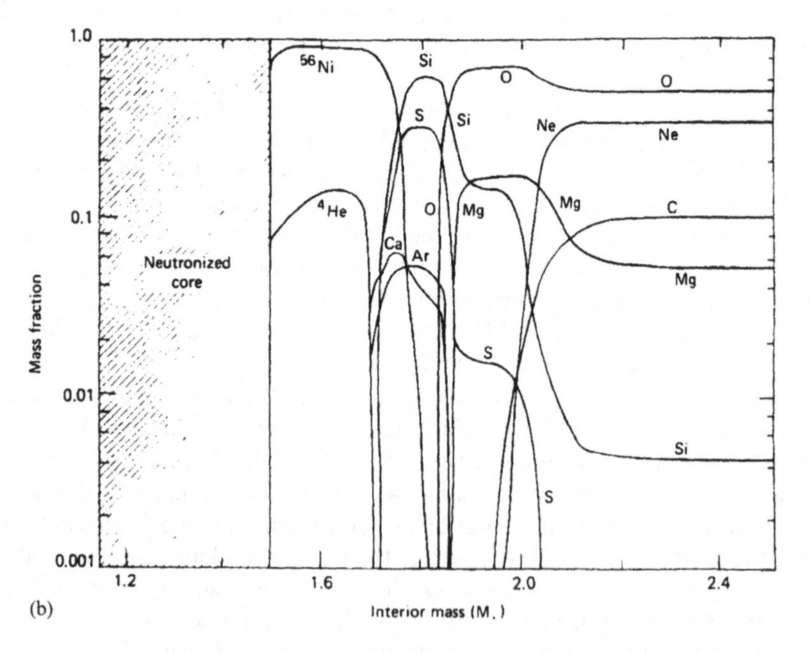

(b)

Figure 10.3. Supernovae Models of 25 Solar Masses. Woosley and Weaver (1982)

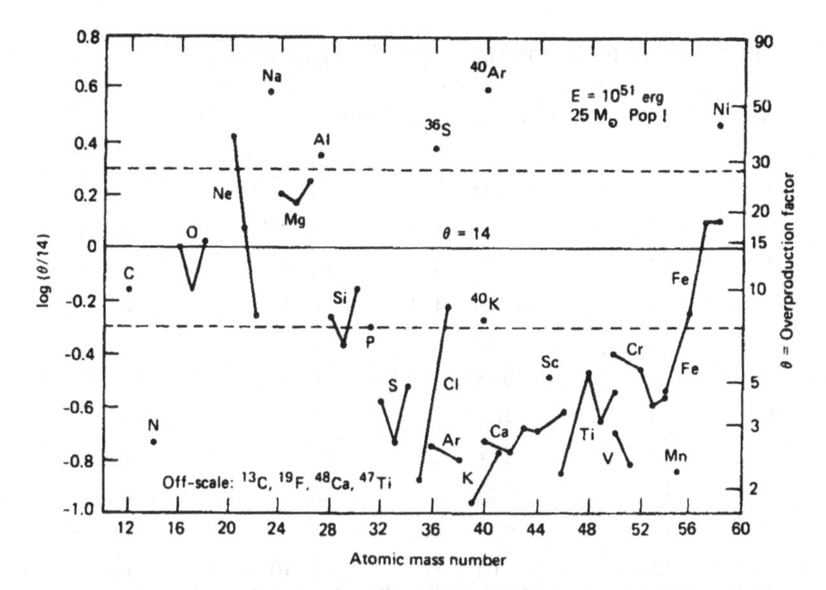

Figure 10.4. Comparison of SAD with Ejected Abundances from a 25-Solar-Mass Star. The vertical axis is the SAD abundance *relative to* the abundances produced by the supernova model, normalized to 14 for ^{16}O. Species with points above the horizontal, full line are *overproduced* by the model. Underproduction of the species from ^{32}S to ^{56}Fe was attributed by Woosley and Weaver (1982) to synthesis of these nuclides in higher-mass stars.

efforts to explain the SAD, the abundances are attributed to a *mixture* of supernovae of several types (see Nomoto, Shigeyama, and Tsujimoto 1991).

We must now discuss briefly the Type I supernovae models, which involve the ignition of a nuclear fuel after mass transfer from a companion star. The progenitors of these supernovae are therefore white dwarf stars in which the majority of the hydrogen either has been processed, or is processed *in situ* prior to explosion. The previously quiescent white dwarf configuration achieves instability by the transfer of mass from the companion. The ejected material is processed under temperatures and densities that lead to near-equilibrium abundance distributions.

The Type I models share with the carbon detonation model of Arnett the fact that the major product is ^{56}Ni which decays to ^{56}Fe. According to Nomoto *et al.* (1991), slightly more than half of the iron in the SAD came from Type I supernovae, the remaining fraction coming from Type II. The precise numbers here are uncertain. This question is discussed further by Timmes (1991).

10.5 The Problem of the Light Nuclei: Li, Be, and B

Before we turn to a discussion of the synthesis of elements beyond the iron peak, we must deal with the three nuclear "fragments" which are essentially bypassed in helium burning ($3^4\text{He}\rightarrow^{12}\text{C}$). Their abundances are strikingly low in the SAD

Table 10.3. *Isotopic Abundances of the Light Elements*

Element	A	% Abundance	Process	Abundance	
				ϵ(Cameron 1982)	ϵ(A&G 1989)
H	1	≈ 100		$2.7 \cdot 10^{10}$	$2.8 \cdot 10^{10}$
	2		U?	$4.4 \cdot 10^5$	$9.5 \cdot 10^5$
He	3		U?	$3.2 \cdot 10^5$	$3.9 \cdot 10^5$
	4	≈ 100	U,H	$1.8 \cdot 10^9$	$2.7 \cdot 10^9$
Li	6	7.5	x	4.4	4.3
	7	92.5	x,H,U	55.6	52.8
Be	9	100.	x	1.2	0.73
B	10	19.6	x	1.8	4.22
	11	80.4	x	7.2	17.0
C	12	98.9	He	$1.1 \cdot 10^7$	$1.0 \cdot 10^7$
	13	1.1	H	$1.2 \cdot 10^5$	$1.1 \cdot 10^5$

(Figure 1.1). Beryllium, by number, is some seven orders of magnitude less abundant than carbon. These deficiencies were duly noted by the incomparable Goldschmidt (Goldschmidt 1954), who had already discussed the problem in 1923! He suggested a common origin for these elements with the low abundance explained in terms of "some instability in certain nuclear processes." Suess and Urey (1956) wrote: "The low abundance of these three elements can easily be understood as a consequence of their instability at high stellar temperatures and their possible thermonuclear reactions with protons."

Because the combination lithium, beryllium, and boron will occur so frequently in our discussion, we shall use the abbreviation 'LiBeB' to refer to these elements collectively. Similarly, we shall write 'CNO' to mean carbon, nitrogen, and oxygen.

Temperatures of the order of a few million degrees are sufficient to destroy LiBeB by reactions with thermal protons. As may be seen from Table 9.3(a–d), these temperatures are exceeded in typical stellar interiors.

Table 10.3 gives the *isotopic* abundances of the stable light elements from Cameron (1982) and Anders and Grevesse (1989). The differences in the last two columns may arguably be taken as an indication of the uncertainties of the SAD determinations. The stable isotopes of carbon are included for comparison. The abundances are given with respect to that of silicon at 10^6.

The abundance contrast with ^{12}C and ^{13}C is apparent. Note the abundance of ^{7}Li, which is several times larger than that of ^{11}B, the next most abundant of the LiBeB isotopes. This is an indication of a history not shared with the other isotopes, to which we shall return. In Column 4, we give the nuclear processes primarily responsible for the production of the isotopes following Cameron (1982). 'U' means the species is thought to have emerged from the birth of the universe. 'H', and 'He' are for stellar hydrogen and helium burning respectively. We shall discuss these possibilities later in the chapter. B^2FH attributed the origin of the light nuclei of LiBeB and deuterium to an unknown or *x*-process. Let us momentarily set aside the cosmological synthesis, and consider the question of the abundance of those light nuclides in Table 10.3 attributable to other processes.

We first review briefly the geochemical behavior of LiBeB. The ionic radii of these elements, in valence states +1, +2, and +3 respectively, are 0.68, 0.35, and 0.23 angstroms. If these radii are compared with those from Figure 5.7, it can be seen that extensive substitution is not expected for these ions with any of the major minerals of the olivine, pyroxene, or feldspar families, with the exception of Li^+ for Mg^{2+}. Beryllium and boron have a proclivity to the formation of complex anions. Of these, the borate borax ($Na_2B_4O_5(OH)_4 \cdot 8H_2O$) was made famous by the TV classic *Death Valley Days*.

The net result of the incompatibility with major minerals is twofold. First, no one of these three light elements is a major element of the earth's crust, and second all three are somewhat enriched in the crust because their ionic incompatibility leads to concentration in residual melts. We note in Figure 5.1 that the deficiency of LiBeB in the earth's crust is only four orders of magnitude with respect to silicon compared with five powers of ten that separate the same elements in the SAD.

We can therefore expect extensive fractions of these light elements whenever (magmatic) geochemical processes come into play. Compilers of the SAD have relied on the CI meteorites to be unfractionated for the refractory elements, that is all elements other than H, C, O, N, and the noble gases including helium. The situation for LiBeB is not simple for an abundance compiler because LiBeB are variable (by factors of 2–5) among the chondrites. Indeed, there was no beryllium abundance determination for a C1 at the time of the compilation of Anders and Ebihara (1982). We refer to their discussion of eclectic selections of the available determinations.

The general rules that account for the odd–even effect are broken for these light nuclei (see Table 7.3). We have two stable isotopes of lithium and boron, and only one of the even-*Z* beryllium (see Table 10.3). ^{10}Be is nearly stable – its half-life is $1.6 \cdot 10^6$ years. The low beryllium abundance prevents this nuclide from being a significant heat source in the early solar system.

In Table 7.3, we can see that lithium, and to a lesser extent, beryllium and boron are depleted in the sun with respect to the SAD. This is reasonably interpreted as due to destruction by burning in the solar envelope.

The convection zone in the standard solar model (Bahcall 1989) extends down to a depth of 0.26 solar radii, at which the temperature is about 1.9×10^6 K. The density is about 0.15 gm/cm^3. New information of relevance has come from the domain of *helioseismology* (Deubner and Gough 1984, Foing 1991), which indicates the convection zone should be deeper than in earlier models.

At the base of the solar convection zone, the lifetime of ^{7}Li is much greater than the present age of the sun. Since ^{7}Li is the dominant isotope, we would therefore not expect the lithium abundance to have been modified by thermal protons during the sun's lifetime. Unfortunately, this conclusion cannot be firm because our knowledge of stellar hydrodynamics is seriously flawed. Indeed, Lebreton and Maeder (1987) concluded that turbulent mixing could lead to the observed depletion of lithium.

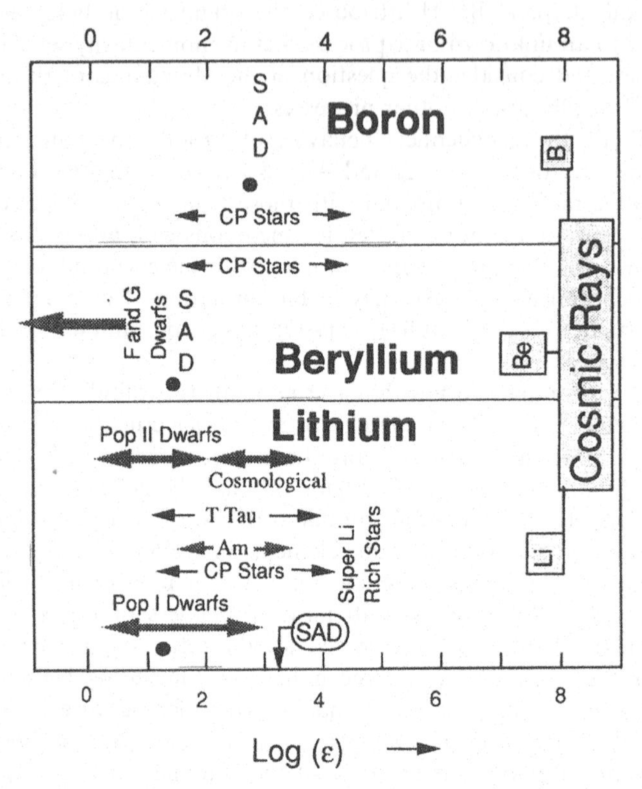

Figure 10.5. Abundances of Li, Be, and B.

The theory upon which this mixing was calculated, however, is only of order of magnitude or dimensional validity. It is called the *mixing length theory* (see Mihalas 1978).

The lithium abundances in main-sequence dwarf stars have long been interpreted as an indicator of age (§7.4). Stellar observations (Boesgaard and Steigman 1985) show that the *maximum* lithium abundance observed in stars is compatible with the logarithmic value 3.3 determined from meteorites. The values scatter over several orders of magnitude, with the lower abundances belonging to old stars.

Figure 10.5 shows the LiBeB abundances in a variety of stars as well as in the SAD. Abundances for the sun itself are shown by the filled circles (Anders and Grevesse 1989). Certain chemically peculiar (CP) stars of the upper main sequence have the ranges indicated (Sadakane, Jugaku, and Takada-Hidai 1985). Some of the lithium abundances for CP stars extend into the domain of the super Li-rich stars indicated separately (Faraggiana *et al.* 1986). We also show super Li-rich stars. The Am Li abundances are from Burkhart and Coupry (1991), the T Tauri's from Magazzù, Rebolo, and Pavlenko (1992). For more details, see Boesgaard and Steigman (1985), Rebolo (1990), and Chapter 12.

The *cosmic rays* are energetic particles first detected from earth and balloon experiments, but now investigated primarily from space satellites (Simpson 1983, Meyer 1989). A distinction is made between those energetic particles that originate in the solar atmosphere (*solar*) and those that come from our Galaxy (*galactic*). For the present, we shall be concerned exclusively with the latter.

The abundances of LiBeB in the cosmic rays are indicated approximately in Figure 10.5 by the squares enclosing the chemical symbols. The numerical values are from Table (2) of Simpson (1983). All three elements have similar abundances, although, interestingly, the odd-Z elements lithium and boron have *slightly higher abundances* than the even-Z beryllium. We see from Figure 10.5 that the abundances of LiBeB in the galactic cosmic rays are many orders of magnitude greater than in any of the other sources including the SAD. The reasons for this will be addressed in the following sections.

Note that typical energies range from hundreds to thousands of MeV, and that the distribution of energies per nucleon approximates a power law for energies over roughly 10 GeV (see Figure 10.6).

Some details of cosmic-ray abundances are shown in Figure 10.6. In (a), we see that the low points between the abundant CNO elements and the iron peak are filled in the cosmic-ray abundances. This filling in is readily understood in terms of spallation reactions of the more abundant cosmic-ray species with interstellar hydrogen (see below). Energy spectra are shown in (b). At low energies these spectra are strongly influenced by solar activity. Several epochs are shown for the various species, to illustrate the magnitude of this effect.

10.6 Cosmic Rays and Abundances of LiBeB

It is usually assumed that the galactic cosmic rays are confined by interstellar magnetic fields whose magnitude is of the order of $3 \cdot 10^{-6}$ gauss. Even though this is a very small field, a straightforward calculation shows that typical cosmic rays will spiral around the field lines with radii that are *much* smaller than the thickness of the galactic plane. These *Larmor* radii are given by the relation $\gamma m v_{\perp} c/(qB)$, where $v_{\perp}$ is the velocity component perpendicular to the field B, q is the charge, and γ is the relativistic factor (see Jackson 1975). The velocity of light, c, appears when *Gaussian cgs units* are used.

The relatively abundant CNO nuclei in the cosmic rays "see" the interstellar hydrogen as highly energetic bombarding particles. How many collisions per second are these nuclei expected to experience? In order to simplify this problem, we shall work with "mean" values, shamelessly averaging over the energy spectrum of particles and enormous fluctuations in the density of the interstellar medium.

Let us work with a typical particle density of 1 hydrogen atom per cm^3, and a "typical" spallation cross section of $<\sigma>= 20$ millibarns (20×10^{-27} cm^2). We shall assume that the energy threshold for the reactions is 20 MeV, so we approximate all of the curves in Figure 10.7 by a step function that is zero for energies below 20 MeV, and equal to 20 millibarns for all higher energies. At the threshold of 20 MeV the relative proton velocity is about 0.2c so we shall not be far wrong if we assume the relative velocity is $c = 3 \times 10^{10}$ cm/sec throughout. The *collision frequency* ν_{coll},

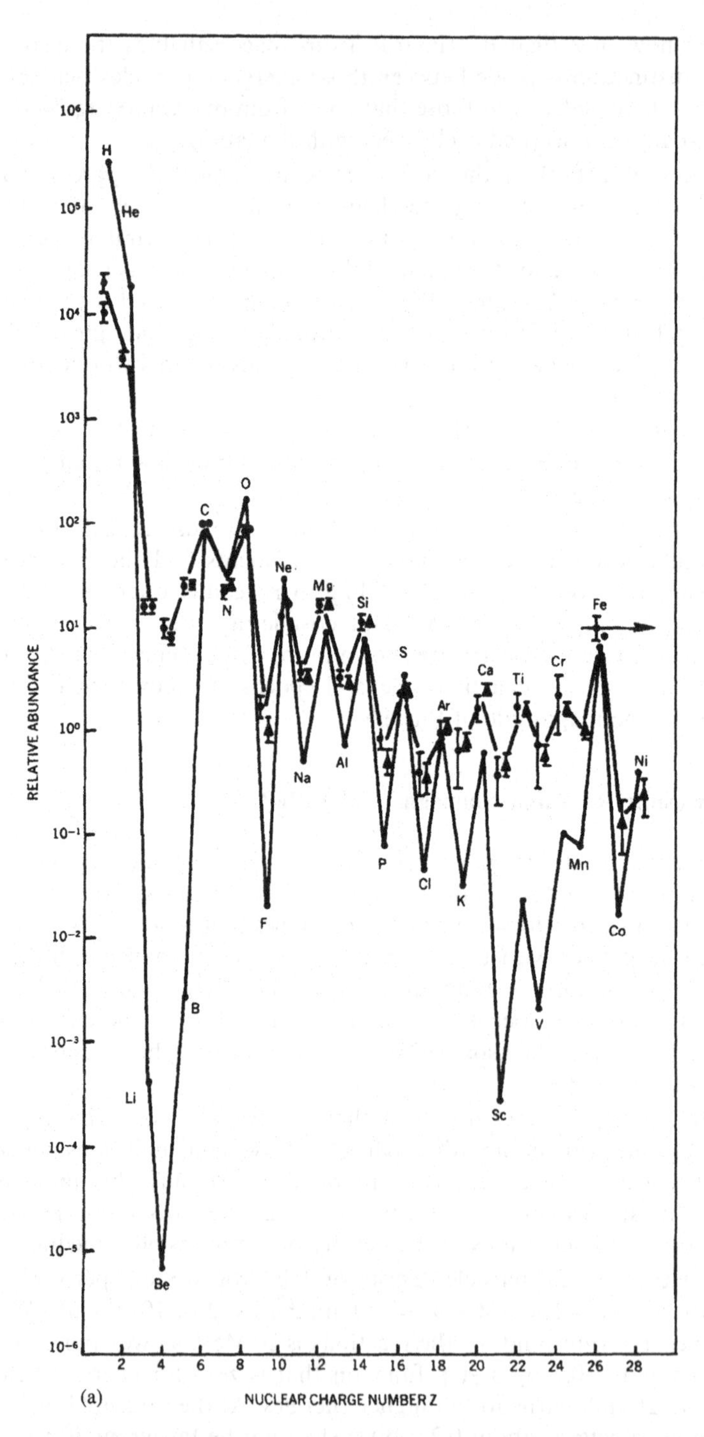

Figure 10.6. (a) Abundances in Galactic Cosmic Rays and the SAD (points without error bars),

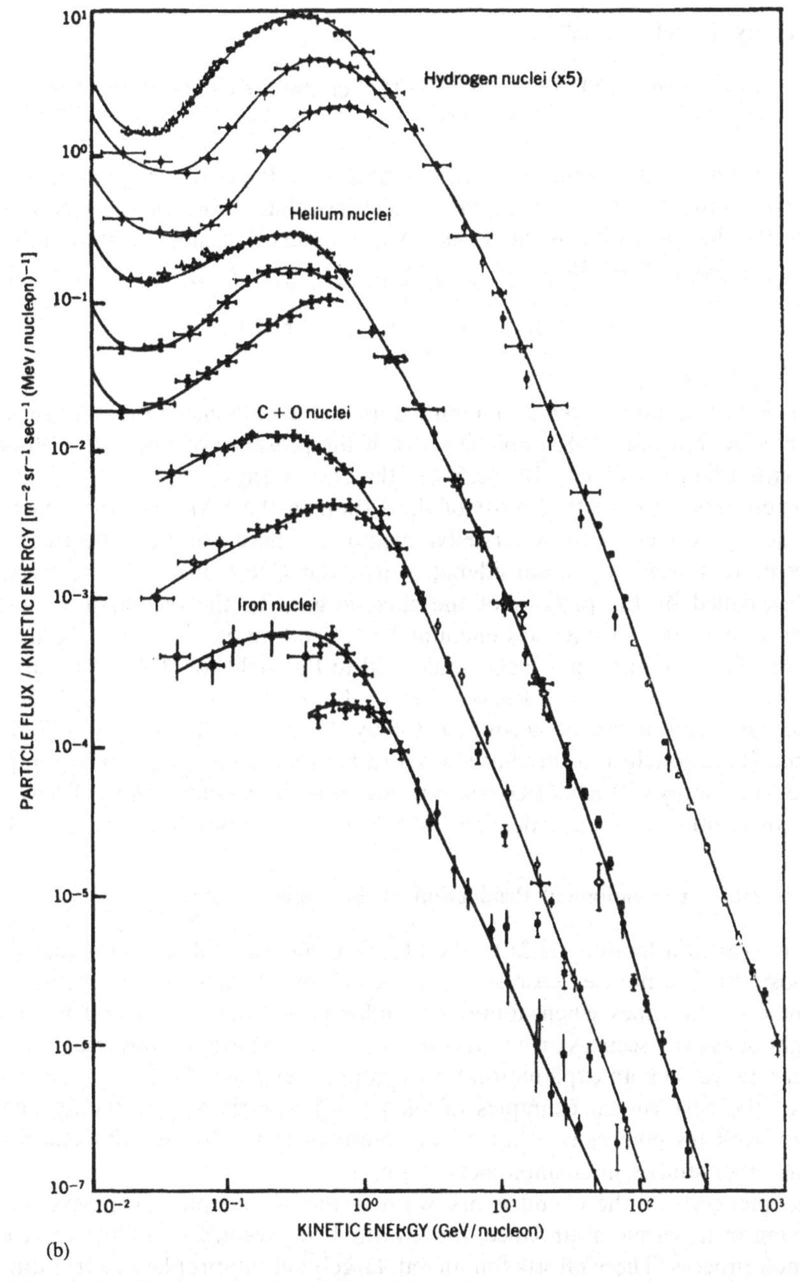

Figure 10.6. (b) Galactic Cosmic-Ray Fluxes For Various Species. Part (a) is from J. A. Simpson (1983). Reproduced with permission from *Annual Reviews of Nuclear and Particle Science*, Vol. 33, ©Annual Reviews, Inc., 1983. Open and closed circles are for energy ranges 1000–2000 and 70–280 MeV/nucleon, respectively. SAD abundances are shown as diamonds. Part (b) is from Meyer, Ramaty, and Webber (1974), courtesy of the American Institute of Physics. Division of the curves for low energies results from different sampling epochs.

is given by the relation (§9.5)

$$\nu_{\text{coll}} = 1\ \text{cm}^{-3} \cdot 20 \cdot 10^{-27}\ \text{cm}^2 \cdot 3 \cdot 10^{10}\ \text{cm/sec} = 6 \cdot 10^{-16}\ \text{sec}^{-1}. \tag{10.13}$$

We see from Figure 10.6(a) that the abundances in the cosmic rays of the fragments LiBeB are crudely one order of magnitude less than those of the CNO "source" nuclei. We thus ask, what is the necessary time interval Δt necessary to multiply ν_{coll} in order to get 0.1? We have

$$\Delta t \approx 0.1/(6 \cdot 10^{-16}\ \text{sec}^{-1}) = 1.67 \cdot 10^{14}\ \text{sec}, \tag{10.14}$$

or $\Delta t \approx 5.3 \cdot 10^6$ years.

This is of the same order of magnitude as the "confinement times" discussed, for example, by Simpson (1983) on the basis of the presence of unstable isotopes (e.g., ^{10}Be with a half-life of $1.5 \cdot 10^6$ years) in the cosmic rays.

It is entirely possible that most of the LiBeB in the SAD was manufactured by spallation processes in the interstellar medium. There are basically two kinds of processes that need to be considered. First, the CNO nuclei of the cosmic rays will be spalled by the protons of the galactic gas. In the second mechanism, the fast protons of the cosmic rays encounter the "stationary" CNO of the interstellar medium. The spallation products would add to the yield for the first mechanism.

According to Audoze and Reeves (1982), about 30% of the current LiBeB in the interstellar medium can be accounted for by the first mechanism, and 70% by the second. These conclusions are based on a rather elaborate model, and extrapolation of the low-energy (20 MeV) cosmic-ray fluxes to interstellar space. Reeves (1991) gives some more recent and detailed estimates for the abundances of LiBeB.

10.7 Cosmological Production of the Light Elements

We have called attention (§1.2) to the old idea that virtually all stars had the same composition. The phrase "cosmic abundances," for which we have substituted SAD, is a relic of the times when abundance differences among stars and nebulae were thought of as excrescences on a universal law. If one believes in universal abundances, it is natural to seek an explanation for them in some kind of cosmological synthesis. One of the best-known examples of such work was the α–β–γ theory of George Gamow and his coworkers. Alpher and Herman (1972) review this early work as well as other studies in cosmological synthesis.

The pioneers in these endeavors wanted the whole pie, as people often do. Believing in a cosmic abundance, they wanted to produce *all* of the elements in a common process. These efforts foundered, largely on the problem of the difficulty of bridging the gaps at mass numbers 5 and 8, for which there are no stable nuclides, and on the problem of getting a high enough density of neutrons to accomplish the task. There may have been ways around some of these difficulties, but in the 1960s, the enormous success of stellar nucleosynthesis largely overshadowed its cosmological counterpart. B^{2}FH wrote that "It seems probable that the elements all evolved from hydrogen, since the proton is stable while the neutron is not." It should

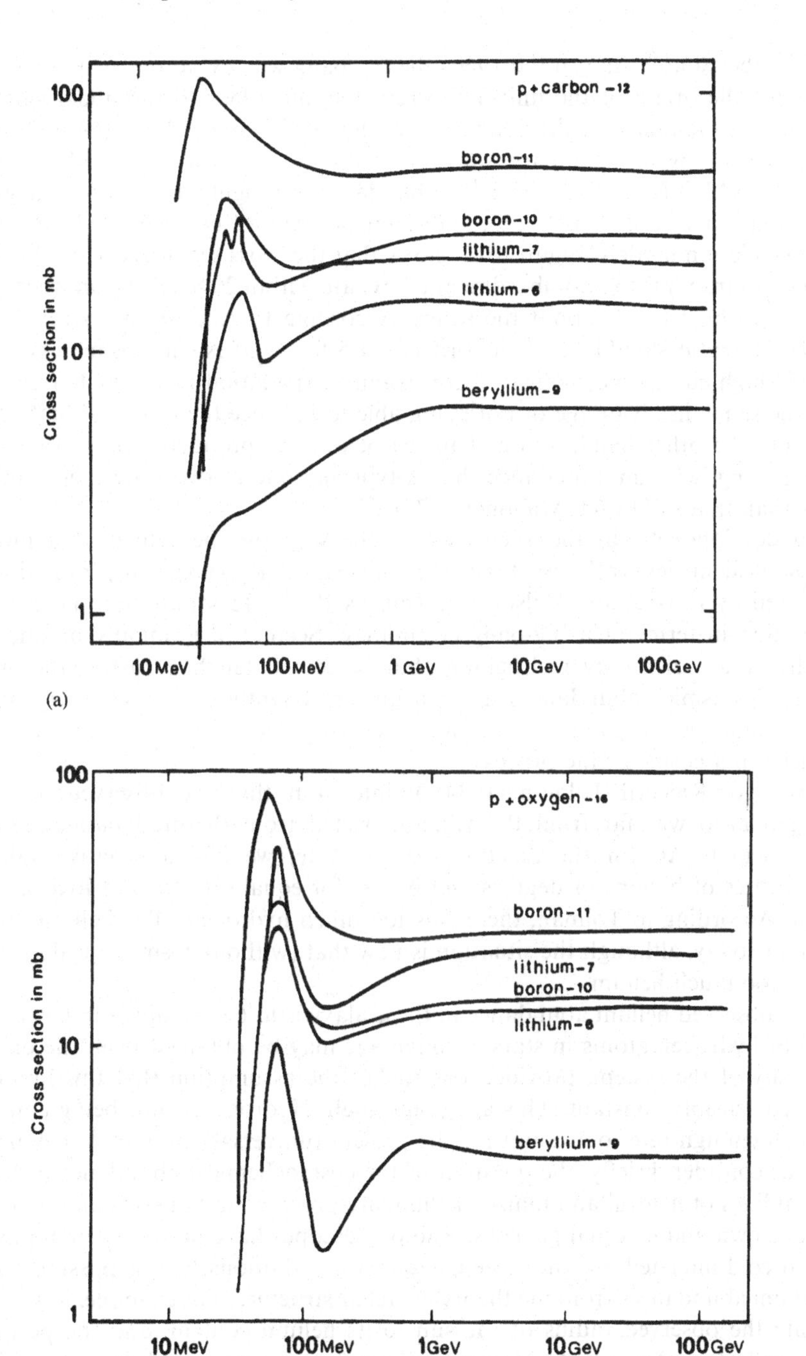

Figure 10.7. Spallation Cross Sections for Production of Light Nuclei: (a) Proton + ^{12}C, (b) Proton + ^{16}O. Courtesy J. Audoze and H. Reeves.

be remembered also, that in the 1960s the big bang was by no means the consensus theory for the origin of the universe. From the late 1950s to the mid-1960s many astronomers thought that the Galaxy was originally composed of pure hydrogen.

Another strong argument for some cosmological synthesis is the abundance of ^{7}Li in the SAD (Table 10.3). The ^{7}Li abundance is roughly an order of magnitude higher than that of ^{6}Li. Such a large discrepancy cannot be accounted for in terms of the spallation models. Figure 10.7 shows that the cross sections for production of ^{6}Li and ^{7}Li from proton bombardment of ^{12}C are within 20% of one another. Yields for protons on ^{16}O are about the same. According to Audoze and Reeves (1982) the ^{7}Li/^{6}Li ratio should be of the order of 1.5 for synthesis in cosmic rays. Other possible high-energy mechanisms – for example, spallation in stellar flares – suffer from the same disadvantage of not being able to produce the observed high ^{7}Li/^{6}Li ratio. On the other hand, as we shall see below, ^{7}Li production is expected from the Big Bang, with an abundance that is typically one to two orders of magnitude larger than that of ^{6}Li (cf. Wagoner 1973).

Two developments in the 1960's paved the way for the return of a modified cosmological nucleosynthesis. First, the universal 3 K background radiation was discovered by Penzias and Wilson (see Penzias 1972). This radiation has a natural explanation in terms of a big-bang cosmology. Second, it gradually became clear that the notion of a universal helium abundance was tenable. Just as the original notion of "cosmic" abundances made a universal synthesis attractive, a universal helium abundance could be taken to indicate that at least the helium may have been produced in a cosmos-wide process.

Long ago, Richard Tolman (1934) pointed out that the observed helium-to-hydrogen ratio was far from the value expected from thermodynamics. Indeed, if we calculate ΔG for the reaction 4H $\rightarrow$ ^{4}He, we find a negative value for temperatures of billions of degrees and below, for equal pressures of hydrogen and helium. According to Tolman, there was too much hydrogen! The helium problem is with us today, although the situation is now that (without cosmological synthesis) there is too much helium.

If the observed helium abundance in the Galaxy is to be explained in terms of the fusion of hydrogen atoms in stars, a rough age may be obtained from the observed luminosity of the system, provided one makes the assumption that this luminosity remained sensibly constant. This age is too high. Hydrogen is not being consumed at a high enough rate to account for the present (universal) amount of helium.

Let us consider, briefly, the question of the cosmochemical abundance of helium. Spectral lines of neutral and ionized helium are observed in hot stars and nebulae in both our own and external galaxies. Radio telescopes have detected helium in both hot and cold interstellar clouds (see Chapter 13). Additionally, it is possible to infer the helium abundances from the theory of stellar structure. For example, it is possible to relate the observed radius of the sun to its helium content, and the periods of certain pulsating stars can be shown to depend on the helium abundances. Finally, we note that the helium abundances in cosmic rays and lower-energy particles from the sun have been measured in space experiments.

None of the methods of determining helium abundances can be accepted uncritically as giving a measure of "the" cosmic abundance. Indeed, there are some objects where the helium abundances are palpably non-cosmic, that is to say, they differ greatly from the average. However, it is possible to make *ad hoc* but nevertheless plausible arguments that these cases are not relevant to the cosmic helium abundance. In some stars, for example, anomalous helium abundances may be explained in terms of fractionation in the atmospheres of stars whose bulk He/H ratios are cosmic (see Chapter 12).

Pagel (1991) discusses abundances of LiBeB in the context of cosmological nucleosynthesis. At least since the work of Kunth and Sargent (1983), external galaxies have been preferred by many as indicators of the primordial helium abundance. High-excitation systems, that is, systems with high ratios of [OIII] to [OII] lines, offer certain analytical advantages. We will give some details in §15.4.

We must postpone a discussion of abundance methods, their results and uncertainties, to later chapters. For the present, we assume that there is reasonable evidence for a primordial helium abundance which may be explained within some cosmological context.

Models of the early universe are currently a subject of active investigation and considerable uncertainty, but useful insight may be obtained from a Newtonian model. If we apply Newton's laws to an elementary volume of mass m at the surface of a (constant) spherical mass M_r of radius r, we have

$$m\ddot{r} = -GmM_r/r^2. \tag{10.15}$$

The quantity $\dot{r}$ is readily seen to be an integrating factor, since $\dot{r}\ddot{r}$ is the time derivative of $\dot{r}^2/2$. Thus

$$(1/2)\dot{r}^2 = GM_r/r + \text{constant}. \tag{10.16}$$

Whatever the value of the constant, it will become increasingly less important as r decreases, so we drop it from further consideration. Equation 10.16 may be integrated once again by elementary means. We obtain

$$\frac{2}{3}r^{3/2} = \sqrt{GM_r} \cdot t, \tag{10.17}$$

where a possible constant of integration has been set equal to zero under the assumption that $r = 0$ at $t = 0$.

We may now use $M_r = 4\pi r^3\rho/3$ to obtain

$$\rho = \frac{\text{constant}}{Gt^2}. \tag{10.18}$$

This relation has the same functional form as that used by cosmologists to describe the early universe. The constant, however, is not correctly given by our Newtonian development.

In cosmology, it is customary to eliminate the length variable, r, and introduce a dimensionless scale factor, defined in terms of the length at some standard epoch at which r is equal to r_0, say. Thus $r = R(t)r_0$. We must make an additional assumption to use Equation 10.18 to describe events, since only one state variable, the density, is specified as a function of time. A natural assumption is to assume that the changes are adiabatic, that is, no external heat is supplied to any volume under consideration. If the only relevant work done by the system is P–V work, the first law of thermodynamics, Equation 4.7, then gives $dE + PdV = 0$, a condition that can be used in addition to the equation of state to eliminate a variable so that a single state variable (plus number density or concentration variables) is sufficient to describe the system.

The pressure of black body radiation is one third its energy density, usually written

$$u = \frac{E_{\text{Rad}}}{V} = a \cdot T^4, \tag{10.19}$$

where a is the *radiation constant* equal to $7.57 \cdot 10^{-15}$ erg $\cdot$ cm^{-3} $\cdot$ deg^{-4}. For an adiabatic change, then, we have

$$dE_{\text{Rad}} = -\frac{1}{3} \cdot \frac{E_{\text{Rad}}}{V} \cdot dV, \tag{10.20}$$

which may be integrated to give

$$\frac{E_{\text{Rad}}}{V} = \text{constant} \cdot \frac{1}{V^{4/3}}. \tag{10.21}$$

Since $V \propto r^3$, Equation 10.21 gives the result that the energy density of radiation depends inversely on the fourth power of the length or the scale factor R. On the other hand, the density of non-relativistic matter is simply equal to particle number density multiplied by mc^2. The particle number density clearly depends on the inverse *cube* of the scale factor. It is therefore clear that for sufficiently early epochs, or equivalently, for sufficiently small scale factors, the universe will be dominated by radiation. This is in fact the situation at the time cosmological synthesis of the light elements is thought to take place. The density that appears in Equation 10.18 is therefore the density of radiation and relativistic particles.

The density of radiation (and neutrinos) is given by Equation 10.18 as a function of time, but how are we to determine the number densities of protons and neutrons with which to begin the calculations of nucleosynthesis?

One way to determine the density of matter would be to measure the present number density and form the ratio with the present number density of photons. It is straightforward, if not immediately obvious, that the latter will scale inversely as R^3, just as for matter. We already know that $u = E_{\text{Rad}}/V$ depends on R^{-4}. The number density of photons may be determined from the Planck formula for energy density of black body photons. The ratio $u_\nu/h\nu$ gives the *number* density of photons per unit frequency interval. Thus, the total number density of photons, n_{p}, is

$$n_{\text{p}} = \int_0^\infty \frac{1}{h\nu} \frac{8\pi h\nu^3}{c^3} \cdot \frac{d\nu}{\exp(h\nu/kT) - 1} = (k/hc)^3 \cdot 16\pi\zeta(3)T^3. \tag{10.22}$$

The integral is not elementary but can be found in a good table (e.g., Dwight 1961, Abramowitz and Stegun 1964, see their formula 27.1.3). Here, $\zeta(3)$ is the Riemann zeta function with argument 3; $\zeta(3) \approx 1.202\,057$.

We need only the fact that $n_p \propto T^3$. Clearly

$$n_p \propto \frac{u}{u^{1/4}} \propto R^{-3}. \tag{10.23}$$

The gurus of this field speak of the *entropy-to-baryon* ratio, meaning thereby (essentially) the entropy of radiation per nucleon (proton or neutron). Thus, a determination of the present photon-to-baryon ratio will enable us to calculate the matter density during the early universe, and allow us to make calculations relative to nucleosynthesis.

We leave to the student the straightforward exercise to show that the entropy of radiation depends linearly on the number density of photons as derived above, and therefore on T^3. Indeed, the entropy *per unit volume*

$$\frac{S}{V} = \frac{4}{3} \cdot aT^3. \tag{10.24}$$

Baryons are defined to be protons and neutrons and those heavier fermions (Chapter 4) that (eventually) decay into protons. Therefore, to a high approximation, the (radiation) entropy per baryon is the same as the entropy per nucleon.

It is possible only in principle to determine the present cosmic nucleon density from observations. Enormous uncertainties are currently extant as a result of apparent departures from the laws of gravitation that are seen in observations of galaxies and aggregates of galaxies (§16.10). While the question is still open, the consensus opinion is that the laws of gravitation are probably not violated, and the apparent departures are caused by the presence of dark matter – also referred to as missing mass. We shall return to this question in Chapter 16. For the present, we note that workers in this field have chosen to invert the problem, and let the entropy per baryon ratio be a free parameter in calculations of big-bang nucleosynthesis.

Once the entropy or photon number density per baryon is specified, a more accurate version of our Equation 10.18 allows us to predict the density of particles as a function of time. During cosmic nucleosynthesis, the universe is still dominated by radiation, that is, the energy density of radiation dominates that of matter. The universe expands, essentially as an adiabatic photon gas, with the added complication of the presence of neutrinos which can modify the thermodynamics of the expansion (see Figure 10.9 below).

The calculations of nucleosynthesis during these early epochs are similar to those for that occurring in the interiors of lower main-sequence stars which burn hydrogen on the proton–proton chains. A major difference is that the calculations have a time-dependent particle density that is dictated by an analogue of Equation 10.18 and the assumed entropy per baryon. There are also free neutrons present that can react directly with protons to form deuterium. In the early epochs, electron–positron pairs can decay to produce radiation. Pioneering calculations were presented by Wagoner, Fowler, and Hoyle (1967). Figure 10.8 is taken from their paper. Nuclear species included in their calculations are shown as boxes, and reactions are indicated by arrows. Weak interactions (β-decays) are shown as dashed arrows.

It is customary to present the results of such calculations as a function of the free parameter, which can be either the entropy-per-baryon or the baryon-to-photon

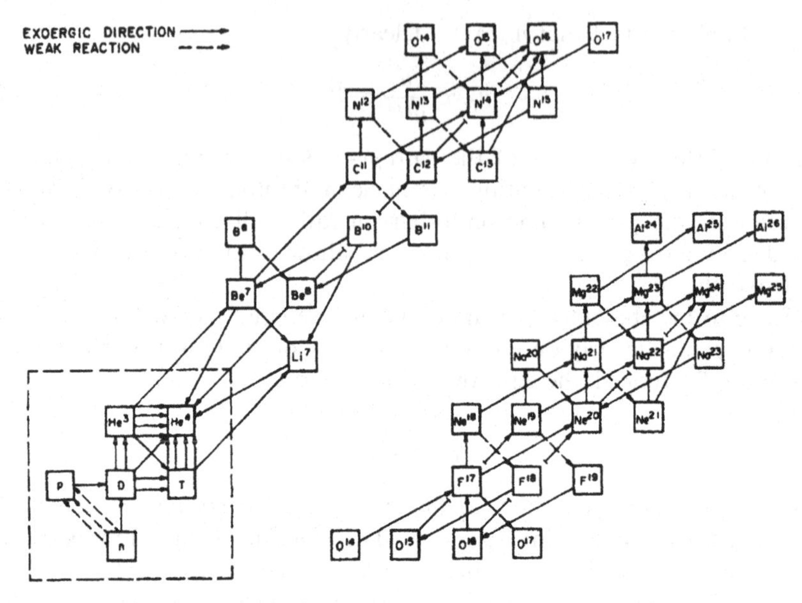

Figure 10.8. Reaction Network for Cosmic Nucleosynthesis (Wagoner, Fowler, and Hoyle 1967).

ratio (Figure 10.9). Traditionally, the symbol η is used for the baryon-to-photon ratio. Since this is a small number, $\eta_{10} = 10^{10}\eta$ is often used. The primordial helium abundance, Y_{p}, is represented as the *mass fraction* of helium. If hydrogen and helium are the dominant species, then

$$Y_{\mathrm{p}} = \frac{\epsilon_{\mathrm{H}} \cdot 1.008 + \epsilon_{\mathrm{He}} \cdot 4.003}{1.008 + 4.003}, \tag{10.25}$$

where the ϵ's are the abundances by number. The SAD values of the hydrogen mass fraction X, and helium mass fraction Y, according to Anders and Grevesse (1989), are 0.708 and 0.273 respectively.

Figure 10.9 shows predictions of several species at the end of cosmological nucleosynthesis. We are grateful to Lawrence Kwano for permission to use the code NUC123 which generated the plotted values (cf. Kwano, Schramm, and Steigman 1988). The results are generally similar to those produced by Wagoner (1973) or Krauss and Romanelli (1990).

Boesgaard and Steigman (1985) gave the following approximate relation for the primordial helium mass fraction as a function of the uncertain parameter ν, the number of kinds of neutrinos N_ν, and (interestingly) the neutron half-life τ_n (in minutes), which until recently was uncertain:

$$Y_{\mathrm{p}} = 0.230 + 0.011 \cdot \ln(\eta_{10}) + 0.013(N_\nu - 3) + 0.014(\tau_n - 10.6). \tag{10.26}$$

Uncertainties in the possible number of kinds of neutrinos as well as the neutron half-life have been greatly reduced since the Boesgaard–Steigman review (see the

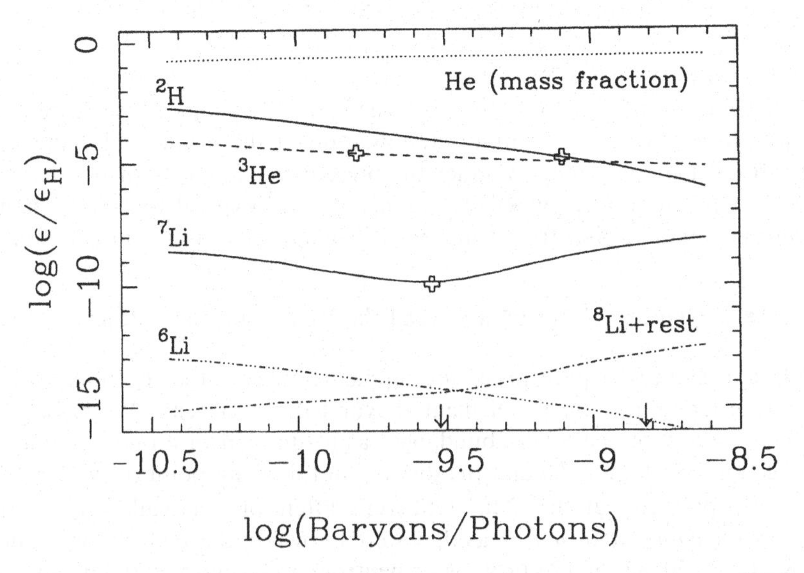

Figure 10.9. Light Nuclide Production as a Function of the Uncertain Photon-to-Baryon Ratio. The crosses are selected observational points from Spite *et al.* (1987) for Li, Bania, Rood, and Wilson (1987) for ^{3}He, and Spitzer (1982) for ^{2}H. These are representative observational points, but are not critically evaluated from the point of view of cosmology (see text). The arrows indicate the range of the baryon-to-photon ratio according to Reeves (1991), whose "subjective feeling" probably gives as good an estimate as is currently available.

references in Trimble 1991, her §3.2). This leaves only η to determine the primordial helium. Stars are made of baryons, and we know something about stars. Galaxies are made of stars and gas, and we think we know something about the latter (Chapters 14 and 15). We observe galaxies, and know something of their distances, so it would seem that we might estimate the number density of baryons in the universe. Tammann (1982), for example, gave $3 \cdot 10^{-7}$ cm^{-3} for the baryon density, N_B. The number density of photons is given directly by 10.22, and if we assume the cosmic background radiation is the dominant source of photons, then $T = 2.736$ (cf. Silk 1991), and we obtain $\eta = 7.2 \cdot 10^{-10}$.

If we use $\tau_n = 10.4$ and $N_\nu = 3$ in Equation 10.26, we find $Y_p = 0.249$. The Kwano code gives $Y_p = 0.245$ at the center of Reeves's "subjective-feeling" range for η. These numbers are surprisingly close to the value 0.23 for which Pagel (1991) makes a good case. However, the determination of N_B involves not only the distance scale of the universe (the Hubble constant H_0), but the role of dark matter (§16.10) and large-scale density fluctuations (§16.1).

It is also very difficult to know how to interpret the observed values of the deuterium, ^{3}He, and lithium abundances. Not only are there formidable observational uncertainties, but all three species may be destroyed in stars. Enormous fractionations occur for deuterium (§14.2, Zinner 1988). More authoritative selections of observational points than those shown in Figure 10.9 may be found in a number of papers (e.g., Pagel 1991, Reeves 1991).

While the uncertainties are considerable, the stakes are high. If the current value of η can be fixed, let us say, from a determination of the primordial deuterium or lithium ratio, it may determine whether the present mass density of the universe is sufficient for closure, that is, whether the universe will expand forever, or if it will reach some maximum radius from which it will collapse once more into a singularity.

10.8 Synthesis of Nuclides Beyond the Iron Peak: the *s*-Process

The SAD abundances require processes supplementary to the reactions discussed thus far in order to account for elements beyond the iron peak. We have already discussed the graph of the SAD abundances as a function of atomic number A in this connection (see Figure 7.6 and the discussion following point (6) within §7.4).

In §9.3, we pointed out that Maria Mayer's attention was called to the magic numbers by the *cosmic abundances*, as they were known in the 1940's. Those nuclides with 2, 8, 20, 28, 50, 82, or 126 protons or neutrons were more abundant than their neighbors, and this fact led Mayer to the shell theory of the atomic nucleus. This development led to what was called the *s*-process by B^{2}FH.

Two general schemes have been proposed for the buildup of elements beyond iron. One of these involves the addition of neutrons to *seed* nucleons, while the other, discussed by Tsuruta and Cameron (1965), represented an extension of nuclear statistical equilibrium to density domains much higher than those considered in §10.4.

If the buildup of the heavy elements could be attributed to neutron addition, the relationship between abundance and neutron capture cross section was surely relevant. All other things being equal, those species with small cross sections would be less likely to be destroyed by neutron addition, and their abundances would accumulate. In order for this scheme to be studied, the neutron synthesis must naturally involve *those* nuclei for which the cross sections were measured. These particular nuclei lay mostly along the valley of β-stability.

One could imagine very rapid neutron addition, which would lead to species far from the valley of β-stability. Such rapid addition, called the *r*-process, will be discussed in the following section, along with equilibrium at very high densities. These are illustrated in Figure 10.10 from Käppeler, Beer, and Wisshak (1989).

The figure shows a portion of a nuclide chart (Z vs. N). The heavy line shows the path followed by the *s*-process, which leads through the valley of β-stability. This path can branch (cf. Figure 9.9). The *r*-process (see below) populates a region well below the valley of β-stability, and the neutron-rich species β-decay. Nuclides labeled 'r' cannot be reached by the *s*-process, while those labeled 's' are shielded from the *r*-process by other stable nuclei. The inset is a schematic abundance curve showing the features described in §7.4.

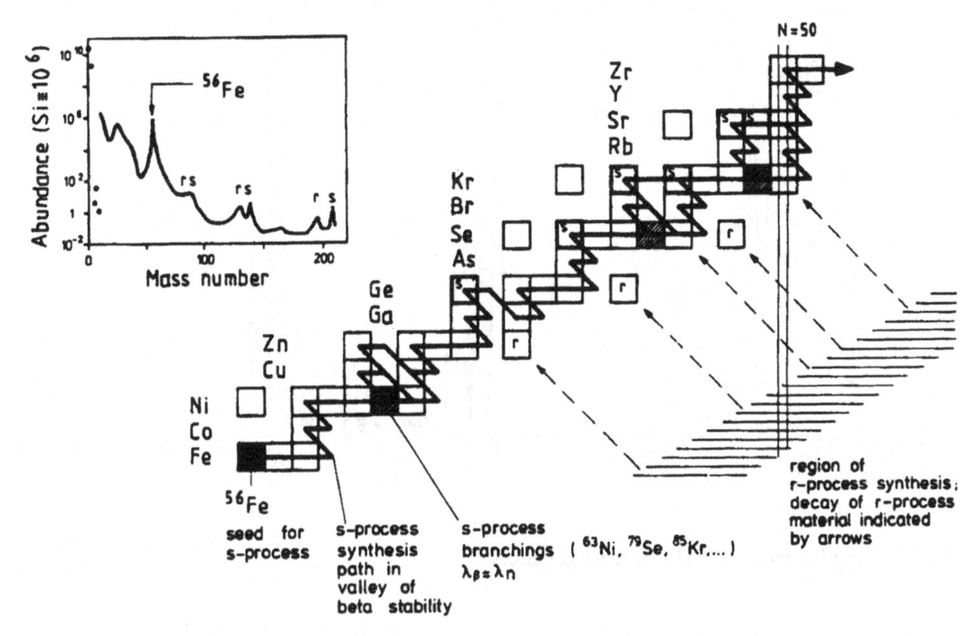

Figure 10.10. Nucleosynthesis beyond the Iron Peak. Reproduced with permission from *Reports of Progress in Physics*, Vol. 52, 1989. Courtesy of F. Käppeler.

We turn now to slow neutron addition or the *s-process*. The classical *s*-process of B^2FH meant "slow" with respect to β-decays. In the simplest approximation, any β-decay that could occur, would occur. The *s*-process path led along the valley as suggested by Figure 10.10. Note that the *s*-process will bypass the nuclides that have been labeled 'r.' We have, therefore, *s-only* nuclides. In principle, these *s*-only isotopes could be reached by β^+-decays from the proton-rich side of the valley of β-stability. However, there is essentially no evidence that the abundances of the *s*-only isotopes have been influenced by such contributions, and we shall ignore them in what follows.

The proton-rich isotopes are to the left of the valley of β-stability. They are accounted for by the *p-process* of B^2FH, which we shall discuss in §10.10.

There are certain places in the valley of β-stability where the *s*-process path is not unique. For example, ground-level ^{152}Eu will decay to ^{152}Sm 73% of the time, and to ^{152}Gd the remaining 27%. If we ignore such difficulties, we can write a series of very simple linear differential equations in which each nuclide has a unique predecessor and successor. Thus, for the *i*th species,

$$\frac{dN_i}{dt} = -N_n N_i < \sigma_{i \to i+1} v > + N_n N_{i-1} < \sigma_{i-1 \to i} v > . \tag{10.27}$$

Here, N_n represents the number density of neutrons, and v is the relative velocity of the neutron *projectiles* and their targets. The angular brackets indicate that

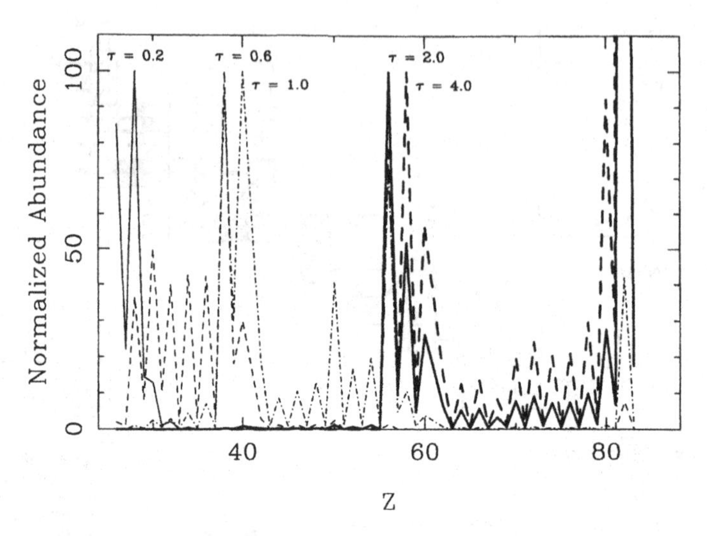

Figure 10.11. CFHZ *s*-Process Calculations.

the product of the cross section and relative velocity has been averaged over the Maxwellian velocity distribution (Problem 3 of Chapter 4).

A system of linear equations, together with boundary conditions $N_i(t = 0)$, may be formally solved by the method of Laplace transforms. Clayton (1968) works out some *s*-process solutions using this technique, which is now primarily of heuristic value. Network calculations, of the kind we have already discussed, are used in most modern work.

Until the introduction of the reaction network in slow neutron addition, most calculations were based either on an approximation introduced by Clayton, Fowler, Hull, and Zimmerman (1961, henceforth CFHZ), or a solution found by Clayton and Ward (1974). The latter is an exact mathematical solution under restricted conditions.

Figure 10.11 shows calculations based on the CFHZ method. Five solutions are displayed, for a number of exposures. The curves are labeled by the parameter τ, where

$$\tau = \int_0^t N_n v_T \cdot dt, \tag{10.28}$$

and the quantity v_T is defined such that $< \sigma v >=< \sigma > \cdot v_T$. The units of τ in the figure are inverse millibarns. All solutions are displayed as a function of atomic number Z, and they are normalized so that the highest abundances (apart from those of Pb and Bi) are set at 100. The two highest exposures, $\tau = 2.0$ and 4.0, are shown with thick solid and dashed lines. Notice how the low neutron exposure ($\tau = 0.2$) moves the iron peak over to nickel.

CFHZ labeled their results with the parameter n_c, the average number of neutrons captured per seed nucleus. This quantity depends on all of the relevant cross sections. Clearly n_c is a monotonic function of τ. Indeed, if all of the cross sections had the

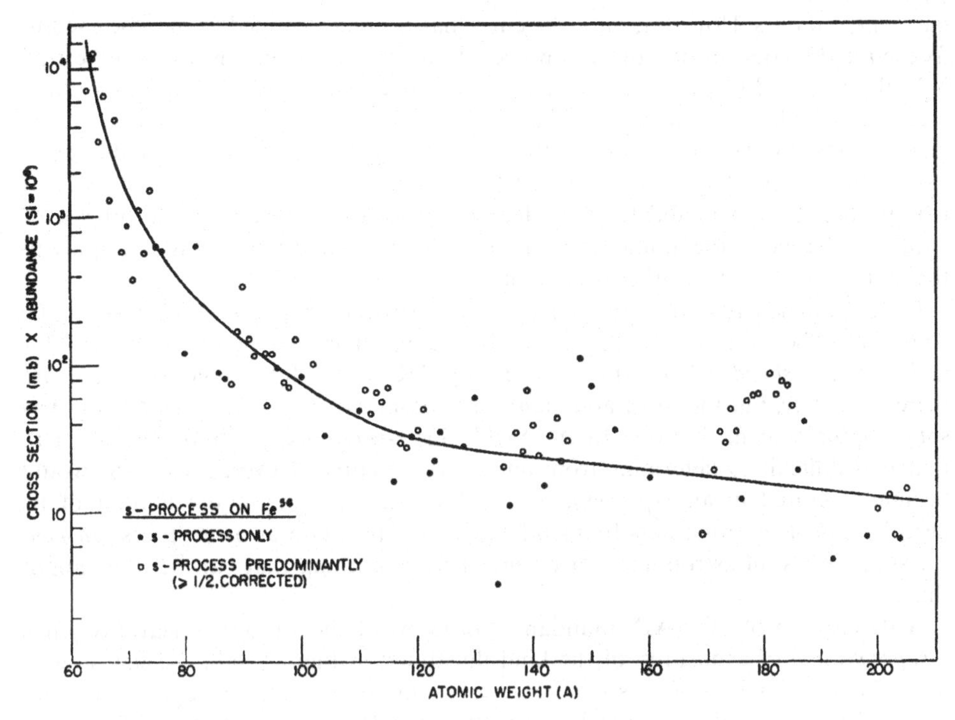

Figure 10.12. The Original $N\sigma$ vs. A Curve of B²FH.

same value, say 80 millibarns, then $\tau = 1.25\times\ 10^{-2} n_c$ millibarn^{-1}. Conversely, the number of neutrons captured per seed ^{56}Fe is 80τ millibarn^{-1}.

For $\tau = 2.0$ millibarn^{-1} the mean mass of the resulting nuclide would be $56+160 = 216$, already past the mass of the last α-stable nucleus ^{209}Bi. Our calculations cannot be realistic for the largest exposures, especially at the high-mass end.

One can see the effects of the neutron magic numbers which occur near $A = 88$ ($Z = 38$ and 40), $A = 140$ ($Z = 56$ and 60), and $A = 208$ ($Z = 82$). Notice how single neutron exposures can lead to very large overabundances of a small number of elements. Most cosmic abundances of heavy elements do not show such isolated peaks, although a few unusual cases are known: Bond, Luck, and Newman (1979), for example, discuss the "extraordinary" strength of the Sr II resonance lines in the R CrB star U Aquarii.

In B²FH it was argued that the nuclides that were created by the *s*-process should fall along a smooth curve when the product $N\sigma$ was plotted vs. atomic number A. The rationale for this was the logical notion that the abundances of *s*-processed nuclei should be inversely proportional to the cross sections. The detailed character of the plot would depend on the exposure, as we have seen from the CFHZ plots which, incidentally, were *not* available at the time of B²FH's seminal work. We show their plot for the Suess–Urey abundances, the SAD of their time, in Figure 10.12.

The shape of the expected $N\sigma$ vs. A curve must be determined empirically *from the s-only isotopes.* However, once the curve has been determined, it may be used to fix the *relative* contribution of the *s*-process to those species that may be synthesized in both the *r*- and *s*-processes. The (r, s)-isotopes (see Figure 10.10) that fit the curve must be considered *predominantly* due to the *s*-process. B^2FH assigned all of the heavy (trans-iron-peak) nuclides to the *r*-, *s*-, or *p*-processes or a combination, e.g., "*ps*." This practice has been continued by modern workers (cf. Cameron 1982). It is important to understand that the decision to assign a nuclide to one process or another is based on the abundances *in the SAD*, and would not necessarily apply to the abundances in some other cosmic object.

B^2FH assigned (see the Appendix to their paper) ^{86}Sr, ^{87}Sr, and ^{88}Sr to the *s*-process. The proton-rich ^{84}Sr is not tabulated. The two lighter isotopes, ^{86}Sr and ^{87}Sr, are *shielded* from the *r*-process by ^{86}Kr and ^{87}Rb respectively, and are therefore *s-only*, but the most abundant, neutron-magic ^{88}Sr, is unshielded, and has some *r*-process contribution. In the SAD, this *r*-process contribution is so small that it is difficult to determine from an $N\sigma$ vs. A curve. Nevertheless, ^{88}Sr would be manufactured in an *r*-process, in an abundance roughly equal to that of its neighbors in A. Astronomers frequently speak of strontium as an *s*-process *element* – I suppose, as an astronomer once said, there is nothing that can be done about this.

A description of the SAD abundances in terms of the *s*-process clearly requires a *superposition* of exposures, of the kind illustrated in Figure 10.11. Seeger, Fowler and Clayton (1965) superposed CFHZ solutions to produce the curve shown in Figure 10.13(a). The theoretical curve corresponds to a distribution of neutron exposures τ that is exponential in form, viz.

$$\rho(\tau) = G \cdot \exp(-\tau/\tau_0). \tag{10.29}$$

CFHZ solutions for various τ were multiplied by the weighting function $\rho(\tau)$ and added together. Seeger *et al.* noted that an acceptable fit could actually be obtained with a superposition of only three exposures.

It has been generally assumed that the appropriate distribution function for the SAD is an exponential one, although it is recognized that the observations will accommodate other forms. Clayton and Ward (1974) showed that the resulting abundances from an exponential distribution had a relatively simple analytical solution. Their work has been used by astronomers in an attempt to establish the parameters of the *s*-process thought to be responsible for certain *stellar* abundances. We shall return to this point in Chapter 12.

There is little doubt that improved abundances and nuclear cross sections have strengthened the foundation of the *s*-process. The fit by Käppeler *et al.* (Figure 10.13(b)) uses three exponential exposures, representing what the authors call a main, a strong, and a weak component. Their methods now integrate Equation 10.27 numerically, and allow for the production of a few species off the classical *s*-process path. The dips in the full curves reach such nuclides.

The parameter τ_0 for their main component was 0.30 millibarn^{-1}. Clayton and Ward (1974) suggested $\tau_0 = 0.27$ millibarn^{-1} reproduced the solar *s*-process abundances. Thus, the broad picture for the *s*-process contribution to the solar

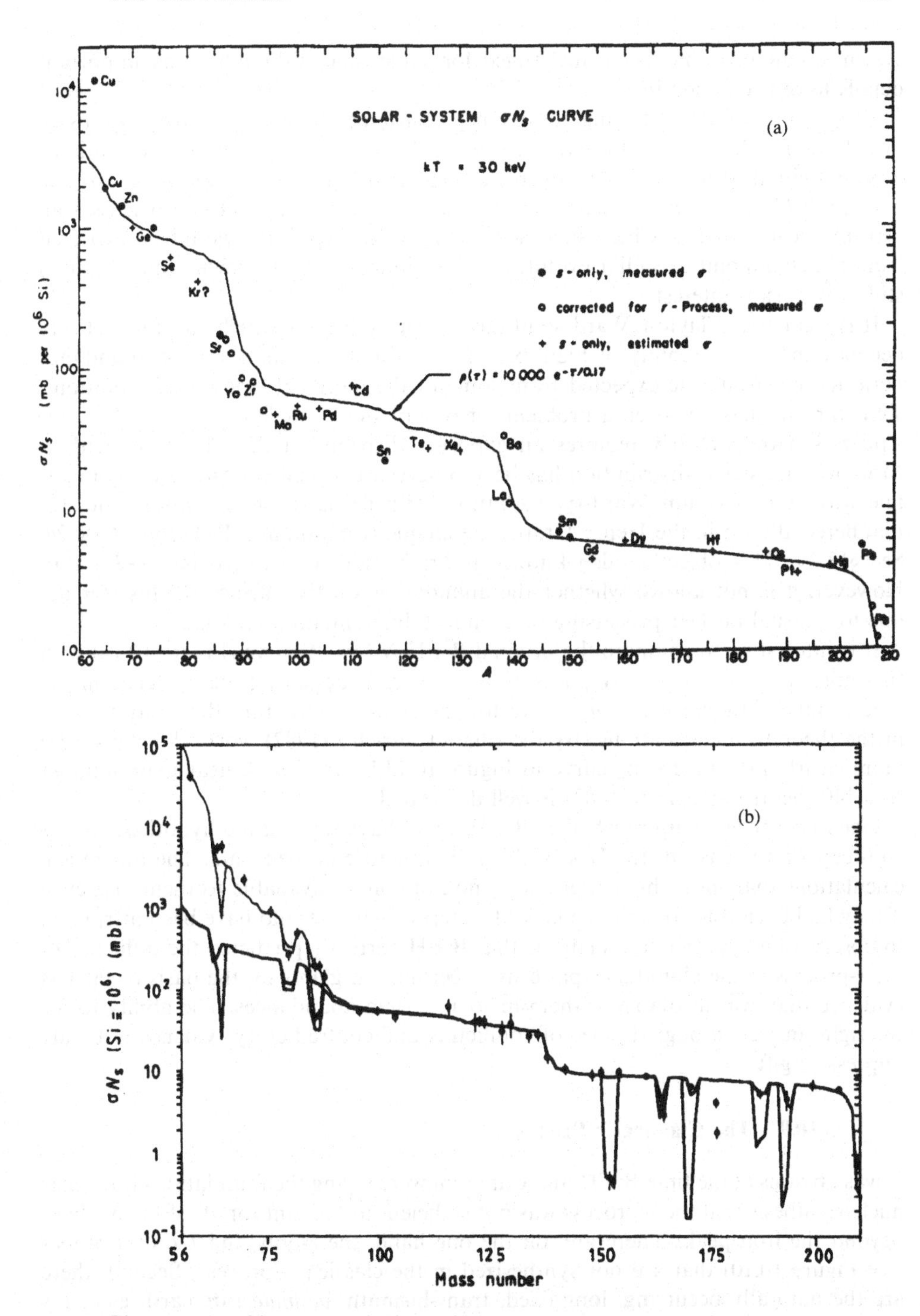

Figure 10.13. Observed and Theoretical $N\sigma$ vs. A Curves. (a): Seeger *et al.* (1965). (b): Käppeler, Beer, and Wisshak ©1989 IOP Publishing, Ltd. See text for discussion.

system abundances has been unchanged for some time, although many important details have been filled in.

The nature of the resulting abundance distribution from a neutron *exposure* distribution such as that of Equation 10.29 should be carefully noted. It is, essentially, a probability distribution for exposures τ, and according to it, *low exposures are the most probable and high exposures are the least probable.* Ulrich (1982) discusses an astrophysical model in which such a distribution of exposures would be expected from the standpoint of stellar evolution. This model is now somewhat dated, but is still of historical interest.

It is clear that Clayton–Ward solutions, or any solutions which call for a steady decrease in the probability of high exposures, cannot give the kinds of abundance patterns that would be expected from some of the more extreme CFHZ solutions. Thus far, this has not been a problem for observational nucleosynthesis. The *solar* s-process pattern clearly requires an exposure distribution that decreases with τ. Moreover, no stellar distribution has been observed that is demonstrably incompatible with some Clayton–Ward-type solution, with perhaps one exception: the star numbered 101065 in the Henry Draper Catalogue (Cannon and Pickering 1918–24. See §13.2). This object is also known as Przybylski's (see Przybylski 1977) star. However, it is not known whether the abundances on the surface of this star are due to unusual nuclear processing or chemical differentiation (§13.4.3).

It is interesting to compare the original B^2FH $N\sigma$ vs. A curve (Figure 10.12) with the more modern versions (Figures 10.13). There is actually little evidence in the data points of the Seeger *et al.* curve for the step-like structure that may be seen in the theoretical calculations. By the time of Ulrich's (1982) work, the steps were more clearly present. In the curve in Figure 10.13(b), the break near mass number $A = 140$ (neutron-magic $N = 82$) is well delineated.

One may speculate upon whether B^2FH might have *rejected* the hypothesis of the s-process on the basis of the *lack* of this step structure in their data. The theoretical calculations that show this pattern were not, of course, available to them, but even if they had been it is possible the lack of a step structure would have been attributed to observational scatter. We shall see that B^2FH correctly predicted the influence of the s-process in the abundance patterns of certain red giants on the basis of far less evidence than was available to them in solar system abundances. The ability to see the right answer through a maze of conflicting and contradictory evidence is a truly impressive gift.

10.9 The Classical r-Process

It was obvious at the time B^2FH and Cameron were laying the foundations for stellar nucleosynthesis that the s-process was not sufficient to account for all of the nuclides beyond the iron peak. There are, on the one hand, the r-only and p-only isotopes (see Figure 10.10) that are not synthesized in the classical s-process. Second, there are the naturally occurring, long-lived, trans-bismuth, *actinide* rare-earth elements thorium and uranium beyond the s-process path.

Those neutron-rich isotopes that are not included in the s-process can be synthesized in a process of rapid neutron addition. In the classical r-process, neutrons are added on a time scale that is much shorter than the β-decay rates of species near the valley of β-stability. Why does one omit intermediate neutron capture rates? It turns out that much interesting modern work has been concerned with intermediate rates. We shall discuss some of them in due course. However, the gross properties of the *solar system abundances* may be accounted for in terms of the extreme r- and s-addition rates.

The existence of separate abundance peaks associated with the neutron magic numbers is strong evidence that two distinct kinds of exposures occurred. These peaks are seen most clearly on the plot of abundance vs. A, in the mass number ranges about 124–142 and 188–209 (Figure 7.6). It will be apparent after we have discussed the classical r-process that intermediate exposures would fill in the region between these peaks. One may now consider the extent to which such filling has in fact occurred, and use it as an indication of the possible influence of intermediate neutron addition rates, but the gross features of the abundance curve are, remarkably, accounted for by two vastly different processes.

Perhaps the most comprehensive discussion of the r-process following those of B^2FH and Cameron was given by Seeger, Fowler, and Clayton (1965). Their work has now been superseded by reaction network calculations (Mathews and Ward 1985; Cowan, Thielemann, and Truran 1991). We shall, however, follow the discussion in Seeger *et al.* for heuristic insight into the results of the more modern work.

If one assumes that the s-process contribution to the heavy elements is understood, it is then possible to subtract it to obtain the r- and p-contributions. Since the latter is minor for those species that can be reached by the r-process, we obtain effectively the r-process abundances on subtraction of the s contribution. This is shown in Figure 10.14. The data are from Käppeler, Beer and Wisshak (1989). The asterisks mark r-only isotopes. The chemical symbols refer to these nuclides.

One sees clearly the peaks associated with the neutron magic numbers $N = 82$ ($A \approx 130$), and 126 ($A \approx 195$). In both Figures 7.6 and 10.14 the first r-process peak ($N = 50, A \approx 80$) is less marked. An intermediate "hump" in the domain of the lanthanide rare earths ($A \approx 160$) is still something of a puzzle.

In the classical r-process, neutrons are added at a rapid rate to species with a given Z until an equilibrium is established between (n, γ) and (γ, n) reactions. The neutron densities are of the order of $\approx 3 \cdot 10^{25}$ cm^{-3} and temperatures 1–2.4 $\times 10^9$K. Under these conditions the (n, γ) and (γ, n) reactions are much faster than the β-decay rates. The r-process path may be located at every Z by assuming equal numbers of isotopes in equilibrium with neutron numbers N and $N + 1$. This is a simple problem provided the relative neutron binding energies are known.

The neutron binding energies are of course directly related to the nuclear masses, but in the case of the conditions postulated, one must rely on nuclear theory to supply the information, as the necessary masses have not been measured in the laboratory. We see now the key role played by the mass formula in the theory of nucleosynthesis.

The elementary nuclear mass formula 9.5 is insufficient for the purposes of nucleosynthesis because it does not include the effects of shell closings. Seeger and his colleagues added terms to account for the shell effects as well as terms to account

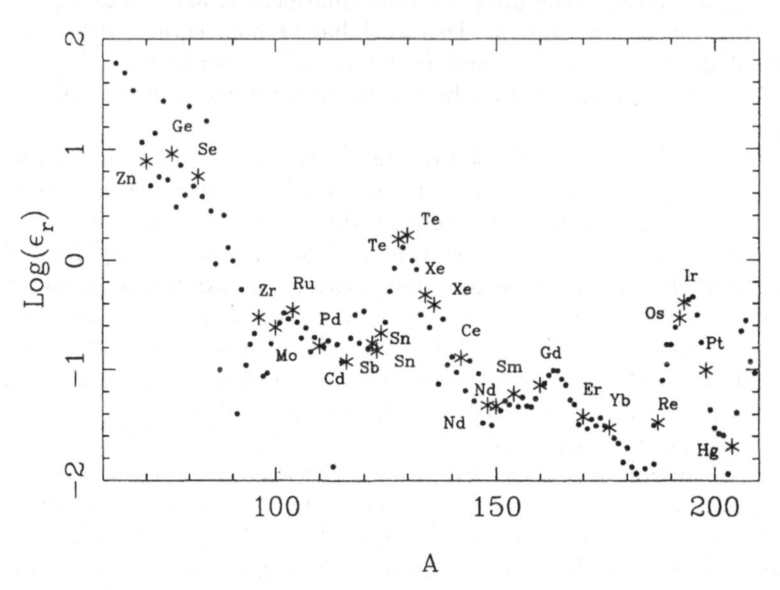

Figure 10.14. The r-Process Contribution to the SAD. Courtesy of Käppeler, Beer, and Wisshak (1989).

for deformations from sphericity which were considered as possibly responsible for the enigmatic "hump" near $A = 160$. The deformations occur because of effects described by *collective models* of the nucleus, which are based on the motions of groups of nucleons. The resulting energy states resemble those of molecular rotation and vibration, and, of course, depart from spherical symmetry. The collective effects are most apparent between the shell closings.

For a given element the (n, γ) equilibrium establishes the position Seeger *et al.* called the wait point. Nuclei leaked away from this position by β-decay, and moved along to the wait point for $Z + 1$.

The assumption of nuclear equilibrium did not fix the wait point uniquely. There is a range of N values for a given Z, within which the value corresponding to the wait "point" is merely the most probable.

For neutron magic numbers, the wait points have the same value of N for several consecutive Z's, because of the stability of the N-magic nuclei. This is the reason why the r-process path runs vertically in Figure 10.15 where N is magic.

If one ignores the breadth of the wait points, a well-defined r-process path is specified, with the time rate of change of any species given by the simple relation

$$\frac{dN_Z}{dt} = \lambda_{Z-1}N_{Z-1} - \lambda_Z N_Z + (\text{fission sources}). \tag{10.30}$$

The λ's are the β-decay rates.

Theoretical r-process solutions from Cameron *et al.* (1983) are shown in Fig-

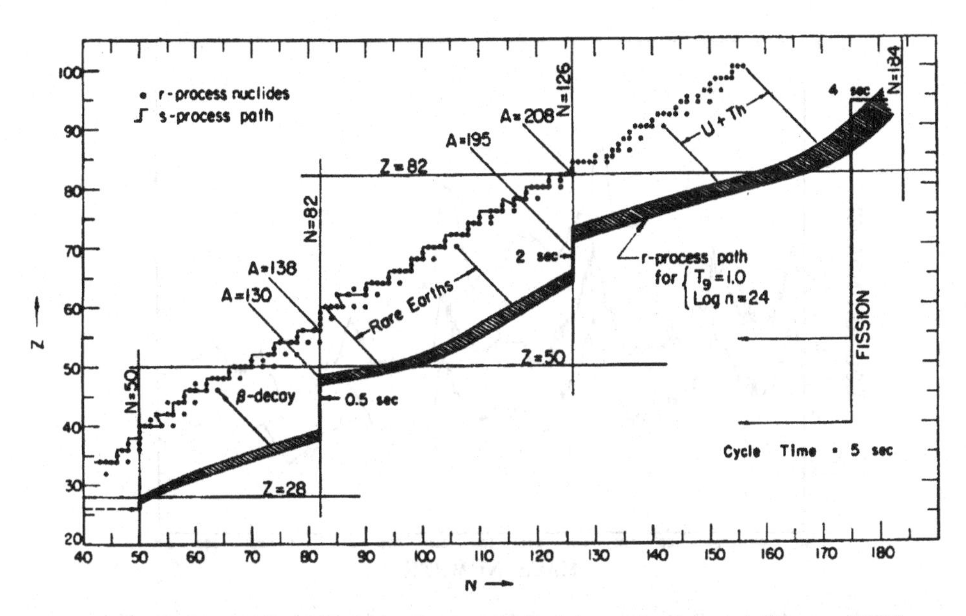

Figure 10.15. The Classical *r*-Process Path. Seeger, Fowler, and Clayton (1965).

ure 10.16. The upper curve is the SAD, while the two lower curves are *steady-flow solutions,* obtained by setting the time derivatives in 10.30 equal to zero. The middle curve gives the abundances along the *r*-process path, while the lower shows the results after the species have decayed to the valley of β-stability.

The use of reaction network solutions removes the need to worry about the width of the wait points or the nature of the freeze-out, that is, what happens *as* the neutron number density decreases from *r*-process values toward zero. It is then possible to investigate intermediate neutron addition rates, called *n*- or *r*(*n*)-processes. Solutions have been carried out through the region of the putative neutron magic number, $N = 184$, which has led to so much speculation about an island of nuclear stability near $A = 298$. It is possible that the fission of neutron-rich, super-heavy species could account for the abundance hump of the SAD near $A = 160$.

These questions were discussed by Schramm (1982), who summarizes one of the global puzzles of the *r*-process: its site. Mathews and Cowan (1990) argued for Type II supernovae.

Keen interest attaches to the *r*-process today because of questions of *nucleocosmochronology,* the dating the age of the cosmos with radioactive nuclei. We shall return to this question in later chapters. We only point out here that critical input into the determinations is the relative *production ratios* of the species. For those nuclides heavier than bismuth, the *r*-process, in some modern guise, is the production mechanism of choice. Fowler's (1987) Milne Lecture on nucleocosmochronology discusses *r*-process calculations.

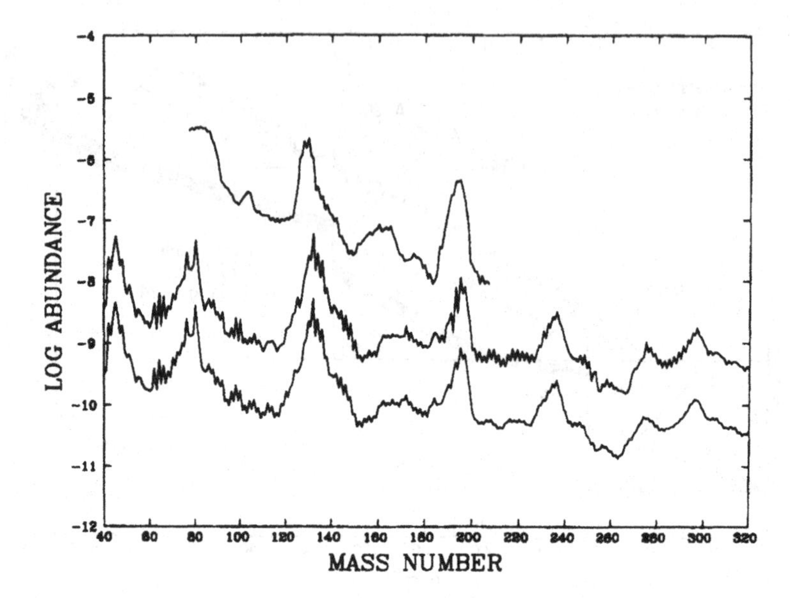

Figure 10.16. *r*-Process Solutions. Upper curve: SAD (Cameron 1982). The middle and lower curves are theoretical calculations using the steady-flow approximation. The assumed temperature was $3 \cdot 10^8$ K, with a neutron number density of 10^{20} cm^{-3}. From Cameron *et al.*, *Astrophysics and Space Science*, **91**, 221 (1983), reprinted by permission of Kluwer Academic Publishers.

10.10 Additional Mechanisms for the Production of Elements Beyond the Iron Peak

We have already mentioned the *p*-process, responsible for those proton-rich species that cannot be accounted for by either the *r*- or the *s*-process. The *p*-process nuclides are generally an order of magnitude or more *less* abundant than their *r*- and *s*-congeners, as shown in Figure 10.17 from Cameron (1982).

It is generally agreed that the *p*-nuclei originate in *secondary* processes, acting on the species already synthesized in the *r*- and *s*-processes. This would explain the rough parallel of the abundance curves in Figure 10.17. Even the abundance humps in the *r*- and *s*-nuclide abundances near $A = 90$ and 130 appear to be mimicked in the *p*-distribution at slightly larger values of *A*. If the *p*-nuclides are created from *r*-and *s*-nuclides, by (p, γ), (p, n), or (γ, n) reactions, such a parallelism would be expected. Current ideas by Prantzos *et al.* (1991) or Woosley and Howard (1990) involve (γ, n) reactions during the passage of a shock wave through the oxygen–neon shells of Type II supernovae. Lambert (1992) has reviewed the origin and abundances of *p*-nuclei. He also favors (γ, n) reactions as the source but suggests supernovae of Type Ia (§10.11) as the site.

Unfortunately, there are no elements whose abundances derive primarily from the *p*-process. By contrast, we know that strontium is dominated by the *s*-process, and

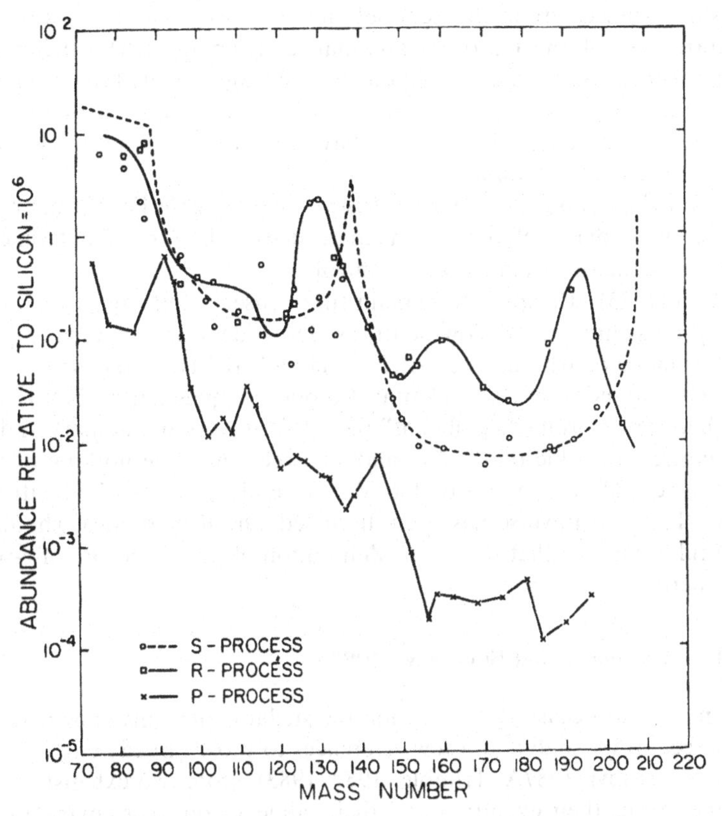

Figure 10.17. SAD Abundances for *r*-, *s*-, and *p*-processes. Cameron (1982) with permission.

europium by the *r*-process. Fascinating new stellar abundance work is beginning to show different histories for *r*- and *s*-process production. We are in a very poor position to attack similar questions regarding the *p*-process. Virtually all of our information comes from the isotopic anomalies in meteorites (Lee 1988).

Figure 10.17 illustrates one of the most puzzling of questions of stellar nucleosynthesis. Why are abundances of *r*- and *s*-nuclides correlated? It is clear that the *s*-process cannot take place under the explosive, high T–ρ conditions that must apply to the *r*-process. Violent events, on a time scale less than a day, are incompatible with the assumption of free β-decay. We suggested the *r*-process occurs in supernovae explosions, while the *s*-process goes on in red giants which are hydrostatically burning helium. How is it that these totally different processes manage to produce comparable abundances of *r*- and *s*-species over such a wide range of nuclear masses?

Table 10.4 gives the relationship between abundances derived for the *r*-process by subtracting the *s*-process component (Käppeler, Beer, and Wisshak 1989). The column labeled ϵ(A&G) are the SAD abundances from Anders and Grevesse (1989).

The *r*-process abundances are in the next column. The last column gives the result of formal subtraction of the *r*-process abundances from the SAD, therefore the *s*- *plus* *p*-components. Since the *p*-component is usually small, this is effectively the *s*-contribution in most cases. Most of the entries in the last two columns are uncertain in the *second* figure, although we have retained three figures when they appeared in the original tabulation.

Note that the lightest and the heaviest elements in the table have similar *r*- and *sp*-abundances, as do many of the intermediate ones. The *largest* differences are rarely as much as a factor of ten for any element.

Suess and Zeh (1973) attempted to explain these observations by assuming post-exposure of *r*-process nuclides to slow neutrons. Their ideas were never accepted by mainstream astronomers, who have chosen, along with B^2FH, to regard the "correlation" of *r*- and *s*-abundances as accidental. No one can question the extraordinary knack B^2FH had for choosing "significant" results from very preliminary and noisy data, so it is entirely possible that the *r*- and *s*-correlations hide nothing of fundamental importance. Still, one may wonder. Very recently, synthesis by neutron-rich pockets within the early universe has been discussed, and if these ideas should turn out to be fruitful, it may be that the Suess–Zeh notion of post-exposure of *r*-process products has merit.

10.11 Supernovae and Supernova 1987A

We have had numerous occasions to mention the stellar explosions known as supernovae. In this section we shall add a few references to those already cited, and then briefly turn to Supernova 1987A. Trimble (1982, 1983) wrote two exhaustive review articles on supernovae, their evolution and their influence on their environments as well as astronomical thought. More recent references follow below.

The last (observed) supernova in our own Galaxy was seen by Kepler in 1604. From a variety of observations, we expect several supernovae per century in our Galaxy; certainly we are overdue for one. Supernova 1987A exploded not in our own Galaxy, but in a small "satellite" known as the Large Magellanic Cloud. There is an enormous literature on this object. The reader might start with Arnett *et al.* (1989), Petschek (1990), Wheeler (1991), or Woosley (1991).

We present a classification scheme following Harkness and Wheeler (1990, see also Gaskell *et al.* 1986). It is based primarily on the spectrum. Branch (1990) has assembled references to the spectra of supernovae of various types, going back to 1885.

SUPERNOVAE CLASSIFICATION

Type I – Spectra near maximum light show no hydrogen
- (Ia.) Silicon lines present
- (Ib.) No silicon, helium present, strong [O I] in late spectra
- (Ic.) No silicon, no helium

Type II – Hydrogen in maximum-light spectrum
- (IIL.) *Linear* light curve decline from maximum

Table 10.4. *Elemental, ps- and r-Abundances (Si = 10^6)*

Z	El	ϵ(A&G)	$\epsilon(r)$	$\epsilon(sp)$
33	As	6.56	5.33	1.23
34	Se	62.1	44.1	18.0
35	Br	11.8	4.64	7.16
36	Kr	45.0	22.7	22.3
37	Rb	7.09	2.89	4.20
38	Sr	23.5	2.5	21.0
39	Y	4.64	1.31	3.33
40	Zr	11.4	2.04	9.36
41	Nb	0.698	0.11	0.588
42	Mo	2.55	0.64	1.91
44	Ru	1.86	1.11	0.75
45	Rh	0.344	0.289	0.055
46	Pd	1.39	0.771	0.619
47	Ag	0.486	0.435	0.051
48	Cd	1.61	0.65	0.96
49	In	0.184	0.119	0.065
50	Sn	3.82	1.414	2.41
51	Sb	0.309	0.305	0.004
52	Te	4.81	4.04	0.77
53	I	0.90	0.85	0.05
54	Xe	4.7	4.01	0.69
55	Cs	0.372	0.316	0.056
56	Ba	4.49	0.61	3.88
57	La	0.446	0.111	0.335
58	Ce	1.14	0.26	0.88
59	Pr	0.167	0.095	0.072
60	Nd	0.828	0.42	0.41
62	Sm	0.258	0.178	0.080
63	Eu	0.097 3	0.090 7	0.006 6
64	Gd	0.330	0.27	0.06
65	Tb	0.060 3	0.054 6	0.005 7
66	Dy	0.394	0.359	0.035
67	Ho	0.088 9	0.081 8	0.007 1
68	Er	0.251	0.211	0.040
69	Tm	0.037 8	0.032	0.005 8
70	Yb	0.248	0.163	0.085
71	Lu	0.036 7	0.030 9	0.005 8
72	Hf	0.154	0.079 9	0.074 1
73	Ta	0.020 7	0.013 3	0.007 4
74	W	0.133	0.062 5	0.070 5
75	Re	0.051 7	0.047 4	0.004 3
76	Os	0.675	0.651	0.024
77	Ir	0.661	0.650	0.011
78	Pt	1.34	1.30	0.04
79	Au	0.187	0.176	0.011
80	Hg	0.34	0.15	0.19
81	Tl	0.184	0.053	0.131
82	Pb	3.15	0.62	2.53
83	Bi	0.144	0.093	0.051

(IIP.) Light curve shows a *plateau*
SN 1987A – unique light curve
SN 1987K – later spectrum resembled Type I

The outburst on 23 February, and subsequent behavior of Supernova 1987A, were in general agreement with expectations. For decades, supernovae played the role of an astronomical *deus ex machina*, providing the energy or extreme conditions that were necessary to explain the *r*- or the *e*-process, and a variety of other astronomical puzzles such as the pulsars. Possibly, the most gratifying result from the study of 1987A was the detection of neutrinos from the event (Burrows 1990). These were observed by several stations, and arguably agreed both in energy and particle flux with values that could be accommodated with current theory. This is not to say that all observations are understood. In particular, data from the Mont Blanc site cannot be accommodated. However, the general consensus is that current supernova models must be basically correct.

The light curve could be successfully modeled, but only with the addition of energy from the radioactive decay of ^{56}Co, expected from the doubly magic parent ^{56}Ni. In order for this energy to be supplied at the appropriate rate, it was necessary to introduce departures from spherical symmetry – radial mixing – in the expanding envelope (Arnett, Fryxell, and Müller 1989). Some of the ^{56}Co could be confirmed by observations of γ-radiation with energies 0.847 and 1.238 MeV, corresponding to electric dipole transitions from the first two excited states of ^{56}Co (Matz *et al.* 1988).

Jeffery and Branch (1990) give a very detailed account of fits of theoretical to observed spectra of Supernova 1987A in the wavelength region 3900–7000 Å. They discuss observations from 25 February, soon after the outburst, until early July. Only modest departures from solar abundances were necessary. Interestingly, one significant modification was for the element barium, whose dominant isotope in the solar system is attributed mostly to the *s*-process. The Na D lines were also stronger than predictions based on solar abundances. The D lines are known to behave capriciously in certain otherwise normal stars. Danziger *et al.* (1991) have made a comparison of observed and predicted abundances for SN 1987A.

10.12 Problems

1. Calculate the number density of ^{12}C* nuclei *in equilibrium* for a density of helium of 10^5 gm/cm^{-3} and a temperature of 10^8 K. The excited nucleus is 7.654 MeV above the ground state. *Hint:*Combine two dissociation equations. The relevant statistical weights (§10.4) are all unity, and the partition functions may be assumed equal to the statistical weights of the ground states. Why? The energy in the final exponential term is $\Delta(^{12}\mathrm{C}) + 7.654\mathrm{MeV} - 3\Delta(^4\mathrm{He})$.
2. Estimate the rate of the 3α process for the above conditions in the following way. Of the ^{12}C* produced, a fraction Γ_α/Γ breaks up into $^8\mathrm{Be} + \alpha$, while a much smaller fraction Γ_γ/Γ emits a γ-ray and stabilizes to ^{12}C. Here (see Rodney and Rolfs 1988) $\Gamma_\alpha = 8.90$ eV, while $\Gamma_\gamma = 0.003\,67$ eV, and $\Gamma = \Gamma_\alpha + \Gamma_\gamma$. The rate of formation of ^{12}C is then the number density of excited nuclei times the *rate* Γ_γ in sec^{-1}. (To convert the Γ's to sec^{-1}, divide by *h*.) Note that the half-life of

the 7.654 MeV level that corresponds to the *total* Γ ($0.05 \cdot 10^{-15}$ sec or 0.05 fs) is given in the nuclear data tables.

3. Calculate the radius of gyration of a 10^{12} eV proton in a magnetic field of 10^{-6} gauss. Give your answer in *parsecs* (see §13.3, Table A1, Appendix), and compare that value with a typical thickness of the plane of our Galaxy of several hundred parsecs.
4. Calculate the tin isotopes that are in equilibrium for $n = 1 \cdot 10^{24}$ cm^{-3} and $T = 1 \cdot 10^{9}$ K. You will need to assume a mass formula in order to find the relevant neutron binding energies. Explain your choice of parameters. The "wait point" for the nuclide chart is the (Z, N) value where equilibrium holds. Is it realistic to consider it to be a *point*?
5. Seeger *et al.* give the following simple formula for the decay constant λ_β in sec^{-1}:

$$\lambda_\beta = \frac{10^{-5}}{18 \cdot \ln(2)} \cdot \frac{W_0^6}{\Delta} \text{ sec}^{-1}, \tag{10.31}$$

where W_0 is given by

$$W_0 = NM(Z, A) - NM(Z + 1, A) - M_e. \tag{10.32}$$

We assume here "bare" nuclear masses, *NM*, so orbital electron binding energies do not enter. The quantity Δ is the average level spacing in the daughter nucleus. This density is a property of the individual nuclei, of course, but Seeger *et al.* used some overall approximations. For example, for spherical, odd-*A* nuclei, they assumed $\Delta = 5 \cdot \exp(A/290)$. Use these figures to estimate the β-decay rate at the wait point found in the above problem. Compare your answers to the information given in Figure 10.15.

6. Collective nuclear effects appear between closed shells. Indeed, the shell model is essentially incompatible with them. Explain.
7. Locate the levels in ^{56}Co fed by decays of ^{56}Ni with the help of tables such as those of Lederer and Shirley (1978). Find the observed 0.847 and 1.238 MeV transitions. Would you expect additional γ-ray lines? Can you find evidence for them in the observations of Matz *et al.* (1988)?
8. Examine the isotopic abundances on a nuclide chart and suggest elements that would have a large *p*-process contribution. (Hint: Molybdenum has an abundant *p*-only isotope. Can you find others?)

11
Atomic and Molecular Spectra

11.1 Introductory Remarks

The philosopher Auguste Comte (1798–1857) asserted that man would never know the chemical composition of the stars. It is therefore ironical that Gustav Kirchhoff (1824–1887) discovered the laws of spectroscopy at about the same time as Comte's death. With the help of the principles articulated by Kirchhoff we now claim knowledge of the composition not only of the nearby stars, but of galaxies so distant that it has taken a substantial fraction of the age of the universe for their light to reach us.

In this chapter we will review the laws of atomic and molecular spectroscopy that enable us to analyze the electromagnetic radiation from space. Naturally, we cannot give a complete account of these rather complicated topics. There is only space to highlight the nomenclature, and in some cases provide heuristic insight into the more important formulae.

The following chapters will deal with the application of atomic and molecular physics to chemical analyses of stars and stellar systems, and interstellar material.

11.2 Atomic Spectra: The Nomenclature of *LS* Coupling

The identification of spectral lines in a star is done with the help of certain reference volumes, the most important of which is possibly C. E. Moore's (1972) *A Multiplet Table of Astrophysical Interest.* While the basic work appeared as a series of the publications of the Princeton Observatory, the demand for this material was so great that it has gone through one major revision and innumerable reprints and updates. Often the basic work is referred to as the RMT, for Revised Multiplet Tables, since the 1945 work was a revision of a publication that first appeared in 1933.

The use of the RMT or comparable material presupposes an elementary knowledge of atomic spectroscopy, which we outline without proof here. Herzberg's (1944) *Atomic Spectra and Atomic Structure* is a satisfactory reference, although the reader may wish to consult a more advanced or recent source. For this we suggest Cowan's (1981) *The Theory of Atomic Structure and Spectra,* or Sobelman (1992). Condon and Shortley's (1935) *The Theory of Atomic Spectra* was for many years the astronomer's bible, but it was written before the development of many powerful theoretical and computational techniques, and is now largely of historical interest.

A single electron may be described by four quantum numbers. We may choose n, l, m_l, and m_s, or n, l, j, and m_j (see Table 9.2). The n and l stand respectively for the principal quantum number and the angular momentum quantum number of orbital motion. Projections of the orbital and spin angular momentum on the z-axis give rise respectively to m_l and m_s. The quantum number j specifies the (total) angular momentum of orbital motion plus spin, and m_j the z-component of the total angular momentum. The quantum number l can take values $< n$; those electrons with $l = 0, 1, 2, 3, 4$ are called respectively s-, p-, d-, f-, and g-electrons.

The quantum numbers for the *sum* of the angular momenta of several electrons, which correspond to l, j, and m_j, are designated by the corresponding upper case letters. The subscript is usually dropped from the latter. Thus L, J, and M correspond to quantum numbers of total orbital angular momentum, total orbital-plus-spin angular momentum, and the projection of the total angular momentum vector on the z-axis (in units of $\hbar$ or $h/2\pi$ where h is Planck's constant).

The total angular momentum of an electron with a given l is $\sqrt{l(l+1)}\hbar$. Similarly, the total orbital and spin angular momenta are $\sqrt{L(L+1)}$, and $\sqrt{S(S+1)}$, in units of $\hbar$. Thus the total angular momenta are not *exactly* integral multiples of $\hbar$, while the *projections* m_l and M *are* exactly integral multiples. When spin is involved, as with m_s, m_j, M_S, or M, we can have half-integral multiples of $\hbar$. For convenience, one often speaks of one unit of (total) angular momentum for a p electron, of two units for a D term, etc., ignoring the difference, for example, between $\sqrt{l(l+1)}$ and l.

The most elementary atomic state of an atom (omitting for the moment possible nuclear interactions) is one in which all of the n's and l's, L, J, and M are specified along with a quantum number S, which describes the total electron spin angular momentum. For a single electron, this value is always 1/2, but for two electrons, it may be 0 or 1, depending on the atomic state. There are $2J + 1$ atomic states with the same energy, and this group of states is called a *level*. The quantity $2J + 1$ is called the *statistical weight* of an atomic level. The values of M for a level are the $2J + 1$ integers between $-J$ and $+J$.

The concepts of states, levels, and statistical weights are slightly different in cases where *hyperfine structure* is important (see §11.8).

A group of levels is said to form a term if all of the levels have the same L and S. Different levels within a term have J values that may range by integers from $|J - S|$ to $|J + S|$. If the energy levels within a term increase with increasing J, the term is said to be normal. In an *inverted* term, the highest J belongs to the lowest level.

The value $2S + 1$ is called the multiplicity of the term. Terms for which $2S + 1$ is $1, 2, 3, 4, \ldots$ are called singlets, doublets, triplets, quartets, etc. If L is equal to or greater than S, then the multiplicity is the same as the number of levels in the term, but if L is less than S, the number of levels is $2L + 1$. The multiplicity of the term is by definition $2S + 1$ in either case. Thus a quartet D has four levels, but a sextet P has only three.

This scheme of describing atomic states is known as *LS* coupling or Russell–Saunders coupling. It is only an approximate description. In principle, any atomic state may be accurately described as some mixture of *LS*-coupling states. The atom is then said to be described by intermediate coupling.

At one time it was fashionable for students in astronomy to learn the nomenclature of both LS and jj coupling. In the latter scheme, the individual electron's (vector) orbital and spin angular momenta coupled, viz. $\boldsymbol{j} = \boldsymbol{l} + \boldsymbol{s}$. The $\boldsymbol{j}$'s then add to form a resultant $\boldsymbol{J}$. The textbooks often said that LS coupling was valid for light atoms while jj coupling was more appropriate for complex atoms. Both statements are only approximate, and it has become the custom in modern listings to give preference to LS coupling (see below). We refer to the book by Cowan for additional details.

Electronic transitions between atomic states are known as *components*, or sometimes, because the states are split by the interaction of an external magnetic field with the atom, called the Zeeman effect, Zeeman components. When the magnetic fields are zero, or when Zeeman splitting may be neglected, the components fall very closely together in wavelength, and one speaks of a line. All of the lines that are allowed between two terms are collectively referred to as a multiplet. We thus have the hierarchy: state, level, term, corresponding respectively to component, line, multiplet.

An atomic level is said to have even or odd *parity* according to the sum of the individual l-values. Thus, the electronic configurations s^2, p^4, and s^2p^6 all have even parity, while p, s^2p, pd^2, and df have odd parity. If the parity of a level is odd, it is indicated as a superscript 'o' on the term symbol, e.g., pd^3P^o, or $df\,{}^1D^o$.

In atomic spectroscopy, one speaks of permitted and forbidden transitions. By the former, one means transitions that take place according to selection rules for electric dipole radiation. The most basic of these is the Laporte parity rule, according to which a transition between two odd or two even levels is *forbidden*. The rule $\Delta J = 0, \pm 1$, with $J = 0 \not\leftrightarrow J = 0$, holds independently of the coupling rules (LS, or other). Often, transitions which violate the LS-coupling selection rules are also called forbidden, or sometimes semi-forbidden. These rules are $\Delta L = 0, \pm 1$, and $\Delta S = 0$. The most commonly broken of these rules is the latter, $\Delta S = 0$. The Laporte rule is valid regardless of the coupling conditions.

It is the custom in atomic spectroscopy to write the lower level or term first, whether the corresponding line is in absorption or emission. Forbidden transitions are indicated by the use of brackets. Thus, the transitions between low-lying levels of doubly ionized oxygen with the same parity are called [O III]. The brackets indicate the lines are forbidden for electric dipole radiation. The corresponding *forbidden multiplet* is written ${}^3P$–${}^2D$ (see Figure 15.3).

Astronomers who work in the field of emission-line regions (Chapter 15) sometimes use a single bracket for lines that violate the LS-coupling rule against changes of spin. For example, the line at 1909 Å in doubly ionized carbon, $2s^2\ {}^1S$–$2s2p\ {}^3P$, may be written C III]; the transition is said to be *semiforbidden* or *spin forbidden*. This notation is less useful in the spectra of complex atoms (e.g., of the iron group), where violations of the $\Delta S = 0$ rule of LS coupling are very common, and it is not generally used by atomic or stellar spectroscopists.

Figure 11.1 shows the three lowest terms of O I. These levels are all based on the electron configuration p^4, which is the only open subshell in neutral oxygen. Note that the same terms arise in O III, which has the configuration p^2 (see Figure 15.3). The levels shown are the *only* ones that can occur for p^4 without violating the Pauli principle. Note that the ${}^3P$ term of O I is *inverted*.

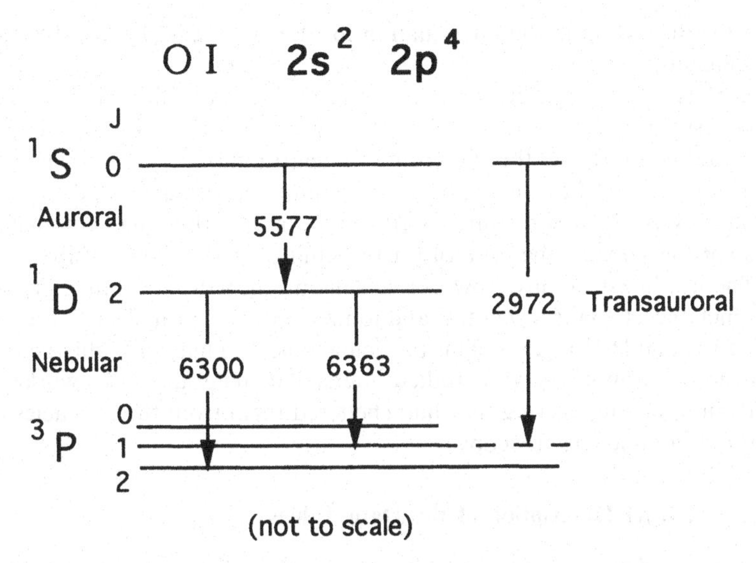

Figure 11.1. Forbidden Transitions in Low-Lying Levels of O I.

The line 5577 Å is called the green *auroral line.* Similarly, lines at 6300 and 6363 Å are called the red auroral lines. By convention, however, the ^1S–^1D transition in either O I *or* O III is called *auroral*, while the ^1D–^3P transitions are called *nebular*. This is because of the strong ^1D–^3P lines at 4959 and 5007 Å that are seen in gaseous nebulae (cf. §§15.1 and 15.3). Occasionally, one speaks of a *transauroral* transition, meaning ^1S–^3P.

In astronomy, wavelengths in angstrom units are often truncated rather than rounded. Thus, the longer of the red auroral lines is called λ6363 here, even though its wavelength is given in the Multiplet Tables as 6363.88. Because many authors do not follow this custom, we *have* occasionally rounded wavelengths in order to follow the work under discussion. Another notation common in astronomy is to indicate the wavelength *in angstrom units* with the λ-symbol. Thus λ5000 means 5000 Å. When several wavelengths are to be given, two λ's may be written, for example, λλ4959 and 5007 means 4959 and 5007 Å.

A large number of atomic energy levels are known, for both neutral atoms and various ionized species. In the years 1949–1958, three volumes by C. E. Moore appeared, known as *Atomic Energy Levels.* For the rare-earth elements, a corresponding volume was published some 20 years later (Martin, Zalubas, and Hagan 1978). Sugar and Corliss (1985) published updated energy levels for the 3*d* or iron-group elements. These volumes contain tabulations of the energy levels, beginning with the ground atomic states and continuing as high in energy as the levels are known. For each level, a description of the atomic states is given. This means the *n* and *l* values of the electrons, and whenever possible, the *LS*-coupling description of the angular momenta.

In some cases *LS* coupling provides so poor a description of energy levels that alternative schemes are used. These are covered by Cowan's text. In addition, we recommend the discussion in the introduction to Martin *et al.*'s (1978) tables on the rare-earth elements.

Spectroscopists use the symbol of the chemical element, followed by a Roman numeral, to designate the spectrum of an neutral atom, singly ionized atom, etc. Thus the symbol Fe I refers to the spectrum of a neutral iron atom. Fe II, Fe III, etc. mean the spectra of Fe^{+}, Fe^{2+}, etc. The Roman numerals are also used to describe atomic energy levels. It is sometimes useful to be aware that the symbol for the spectrum is *not* the same as the symbol for the kind of particle responsible for that spectrum. There is in astronomy, however, a well-entrenched but harmless misusage of this nomenclature. Clouds of neutral and ionized interstellar hydrogen are almost always called H I and H II regions. This particular solecism may probably be blamed on the first radio astronomers who studied interstellar material. These workers had backgrounds in radio engineering that had sheltered them from the delicacies of the nomenclature of atomic spectroscopy.

11.3 A Brief Description of the Data Tables

The nearly unambiguous way to describe an energy level is to give the energy itself, and this is traditionally given in cm^{-1} (see Equation 11.2 below). In general, these figures will have comparable accuracy to wavelength measurements which are typically good to six figures. Sometimes energies overlap very closely, so additional information must be supplied.

Figure 11.2 is an abbreviated listing from Sugar and Corliss's (1985) revision of Moore's *Atomic Energy Levels* for the neutral iron, or Fe I spectrum. The ground state (level) of the atom has the electron configuration $3d^6 4s^2$. The inner electrons are not written explicitly. They occupy closed shells, and also contribute nothing to the net angular momentum of the term. The $4s^2$ electrons also constitute a closed shell which makes no contribution to the net angular momentum. The six equivalent (same n and l) d electrons give rise to the lowest term, the quintet D, whose lowest *level* has angular momentum $J = 4$, and an energy of 0.00 cm^{-1}. Other levels in this term follow in order of increasing energy until we reach the next term, a quintet F. The low terms in Fe I are all inverted.

In iron, as well as in other transition elements, there is a strong competition between the $3d$ and $4s$ shells for the outer electrons. This is shown by the electron configuration of the quintet F, which is $3d^7 4s$. The closeness of the ^5D and ^5F terms reflects the small difference in energy between a $3d$ and a $4s$ electron.

The terms in parenthesis in the Configuration column are called the *parents.* If the $4s$ electron were removed, from the ^5F, the quartet F would be the resulting term of the Fe II spectrum. Thus the ^4F is the parent of the ^5F. More generally, one may look at the term in parenthesis as giving the sums of the orbital and spin angular momenta of the configurations immediately preceding it. Then in the case of the $^7D^o$ term, whose first level is at 19350 cm^{-1}, the $4s$ and $4p$ electrons together give a $^3P^o$, and the $^3P^o$ and ^5D combine to give the $^7D^o$. This is not strict LS coupling.

Since there may be numerous terms with the same L and S but different energies, it was customary to give each term a label. Thus the lowest term with a given L and S was called "a", the next term with the same L and S and parity was called "b",

Fe I

Configuration	Term	J	Level (cm⁻¹)	g	Leading percentages		
$3d^6\,4s^2$	$a\,^5$D	4	0.000	1.50020	100		
		3	415.932	1.50034	100		
		2	704.004	1.50041	100		
		1	888.129	1.50022	100		
		0	978.072		100		
$3d^7(^4$F$)4s$	$a\,^5$F	5	6 928.266	1.40021	100		
		4	7 376.760	1.35004	100		
		3	7 728.056	1.24988	100		
		2	7 985.780	0.99953	100		
		1	8 154.710	-0.014	100		
$3d^7(^4$F$)4s$	$a\,^3$F	4	11 976.234	1.254	98	1	$3d^64s^2\,^3$F2
		3	12 560.930	1.086	98	1	
		2	12 968.549	0.670	98	1	
$3d^7(^4$P$)4s$	$a\,^5$P	3	17 550.175	1.666	99		
		2	17 726.981	1.820	99		
		1	17 927.376	2.499	99		
$3d^6\,4s^2$	$a\,^3$P2	2	18 378.181	1.506	55	32	^{3}P1
		1	19 552.473	1.500	55	32	
		0	20 037.813		55	32	
$3d^6(^5$D$)4s4p(^3$P°$)$	$z\,^7$D°	5	*19 350.892*	1.597	99		
		4	*19 562.440*	1.642	98		
		3	*19 757.033*	1.746	99		
		2	*19 912.494*	2.008	99		
		1	*20 019.635*	2.999	100		
$3d^6\,4s^2$	$a\,^3$H	6	19 390.164	1.163	100		
		5	19 621.005	1.038	100		
		4	19 788.245	0.811	100		
$3d^6\,4s^2$	$b\,^3$F2	4	20 641.109	1.235	71	21	^{3}F1
		3	20 874.484	1.073	71	21	
		2	21 038.985	0.663	71	21	
$3d^7(^2$G$)4s$	$a\,^3$G	5	21 715.730	1.197	88	10	$3d^64s^2\,^3$G
		4	21 999.127	1.051	88	10	
		3	22 249.428	0.756	88	10	
$3d^6(^5$D$)4s4p(^3$P°$)$	$z\,^7$F°	6	*22 650.421*	1.498	100		
		5	*22 845.868*	1.498	99		
		4	*22 996.676*	1.493	99		
		3	*23 110.937*	1.513	99		
		2	*23 192.497*	1.504	99		
		1	*23 244.834*	1.549	100		
		0	*23 270.374*		100		
$3d^7(^4$P$)4s$	$b\,^3$P	2	22 838.318	1.498	92	4	$3d^64s^2\,^3$P1
		1	22 946.808	1.489	79	10	$3d^7(^2$P$)4s\,^3$P
		0	23 051.742		79	12	$3d^7(^2$P$)4s\,^3$P
$3d^6(^5$D$)4s4p(^3$P°$)$	$z\,^7$P°	4	*23 711.457*	1.747	98		
		3	*24 180.864*	1.908	99		
		2	*24 506.919*	2.333	98		

Figure 11.2. Energy Levels for Fe I. Sugar and Corliss (1985)

Fe I—Continued

Configuration	Term	J	Level (cm^{-1})	g	Leading percentages		
$3d^6\,4s^2$	$b\,^3$G	5	23 783.614	1.200	88	10	$3d^7(^2$G$)4s\,^3$G
		4	24 118.814	1.048	88	10	
		3	24 338.762	0.761	88	10	
$3d^7(^2$P$)4s$	$c\,^3$P	2	24 335.759	1.484	90	4	$(^2$D2$)\,^3$D
		1	24 772.017	1.466	81	7	$(^4$P$)\,^3$P
		0	25 091.597		79	12	$(^4$P$)\,^3$P
$3d^7(^2$G$)4s$	$a\,^1$G	4	24 574.650	1.001	90	3	$(^2$H$)\,^3$H
$3d^6(^5$D$)4s4p(^3$P$^\circ)$	$z\,^5$D$^\circ$	4	*25 899.987*	1.502	91	6	$3d^7(^4$F$)4p\,^5$D$^\circ$
		3	*26 140.177*	1.500	91	6	
		2	*26 339.691*	1.503	92	6	
		1	*26 479.376*	1.495	92	6	
		0	*26 550.476*		93	5	
$3d^7(^2$H$)4s$	$b\,^3$H	6	26 105.904	1.165	100		
		5	26 351.039	1.032	98	2	$(^2$H$)\,^1$H
		4	26 627.604	0.811	98	2	$(^2$G$)\,^1$G
$3d^7(^2$D2$)4s$	$a\,^3$D	3	26 224.966	1.335	74	3	$(^2$D1$)\,^3$D
		1	26 406.470	0.731	45	35	$(^2$P$)\,^1$P
		2	26 623.730	1.178	67	18	$(^2$D1$)\,^3$D
$3d^6(^5$D$)4s4p(^3$P$^\circ)$	$z\,^5$F$^\circ$	5	*26 874.549*	1.399	95	4	$3d^7(^4$F$)4p\,^5$F$^\circ$
		4	*27 166.819*	1.355	94	4	
		3	*27 394.688*	1.250	94	4	
		2	*27 559.581*	1.004	95	4	
		1	*27 666.346*	-0.012	95	4	
$3d^7(^2$P$)4s$	$a\,^1$P	1	27 543.004	0.817	62	23	$(^2$D2$)\,^3$D
$3d^7(^2$D2$)4s$	$a\,^1$D	2	28 604.606	1.028	64	16	$(^2$D1$)\,^1$D
$3d^7(^2$H$)4s$	$a\,^1$H	5	28 819.946	1.000	98		
$3d^6(^5$D$)4s4p(^3$P$^\circ)$	$z\,^5$P$^\circ$	3	*29 056.321*	1.657	98		
		2	*29 469.020*	1.835	97		
		1	*29 732.733*	2.487	97		
$3d^6\,4s^2$	$a\,^1$I	6	29 313.003	1.014	100		
$3d^6\,4s^2$	$b\,^3$D	1	29 320.028		88	8	$3d^7(^2$D2$)4s\,^3$D
		2	29 356.740		81	7	
		3	29 371.811	1.326	94	4	
$3d^6\,4s^2$	$b\,^1$G2	4	29 798.933	0.979	62	35	^{1}G1
$3d^6(^5$D$)4s4p(^3$P$^\circ)$	$z\,^3$F$^\circ$	4	*31 307.243*	1.250	94	5	$3d^7(^4$F$)4p\,^3$F$^\circ$
		3	*31 805.067*	1.086	97		
		2	*32 133.986*	0.682	93	5	$3d^7(^4$F$)4p\,^3$F$^\circ$
$3d^6(^5$D$)4s4p(^3$P$^\circ)$	$z\,^3$D$^\circ$	3	*31 322.611*	1.321	90	8	$3d^7(^4$F$)4p\,^3$D$^\circ$
		2	*31 686.346*	1.168	90	8	
		1	*31 937.316*	0.513	91	8	
$3d^8$	$c\,^3$F	4	32 873.619	1.264	92	3	$3d^7(^2$F$)4s\,^3$F
		3	33 412.713	1.066	92	5	
		2	33 765.304	0.677	86	6	

Figure 11.2 (*continued*). Energy Levels for Fe I. Sugar and Corliss (1985)

and so on. Thus the lowest ^3F term, starting at 11 976 cm^{-1}, is called a^3F, and the term starting at 20 641 cm^{-1} was called b^3F by Moore. Sugar and Corliss have added a '2' in order to distinguish it from a much higher ^3F that has the same electron configuration (and therefore the same parity).

This sequence (a, b, etc.) is not rigorously followed throughout the entire set of atomic energy levels, especially in some of the newer compilations. Nevertheless, it has been of value. Terms arising from electron configurations with parity opposite to the ground term were labeled with letters at the end of the alphabet. So the lowest ^5D ($\geq$ 25 900 cm^{-1}) that could be reached by an absorption was called z^5D. The next ^5D is then y^5D (not shown) at ($\geq$ 33 095 cm^{-1}), and so on. The terms with labels z, y, x, etc. have parity opposite to those labeled a, b, c, etc.

The Multiplet Tables have two parts. Part I is called the table of multiplets, while Part II is a finding list, essentially a key to the first part. In the finding list, all wavelengths are listed in order, and a multiplet number is given, as illustrated in Figure 11.3(a) and (b). Figure 11.3(a) is a portion of Part I. Consider, for example, the Fe I line from Multiplet 4, λ3859.913, which may be found in the illustrated portions of both Parts I and II. Each multiplet is assigned a number, and the order of listing is determined by the order of the terms in the atomic energy tables. The first multiplets listed are those that can combine with the ground term of Fe I (a^5D), and they are listed in order of the energy of the upper term.

Over the years since the Multiplet Tables were constructed, a great deal of information on atomic spectra has been accumulating. Much recent material may be found with the help of Cowan's textbook. Commission 14 of the International Astronomical Union, Atomic and Molecular Data, publishes a summary every three years of interim research work. Their compilation appears, along with similar reviews by other Commissions, in the form of Reports which are a part of the *Transactions of the IAU*. These Reports may be the best guide to the literature available at the time of their writing.

Ultimately, workers whose results depend upon the most recent information on atomic data should contact the physicists at the National Institute of Standards and Technology (NIST) in Gaithersburg, Maryland. The Institute was formerly known as the National Bureau of Standards. The appropriate telephone number (August 1994) is 301/975-3200.

11.4 Diatomic Molecules: Rotation and Vibration

Molecular lines have been known in the spectra of the sun and stars from the early days of spectroscopy. In the original four classes of stellar spectra, Angelo Secchi (1878) made use of what are now known to be features of diatomic molecules. In Secchi's Class III, broad absorption features were noticed that were sharp on the violet side, and faded in intensity or were *degraded* to the red. We now know that the features he saw are due to the TiO molecule. In Class IV, the wide absorptions (due to C_2 and CN) were degraded to the violet.

The serious, detailed study of these features was only taken up at about the middle of the twentieth century. Possibly the appearance of books by the Nobel Laureate Gerhard Herzberg is one reason for heightened activity. The impact of his

Laboratory I A	Ref	Int	E P Low	E P High	J	Multiplet (No)
Fe I	I P 7.858	Anal A	List A		Feb 1943	
5166.286	J	4	0.00	2.39	4-5	$a^5D-z^7D^\circ$
5221.43	P		0.05	2.41	3-4	(1)
5247.052	V	1	0.09	2.44	2-3	
5254.956	V	1	0.11	2.46	1-2	
5250.212	V	1	0.12	2.47	0-1	
5110.414	B	10	0.00	2.41	4-4	
5168.901	B	4	0.05	2.44	3-3	
5204.582	J	2	0.09	2.46	2-2	
5225.533	V	1	0.11	2.47	1-1	
•5060.079	T	(1)	0.00	2.44	4-3	
5127.68	P	⊙	0.05	2.46	3-2	
5175.71	P		0.09	2.46	2-1	
4375.932	B	9	0.00	2.82	4-5	$a^5D-z^7F^\circ$
4427.312	B	10	0.05	2.84	3-4	(2)
4461.654	B	8	0.09	2.85	2-3	
4482.171	J	4	0.11	2.86	1-2	
4489.741	B	3	0.12	2.87	0-1	
4347.239	V	(1)	0.00	2.84	4-4	
4405.02	P	⊙	0.05	2.85	3-3	
4445.48	U	(1)	0.09	2.86	2-2	
4471.68	P	(1)	0.11	2.87	1-1	
m4325.74	P	Fe	0.00	2.85	4-3	
4389.244	J	2	0.05	2.86	3-2	
4435.151	J	2	0.09	2.87	2-1	
m4466.57	P	Fe	0.11	2.87	1-0	
4216.186	B	8	0.00	2.93	4-4	$a^5D-z^7P^\circ$
4206.702	J	3	0.05	2.99	3-3	(3)
4199.97	W	1	0.09	3.03	2-2	R
4134.343	V	(1)	0.00	2.99	4-3	
4149.76	P	(1)	0.05	3.03	3-2	
•4291.466	I	4	0.05	2.93	3-4	
4258.320	J	2	0.09	2.99	2-3	
3859.913	B	300R	0.00	3.20	4-4	$a^5D-z^5D^\circ$
3886.284	B	40R	0.05	3.23	3-3	(4)
3899.709	B	30R	0.09	3.25	2-2	
3906.482	B	8	0.11	3.27	1-1	
3824.444	B	50r	0.00	3.23	4-3	
3856.373	B	50r	0.05	3.25	3-2	
3878.575	B	100r	0.09	3.27	2-1	
3895.658	B	25r	0.11	3.28	1-0	
3922.914	B	25R	0.05	3.20	3-4	
3930.299	B	25R	0.09	3.23	2-3	
3927.922	B	30R	0.11	3.25	1-2	
3920.260	B	20r	0.12	3.27	0-1	
3719.935	B	250R	0.00	3.32	4-5	$a^5D-z^5F^\circ$
3737.133	B	150R	0.05	3.35	3-4	(5)
3745.561	J	100R	0.09	3.38	2-3	
3748.264	B	60R	0.11	3.40	1-2	
3745.901	J	40r	0.12	3.42	0-1	
3679.915	B	40r	0.00	3.35	4-4	
3705.567	B	100r	0.05	3.38	3-3	
3722.564	B	50r	0.09	3.40	2-2	
3733.319	B	40r	0.11	3.42	1-1	
3649.304	J	5	0.00	3.38	4-3	
3683.054	G	10	0.05	3.40	3-2	
3707.828	V	20	0.09	3.42	2-1	

Laboratory I A	Ref	Int	E P Low	E P High	J	Multiplet (No)
Fe I continued						
8047.60	O	15	0.86	2.39	5-5	$a^5F-z^7D^\circ$
8204.10	P	⊙	0.91	2.41	4-4	(12)
8310.98	P		0.95	2.44	3-3	
8382.23	P	⊙	0.99	2.46	2-2	
8425.89	P	⊙	1.01	2.47	1-1	
7912.866	E	6	0.86	2.41	5-4	
8075.13	O	4	0.91	2.44	4-3	
8204.93	P	⊙	0.95	2.46	3-2	
8307.61	P	⊙	0.99	2.47	2-1	
8349.05	P	⊙	0.91	2.39	4-5	
8447.63	P	⊙	0.95	2.41	3-4	
6358.692	I	3	0.86	2.80	5-6	$a^5F-z^7F^\circ$
m6462.72	P	Fe	0.91	2.82	4-5	(13)
6547.58	P		0.95	2.84	3-4	
6609.68	P	⊙	0.99	2.85	2-3	
6648.08	P	⊙	1.01	2.86	1-2	
6280.625	I	2	0.86	2.82	5-5	
6400.335	V	(2)	0.91	2.84	4-4	
6498.950	V	5	0.95	2.85	3-3	
6574.238	V	3	0.99	2.86	2-2	
6625.04	V	1	1.01	2.87	1-1	
6221.661	U	(-)	0.86	2.84	5-4	
6353.84	P	⊙	0.91	2.85	4-3	
6464.67	P		0.95	2.86	3-2	
6551.68	P	⊙	0.99	2.87	2-1	
6613.83	P	1	1.01	2.87	1-0	
5956.702	J	(3)	0.86	2.93	5-4	$a^5F-z^7P^\circ$
•5949.35	V	(2)	0.91	2.99	4-3	(14)
•5958.23	P	(2)	0.95	3.03	3-2	
6120.25	P	⊙	0.91	2.93	4-4	
5269.541	I	60	0.86	3.20	5-4	$a^5F-z^5D^\circ$
5328.042	I	50	0.91	3.23	4-3	(15)
5371.493	B	50	0.95	3.25	3-2	
5405.778	B	40	0.99	3.27	2-1	
5434.527	B	30	1.01	3.28	1-0	
5397.131	B	40	0.91	3.20	4-4	
5429.699	B	40	0.95	3.23	3-3	
5446.920	B	40	0.99	3.25	2-2	
5455.613	B	40	1.01	3.27	1-1	
5501.469	B	12	0.95	3.20	3-4	
5506.782	B	18	0.99	3.23	2-3	
5497.519	B	15	1.01	3.25	1-2	
5012.071	B	12	0.86	3.32	5-5	$a^5F-z^5F^\circ$
5051.636	B	10	0.91	3.35	4-4	(16)
5083.342	B	7	0.95	3.38	3-3	
5107.452	J	6	0.99	3.40	2-2	
5123.723	B	6	1.01	3.42	1-1	
4939.690	B	4	0.86	3.35	5-4	
4994.133	B	6	0.91	3.38	4-3	
5041.074	J	7	0.95	3.40	3-2	
5079.742	J	4	0.99	3.42	2-1	
5127.363	B	5	0.91	3.32	4-5	
5142.932	J	6	0.95	3.35	3-4	
5150.843	B	6	0.99	3.38	2-3	
5151.915	J	4	1.01	3.40	1-2	

Figure 11.3. (a) A Portion of Part I of the Multiplet Tables,

first book on molecular spectroscopy was undoubtedly lessened by World War II, but his revision on diatomic molecules (Herzberg 1950) was a well-thumbed bible of any astronomer interested in the study of molecular features. A later introductory volume (Herzberg 1971) is highly recommended.

The Doctoral Dissertation of A. Schadee (1964) provides a historical overview and introduction to the study of molecular features in the sun. Moore, Minnaert, and Houtgast (1966) list some 3500 solar wavelengths due wholly or in part to the diatomic molecules. The species identified at the time of that writing were CH, CN, C_2, NH, MgH, and OH. In his second edition of *Atmospheres of the Sun and Stars,*

FINDING LIST

I A	Type	Element	Multiplet No.	I A	Type	Element	Multiplet No.
3846.803		Fe I	664	3856.16		O II	12
3846.949		Fe I	176	3856.281		Cr I	69
3847.01		Zr I	10	3856.373		Fe I	4
3847.086		F II	1	3856.515		Rh I	7
3847.252		Mo I	8	3856.796		Co I	60
3847.323		V I	7	3857.032		Ce II	158
3847.323		V II	156	3857.18		O II	13
3847.38		N II	30	3857.240		Ce II	127
3847.501		W I	4	3857.26	P	Y II	16
3847.511		Sm II	34	3857.631		Cr I	69
3847.89		O II	12	3857.912		Sm II	28
3848.023		Tm II	2	3858.07		He II	4
3848.194		Y II	72	3858.133		Ti I	176
3848.233		Nd II	19	3858.301		Ni I	32
3848.24		Mg II	5	3858.32		A III	5
3848.29	P	Fe I	224	3858.48	P	Fe I	565
3848.524		Nd II		3858.90		Cr I	138
3848.597		Ce II	36	3859.214		Fe I	175
3848.779		Sm II		3859.24		Mg I	21
3848.983		Cr I	69	3859.26		S II	30
3849.02		La II	12	3859.33		Al II	38
3849.26		Zr I	6	3859.341		V I	44
3849.324		V I		3859.36.	P	Sc II	1
3849.365		Cr I	138	3859.913		Fe I	4
3849.52		Hf II	61	3860.12	P	Fe II	128
3849.534		Cr I	24	3860.13		Cr I	39
3849.58		Ni II	11	3860.15		S II	41
3849.758		V II	33	3860.46		Fe III	109
3849.969		Fe I	20	3860.64		S II	50
3849.987		F II	1	3860.64		S III	5
3850.042		Cr I	69	3860.74	P	Fe I	701
3850.40		Mg II	5	3860.80		Cl II	25
3850.409		V II	11	3860.915		Fe II	
3850.57		A II	10	3860.98		Cl II	25
3850.69		Gd II	2	3861.079		Ti I	
3850.81		O II	12	3861.164		Co I	33
3850.820		Fe I	22	3861.18		Eu II	
3850.825		Pr II		3861.341		Fe I	283,663
3850.93		S II	50	3861.40		Cl II	25
3850.945		Co I	17	3861.60		Fe I	663

Figure 11.3. (b) A Portion of the Finding List. Courtesy of National Institute of Standards and Technology and Princeton University Observatory. See text for discussion.

Aller (1963) reviewed the status of molecular spectroscopy in stars (see his Section 8-12). A symposium on cool stars (Jaschek and Keenan 1985) describes much work on molecular spectra in stars.

Modern literature on astrophysical applications of *stellar* molecular spectroscopy is perhaps best approached by reading the relevant parts of the Reports of Commission 29 of the International Astronomical Union, for example, Lambert (1988), or Gray (1991). IAU Commissions 29 and 45 sponsor a Working Group on Chemically Peculiar Red Giant Stars. This group publishes a newsletter currently distributed by electronic as well as regular mail. The present (summer 1993) editor is Sandra Yorka (1993, yorka@cc.denison.edu).

We now outline the nomenclature of the spectra and structure of diatomic molecules. We must leave technical descriptions to the literature; our approach must be primarily mnemonic and heuristic.

In a first approximation, the diatomic molecule may be considered to be two mass points bound by some force which may be taken to obey Hooke's law. The mass points, which belong to the two atomic nuclei, may rotate relative to the center of mass of the system, and they may vibrate along the line joining the axes. The "Hooke's-law" force is due, of course, to the electronic binding, whose consideration we shall take up shortly.

In the simplest approximation of all, the internuclear distance is considered fixed, and we have what is called the rigid rotator. The motion of one of the masses with respect to the other then follows the same law as in the two-body problem (the hydrogen atom) except that the radial variable is constant. The angular variables, usually called θ and ϕ, are described by the same functions as in any central force problem (electron in the hydrogen atom, nucleon in the shell model).

It will be recalled that these functions, the spherical harmonics, usually called Y_l^m, represented particles with angular momentum l (strictly $\sqrt{l(l+1)}$) in units of Planck's constant h divided by 2π, $\hbar \equiv h/(2\pi)$. In the case of the rigid rotator, it is usual to call the angular momentum quantum number J rather than l, and the projection of the angular momentum vector on the z-axis is usually designated M. Thus, we might write Y_J^M rather than Y_l^m for the solutions, but the properties of these solutions are the same.

Now the classical energy of a rotating, rigid body is given by the square of the angular momentum divided by twice the moment of inertia ($I = mr^2$), which is also called the inertial mass. If r is the separation of the two point masses, then m must be the reduced mass, $m_1 m_2/(m_1 + m_2)$.

The name inertial mass is of useful mnemonic value, since the energy of a free particle with momentum p, which everyone knows, is $p^2/(2m)$. The appropriate quantum mechanical expression for the energy of a rigid rotator is thus the square of the angular momentum divided by $2I$, or

$$E_{\text{rot}} = \frac{J(J+1)\hbar^2}{2I}. \tag{11.1}$$

It is also common in molecular spectroscopy to use reciprocal centimeters (wavenumbers) to specify the energy. We have the simple relation

$$E = h\nu = h\frac{c}{\lambda} = hc\tilde{\nu}, \tag{11.2}$$

where $\tilde{\nu}$ is used for wavenumbers. Thus the energy in wavenumbers may be converted to ergs or joules on multiplication by the product hc. This "energy" usually given in "reciprocal cm" is called the term value in molecular spectroscopy. (The reader should note that "term value" usually has a different meaning in atomic spectroscopy.)

The term values for the rigid rotator are written

$$F(J) = J(J+1)B, \tag{11.3}$$

where the constant B is readily seen from 11.1 and 11.2 to be given by

$$B = \frac{h}{8\pi^2 cI}. \tag{11.4}$$

It is easy to see that the characteristic energies of rotation of diatomic molecules will be much less than those of the excitation of atomic (or molecular) electrons. The moments of inertia I are given by the product of the reduced mass of the two mass points, μ, times their separation r. These separations are of the order of an angstrom, 10^{-8} cm. For comparison, the Bohr radius of the hydrogen atom is about half an angstrom. For the abundant diatomic molecules such as CN, or CO, the reduced mass is (very) roughly 10 atomic mass units ($1.6 \cdot 10^{-24}$ gm).

We therefore have moments of inertia for these diatomic molecules of the order of $10 \cdot 1.6 \cdot 10^{-24} \cdot (10^{-8})^2 = 1.6 \cdot 10^{-39}$. This leads, by Equation 11.4, to $B \approx 1.75\,\text{cm}^{-1}$. The difference in the energy between the rotational level with $J = 0$ and that with $J = 1$ is therefore $2B$, or about $3\,\text{cm}^{-1}$.

A wavenumber of $3\,\text{cm}^{-1}$ corresponds to a wavelength of 0.3cm, or a frequency of 10^{11} Hz (100000 MHz). This is the millimeter region, or the short-wavelength end of the microwave domain. Roughly, the microwave region of the spectrum means wavelengths of a millimeter to 100 cm, or frequencies from $3 \cdot 10^8$ to $3 \cdot 10^{11}$ Hz.

While optical photons have energies of the order of several electron volts, a photon with a frequency of 10^{11} Hz has an energy of about 0.0004 eV.

Because diatomic molecules may vibrate as well as rotate, the formula for the rotational energies is not simply given by 11.3, but includes terms that depend on the vibrational energy state. Let us examine these vibrational states now, first on the assumption that the rotation may be neglected. The molecule then is approximated by a quantum mechanical simple harmonic oscillator (SHO). We shall not work through this problem of introductory quantum mechanics. It leads to quantized energies

$$E(v) = h\nu_{\text{osc}}(v + 1/2), \tag{11.5}$$

where the vibrational quantum number v takes integral values $0, 1, 2, \ldots$. For an SHO, the quantity ν_{osc} is given by

$$\nu_{\text{osc}} = \frac{1}{2\pi} \cdot \sqrt{k/m} \tag{11.6}$$

for a force constant k and mass m.

A useful mnemonic is to think of Planck's postulate for photons, that the energies were multiples of $h\nu$. This gives the result 11.5 without the so-called *zero-point energy*, which we must remember. The relation between classical oscillators and photons is a remarkably close one in which the zero-point energy plays a key role. Unfortunately we cannot pursue it here.

If we express the vibrational energy in wavenumbers, we may write a vibrational *term value*

$$G(v) = \omega_e(v + 1/2). \tag{11.7}$$

The quantity ω_e is an empirically determined constant (see Table 11.1), related to the Hooke's-law-like force which binds the nuclei. For a given electronic state, this quantity changes slightly with the vibrational state of the molecule, making it necessary to introduce higher-order, empirical terms.

For the diatomic molecules, the vibrational energies (in cm^{-1}) are roughly multiples of the ω_e-parameter. A number of these are given in Table 11.1, and it can be seen that they are typically of the order of several thousand cm^{-1}. A wavenumber of 10^3 cm^{-1} corresponds to 100 000 angstroms, 10 microns, or 10 micrometers (μm). This is in the infrared region of the electromagnetic spectrum (about 0.75 μm to 1 mm). The infrared domain also contains the spectra of the vibrations of more complicated molecules.

In the following discussion, we shall make plausibility arguments to justify certain functional forms involving molecular "constants." These quantities are determined empirically from laboratory measurements, and tabulated.

Because the interplay between rotational, vibrational, and electronic structure of molecules leads to a congeries of energy levels and wavelengths it is not customary in molecular spectroscopy to tabulate them all. Instead, measurements of the various wavelengths are made, and the corresponding term values are fitted with the help of interpolation formulae. In the first approximation, these formulae are simply our equations 11.3 and 11.7. The constants B and ω would be different, of course, for each state of electronic excitation, and are therefore tabulated separately (see the references). An abbreviated table of molecular constants appears in Table 11.1, and the reader may find it helpful to refer to it as the various constants are introduced.

We mentioned that the "constant" B must be modified for a vibrating rotator. For an arbitrary vibrational state, one must replace B with B_v, where

$$B_v = B_e - \alpha_e(v + 1/2) + \cdots. \tag{11.8}$$

The subscript "e" may be thought of as implying the equilibrium position for the "$v = 0$" vibrational state. Similarly, B_v means "corrected for vibration." A higher-order correction, depending on the square of $(v + 1/2)$, may be used in 11.8.

In order to fit 11.3 to the data for actual molecules, a higher-order term in J *has* often been used. The complete expression for the rotational term values thus becomes

$$F(J) = J(J + 1) \cdot B_v - J^2(J + 1)^2 \cdot D_v. \tag{11.9}$$

There is a formula for D_v similar to 11.8, involving still another empirical constant β_e, but the equilibrium value D_e is often sufficient.

The vibrational levels are fitted as follows:

$$G(v) = \omega_e(v + 1/2) - \omega_e x_e(v + 1/2)^2 + \omega_e y_e(v + 1/2)^3 + \dots. \tag{11.10}$$

This functional form, with the clumsy notation $\omega_e x_e$ etc., arose for subtle historical reasons (see Townes and Schawlow 1975, §1-2). Data tables include ω_e, $\omega_e x_e$, and (sometimes) $\omega_e y_e$. For very high rotational states, it may be necessary to modify these rotational constants, but we shall not write the relevant formulae here.

We must now deal with the electronic states.

11.5 Diatomic Molecules: Electronic Structure and Wavelengths

In a diatomic molecule, there obviously cannot be spherical symmetry. Axial symmetry remains, and is of major importance in the classification of the electronic *states.* For homonuclear diatomic molecules such as H_2 or C_2, there is an additional symmetry with respect to exchange of the nuclei, which has a fascinating influence on the properties of these molecules. It should be noted that the distinction between the words "state," "level," and "term" which was made for atomic spectra is not typically made in molecular spectroscopy.

It may be useful to give at this point some data on actual diatomic molecules. The symbols in Table 11.1 that have not been explained thus far will be clarified below. The data are from the compilation of Huber and Herzberg (1979). Note that unlike the case of atomic energy levels, the lowest molecular states are listed last.

The component of the electronic angular momentum along the internuclear axis forms the basis for the classification of electronic states. If we were to write down and attempt to solve the wave equation for a single particle subject to a potential with two centers, we would find that the ϕ-dependent part could be separated, and would have solutions $\propto \exp(im\phi)$, where $m = 0, \pm 1, \pm 2, \ldots$. This is the same as in the atomic case. The equations describing the remaining two coordinates depend only on m^2, and not on m itself. This leads to electronic states independent of the *sign* of m. It is therefore customary to introduce

$$\lambda = |m|, \tag{11.11}$$

and use this parameter to describe the individual electronic wave functions.

The electronic states themselves are called σ, π, δ, ϕ, ..., for $\lambda = 0, 1, 2, 3, \ldots$. The solutions for the molecule as a whole are described by the *sum* of the orbital angular momentum vectors, $\boldsymbol{L}$. The absolute magnitude of the projections of $\boldsymbol{L}$ (in units of $\hbar$), usually called M_L, are the basis for the classification of molecular electronic states. The quantum number Λ is used for states with a given $|M_L|$. Those with $\Lambda = 0, 1, 2, 3, \ldots$ are called Σ, Π, Δ, $\cdots$.

The electronic orbital angular momentum $\boldsymbol{L}$ may have a z-component that points toward one or the other of the two nuclei. While it is the *projection* of this angular momentum that determines the electronic state, Σ, Π, Δ, etc., these two possible projections manifest themselves in quite different ways, for $\Lambda = 0$ (Σ-states), and for nonzero Λ-values. If Λ is nonzero, the states with the two projections have very nearly the same term value, that is, they are *degenerate.* This degeneracy can be lifted by molecular rotation, giving rise to a phenomenon known as Λ-*doubling*. In Σ states, the corresponding ambiguity usually causes widely separated states (see Problem 6). In one of these Σ states, the electronic orbital wave function will change sign on reflection in a plane containing the internuclear axis, while the other will not. The former state is called Σ^- and the latter Σ^+. These pluses and minuses apply to the *electronic* orbital wave-functions of Σ states only.

The total electronic angular momentum of spin determines the multiplicity of the states just as for atoms. The corresponding quantum number is also designated S. Thus, if $S = 1/2$, the molecular states are doublets, and written $^2\Sigma, ^2\Pi, ^2\Delta$, etc. The electronic spin sometimes has a large effect on the energy levels of the diatomic molecule. The electronic term value for a given S and Λ may be approximated by

$$T = T_0 + A\Lambda\Sigma, \tag{11.12}$$

where A is a constant that depends on the molecule, and Σ here is *the projection of the total electronic spin vector* $\boldsymbol{S}$ *on the internuclear axis.* The same symbol is used with the meaning $\Lambda = 0$, and the distinct meanings must simply be remembered. Unlike atomic states, molecular states have their full multiplicity, so there would be *four* $^4\Pi$ states. The algebraic sum of Λ and Σ *may* be appended to the term symbol. In the case of the $^4\Pi$, we would have $^4\Pi_{5/2}$, $^4\Pi_{3/2}$, $^4\Pi_{1/2}$, $^4\Pi_{-1/2}$, since for a Π state, Λ is always 1, while Σ ranges from $+(3/2)$ to $-(3/2)$. The quantum number $\Omega = |\Lambda + \Sigma|$ characterizes the projection of the total electronic angular momentum (orbit + spin). It is *usually* defined as an absolute value (Herzberg 1950, Eq. V,7; Townes and Schawlow 1975, see p. 177), but not always (Herzberg 1971, Eq. 48)!

The constant A in 11.12 may either be positive or negative, so that, for example, the $^4\Pi_{5/2}$ may lie either highest or deepest in energy of $^4\Pi$ states. Similar to the nomenclature of atomic spectroscopy, one speaks of regular or inverted multiplets, and the symbols r and i are often inserted as a part of the electronic state designation (see e.g., the ground states of TiO and OH in Table 11.1). For TiO, the energy splitting due to spin is appreciable.

The various angular momenta that influence the energy levels of a diatomic molecule may be combined in several ways. The situation is very much like that of the coupling schemes in atomic spectroscopy, where LS, jj, or some other scheme may provide a convenient description of atomic energy levels. In molecular spectroscopy, the coupling schemes are known as *Hund Cases*, and the most important of these are Case (a) and Case (b). These cases are well explained by Herzberg (1950, §V.2). Unfortunately, some of notation has changed since Herzberg's book was written. There are also notational differences between Herzberg (1971) and the microwave spectroscopists. He calls the angular momenta of nuclear end-over-end rotation $\boldsymbol{R}$, while Townes and Schawlow (1975) and Gordy and Cook (1984) call them $\boldsymbol{O}$.

For Σ states, Λ is zero, so the vectors $\boldsymbol{O}$ or $\boldsymbol{R}$ and $\boldsymbol{N}$ (see below) coincide. We show the coupling schemes for Cases (a) and (b) in Figure 11.4.

Cases (a) and (b) may be distinguished respectively by the relative importance (energies) of electronic spin–orbit interaction, and the rotational energies of the nuclei. Case (a) is the most common, apart from Σ states, where $\Lambda = 0$ in Equation 11.12, and there can be no spin–orbit energy. Herzberg (1950) discusses additional coupling cases which we cannot treat here. In general, intermediate coupling is needed to accurately describe molecular states.

The angular momentum in Case (b) apart from electronic spin is now called $\boldsymbol{N}$. It was called $\boldsymbol{K}$ by Herzberg (1950, Figure 100).

Electronic terms are labeled by letters. The recommended scheme (Jenkins 1953) is as follows. The ground term is called X, while higher terms with the same multiplicity are labeled with upper-case letters, $A, B, \ldots$. Other terms are labeled with lower-case letters, $a, b, \ldots$. Obviously, newly found terms may not fit into the recommended scheme.

Table 11.1. *Selected Constants for Diatomic Molecules. (Units are* cm^{-1} *except as indicated.)*

State	Te	ω_e	$\omega_e x_e$	B_e	α_e	D_e	r_e(Å)
		$^1H_2$	$D_0^0 = 4.4781_3$ eV				
$C\,^1\Pi_u 2p\pi$	100 089.8	2 443.77	69.524	31.362_9	1.664_7	2.23	1.03
$B^1\Sigma_u^+ 2p\sigma$	91 700.0	1 358.09	20.888	20.15_4	1.184 5	1.625	1.29
$X\,^1\Sigma_g^+ 1s\sigma^2$	0.0	$4\,401.21_3$	121.33_6	60.853_0	3.062_2	4.71	0.74
		$^{12}C_2$	$D_0^0 = 6.21$ eV				
$d^3\Pi_g$	20 022.50	1 788.22	16.440	1.752 7	0.016 08	6.7_4	1.27
$c^3\Sigma_u^+$	$13\,312._1$	$1\,961._6$	$13._7$	1.87			1.23
$A^1\Pi_u$	8 391.00	1 608.35	12.07_8	$1.616\,3_4$	$0.016\,8_6$	6.44	1.32
$b^3\Sigma_g^-$	$6\,434.2_7$	$1\,470.4_5$	11.1_9	1.485_2	0.016 34	6.22	1.37
$a^3\Pi_u$	716.2_4	1 641.35	11.67	$1.632\,4_6$	0.016 61	6.44	1.31
$X^1\Sigma_g^+$	0.0	1 854.71	13.34_0	$1.819\,8_4$	$0.017\,6_5$	6.92	1.24
		$^{12}C\,^{14}N$	$D_0^0 = 7.7_6$ eV				
$D^2\Pi_i$	54 486.3	$1\,004.7_1$	8.7_8	1.162	0.013	7.	1.50
$a^4\Sigma^{(+)}$	(32 400.)						
$B^2\Sigma^+$	25 752.0	2 163.9	20.2	1.973	0.023	[6.6]	1.15
$A^2\Pi_i$	9 245.28	$1\,812.5_6$	12.60_9	1.715 1	0.017 08	5.93	1.23
$X^2\Sigma^+$	0.0	2 068.59	13.087	$1.899\,7_4$	$0.017\,36_9$	6.40	1.17
		$^{12}C\,^{16}O$	$D_0^0 = 11.09_2$ eV				
$a'^3\Sigma^+$	$55\,825.4_9$	1 228.60	10.468	1.344 6	$0.018\,9_2$	6.41	1.352 3
$a^3\Pi_r$	48 686.70	$1\,743.4_1$	14.3_6	1.691 24	0.019 04	6.36	1.205 74
$X^1\Sigma^+$	0.0	2 169.814	13.288 3	1.931 3	0.017 5	6.121	1.128
		$^{16}O\,^1H$	$D_0^0 = 4.392$ eV				
$X\,^2\Pi_i$	0.0	$3\,737.76_1$	84.881_3	18.910_8	0.7242	19.38	0.969 66
		$^{48}Ti\,^{16}O$	$D_0^0 = 6.87$ eV				
$C\,^3\Delta_r$	19617.0	838.26	4.76	0.489 89	0.003 06	6.7	1.69
	19 525.5						
	19 427.12						
$B^3\Pi_r$	$16\,331._3$	875.	5.	[0.506 17]		[6.86]	[1.67]
	$16\,315._1$						
	$16\,293._5$						
$b^1\Pi$	$a + 11\,322.0_3$	[911.20]	(3.7_2)	0.513 37	$0.002\,9_1$	6.1	1.65
$A^3\Phi_r$	14 431.0	867.78	3.942	0.507 39	0.003 15	6.92	1.66
	14 262.8						
	14 089.91						
$E^3\Pi$	12 025.	$924._2$	5.1				
$d^1\Sigma^+$	$a + 2\,215.6$	[1 014.6]	(4.6_4)	0.549 22	0.003 37	[6.0]	1.60
$a^1\Delta$	a	[1 009.3]	3.9_3	0.537 60	0.002 98	5.9	1.62
$X^3\Delta_r$	197.5	1 009.02	4.498	0.535 41	0.003 01	6.03	1.62
	96.4						
	0.0						

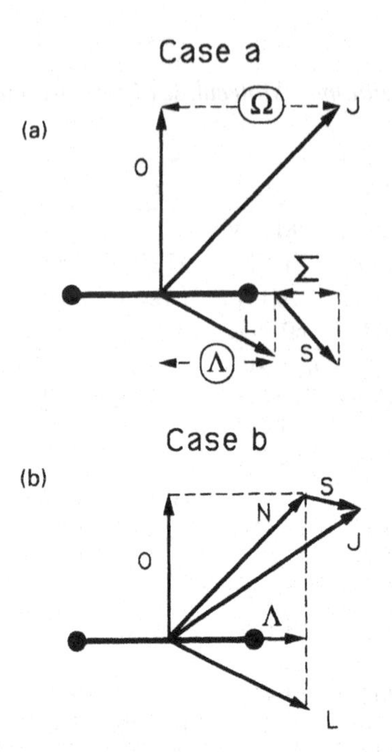

Figure 11.4. Hund's Cases (a) and (b).

For homonuclear diatomic molecules, the states are classified depending upon whether the electronic, orbital wave functions change sign on a reflection at the point of symmetry on the internuclear axis. If the wave function changes sign, the states are called *u*, for ungerade, which means uneven in German. If the wave function does not change sign, the state is called *g*, or *gerade*, meaning even.

We have now assembled sufficient nomenclature for an understanding of the data given by Huber and Herzberg (1979). Table 11.1 presents a sample of the information that is available for the molecules H_2, C_2, CN, CO, OH and TiO. Let us now turn to the use of such data in calculation of the wavelengths in molecular bands.

To a fair approximation, the energy of a molecular level of a diatomic molecule may be written as the sum of rotational, vibrational, and electronic terms:

$$T = T_e + G(v) + F(J), \tag{11.13}$$

where the $G(v)$ and $F(J)$ are discussed in §11.4.

According to the adopted notation of molecular spectroscopists, lower states are designated by double primes, and upper ones by single primes. In specifying a transition, the upper state is always written first. Neither of these conventions is followed in atomic spectroscopy.

The wavenumber for a transition in a diatomic molecule is thus $\tilde{\nu} = T' - T''$. Let us hold the electronic and vibrational energies fixed, and examine the structure imposed by the rotational energies on a series of lines. The resulting structure is called a band.

Let us also ignore any possible symmetries that might be imposed by the electronic structure, and suppose the rotational wave functions are adequately described by the rigid rotator. The selection rule on J is then the same as that on l for an atomic electron, since both wave functions are the spherical harmonics: $Y_{l\,\mathrm{or}\,J}^{m\,\mathrm{or}\,M}$. Thus, the selection rule on J is $\Delta J = 0, \pm 1$, with the additional proviso that $J = 0 \nleftrightarrow J = 0$.

The wavenumbers for a transition may then be written

$$\tilde{\nu} = \tilde{\nu}_0 + J'(J'+1)B' - J''(J''+1)B'', \qquad (11.14)$$

where the $\tilde{\nu}_0$ contains differences due to electronic and vibrational energies. Transitions where $J' = J'' + 1$ are said to fall in the R branch, while those with $J' = J'' - 1$ lie in the P branch. For lines in either branch, we may eliminate one or the other of the J values in 11.14, since there is a simple relation between them. It turns out that a clever substitution allows us to combine both branches in a single expression, with one running variable.

For the R branch, put

$$J' = J'' + 1 = m. \qquad (11.15)$$

The possible (emission) transitions are then, writing the upper J-value first, $(1 \to 0)$, $(2 \to 1)$, $(3 \to 2)$, etc., which correspond to $m = 1, 2, 3$, etc. Likewise with the P branch, put

$$J'' = J' + 1 = -m. \qquad (11.16)$$

The possible (emission) transitions, with the upper J written first, are then $(0 \to 1)$, $(1 \to 2)$, $(2 \to 3)$, etc. The corresponding values of m are $-1, -2, -3$, etc. The reader may verify that *either* 11.15 or 11.16 leads to the expression

$$\tilde{\nu} = \tilde{\nu}_0 + (B' + B'')m + (B' - B'')m^2. \qquad (11.17)$$

Positive values of m correspond to R branch transitions with negative ones giving P branch transitions. Let us plot m as the ordinate and $\tilde{\nu}$ as the abscissa. Then clearly Equation 11.17 is a parabola, which opens to the right or to the left depending on the sign of the difference $(B' - B'')$.

Two parabolas are plotted in Figure 11.5. They illustrate the structure of the $v' = 0$, $v'' = 0$ band of the $B^2\Sigma^+ \to X^2\Sigma^+$ transition in $^{12}C^{14}N$ and its isotopic congener $^{14}C^{14}N$. The curves were generated with the rotational constants from Table 11.1. We will discuss the isotopic effect in a moment.

Note that $\tilde{\nu}$ increases to the left, so wavelength increases from left to right, which is the standard way in which astronomical spectra are presented. Consider first only the solid line. For positive values of m, $\tilde{\nu}$ increases (left, in the plot). For negative m, $\tilde{\nu}$ first decreases (right), but finally increases (left). This can only happen if the quantity $(B' - B'')$ is *positive*. At the bottom of the plot, short vertical lines are drawn for integral values of m to illustrate the band wavelengths. The lines in this band therefore crowd together on the red side of the structure forming what is called a band head. The band is said to be *degraded to the violet*. This was the situation, one may recall, with Secchi's Class IV, and we now know that the features he saw were

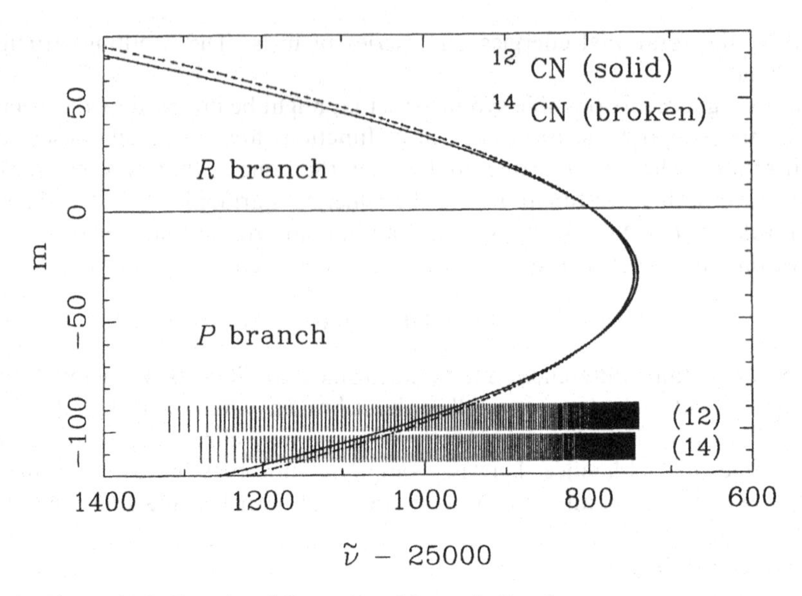

Figure 11.5. Rotational Structure of Isotopic Bands

due to C_2. He saw bands from what are known as the Swan bands, $d^3\Pi_g - a^3\Pi_u$. In the Type III stars, Secchi saw the α-system in TiO, $C^3\Delta_r - X^3\Delta_r$. We will describe the naming of band systems below.

All of these molecular states appear in Table 11.1; $B' - B''$ is positive for CN and C_2, and negative for TiO.

When $B' - B''$ is negative, the parabola will open to the right (decreasing $\tilde{\nu}$), and the band head will occur for positive m, that is, in the R branch. The head will fall at the violet side of the absorption, and the band will be degraded to the red.

Astronomical and molecular spectroscopists often refer to *named* bands. We have mentioned the α-system of TiO, and the *Swan* bands of C_2. Figure 11.5 illustrates a band in the *Violet System* of CN. Huber and Herzberg's (1979) tables give many of these designations in the columns labeled "Observed Transitions." The Swan bands of C_2 are identified on p. 112 of their book by 'd ↔ a'. Here the 'd' and 'a' are used as labels for the $^3\Pi_g$ and $^3\Pi_u$ respectively. We have used Huber and Herzberg's labels throughout our Table 11.1.

Other good sources for the names of band systems are Pearse and Gaydon's (1976) *The Identification of Molecular Spectra* and Rosen (1970).

Let us now return to the broken line in Figure 11.5, which also illustrates the effect of *isotopes* on band structure. The situation can occur with *isotopic molecules* such as $^{13}C\,^{12}C$, and $^{12}C_2$, or $^{12}C^{14}N$ and $^{13}C\,^{14}N$. The atomic masses enter into the rotational constants B, and the positions of the rotational lines change slightly if a diatomic molecule is composed of different isotopes. For $^{14}C^{14}N$, the rotational constants for Figure 11.5 were generated by multiplying the B's for $^{12}C^{14}N$ by $\mu/\mu_{(14)}$, the ratio of the reduced masses of the respective molecules. This effect provides one of the few sources of information on isotopic abundances of elements in cosmic sources beyond the solar system (see Jorissen, Smith, and Lambert 1992, Chapt. 14).

11.6 Rotational Structure of Symmetrical Top Molecules

The US National Bureau of Standards published a five-volume series entitled *Microwave Spectral Tables*, beginning with the work of Wacker, Mizushima, Peterson, and Ballard (1964). The material in these volumes is of general interest to the student of molecular spectra, and we shall have occasion to refer to individual volumes below.

The symmetrical top is one of the few problems that can be solved completely in both classical and quantum mechanics without resort to perturbation theory. While only a few molecules are accurately described as symmetrical tops, many may be satisfactorily described as perturbed symmetrical tops. It is therefore convenient to begin our discussion with the symmetrical top.

The classical expression for the rotational kinetic energy of a rigid body is

$$E = \left(\frac{J_x^2}{2I_{xx}} + \frac{J_y^2}{2I_{yy}} + \frac{J_z^2}{I_{zz}} \right). \tag{11.18}$$

Here the I's are the principal moments of inertia; the *products of inertia* vanish for a symmetrical top. Moreover, we have either $I_{xx} = I_{yy} > I_{zz}$ for a *prolate* (North American football) top, or $I_{xx} = I_{yy} < I_{zz}$ for an *oblate* (spinning planet) top. So

$$E = \frac{J_z^2}{2I_{zz}} + \frac{1}{2I_{xx}} \left(J_x^2 + J_y^2\right). \tag{11.19}$$

Using J for the *total* angular momentum, we can write

$$J_x^2 + J_y^2 = J^2 - J_z^2, \tag{11.20}$$

whence

$$E = \frac{J_z^2}{2I_{zz}} + \frac{1}{2I_{xx}} \left(J^2 - J_z^2\right). \tag{11.21}$$

We cannot present the quantum mechanical derivation of the quantized energy levels here. In a typical derivation, the expression for the kinetic energy is transformed from its form in rectangular Cartesian coordinates to the equivalent form in terms of three Euler angles, e.g., θ, ϕ, and χ. The resulting equation may be factored into three second-order, linear differential equations that are solved by power series methods. The subtle part of this procedure is to obtain the wave equation with the Euler angles as independent variables.

The interested reader may find a derivation in the book by Margenau and Murphy (1956, p. 368). A detailed solution, although not a derivation of the wave equation itself, may be found in the standard reference by Pauling and Wilson (1935). Surprisingly, the well-known reference by Edmonds (1960) contains a conceptually flawed derivation of this equation. The solution is usually written

$$\psi = \Theta_{JKM}(\theta) \exp(iK\chi) \exp(iM\phi). \tag{11.22}$$

The functions $\Theta_{JKM}(\theta)$ are essentially hypergeometric functions, the so-called ${}_2F_1(a,b,c;x)$'s, power series in x. The particular ${}_2F_1$'s that form the solution to the symmetric top are closely related to the *Jacobi polynomials*, described, for example, by Abramowitz and Stegun (1964) and in books that deal with the special functions of mathematical physics. Unlike the *confluent* hypergeometric functions of the hydrogen atom, the Jacobi polynomials contain *three* parameters, which we can take to be J, K, and M. J describes the total angular momentum, and K its projection on the z-axis. The quantum number M is a projection of the angular momentum on a z-axis that is *fixed* in the top. For the symmetric top, the energies do not depend on M.

We must now resort to a heuristic argument, and *assume* without proof, but using the analogy of atomic electron orbits, that the *total* angular momentum is quantized with values whose *square* is $\hbar 2J(J+1)$, where J is zero or a positive integer. Similarly, we assume that the z-component of the angular momentum is quantized, with values $\hbar K$, where $K = 0, \pm 1, \pm 2, ..., \pm J$, etc. The quantum number M likewise takes on values $0, \pm 1, \pm 2, ..., \pm J$, etc., but the *symmetric top* is completely degenerate with respect to M. The top is also degenerate with respect to the *sign* of K. The solutions are thus $(2J+1)$-fold degenerate ($K = 0$), and $(4J+2)$-fold degenerate ($K \neq 0$).

The correct quantum mechanical expression for the energy levels of a symmetrical top is thus

$$E = \hbar^2 \frac{K^2}{2I_{zz}} + \frac{\hbar^2}{2I_{xx}} \left[J(J+1) - K^2 \right]. \tag{11.23}$$

As in the case of diatomic molecules, it is customary to write the energy of symmetric top molecules in wavenumbers. Accordingly, molecular constants A, B, and C are defined (as for diatomic molecules). By international convention, these constants are always defined in such a way that $A > B > C$. Because of the naming convention, for the rotational constants, it has been necessary to introduce new expressions for the moments of inertia in 11.24 below. I_{AA} is the *smallest* of the three principal moments of inertia, and I_{CC} the largest. Thus

$$A = \frac{\hbar^2}{2I_{AA}hc}, \quad B = \frac{\hbar^2}{2I_{BB}hc}, \quad C = \frac{\hbar^2}{2I_{CC}hc}. \tag{11.24}$$

We must also write the expression for the energies in cm^{-1} or the *term values* separately for oblate and prolate symmetrical top molecules. We have

$$F(J,K) = AJ(J+1) + (C-A)K^2, \tag{11.25}$$

for the oblate case, or

$$F(J,K) = CJ(J+1) + (A-C)K^2, \tag{11.26}$$

for the prolate case.

Rotational energy level diagrams for symmetrical tops are usually plotted with the energy increasing with the vertical direction, along with the total angular momentum J, while the absolute magnitude of K is plotted increasing to the right. Since A is

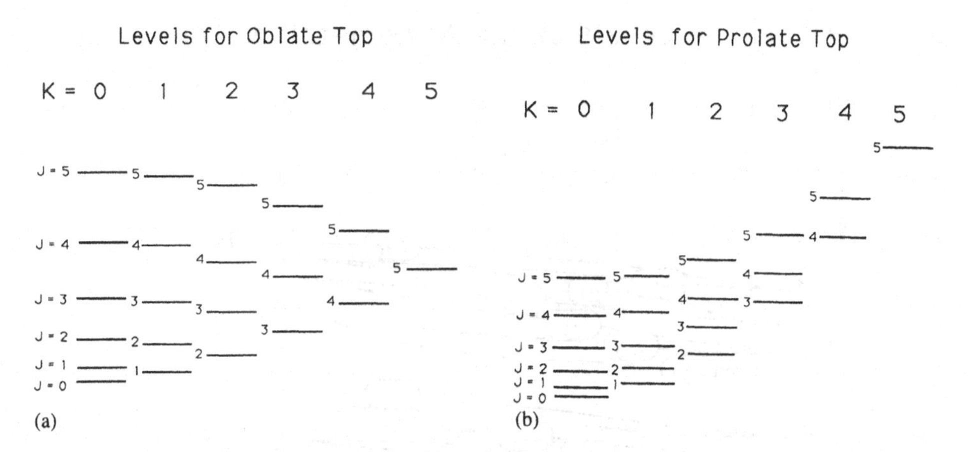

Figure 11.6. Rotational Energy Levels for Symmetrical Top Molecules: (a) Oblate; (b) Prolate.

always larger than C, it is clear from Equations 11.25 and 11.26 that for oblate tops, the energy *decreases* with increasing $|K|$, while for prolate tops, the energy *increases* with $|K|$. This is illustrated in Figure 11.6.

Selection rules for electric dipole transitions among the energy levels for the pure symmetrical top are simple. They are $\Delta J = 0, \pm 1$, $\Delta K = 0$, with $J = 0 \nrightarrow J = 0$. The selection rules for real molecules are considerably more involved as a result of additional degrees of freedom in the form of vibration, electronic orbital motion, and electronic and nuclear spin. These are discussed in standard works, such as the books by Townes and Schawlow (1975) or Herzberg (1966). Most working cosmochemists hardly need to carry these complicated rules in their heads. Identified interstellar microwave frequencies are tabulated, for example by Lovas (1992). Volume V of the NBS *Microwave Spectral Tables* (Cord, Lojko, and Peterson 1968) includes a listing by frequency of microwave transitions for some 260 polyatomic molecules.

11.7 Rotational Structure of Asymmetrical Top Molecules

Most polyatomic molecules are *not* symmetrical tops, and closed-form solutions such as 11.22 are not known. Nevertheless, the asymmetrical top may be regarded as a "solved" problem, the solution being achieved with the help of perturbation theory. The wave functions for the asymmetrical tops are written as series expansions in the complete, orthogonal functions of the symmetrical top. The perturbation can remove the degeneracy that is associated with the sign of the angular momentum quantum number K of the symmetrical top. The wave functions and consequently the energies share the properties of both the prolate and oblate symmetrical top. The energy levels may be labeled by three numbers, rather than only two, as for the symmetrical top. The numbers may be chosen to correspond to the nearest values of K for the prolate and oblate tops respectively (see Figure 11.7).

The *Microwave Spectral Tables* (Cord, Peterson, Lojko, and Haas 1968) use J, K_{-1},

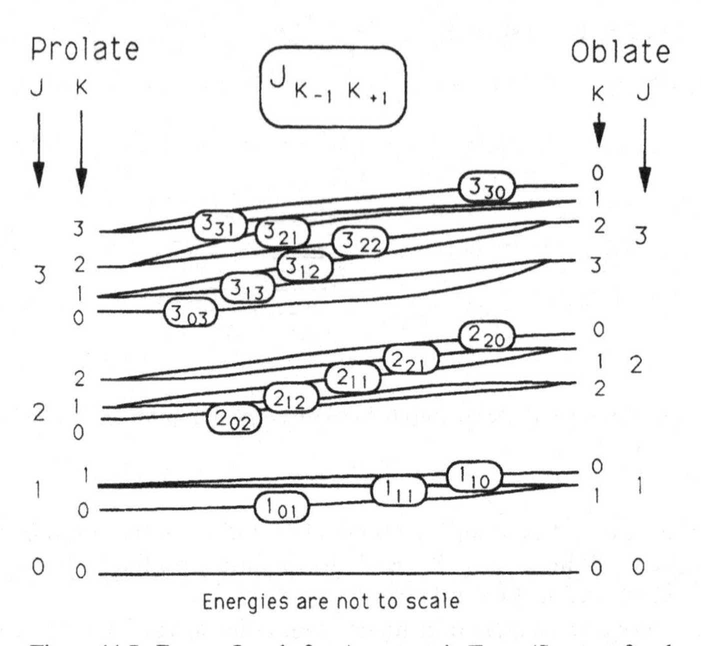

Figure 11.7. Energy Levels for Asymmetric Tops. (See text for description).

and K_{+1}. If the top is "nearly" prolate, then the energy levels are close to those of the prolate top with K having the value K_{-1}. Similarly, if the top is nearly oblate, the energies are close to those calculated for the oblate symmetrical top using $K = K_{+1}$.

Figure 11.7 illustrates the scheme for labeling energy levels for asymmetric rotators. Suppose the top is very nearly prolate. Then $B \approx C$, and we can expect that Equation 11.25 would give approximately correct answers for the energy levels if we took an average of B and C. Indeed, while we must refer to the references for a proof, it is possible to write the rotational term values for *any* asymmetric top with the help of an *asymmetry parameter*

$$b = \frac{C - B}{2\,[A - 0.5(B + C)]}. \tag{11.27}$$

Clearly $b = 0$ for the prolate top ($C = B$), and it is -1 for the oblate top. We can not prove it here, but one may then calculate a quantity W_τ that is a function of b such that W_τ plays the role of K in the symmetric top:

$$F(J, \tau) = \frac{1}{2}(B + C)J(J + 1) + \left[A - \frac{1}{2}(B + C)\right] W_\tau. \tag{11.28}$$

Townes and Schawlow (1975) as well as Herzberg (1945) give a series of algebraic equations for the parameter W_τ. For each J, there are a sufficient number of equations for $2J+1$ real roots, and the parameter τ may be used to label each of these roots. Indeed, we may take $\tau = K_{-1} - K_{+1}$. Then τ runs from $-J$ to $+J$ just as M does for the symmetrical top. It may be simplest to think of the τ-parameter as a label for the new levels that appear when the $(\pm K)$-degeneracy of the symmetrical top is lifted by the *asymmetry*.

Equations 11.27 and 11.28 are convenient when the asymmetrical top is more nearly prolate than oblate ($B \approx C$). Equivalent relations may be found in Herzberg or Townes and Schawlow for the nearly oblate top.

The theory of asymmetrical tops presented thus far is sufficient for us to understand the *nomenclature* of references such as those of Lovas (1992). In general it is not sufficient for us to accurately calculate the energy levels and microwave transition frequencies. Even in the case of molecules that are in their lowest vibrational level, *centrifugal distortions* introduce significant changes in the rotational constants with increasing J. For water, one of the worst cases, the centrifugal distortions cause significant deviations from the calculated energy levels even for $J = 1$!

11.8 Nuclear Effects in Atomic and Molecular Spectra: Hyperfine Structure

In this section we shall first review several ways in which properties of the atomic nucleus manifest themselves in atomic spectra. We then consider similar effects in molecular spectra.

The nucleus manifests itself in atomic spectra in three distinct ways. First, the reduced mass of the electron,

$$\frac{1}{\mu} = \frac{1}{M} + \frac{1}{m_e}. \tag{11.29}$$

depends slightly on the value of the mass of the nucleus M, and this has an influence on the atomic energy levels. One easily derives from Bohr theory, or from dimensional analysis (see Problems below), that the energy levels depend on the first power of the reduced mass. From Equation 11.29 it is clear that M will be important only for the lightest nuclei.

For atoms with more electrons, there are two terms in the isotopic mass effect, which may be derived following Clark (1984) by considering the kinetic energy of the nucleus in the center-of-mass system of the atom. If we choose our coordinate system at this center of mass, then the momenta of the electrons, $\boldsymbol{p}_i$, and the nucleus, $\boldsymbol{P}$, must obey the relation

$$\boldsymbol{P} = -\sum_i \boldsymbol{p}_i. \tag{11.30}$$

We may therefore write the *kinetic energy of the nucleus* as

$$T_N = \left(\frac{1}{2M}\right) \cdot \sum_i \boldsymbol{p}_i^2 + \left(\frac{1}{M}\right) \cdot \sum_{i,j>i} \boldsymbol{p}_i \cdot \boldsymbol{p}_j. \tag{11.31}$$

These terms are small, because of the factors $(1/2M)$, and may be taken as perturbations. The first term arises in Bohr theory and gives what is called the normal

Table 11.2. *Isotope Shifts for* $\lambda(^4\mathrm{He\ I})-\lambda(^3\mathrm{He\ I})$.

Wavelength	$\Delta\lambda$(Å)	Wavelength	$\Delta\lambda$(Å)
3819	0.08	4713	0.07
3867	0.08	4921	0.33
4026	0.08	5015	0.21
4121	0.07	5875	0.04
4388	0.28	6678	0.50
4472	0.07		

mass shift. As we can see from 11.29, μ increases with the nuclear mass M. The levels of the more massive isotope lie deeper in their potential wells. The absolute (negative!) energy difference of the excited levels of isotopes is smaller than that of the lower levels, since all excited levels converge (from negative) to zero potential energy. The net result is that the normal isotope effect displaces the wavelengths of the heavier isotope toward the violet. The wavenumber differences $\Delta\tilde{\nu}$ for the normal isotope shifts are defined to be positive: $\tilde{\nu}(\text{heavy}) - \tilde{\nu}(\text{light}) > 0$.

The second term in 11.31 gives rise to the *specific mass shift*, which can be positive or negative. It may be calculated or measured in the laboratory (see Clark 1984, and references therein).

Hydrogen and some of the helium lines have measurable isotopic shifts. Moore (1949) lists separate energy levels for deuterium and tritium, and her Multiplet Tables list a few Balmer and Paschen lines (see §15.2) for deuterium. Because of the typical great breadths of atomic hydrogen lines, the relatively minor wavelength shifts that result from a small admixture of deuterium in a stellar spectrum are masked. ^{3}He, on the other hand, was discovered in the optical spectrum of the star 3 Cen A by Sargent and Jugaku (1961). This early B star belongs to a class described by Hartoog and Cowley (1979). We reproduce their table of isotopic shifts here (Table 11.2; see also Fred *et al.* 1951).

Considerable interest has attached to observations of the light elements, lithium, beryllium, and boron (§§10.5–10.7), because of a variety of connections their abundances have with stellar evolution, nucleosynthesis, and cosmology. Little information is presently available about isotopic abundances of these elements from their *atomic* spectra. Most attention has been devoted to the Li I doublet ($^2S_{1/2}$–$^2P_{1/2,3/2}$) near λ6707. The isotope shift $\lambda(^6\mathrm{Li})-\lambda(^7\mathrm{Li}) \approx 0.16$ Å, given, for example, by Hughes (1955), is quite large with respect to the wavelength accuracy of single lines with high-resolution spectrographs (≈ 0.01 Å). The situation is complicated because the line is a doublet with other atomic lines nearby (cf. Hobbs 1985). Thus far there is only fragmentary evidence for non-SAD abundance ratios of $^6\mathrm{Li}/^7\mathrm{Li}$ from stellar absorption lines (see Smith 1993).

Nuclear volume effects manifest themselves in heavier atoms. Here, the finite size of the nucleus causes departures from a strict $(1/r)$-potential, and this is reflected in the atomic energy levels. The magnitude of the shifts depends on the amount of overlap of the electronic wave function with the nucleus, and it is therefore largest for s-electrons.

Two transitions that are particularly sensitive to the nuclear volume effects are $5d^7 6s^2\ ^4F_{5/2}$–$5d^8(^3F)6p\ ^4D^o_{5/2}$ of Pt II, λ4046, and $5d^9 6s^{2\,2}D_{5/2}$–$5d^{10}(^1S)6p\,^2P^o_{3/2}$ of Hg II, λ3984. For the Hg II line, the shift from ^{198}Hg to ^{204}Hg is 0.22 Å (cf. Cowley and Aikman 1975). The behavior of these features in stars is discussed by Dworetsky (1986), who predicted the isotope shifts in Pt II later measured by Engleman (1989).

Both the mass and nuclear volume effects are sometimes included in the general term hyperfine structure, although the latter often is restricted to those effects that depend on the nuclear spin. The theory of all of these effects is discussed in many texts, for example, by Cowan (1981, pp. 505–511).

The energy difference in atomic levels belonging to the same term (e.g., ^2P) but with different values of J (e.g., $^2P_{1/2}$ and $^2P_{3/2}$) is referred to as *fine structure*. This splitting arises because of the interaction of the electron's orbital angular momentum $\boldsymbol{L}$ with the spin $\boldsymbol{S}$. Hyperfine structure arises in an entirely analogous way, from nuclear spin.

Atomic nuclei with unpaired nucleons have both integral and half-integral spins. The spin angular momentum vector is designated $\boldsymbol{I}$, and it has the usual magnitude $\sqrt{I(I+1)}\hbar$. $\boldsymbol{I}$ and $\boldsymbol{J}$ form resultants $\boldsymbol{F}$ according to the same rules that $\boldsymbol{S}$ and $\boldsymbol{L}$ add to form $\boldsymbol{J}$, viz.

$$\boldsymbol{F} = \boldsymbol{J} + \boldsymbol{I}. \tag{11.32}$$

If $I = 1/2$ and $J = 2$, the possible values of F are 5/2 and 3/2, where, strictly speaking, we mean $\boldsymbol{F}$ has the magnitude $\sqrt{(5/2)(7/2)}\hbar$, and $\sqrt{(3/2)(5/2)}\hbar$.

Hyperfine splitting for some atomic lines can amount to several tenths of an angstrom, and calculations of stellar line strengths must take it into account. The widths of the hyperfine patterns must be measured, or calculated with a reasonably sophisticated atomic structure code. The relative strengths of the lines within a hyperfine multiplet may be predicted with comparative ease (see Cowley and Frey 1989 and references therein).

When hyperfine structure is large, it may be necessary to consider a level as being defined by L, S, J *and* F rather than just L, S, and J. The corresponding statistical weight is then $2F + 1$. Usually, we can ignore the weights that arise from nuclear spin, because they cancel from ratios that are physically important. When the splitting is large, one must take care to get the weight factors correctly for the atomic levels and in the partition functions for the atoms and ions.

We now turn to a discussion of the effects of nuclear spin on molecular spectra. We have already discussed the influence of isotopic masses on the vibrational spectra of diatomic molecules. Similar effects are observed for polyatomic molecules, but they are naturally much more complicated, and cannot be covered here (see Herzberg 1945, §II.6). Nuclear volume effects, analogous to those we discussed for the atomic Hg II and Pt II lines, may very well be measurable, but are not known to be of cosmochemical interest, and in any case may be hopelessly mixed with other complexities of molecular spectra.

The relatively simple influence of nuclear spin structure on molecules in their ground electronic states is of great contemporary interest because it can be observed in interstellar molecules. In principle, relative isotopic abundances of the elements in molecular clouds may be determined for hydrogen, carbon, oxygen, nitrogen, and a few other elements. Such information is difficult to obtain from optical spectroscopy. We shall discuss observational results in §14.2.

In molecules, we have the possibility of a contribution to the total spin from more than one nucleus. We shall confine ourselves here to the case where only a single nucleus with angular momentum of spin $\boldsymbol{I}$ couples to the total angular momentum of spin $\boldsymbol{J}$ of the molecule. For this simple case, Equation 11.32 above is valid, and we can predict the resulting angular momentum states from the vector model.

The ground electronic states of *most* diatomic molecules are Σ-states. The same statement is true of polyatomic molecules for which this description applies (e.g., linear molecules). It is also typically true that the total angular momentum of spin is *usually* zero. The astrophysically important CN and CH molecules, whose ground electronic states are $^2\Sigma$ and $^2\Pi$, are among a number of important exceptions, which also include OH (see Table 11.1 and below).

The most important interstellar microwave molecular lines are arguably those due to rotational transitions in the ground electronic ($^1\Sigma$, and vibrational $v = 0$) states of the CO molecule. The most abundant isotopic species of carbon and oxygen, ^{12}C and ^{16}O, have spinless nuclei. The rotational levels are therefore unsplit by hyperfine structure, and give a single microwave line at 0.26 cm or 115 271 MHz. Recently, transitions from ^{12}C^{17}O have been observed in the infrared source known as B335 (see Chapter 14 for nomenclature of sources). The ^{17}O nucleus has spin 5/2. Applying the ordinary rules for the addition of angular momenta, we expect the ground rotational level to be unsplit with $I = 5/2$. The $J = 1$ level would split into three hyperfine levels, with $I = 7/2, 5/2$, and 3/2. The corresponding lines are listed in Table 11.3.

For ^{12}C^{14}N, the hyperfine structure is already complicated, even for the lowest rotational transition, because of the unpaired electronic spin which makes the ground state a $^2\Sigma$. The structure is shown in Figure 11.8, although not to scale (see Turner and Gammon 1975). The lowest rotational level thus has a total angular momentum J of 1/2, which can couple with the nuclear spin of ^{14}N, $I = 1$, to give $F = 3/2$ and 1/2. The first excited rotational term is split into two sub-levels by the electronic spin, giving $J = 3/2$ and 1/2. *Each* of these is split by the nuclear spin. The $J = 3/2$ is split into three levels with $F = 5/2, 3/2$, and 1/2, while the $J = 1/2$ is split into two levels with $F = 3/2$, and $F = 1/2$ (see Figure 11.8). Of the nine possible transitions, eight are found in the source OriMC-1. They are listed in Table 11.3.

Figure 11.8 also shows the structure of the four lowest hyperfine levels in the OH molecule. This is an example of Λ-doubling, already mentioned in §11.5. The main splitting, giving rise to the two pairs of levels marked + and − are due to this effect. The possible values of Σ, the projection of $\boldsymbol{S}$, are + and $-1/2$, giving rise to doublets, as in the case of CN. Unlike CN, the ground electronic state of OH is a $^2\Pi_{3/2}$. It may be described by *Hund's Case* a, in which $|\Lambda + \Sigma| = \Omega$, the number that appears on the term designation as a subscript. As discussed in §11.5, the degeneracy associated with the direction of $\boldsymbol{L}$ (for a fixed Λ) can be lifted by a perturbation. The Λ-doubling splits the level depending on whether the angular momentum vector points toward the O or the H nucleus.

Table 11.3. *Microwave Molecular Transitions and Hyperfine Structure in CN and CO*

Molecule	Frequency (MHz)	N'	J'	F'	N''	J''	F''
$^{12}C^{16}O$	115 271.204	1	1	1	0	0	0
$^{12}C^{17}O$	112 358.780	1	1	3/2	0	0	5/2
	112 358.988	1	1	7/2	0	0	5/2
	112 360.005	1	1	5/2	0	0	5/2
$^{12}C^{14}N$	113 123.4[a]	1	1/2	1/2	0	1/2	1/2
	113 144.192	1	1/2	1/2	0	1/2	3/2
	113 170.528	1	1/2	3/2	0	1/2	1/2
	113 191.317	1	1/2	3/2	0	1/2	3/2
	113 488.140	1	3/2	3/2	0	1/2	1/2
	113 490.982	1	3/2	5/2	0	1/2	3/2
	113 499.639	1	3/2	1/2	0	1/2	1/2
	113 508.944	1	3/2	3/2	0	1/2	3/2
	113 520.414	1	3/2	1/2	0	1/2	3/2

[a] Not observed in Or:MC–1

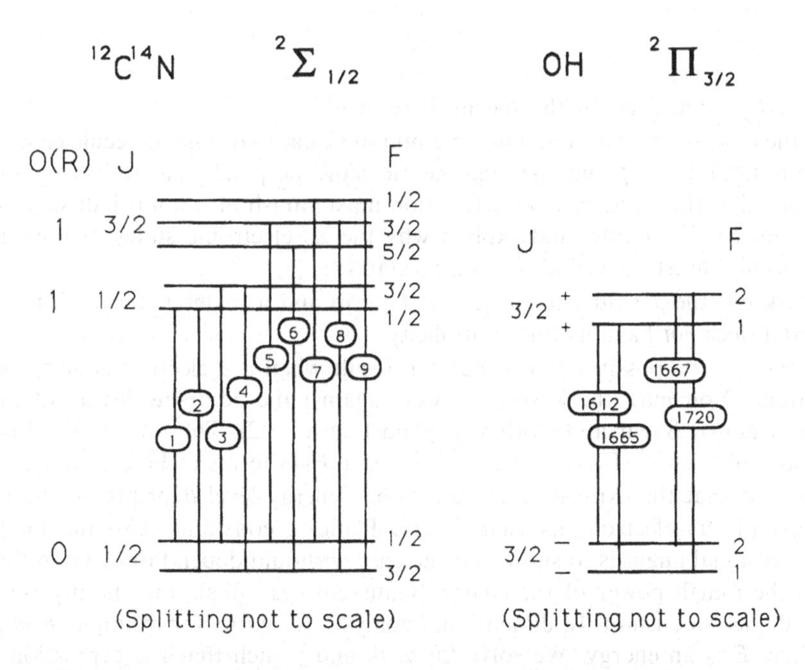

Figure 11.8. Level Structure in CN and OH. For the $^2\Sigma$ state of CN, the quantum number N is equivalent to O. In the older literature, N was called K.

Thus, for $I = 0$, there are *two*, Λ-doubled, $^2\Pi_{3/2}$ levels. These are split by the nuclear spin ($I = 1/2$) of the unpaired proton in H.

The $X^2\Pi$ electronic term of OH carries the subscript i (inverted) in Table 11.1 because the $\Omega = 3/2$ levels lie lower than the $\Omega = 1/2$.

The radio brightness temperatures (see §14.1 and Equation 14.1) observed in certain of these lines are so high, sometimes more than 10^{14} K(!), that they could not possibly be emitted thermally – the lines must be subject to *interstellar maser action.* We must leave a discussion of this phenomenon to the references (e.g., Reid and Moran 1981, Mezger 1991). We only note that the observed intensity patterns in the OH lines depend on the nature of the emitting source, e.g., circumstellar envelopes or molecular clouds.

11.9 Problems

1. What are the allowed terms from the configuration d^2? List the energy levels for the lowest levels for the terms based on the configuration d^2s^2 in Ti I in Sugar and Corliss (1985). Is this configuration completely known in Ti I? Name any missing terms.
2. Prove that the number of (elementary) states belonging to the electron configuration l^w is the binomial coefficient

$$\binom{4l+2}{w}.$$

   Verify your answer for the specific case of p^3.
3. If the electronic wave function for a homonuclear diatomic molecule changes sign on reflection in a plane bisecting the molecule, perpendicular to its axis, one may argue that the electronic wave function must vanish at a point half-way between the nuclei. Elaborate, and explain why the Σ_u electronic states of homonuclear molecules must lie well above the Σ_g states.
4. Work out the J values for a quartet D term and a sextet P term. Note that the sextet does *not* have its full multiplicity.
5. Work out the possible terms that can result when a p electron is added to a ^4F parent. You may check your answers against the complete listing of the Fe I spectrum by Sugar and Corliss (beyond Figure 11.2). The lowest $3d^7(^4F)4p$ level is at 33 095 cm^{-1}. Find as many of the $3d^7(^4F)4p$ terms in Fe I as you can.
6. Assume that the expression for an atomic energy level depends on the reduced mass of the electron, its charge, and Planck's constant. Use the method of dimensional analysis to show that the energy should depend directly on the mass, on the fourth power of the charge (Gaussian cgs units), and the inverse square of Planck's constant. In dimensional analysis, we write, for example, $E = \mu^\alpha h^\beta e^\gamma$, where E is an energy. We solve for α, β, and γ such that the expression on the right will have units of energy.
7. Prove that Equation 11.17 correctly describes both the P and R branches if the substitutions 11.15 and 11.16 are made in 11.14.

8. The four OH levels shown in Figure 11.8 all have $\Omega = 3/2$ and O (the nuclear rotational quantum number) $= 0$. Sketch two families of sets of four levels, one set based on $\Omega = 3/2$, and another on $\Omega = 1/2$. For each value of Ω, you get a new set of four levels as O increases by unity. Note that for the levels shown in the figure, $\Omega = J$, while for the set above it, $J = \Omega + 1$, $\Omega + 2$, etc.

12
The Analysis of Stellar Spectra

12.1 The Identification of Lines in Stellar Spectra

We start our discussion of the analysis of stellar spectra with line identifications. In chemistry, this would be called qualitative analysis. With line identification, we find out which elements and ions are present in stellar atmospheres, leaving the quantitative analysis to subsequent techniques.

The most thoroughly explored stellar spectrum is surely that of the sun. The region from $\lambda\lambda 2935$–8770, essentially the traditional spectrum available from the ground, is described in a volume by Moore, Minnaert, and Houtgast (1966). This work is a revision of a previous study that was, itself, revised from a still older work. No one at the present time needs to begin the study of any stellar spectrum from first principles. It will be possible in virtually every case to find some at least *relevant* identification list for a star whose spectrum is similar in nature to the one for which new identifications are desired.

Many of the classical identification studies were done from a list of measured wavelengths, for which the measurer had supplied eye estimates of the line intensities. In modern work one could always have in addition to the list of wavelengths, a tracing of the spectrum, such as the one shown in Figure 12.1. On tracings such as this, one quickly identifies some of the strongest features, such as the hydrogen lines, the resonance lines of Ca II, called H and K, or the strong lines of iron. In this way a framework is established, and the main job of identification work involves identification of the weaker features. A few identifications are indicated on the figure.

Figure 12.1 was made from spectra obtained at the Dominion Astrophysical Observatory, where the author has frequently been a guest investigator. The spectra, which were originally obtained on photographic plates, were digitized with the help of a microdensitometer, and the results added together to increase the signal-to-noise level. The procedure has been described by Adelman, Cowley, and Hill (1988, see also the references cited). Most modern spectra are obtained with electronic devices that can produce beautiful, low-noise material (Cayrel de Strobel and Spite 1988).

The programs which produce tracings of this kind typically provide a provisional wavelength scale, such as the one shown. In this case, the strong line Hγ is immediately obvious, and an experienced spectroscopist would next search this portion of the spectrum for the strong Fe II line $\lambda 4233$. These two lines would

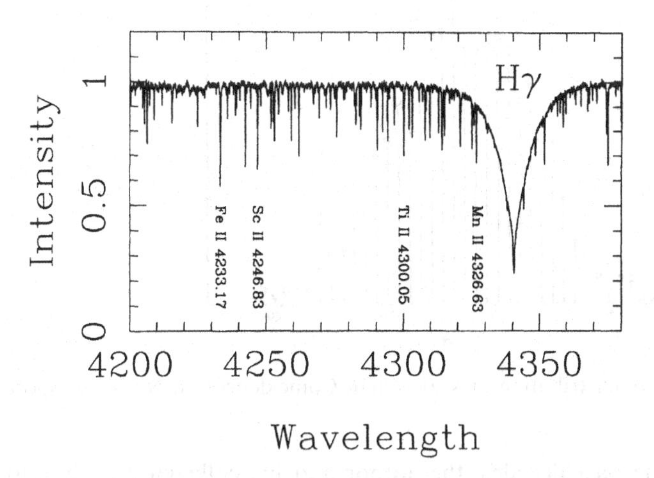

Figure 12.1. A Tracing of the Region Near $H\gamma$ in the Spectrum of ϕ Herculis.

be sufficient to provide a preliminary wavelength scale, if one were not already available. From this point, it is a simple matter to fill in the identifications of the stronger features with the help of an identification list.

Complete line identification studies, in which one attempts to identify every absorption feature, are rarely done at the present time. Most of the current efforts have gone into identification of features suitable for abundance work, or element identifications in chemically peculiar stars. That part of stellar spectra available from ground-based instruments is now rather well explored, in all but a few of the most bizarre objects. However, satellite observations have revealed a new spectral domain in which our current tools are marginally adequate for a complete understanding.

Recent studies have demonstrated that many of the features in the satellite ultraviolet cannot be properly identified because the atoms which almost certainly give rise to some of the absorptions have not been well enough studied in the laboratory! This situation has improved dramatically in the last several years as a result of efforts from Johansson and his colleagues at Lund (Johansson and Cowley 1988, Martin 1992) and the colossal work of R. L. Kurucz (1991a).

The most straightforward way of doing detailed identification work is to find some similar star for which identifications have already been made. For many late-type stars, the solar spectrum may serve as a standard. One then compares two tracings of the standard and program star, hopefully with the same spectral dispersion, and usually there is a one-to-one correspondence of the features, so the identifications may simply be transferred. This procedure works well for many stars. For virtually all objects, this is the best way to get started.

If one has an unusual spectrum, and no good identification study in a similar star, it is useful to have a procedure that will indicate which atomic spectra are present

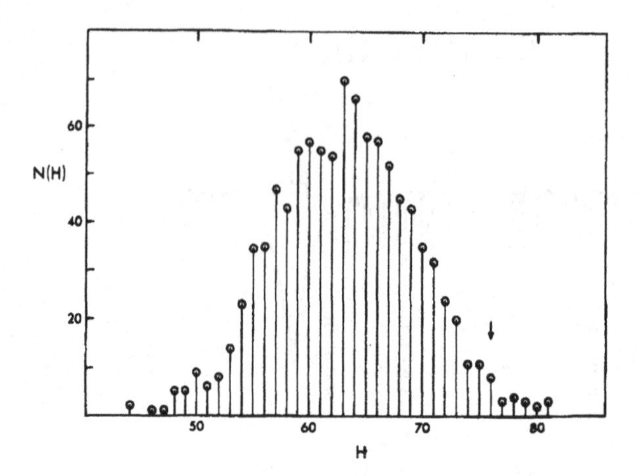

Figure 12.2. A Distribution of Wavelength Coincidences on Nonsense Spectra.

in the star. Over the past decades, the author and his colleagues have automated a method by which this may be quickly accomplished. The method consists of a search for wavelength coincidences with virtually all conceivable atomic spectra for which laboratory wavelengths are available. At the present time, in surveys of stellar spectra, we make use of some 400 line lists containing atomic spectra of weak and strong lines of atoms in various stages of ionization.

The coincidences are analyzed by a statistical procedure, so the method is called wavelength coincidence statistics, or WCS. The method is basically a test against the null hypothesis that H coincidences of laboratory and stellar lines arose by chance. Let us suppose the laboratory list contains N lines, and H coincidences were found. We then make up n sets of N "nonsense" lines, by adding or subtracting a random number from the true position. Searches are then made for coincidences with these lines. We can then form a mean number of coincidences $< H_N >$ on the nonsense lines, and compute the variance $\sigma^2 =< (H_N - < H_N >)^2 >$. The ratio $(H - < H_N >)/\sigma$ then tells the number of standard deviations away the hits on the laboratory wavelengths are from the hits on the nonsense wavelengths. Equally valuable is the quantity we call p, which is the fraction of times H or more hits were obtained on the nonsense wavelengths. The confidence level at which we may reject the null hypothesis is $(1.0 - p) \times 100\%$.

Figure 12.2 shows an example of the distribution of coincidences on 1000 sets of nonsense wavelengths, with the coincidences H on the laboratory set indicated by an arrow. The confidence level at which the null hypothesis may be rejected is about 95% in the example shown. In survey work it is wise to ignore such marginal cases, and concentrate on those with 99.99% confidence or higher, since those are usually the atomic species providing the majority of the atomic absorption lines.

Gulliver and Stadel (1990) describe a program for making line identifications that combines traditional techniques with coincidence statistics. This powerful program is recommended for all future identification work.

WCS does not help much for species with a few strong lines, such as Ca II or Sr II.

Often, these lines are so strong that their identifications are unmistakable, but when there are only one or two weak stellar features that coincide in wavelength with the strongest expected atomic lines, there is very little any method can do to allow one to reject the obvious null hypothesis. Of course, one may try to obtain observations in another wavelength region, where the atomic lines are stronger. Indeed, this idea provided the great hope for the information obtained from the satellite ultraviolet, especially the rich data sources of the archives of the International Ultraviolet Explorer (IUE).

Many of the elements on the right-hand side of the periodic table, such as the halogens, and the second spectra of potassium and sodium, have strong lines in the ultraviolet. Unfortunately, the lines are often totally overwhelmed by a congeries of lines from commonplace elements such as iron and chromium, and it has not been easy to "fill in" the missing stellar abundances with the help of this region. Research is currently being devoted to this problem which will be even more acute as data accumulate from the Hubble Space Telescope.

12.2 Details of Identification Work

Let us suppose that one has made certain identifications of the more obvious features in the spectrum of a star with the help of the procedures mentioned above. The preliminary work will have revealed a number of atomic spectra that are surely present in the star, usually Fe I and II, Cr II, Ti I and II, etc., and if WCS trials have been made, one also knows other species that *may* be present.

We shall also suppose that accurate wavelengths are available. While some remarkably good identification work has been done using measurements from tracings of spectra, there is now no reason to proceed without accurate wavelength measurements. Modern spectra are invariably digitized, and programs exist to measure wavelengths automatically from digitized spectra. Large reduction packages are now in use at observatories and most of them will do wavelength measurement. These programs work very well, and have had some success with the measurement of blends that reveal themselves as inflections in the wings of stronger lines.

Cowley and Adelman (1983) gave a résumé of such programs available at the time. The field moves so quickly that the modern worker needs to consult online (electronic) sources. The American Astronomical Society (1993, see pp. 17ff) sponsors a working group on Astronomical Software whose chairman may be contacted for advice on reduction packages. A very large collection of software is available through the Collaborative Computational Project No. 7 (see Jeffery 1993). The major world observatories maintain reduction packages, and publish manuals which are referenced in the general facilities descriptions (e.g., Schwarz and Melnick 1989, Crawford 1987). An older discussion of programs may be found in the section of Cowley and Adelman (1983) on New Reduction Programs.

Let us consider the case where one is attempting to identify *every* absorption in a limited spectral region in which a few preliminary identifications are already available. This task can surely be speeded up by the program by Gulliver and Stadel (1990). For late-type stars, types F through K, a reasonable start can be made by simply transferring solar identifications. The resolution of the solar spectrum is typically much higher than that of stellar spectra, so that one has the problem of assigning several solar wavelengths to blended stellar features. This is a tedious problem, but not a difficult one in principle.

Differences between the solar spectrum and that of the star may usually be accounted for in terms either of the respective stellar temperatures, which change the relative intensities of atomic lines, or especially of molecular features. For the majority of F, G, and K stars, there rarely occur features that require one to assume the presence of totally new atomic species. For the most part, the solar features are qualitatively similar, and only quantitatively different. Rarely, a weak solar rare-earth line, for example, coincides in wavelength with a very strong stellar feature. Usually when this occurs, it is for a star whose chemical peculiarities are already known.

Suppose, now, that one must deal with such a chemically peculiar star, for example, a barium star, or an upper main-sequence chemically peculiar (CP) star. Again, the best start would be to transfer identifications from some previous identification list. Once this has been accomplished, there remains the very tedious task of checking the individual features. Certainly with the CP stars, there will often be a sizable fraction of the features that require *ab initio* identifications.

For stars whose spectra are known to contain rare-earth lines, there is one chief source that one must use to supplement the Multiplet Tables – NBS Monograph **145** (Meggers, Corliss, and Scribner 1975). Another useful table appears in the *HCP* (see also Reader and Corliss 1991).

For the CP stars, it is often necessary to make use of the Tables of Kurucz and Peytremann (1975), which contain many predicted lines of iron-peak elements. The latter do not contain intensities directly, but they do contain transition probabilities from which intensities may be calculated. Kurucz's (cf. Kurucz 1991a, b) monumental efforts to improve the astronomer's atomic data base have made it possible to do realistic analytical work in the very crowded satellite ultraviolet.

Ultimately, line identifications are based on coincidences in both wavelengths and intensities. The statistical method mentioned above is capable of identifying a species as a whole, but *not* individual features. To identify these, one must consider simultaneously a number of possible identifications with lines whose relative laboratory intensities are known. One then asks if the stellar wavelengths and line strengths are consistent with the identification of the group as a whole. In past work, it was traditional to consider the multiplet as a source of atomic lines whose relative intensities were reliably known. One could attempt to identify whole multiplets, or at least the stronger lines from multiplets. A sure indication of a wrong identification was the failure to find "stronger" lines from the same multiplet as some candidate for identification.

The preeminence given to the multiplets in the classical identification work of stellar spectroscopy arose because of the notorious unreliability of older laboratory estimates of relative intensities from one multiplet to another. Often results from different light sources were used in compilations such as the Multiplet Tables. Many laboratory sources used today are nonthermal, and some very highly excited lines can appear with unusual strength. However, the relative intensities within multiplets are still much more consistent from one laboratory source to another, or from a laboratory source to a star.

NBS Monograph **145** (Meggers, Corliss, and Scribner 1975) has a list of photometrically estimated intensities, whose accuracy is superior to those generally listed in spectroscopic papers. We must refer to the preface of that work for further details. In the case of rare-earth lines it is sensible to use the intensities of Monograph **145** rather than multiplets. When the temperatures of the stellar atmosphere and the NBS source are very different, one can take the excitation energies into account. There are a few multiplets for rare-earth lines, and some are even in the Multiplet Tables. These multiplets may all be taken into account as time permits, but for many lines the departure from *LS* coupling is so severe that one must delve reasonably far into the vagaries of the particular atomic species to make use of these non-*LS* multiplets, and the work is simply too time consuming. The Monograph **145** intensities serve much the same purpose with minimal work.

The MIT Wavelength Tables (Harrison 1939) must also be mentioned. There are often faint lines in these tables that do not appear in the other sources. Dworetsky's (1969) identification of platinum in the mercury–manganese stars (a variety of CP star) was made with the help of this source. However, the faint lines in the MIT tables are often unclassified, for example 'Ce' might be given, and we would know nothing of the stage of ionization or excitation.

Ground-based spectra of normal stars have very few absorption features that cannot be identified with the help of the sources we have mentioned. For the chemically peculiar stars, especially for the upper main-sequence objects, identifications are more difficult. Great progress has been made possible by the work of Kurucz (1991b). Detailed analyses of spectra obtained in the satellite ultraviolet were almost impossible before his work. With these spectra one must often deal with low signal-to-noise levels as well as severe blending.

Stellar rotation can severely hinder identification work. The author has tried to avoid stars that do not have very sharp lines, corresponding to rotational velocities ($v \sin i$'s) of 10 km/sec or less. Certainly many identifications have been made at an intrinsically lower resolution.

If the stellar spectrum is simple enough, it is possible to identify the stronger features, even at reasonably high rotational velocities. Occasionally one has no choice but to measure and analyze these strong features. However, whenever it is possible, one should avoid working with material whose intrinsic resolution is very low.

In real stars with rich spectra, myriad weak features blend with the noise in a most unfortunate way. This can lead to gross errors sometimes, because of the emphasis that abundance work must place on the weakest lines, as we shall explain below.

12.3 The Analysis of Stellar Spectra: Overview

Stellar abundance work is based on a postulate that is seldom articulated: if the calculated properties of a theoretical model match the observed properties of a real object, then the parameters that describe the model also describe the real object. This implicit assumption is common to all techniques of this kind (cf. Chapter 15).

Of course, we can never observe "all" of the properties of a real star, nor can or do we calculate all of the features that belong to a stellar model. Any piece of analytical work therefore makes the additional assumption that "enough" properties can be matched in the model and its real counterpart.

With these matters confessed, let us now outline the methods by which stars are analyzed in practice. We begin with the most favorable or ideal cases, and then discuss simplifications brought about by such practical matters as the availability of observational material, telescope time, computer time, access to computer codes, etc., etc. The analysis of stellar spectra has two traditional phases. First a numerical model atmosphere is constructed, and second, the emitted spectrum belonging to this model is calculated. The calculated spectrum is then compared with the observed spectrum.

For *most* stars, three parameters specify a model atmosphere: the effective temperature, T_e, the surface gravity, g, and the abundance. "The abundance" is not one, but many parameters. But for a surprisingly large fraction of the stellar population, one may scale all element-to-hydrogen ratios by the same parameter, and match the stellar abundances to within the uncertainties of the method. This is why stellar astronomers have traditionally spoken of abundance(s) as though only one parameter were involved.

This simplistic situation is rapidly eroding because of modern emphasis on chemically peculiar objects as well as improved abundance techniques that make the investigation of small abundance differences (≤ 0.2 in the logarithm) meaningful.

It is usually possible to assume that the stellar atmosphere may be approximated by plane-parallel layers in hydrostatic equilibrium. Then

$$dP = -g\rho dz, \tag{12.1}$$

where P is the gas pressure, g the surface gravity (cm/sec^2), ρ the density, and z a physical height measured with respect to some arbitrary zero.

Construction of the model consists of integrating this equation to obtain a temperature–pressure–depth relation subject to certain boundary conditions. The boundary conditions that are imposed are that there are no energy sources in the atmospheric layers. Alternatively, we keep the energy flux (§12.4) constant. In practice, only radiative and convective energy fluxes are considered, although at the uppermost regions of some stellar atmospheres other modes of transport, such as hydromagnetic waves, are surely relevant.

The most important mode of energy transport is radiation, since we see the surfaces of the stars. It is therefore practical to cast the model atmosphere problem in terms of an optical depth τ rather than a physical depth. Optical and physical depths are related by the equation

$$d\tau = -\kappa dz, \tag{12.2}$$

where κ is the opacity, which we shall discuss in some detail in §12.4. By convention, τ increases downward, toward the center of a star, while the height (z) increases outward.

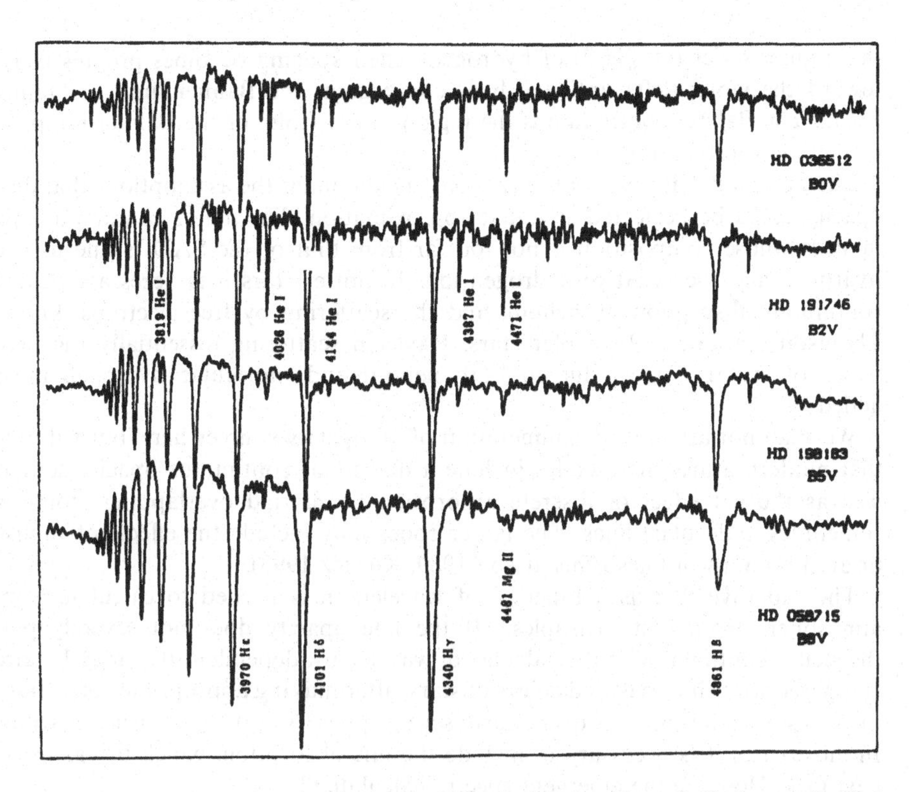

Figure 12.3. Discrete and Continuous Absorption by Neutral Hydrogen in Hot Stars. From Jaschek and Jaschek (1987) with permission. The spectra represent a temperature sequence from the upper star, whose surface temperature is roughly 30 000 K to the lower, where it is some 15 000 K. Note the (high-excitation) helium lines are strongest in the spectra of the hotter stars.

If the opacity were due to N absorbers per cm^3 with cross section α (cm^2) then κ would simply be $N\alpha$. (In stellar atmospheres, cross sections are called α, while in interiors or nuclear physics they are called σ or q. In stellar interiors it is common to write the opacity *per gram* so that the relation corresponding to 12.2 looks like $d\tau = -\kappa\rho dz$. To quote a well-known astronomer once again, there is nothing we can do about this.

The opacity in stellar atmospheres may be divided into two categories, discrete and continuous. Discrete opacity essentially means spectral lines. Because of natural line width, Doppler motions, etc. the absorption is spread over a range of wavelengths, but this range is usually rather small, 1 Å or less, except for the stronger lines. Continuous absorption arises from bound–free or free–free transitions. For this kind of absorption, the opacity changes slowly with wavelength, so long as one is not close to the wavelength limit, between discrete and continuous absorption. This boundary is illustrated in Figure 12.3.

The spectra of four stars are shown in the figure. The strong lines are due to the Balmer series (cf. §15.2) of hydrogen. Their spacing becomes progressively less toward the violet (left) until the lines converge at the Balmer limit or "jump" at λ3647. The Balmer continuum is the region to the violet of the Balmer jump, at the extreme left in the plots.

The first model atmospheres were constructed under the assumptions that discrete opacity could be neglected as a first approximation. The major continuous opacity in stars whose composition is not too far from that of the SAD is due to neutral hydrogen and the negative hydrogen ion. In hotter stars it is necessary to include continuous absorption by helium, and the scattering by free electrons, known as Thomson scattering. In cooler stars, Rayleigh scattering (essentially the extreme wings of spectral lines) due to both neutral and molecular hydrogen must be added.

We shall not attempt an enumeration of all opacity sources here, but will remark that modern atmosphere codes include a dozens of continuous opacity sources as well as the net effect of discrete absorption by myriad overlapping atomic (and sometimes molecular) lines. The larger codes may include the effects of thousands or even millions of lines (Gustafsson 1989, Kurucz 1993).

The radiative flux as a function of wavelength is needed to calculate a model atmosphere from first principles. If the line opacity does not severely perturb the stellar continua, then the calculated, wavelength-dependent flux may be directly compared with the measured colors of stars, after making appropriate corrections for the transmission functions of the earth's atmosphere and the measuring instruments. In most cases it is necessary to include the line absorption, which brings us to the next task. How are the emergent spectra calculated?

Before we discuss the calculation of the line spectra of stars, we must mention one enormous simplification that is often made in the construction of stellar model atmospheres and their emergent spectra – the assumption of local thermodynamic equilibrium or LTE. If this assumption can be made, then all of the thermodynamic properties of the gas in a stellar atmosphere are described by the temperature and pressure alone. At a given physical or optical depth, the excitation, ionization and dissociation of atoms and molecules are given by thermodynamic relations such as the ones presented in Chapter 4.

If the LTE postulate cannot be made, calculations must be based on the postulate of a statistically steady state, that is, one assumes the populations of atomic energy levels as well as the states of ionization are constant in time. In order to arrive at the occupation numbers of atomic states, it is necessary to consider simultaneously all collisional and radiative rates that pertain to each atomic energy level and solve the resulting rate equations in conjunction with the expression of hydrostatic equilibrium of the atmosphere and the boundary condition of constancy of flux.

Since there are (in principle) an infinite number of atomic energy states, it is obvious that a program of this kind may only be carried out in some approximation, and for a number of years the numerical problems alone precluded significant progress. We cannot detail here the history of the work in this area which eventually came to fruition in the 1960s with a series of brilliant successes, especially for the hotter stars. Nor can we give proper credit to the pioneers in this field, R. N. Thomas, R. G. Athay, J. T. Jefferies, L. H. Auer, and D. Mihalas. Mihalas's (1978) *Stellar*

Atmospheres gives an authoritative account of the methods of non-LTE calculations. A more recent review is that of Kudritzki and Hummer (1990).

It should be remarked that the requirements of modeling spectra by non-LTE methods are so formidable, that the technique is still not routinely applied. Usually non-LTE methods are used in cases where LTE is known or expected to fail. Anderson (1989) has introduced statistical methods, which will allow inclusion of thousands of lines in a non-LTE calculation.

In principle, the calculation of the emergent line spectrum in LTE follows the same scheme as the calculation of the continuous spectrum. The only difference is that line opacity is added to the continuous opacity. However, while the number of important continuous absorbers is relatively small the number of line absorbers is enormous. The task of assembling the atomic line absorbers from the literature is a formidable one.

Martin (1992) has summarized relevant atomic data sources, including databases and data files. The paper by Johansson and Cowley (1988) provides an introduction to the literature on complex atoms. We also mention a book by Wynne (1984) containing a section on "Routes to data" prepared by G. Martin of the US National Bureau of Standards. These sources must be supplemented by recent reports of IAU Commissions 14 (Atomic and Molecular Data) and 36 (Theory of Stellar Atmospheres). Some of the papers on atomic and molecular spectra from the 1991 IAU meeting have been collected in a useful volume by Smith and Wiese (1992). Fortunately, Kurucz (see Kurucz 1993) has done most of the relevant work. The modern spectroscopist need only consult the literature for the latest work, and for exotic species not yet treated by Kurucz. Among the more interesting examples of the latter are third spectra of (*4d*)-elements (e.g., Y, Zr, Redfors 1991) and the rare-earth elements.

The accurate calculation of abundances from stellar spectra involves a large number of steps, all of which must be done correctly. Fascinating papers have been published in which abundances were derived from features that were not correctly identified! Even if the astronomical observations are properly made and analyzed, one can come to grief because of inaccurate or carelessly assembled physical data. For many years, the solar iron abundance was thought to be nearly an order of magnitude lower than that obtained for meteorites. Then in the late 1960s careful laboratory work showed that the transition probabilities used for the Fe I lines in abundance calculations were seriously in error. Use of substantially improved atomic data has led to the rather satisfactory (but not perfect!) situation illustrated in Figure 7.7.

The problem of stellar abundances has always reminded the author of the old saying that a chain is only as strong as its weakest link. Time and again workers have brought powerful new techniques to this field only to squander their advantages by too casual a treatment of some critical "link" in the chain.

One of the most serious problems of this kind is still the matter of the hydrodynamics of stellar surfaces, which is not well understood. Nevertheless, it is clear that the gas in virtually all stellar atmospheres is in a state of motion, and some account must be taken of this fact if the emergent spectra are to be properly calculated. We

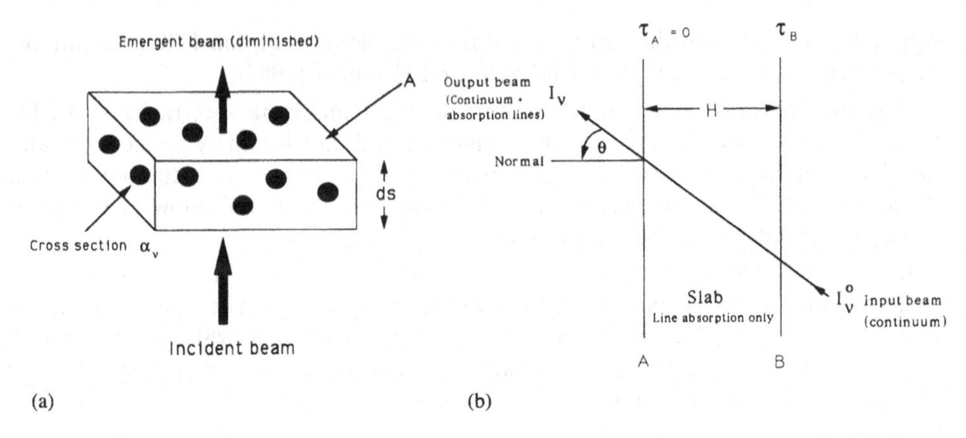

(a) (b)

Figure 12.4. Illustration of Radiative Transfer.

shall illustrate this with the help of a heuristic model to be presented in the next section.

12.4 The Slab Model

In this section we shall develop a simple analytical model that may be used to illustrate the emergent spectrum of stellar atmospheres. The same model is useful to describe interstellar clouds and other cosmic sources.

We begin with the equation of radiative transfer. This simple differential equation describes the change in a variable used to describe the radiation field known as the specific intensity, I_ν. I_ν gives the energy per second carried by photons in a unit frequency interval and in a unit solid angle across unit area. The total energy per unit frequency interval per cm^2 is called the flux, and designated πF_ν by workers in stellar atmospheres. Radio astronomers use a different notation and nomenclature (NB). By definition

$$\pi F_\nu = \int I_\nu \cos(\theta)\, d\omega, \tag{12.3}$$

where the integral is carried out over the sphere. The solid angle increment $d\omega = \sin(\theta) d\theta d\phi$. The cosine factor arises because the flux is calculated for a specific direction, for example, $\theta = 0$, in some coordinate system. For astronomical sources for which we can observe some angular size, such as the sun, or some interstellar clouds, the specific intensity may be – at least approximately – observed. In the case of stars, we observe essentially point sources, and the radiation received is from an entire hemisphere. It is therefore appropriately described as their flux. We shall develop the transfer equation and its solution for the specific intensity. The results may be easily modified for the case of the flux.

Consider the passage of radiation through a volume element with surface area A and thickness ds (Figure 12.4(a)). Let us further suppose that there are N absorbers per cm^3 with cross section α_ν per absorber. Then the fraction by which the beam is diminished by the absorbers is given by the fraction of the area A that is obscured by the absorbers. It may be helpful to think of the absorbers as opaque spheres which block the light beam. So the ratio (dI_ν/I_ν) is the (blocked area ÷ total area). If we introduce a negative sign to indicate I_ν is decreased, and multiply by I_ν, we have

$$dI_\nu = -\frac{AdsN\alpha_\nu}{A} \cdot I_\nu. \tag{12.4}$$

We must now add to this an amount of specific intensity that originates within the volume. We do this in a formal way by introducing a volume emission coefficient j_ν. This is defined to be equal to the specific intensity per unit volume that arises within the volume. Contributions to j_ν may come from spontaneous and stimulated emissions from atoms within the volume, as well as radiation that enters the volume from outside and is scattered into the beam. For the present, we simply write this contribution as $+j_\nu \cdot Ads$.

The equation for the change of I_ν within the volume element Ads is thus

$$A \cdot dI_\nu = -\kappa_\nu I_\nu Ads + j_\nu Ads, \tag{12.5}$$

where we have also substituted κ_ν for the product $N \cdot \alpha_\nu$, the number of absorbers times the cross section. The surface area A cancels from all terms. We may also use the geometric relation between the line element ds, in an arbitrary direction, and the distance normal to the slab surface, that is, $ds \cdot \cos(\theta) = dz$, or $dz = ds \cdot \sec(\theta)$ (Figure 12.4(b)). The equation of radiative transfer is then

$$dI_\nu = -\kappa I_\nu \sec(\theta) dz + j_\nu \sec(\theta) dz. \tag{12.6}$$

We now change the independent variable from physical depth z to optical depth τ, with the help of Equation 12.2 Let us define the ratio of the absorption to the emission coefficient as a new quantity, S_ν, which is called the source function. If we put $\mu = \cos(\theta)$, the equation of radiative transfer becomes

$$\mu \frac{dI_\nu}{d\tau} = I_\nu - S_\nu. \tag{12.7}$$

This is a linear, first order differential equation, so an integrating factor can always be found. One may use standard methods (or simple inspection) to find that $\exp(-\tau/\mu)/\mu$ is the desired integrating factor. We then solve Equation 12.7 subject to the boundary conditions illustrated in Figure 12.4(b): an incident specific intensity I_ν^0 is present at the boundary B, and the optical depth is zero at the boundary A. This means that we shall only be considering the transfer of radiation within the slab, from A to B.

Move I_ν to the left of Equation 12.7, and apply the integrating factor. The result is

$$\int_{\tau_A}^{\tau_B} d\left(I_\nu \exp(-\tau/\mu)\right) = -\int_{\tau_A}^{\tau_B} S_\nu \exp(-\tau/\mu) d\tau/\mu. \tag{12.8}$$

Now let $\tau_A = 0$, meaning no absorption or emission beyond the surface at A. This could represent, for example, the "top" of a stellar atmosphere. For the special case of a constant source function S_ν, we may perform the trivial integration to obtain

$$I_\nu = S_\nu \left[1 - \exp(\tau_B/\mu)\right] + I_\nu^0 \exp(-\tau_B/\mu). \tag{12.9}$$

Equation 12.9 provides a useful heuristic approximation to the emergent specific intensity from a stellar atmosphere. We shall use it to calculate spectral line profiles, and to illustrate virtually all of the tribulations that arise in the far more elaborate numerical models.

In order to make use of Equation 12.9 to treat spectral lines we assume that a stellar atmosphere may be approximated by a uniform slab (T, P, S_ν, etc. are constant). This cool gas lies over a hotter "photosphere" from which only continuous radiation is emitted. The photosphere then provides the specific intensity I_ν^0. We further assume that only discrete or line absorption takes place in the slab.

This simple model was used extensively in early stellar abundance work. The astronomers referred to the slab as the reversing layer because of a phenomenon that could be seen at the time of a solar eclipse. Moments before totality, when only a thin crescent of light from the sun could be seen, spectrograms revealed an emission-line spectrum that was very nearly the inverse of the absorption spectrum obtained from the center of the solar disk. The bright-line spectrum from this crescent was called the flash spectrum, a term still in use.

We designate the depth of a spectral line by a lower-case r (see Figure 12.5). The notation follows that of the German astronomer A. Unsöld (1955), whose classic *Physik der Sternatmosphären* is one of the finest scientific monographs ever written. We take

$$r_\nu = \frac{I_\nu^0 - I_\nu}{I_\nu^0}. \tag{12.10}$$

Since we assume no continuous opacity in the slab, I_ν^0 is the intensity in the continuum, that is, when there is *no* line opacity.

If we insert Equation 12.7 into 12.8, we obtain, after some elementary algebra,

$$r_\lambda = \left[1 - S_\lambda/I_\lambda^0\right]\left[1 - \exp(-\tau_B/\mu)\right]. \tag{12.11}$$

We indicate a wavelength dependence rather than the corresponding frequency dependence of the quantities I and S. Readers should satisfy themselves that this is legitimate! Because of the use of grating spectrographs, the observed spectra are closely linear in wavelength, so r_λ is more suitable than r_ν. But it is common for astronomers to use both frequency- and wavelength-dependent quantities. The transformation from one to the other may cause the novice some trouble, but the difficulty is not profound.

The optical depth at the bottom of the slab, τ_B, is by assumption due only to line (discrete) absorption. Because the slab is uniform, we have from Equation 12.2

$$\tau_B = \int_{z=H}^{z=0} -\kappa dz = \kappa H. \tag{12.12}$$

A rough estimate of the thickness H of the reversing layer may be made from the following considerations. We see deeper into the atmosphere in the continuum, where the opacity is lower than at the wavelengths of lines. In a real atmosphere, the continuous opacity is finite in the domain where the bulk of the line absorption is formed, and not zero, as in our slab model. How far down (in cm) do we see in the continuum of a real atmosphere?

Let us apply Equation 12.8 to a semi-infinite atmosphere ($\tau_B = \infty$, $\tau_A = 0$). Then

$$I_\nu = \int_{\tau=0}^{\tau=\infty} S_\nu \exp(-\tau/\mu) d\tau/\mu. \tag{12.13}$$

One readily sees that the integrand will decrease rapidly for values of τ much greater than unity, so long as the source function S_ν does not increase exponentially, a situation that can be safely ruled out in almost all instances. It is instructive to postulate a linear source function, $S_\nu = a + b\tau$, where a and b are constants. Then

$$I_\nu = a + b\mu, \tag{12.14}$$

which means, for example, if we are looking at the center of the disk of the sun ($\mu = 1$), we see an intensity that corresponds to the value of the source function at optical depth unity.

It is approximately correct to say the radiation that emerges from an optically thick source comes mostly from optical depth unity. Consequently, a reasonable estimate, to within a factor of two or three, is $\kappa(\text{continuum})H = \tau \approx 1$, or $H = 1/\kappa(\text{continuum})$. For most stars, we may make a reasonable estimate of $\kappa(\text{continuum})$, from preliminary values of the physical conditions in the atmospheres.

The factor $[1 - S_\lambda/I_\lambda^0]$ which appears in Equation 12.11 is important. We see readily that r_λ approaches this value in the limiting case of arbitrarily large τ_B. In our slab model, then, we define

$$r_0 \equiv \left[1 - S_\lambda/I_\lambda^0\right] \tag{12.15}$$

as the "limiting line depth." Experience shows that most lines are formed as though there were some limit to their central depth, which is *not* unity. Indeed, ground-based spectra of hot stars show that even the strongest lines rarely have depths much greater than 0.5. The quantity r_0 is a slow function of wavelength. While it is a useful quantity for the description of the behavior of *most* stellar absorption lines, there are well-known cases where the lines are unusually deep. In the solar spectrum, the Na I and Ca I lines are deeper than the value of r_0 that is appropriate for the bulk of the metal lines. This is currently understood in terms of the modern theory of line formation.

If we regard r_0 as a quantity to be fixed by observation, the line depth, as a function of wavelength, is specified by the line opacity κ. This quantity is (see Equations 12.2ff)

$$\kappa = N_n \alpha(\Delta\lambda), \tag{12.16}$$

where N_n is the number of absorbers per cm^3 capable of forming the line, and $\alpha(\Delta\lambda)$ is what stellar astronomers often call the atomic absorption coefficient (per atom). It is the cross section (in cm^2) for absorption by an atomic transition in a photon–atom collision. The quantity $\Delta\lambda$ indicated as the argument of α is the displacement from the line center.

No single function can be written that will describe all atomic line absorption profiles. Hydrogen and helium lines, particularly, have rather complicated profiles, and many stellar abundance codes compute $\alpha(\Delta\lambda)$ for the hydrogen lines by table

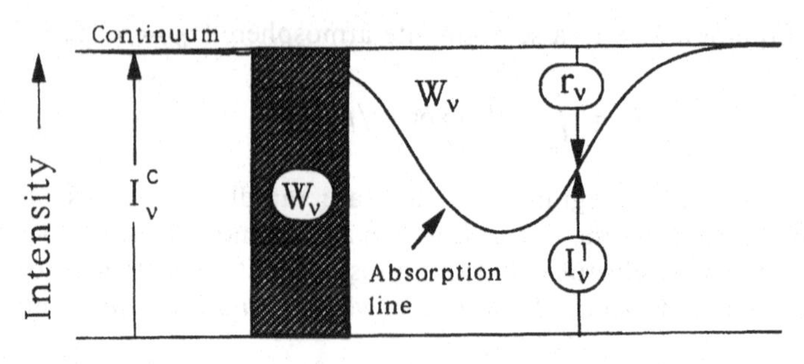

Figure 12.5. Definition of the Equivalent Width.

lookup. They might use, for example, the numerical results of Vidal, Cooper, and Smith (1971). However, for the majority of lines from which abundances may be calculated there is a satisfactory relation that describes the shape of the atomic line absorption cross section in the rest frame of the atom. We may write

$$\alpha(\Delta\lambda) = \frac{\alpha_0}{\Delta\lambda^2 + (\gamma_\lambda/2)^2}. \tag{12.17}$$

The functional form given by Equation 12.17 is known by a number of names. It may be called a damping profile, a dispersion curve, or a Lorentzian. The "curves and surfaces" section of the Chemical Rubber Company's *Standard Mathematical Tables* calls this function the Witch of Agnesi, after the Italian mathematician Maria Gaetana Agnesi (1718–1799).

The quantity α_0 is a constant that can be used to normalize the profile. The damping constant, γ_λ, gives the full width at half maximum of the line absorption cross section. In Equation 12.17, γ_λ has units of length, just like $\Delta\lambda$. It is also common to write this quantity in units of circular frequency, $\omega = 2\pi\nu$. Then $\gamma_\lambda/\lambda = \gamma/\omega$. The name "damping constant" derives from the fact that the amplitude of a charged, classical oscillator will decay exponentially ($\propto \exp(-\gamma t)$) due to the emission of dipole radiation, as is shown in many books (e.g., Cowley 1970). We shall give some details concerning the line absorption coefficient in the following section.

It is convenient to define the equivalent width of a spectral line by means of the relation

$$W_\lambda = \int_{\Delta\lambda=-\infty}^{\Delta\lambda=+\infty} r_\lambda d(\Delta\lambda). \tag{12.18}$$

The equivalent width is the area of the profile defined with respect to a continuum that has been normalized to unity. Therefore, the quantity W_λ is the width of a totally black ($r_\lambda = 1.0$) strip whose area is the same as the area of a given profile, as shown in Figure 12.5. We now note an important point to which we shall return several times: *the equivalent width, or total absorption by weak lines, is independent of the shape of the absorption profile.*

12.5 Details of the Line Absorption Coefficient

The factors that determine the proper functional form of the line absorption coefficient may be conveniently divided into those that are intrinsic to the atom and its rest frame, and those due to the Doppler effect as a result of motion of the atoms with respect to the observer. We consider the former in this section, and will take up the important and often controversial matter of Doppler motions in §12.6.

First, we give a heuristic derivation of the normalization constant α_0 of Equation 12.17. In this derivation, it is convenient to integrate over frequency rather than wavelength. Let us use the notation $\Delta\lambda = (\lambda - \lambda_0)$, and $\Delta\nu = (\nu - \nu_0)$, where the subscript '0' designates line center. The cross section α is *not* a mathematical frequency (or density) function like the Planck or Maxwell–Boltzmann function. The change from $\alpha(\Delta\lambda)$ to $\alpha(\Delta\nu)$ simply involves a substitution of variables using $\Delta\lambda/\lambda = \Delta\nu/\nu$, which comes directly from $\nu = c/\lambda$, with the proviso that $\Delta\nu/\nu$ and $\Delta\lambda/\lambda \ll 1$. Consider then

$$\int_{\Delta\lambda=-\infty}^{\Delta\lambda=+\infty} \alpha(\Delta\nu)d(\Delta\nu). \tag{12.19}$$

The integral is over all $\Delta\nu$ from minus to plus infinity, but for line absorption we use the fact that the cross section is finite for only a small range of frequencies of the order of the quantity γ of Equation 12.17. We thus replace the integral 12.19 by some "average" cross section, times the appropriate frequency interval. For the average cross section, we use πr_e^2, where r_e is the classical electron radius obtained by equating the electrostatic energy e^2/r_e to the rest mass energy of the electron mc^2. For the estimate of the frequency interval, we use the reciprocal of a characteristic time, δT, which we take to be the time for a photon to cross a classical electron radius. The net result, then, is

$$\int \alpha(\Delta\nu)d(\Delta\nu) = \frac{\pi e^2}{mc}. \tag{12.20}$$

For an integral over $\Delta\lambda$, the corresponding relation is

$$\int \alpha(\Delta\lambda)d(\Delta\lambda) = \frac{\pi e^2}{mc^2}\lambda_0^2. \tag{12.21}$$

Both Equations 12.20 and 12.21 are exact, as far as classical electron theory is concerned. Most books on stellar atmospheres (cf. Gray 1976, p. 222) give a detailed development of this classical theory because it provides a simple way of obtaining the shape of absorption lines. Full treatments are naturally given in standard works on electricity and magnetism, such as Jackson's (1975).

The only change that needs to be made to make Equation 12.20 or 12.21 valid according to the quantum theory of absorption by atoms is to multiply the expressions by a factor that is known as the oscillator strength, or f-value. This quantity is traditionally designated f_{nm}, where n and m refer to the appropriate atomic levels.

The f-value is directly proportional to Einstein's probability coefficients A and B (see Cowley 1970, Section 1-5, Aller 1963, Section 4-14). It arose from early attempts to treat atomic absorption with a model consisting of a bound charge – an electric oscillator – that would be forced into vibration by an electromagnetic field. The oscillator would remove energy from a beam of radiation, and then as a result of its motion, reradiate the energy in a characteristic dipole distribution. An atom will do much the same thing, but while the resonant frequency of an oscillator is fixed at $\sqrt{k/m}$, an atom may respond to a number of frequencies, $\nu_{nn'}$, $\nu_{nn''}$, $\nu_{nn'''}$, etc., where n represents the initial level, and n', n'', n''', etc. are possible final levels. To each of these transitions, then, we may say there corresponds a possible oscillator strength, f.

For a free atom, the (full) half-width of the profile, γ_λ, is directly proportional to the oscillator strength, but real atoms are in a bath of particles and fields, and the width of the absorption and emission reflects this. The dominant broadening mechanisms are typically those associated with collisions of various kinds, which are collectively referred to as pressure broadening. Models of atomic absorption must take into account all relevant forms of pressure broadening. Some kinds of pressure broadening also lead (approximately) to Lorentz profiles (Equation 12.17). When this is the case, the γ_λ's may be added linearly and the total profile – radiation plus pressure broadening – remains Lorentzian.

Other mechanisms that act to broaden atomic absorption lines in the atom's rest frame are the Zeeman effect, hyperfine structure, and isotope shifts. The first arises because of magnetic fields, which can be remarkably strong in stars. Hyperfine (§11.8) structure is a result of the interaction of nuclear spin with the radiating electrons. Isotope shifts occur as a result of both nuclear mass and volume differences between isotopes. Some of the meager information that we have on isotopic abundances in stars is obtained by studying the profiles of lines that are sensitive to these effects. We have mentioned the unusual abundances of ^{3}He that occur in the atmospheres of some stars. Related stars display abundance anomalies of the Hg isotopes (Cowley and Aikman 1975, Leckrone 1984).

12.6 Doppler Broadening of Spectral Lines

One of the major sources of line broadening, especially for upper main-sequence stars, is stellar rotation. This effect is purely geometrical, and has already been discussed in §12.2. Stellar rotation is taken into account after the emergent spectrum of a nonrotating star has been computed. The prescription for doing this is straightforward in principle. One computes the emergent intensities for a variety of points over the disk of the star, and adds them after applying the appropriate Doppler shifts. It is less rigorous, but more economical in terms of computer time, to calculate the line profiles in the flux (Equation 12.3), and *convolve* them with a schematic rotational broadening function, whose prescription may be found in the books of Cowley (1970) or Unsöld (1955).

The gas kinetic motion of absorbing atoms is also taken into account by a convolution. The Lorentz profile (Equation 12.13) is convolved with a Gaussian that gives the probability distribution of Doppler shifts. If the Gaussian profile is normalized to unity, its form is

$$f_D(\Delta\lambda) = \frac{1}{\sqrt{\pi}\Delta\lambda_D} \exp\left[-\left(\frac{\Delta\lambda}{\Delta\lambda_D}\right)^2\right]. \tag{12.22}$$

The quantity $\Delta\lambda_D$ is called the Doppler width. Equation 12.22 is easily derived from the expression for the Maxwell–Boltzmann velocity distribution for motion in one dimension, $f(v_z) \propto \exp[-mv_z^2/(2kT)]$. Use $\Delta\lambda/\lambda = v_z/c$, where v_z is the velocity along the line of sight.

It turns out that in most stars, and for most lines, the Doppler width is larger, often by one or more orders of magnitude, than the damping width γ_λ. This means that the Doppler width controls the amount of energy that a moderately strong line can remove from the continuum, as we shall show in detail in the next section.

Gas kinetic motions are not the only cause of Doppler shifts in stellar atmospheres. It is well known that macroscopic motions of solar gases occur in a wide variety of forms described as active prominences, spicules, granules, wiggly lines, acoustical and hydromagnetic waves, shock waves, etc. (see Foukal 1990).

All of these motions could in principle be described by a depth-dependent velocity field. The real velocity field is rigorously the three-dimensional vector function that one sees in hydrodynamics, $\mathbf{v}(x, y, z)$. For purposes of calculating the Doppler effect, one only needs the component along the line of sight. If this velocity field were known, it would be straightforward to incorporate it into the calculations of the emergent intensity. The problem is that we do not understand the hydrodynamics of stellar atmospheres sufficiently well to calculate such velocity fields or even to estimate average quantities from first principles. Unfortunately, the *empirical evidence* for such bulk motions in stellar atmospheres is overwhelming. Simple estimates show that the effects on line profiles *cannot be ignored.*

Does this leave the stellar abundance worker up the proverbial creek without a paddle? The answer is "not quite." First, we may use the total strengths or equivalent widths of either very weak or very strong lines, which will be shown below to be *independent* of the Doppler broadening. Second, we may make empirical estimates of the Doppler broadening due to bulk motions, and include their effects in an approximate but hopefully adequate manner.

It is still useful to account *approximately* for bulk motions in stellar atmospheres by adding a term to the Doppler width, viz.

$$\frac{\Delta\lambda_D}{\lambda} = \frac{1}{c} \cdot \frac{2RT}{\mu} + \xi_t^2. \tag{12.23}$$

The quantity ξ_t is often called the *microturbulent velocity.* In the days when it was thought all stellar atmospheres were in a state of fully developed turbulence, the effect of the motion of elements was modeled by postulating small elements of gas in random motion with respect to one another. If the distribution of these velocities could be assumed to be Gaussian, *and* if they were all optically thin ($\tau < 1$), then we could add the squares of the characteristic widths as in Equation 12.23.

It should be recognized at the outset that this is at best an approximate procedure. We have no basis for assuming that the distribution of velocities of "small elements" will be Gaussian. But it is not unreasonable to assume that the effects of velocity fields may be taken into account in this way. Are the moving "elements" optically thin? This depends on the opacity which for lines is a strong function of wavelength. An element that is quite opaque at line center, may still be transparent enough in the wings to require inclusion in the Doppler width.

Very large elements of gas would act as independent atmospheres. The net effect of their motion would be to change the line profiles, but not the equivalent widths. Such motions are often called *macroturbulent* by workers in stellar atmospheres (Gray 1988).

The fact that the concepts of microturbulence and macroturbulence are approximate does not mean that they lack physical meaning, or that they deserve the pejorative *fudge* so often applied to them. The words "simplification," and "approximation" are more appropriate. If they were *not* used in analytical procedures, it would be necessary to use some other means of accounting for the velocity fields. Some workers have arbitrarily ignored the velocity fields, effectively using $\xi_t = 0$, and derived demonstrably spurious abundances. Others have used much too large a ξ_t. We shall be in a better position to understand the importance of the ξ_t parameter after we have discussed the concept of the curve of growth in the following section.

If it is assumed that the net effect of bulk motions in stellar atmospheres may be accounted for by adding the ξ_t term as in Equation 12.23, it is straightforward to account for both Doppler broadening and the damping profile. We simply need to perform a convolution, or smearing of the two profiles. We must refer to the literature for a discussion of convolution as well as the mathematical details (e.g., Cowley 1970, Section 1-8). The resultant profile must have the same normalization as that given by Equation 12.21. It does not have an analytical form.

$$\alpha(\Delta\lambda) = \frac{\sqrt{\pi}e^2}{mc^2} f_{nm} \frac{\lambda_0^2}{\Delta\lambda_D} \cdot H\left(\frac{\gamma_\lambda}{2\Delta\lambda_D}, \frac{\Delta\lambda}{\Delta\lambda_D}\right), \tag{12.24}$$

where the H-function is defined by

$$H(\alpha, v) = \frac{\alpha}{\pi} \int_{-\infty}^{\infty} \frac{\exp(-t^2)dt}{\alpha^2 + (v-t)^2}. \tag{12.25}$$

This combined profile turns out to be a tedious function to evaluate, and many discussions may be found in the literature. We again refer to texts on stellar atmospheres or spectroscopy for numerical details. We shall discuss the shape of the combined profile in the following section.

12.7 The Curve of Growth for Equivalent Widths

In this section we show how the equivalent width of a line grows as the number of absorbers in the reversing layer increases. We derive heuristic results from our slab model, but it is emphasized that virtually all of the properties obtained from this approximate treatment are closely similar to those obtained with elaborate numerical models. Many astronomers today do not understand the basic ideas of the analysis of stellar photospheres because their professors eschewed the slab model as too simple. There is a difference between simplicity and simplism.

Using Equations 12.9, 12.11, and 12.14, we may write the equivalent width

$$W_\lambda = r_0 \int_{\Delta\lambda=-\infty}^{\Delta\lambda=+\infty} \left[1 - \exp\left(-\tau_B/\mu\right)\right] d(\Delta\lambda), \tag{12.26}$$

where the integral is carried out over the line profile. Assume that there is a single line which is isolated from other absorbers, as in Figure 12.5.

The optical depth τ_B at the bottom of the slab is given by Equation 12.12. Expand the exponential in 12.26 to obtain

$$W_\lambda = r_0 \int \left[\tau_B/\mu - \left(\tau_B/\mu\right)^2/2! + \left(\tau_B/\mu\right)^3/3! - \ldots\right] d(\Delta\lambda). \tag{12.27}$$

For the case where τ_B is small at all wavelengths (and μ is not too small), only the first term in the integral is important, and we have

$$W'_\lambda = \left(r_0 H/\mu\right) \int_{-\infty}^{+\infty} \alpha(\Delta\lambda) d(\Delta\lambda) = \left(r_0 H/\mu\right) \frac{\pi e^2}{mc^2} \lambda_0^2 N_n f_{nm}. \tag{12.28}$$

The integral contains the line absorption cross section, which has a definite value specified by the oscillator strength and atomic parameters independent of the broadening mechanisms. We shall use a prime to indicate the equivalent width of a line calculated with Equation 12.28 – the *weak-line approximation.*

The number of atoms in the nth level of excitation, and tth stage of ionization, per cm^3 capable of absorbing at wavelengths near λ_0, is related to the total number of atoms of the appropriate stage of ionization by the Boltzmann formula

$$N_{n,t} = N_t \frac{g_n}{u_t} \exp\left(-\chi_n/kT\right). \tag{12.29}$$

This notation shows explicitly the stage of ionization (t), while that in Equation 12.28 does not (NB). The quantity g_n is the statistical weight $(2J+1)$ of the absorbing level, u_t the partition function, and χ_n the excitation potential. If H is estimated according to the scheme outlined in §12.4, r_0, λ_0, and μ fixed by observations, and the oscillator strength is known, a single observation of a weak line enables one to determine $N_{n,t}$. Equation 12.29 allows one to determine the total number of absorbers in a given stage of ionization, and the ionization equation will allow calculation of the total number of atoms per cm^3 of the element responsible for a weak absorption line with equivalent width W'_λ. We shall illustrate this in the next section.

We now discuss the behavior of W_λ when the optical depth in the line is not small, and higher-order terms in Equation 12.27 become important. It would be useful to make a plot of the actual equivalent width against the quantity we have called W'_λ, the equivalent width *if* the higher-order terms in 12.27 did not enter. In most studies, W'_λ itself has not been used. A quantity that is directly proportional to it has been substituted. If we take the logarithm of Equation 12.28, and substitute for $N_{n,t}$ from Equation 12.29, the result may be written, after some rearrangement, in the form

$$\log\left[W'_\lambda/\lambda\right] = \log(N_t) + \log(g_n f_{nm}\lambda) - \left(\chi_n/kT\right)\log(e) + \text{constant}. \qquad (12.30)$$

It is important to note that the constant will be the same for all lines arising from the same atomic spectrum, e.g., Fe I, or Fe II, etc. In astronomy, it is common to introduce a reciprocal temperature variable

$$\theta \equiv 5040/T. \qquad (12.31)$$

If χ_n is measured in electron volts, k in cgs units, and T in kelvins, the third term on the right of 12.30 becomes simply $-\theta\chi_n$.

Several points about Equation 12.30 deserve special attention. On the right, the quantities g_n, f_{nm}, and N_t appear only as a product. Therefore, the effect on W'_λ of multiplying any one of the three factors by some value, while holding the other two constant, is independent of the particular factor that is changed – increasing f_{nm} by a factor of ten is the same as increasing N_t by a factor of ten. Since the product $g_n f_{nm} N_t$ occurs in τ_B, this is also true for the equivalent width W_λ, *whether it is weak or not.* This may be seen from the analytical form of Equation 12.26, and a little thought should reveal that even if a depth-dependent, numerical model is used, the product $g_n f_{nm} N_t$ will still enter linearly into the line absorption coefficient. The net equivalent width will again depend on the product of these factors and not their individual values. Finally, we note that the value N_t (atoms per cm^3 in a given stage of ionization) will scale linearly with the abundance of an element ϵ_{EL}. It is therefore common to use the product $g_n f_{nm} \epsilon_{\text{EL}}$ as a single, independent variable of the functional relation for W_λ.

One of the λ_0's has been moved to the left of Equation 12.30, and the subscripts 0 have been dropped. It is common to divide the equivalent width by λ (or λ_0), to produce a dimensionless quantity. A dimensionless quantity may also be produced by dividing by the Doppler width, which has the same units as λ and W_λ.

It is still useful to explore the results of a numerical integration of Equation 12.26. In order to do this, we divide both sides by $r_0 \Delta\lambda_{\text{D}}$, and present the results in the form $\log[W_\lambda/2\Delta\lambda_{\text{D}} r_0]$ vs. $\log[W'_\lambda/2\Delta\lambda_{\text{D}} r_0]$. Since the integrand is symmetrical about zero, we integrate from 0 to ∞, and multiply by 2. This accounts for the appearance of the factor of 2. There results a family of curves as shown in Figure 12.6. The functional form shown in Figure 12.6 is known as a *curve of growth,* because it shows how the equivalent width *grows* as the number of absorbing atoms increases. The parameter α is defined as

$$\alpha \equiv \gamma_\lambda/(2\Delta\lambda_{\text{D}}) = \gamma/(2\Delta\omega_{\text{D}}). \qquad (12.32)$$

The α that appears in Equation 12.32 must not be confused with the absorption cross section $\alpha(\Delta\lambda)$. The relation between γ and γ_λ was discussed following Equation 12.17. We define $\Delta\omega_{\text{D}}$ such that $\Delta\omega_{\text{D}}/\omega = \Delta\lambda_{\text{D}}/\lambda$. The notation for $\gamma/\Delta\omega_{\text{D}}$ is, unfortunately, not universal; the ratio that we have called 2α is often designated $2a$ or a, especially by US authors. The symbols b and β have also commonly been used to denote the Doppler width.

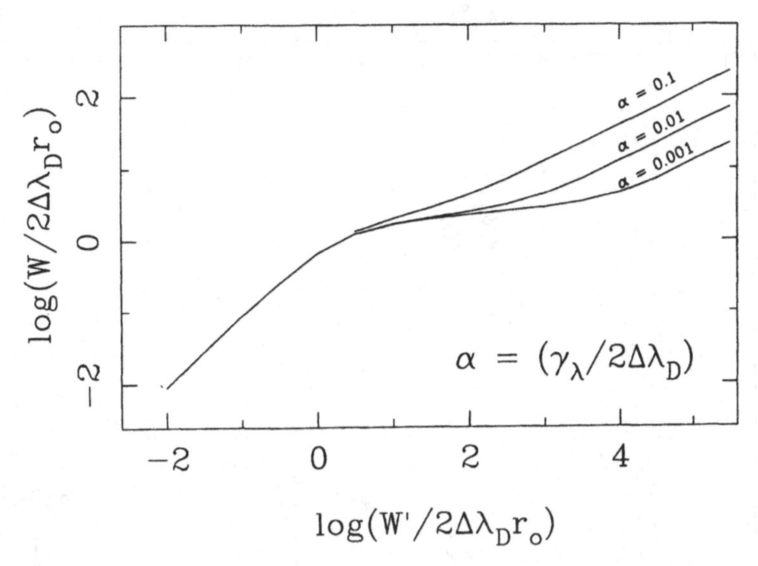

Figure 12.6. A Curve of Growth for the Slab Model.

It is straightforward to interpret the flattening of the curve of growth as a result of the *higher-order* terms in the integrand of 12.27. The eventual increase, with a slope of 1/2, depends on the shape of the line absorption coefficient at large values of $\Delta\lambda/\Delta\lambda_D$. The absorption coefficient is given by Equation 12.24. It is common to use the H-function, defined by 12.25. This $H(\alpha, v)$ is often called the *Voigt* function (pronounced "fokt"). Workers sometimes use a different normalization. The Voigt function as defined here is normalized to $\sqrt{\pi}$. This troublesome transcendental function is discussed in books on stellar atmospheres and spectra.

12.8 Details of the Curve of Growth

A curve of growth for an Fe I line λ4600 of Multiplet 41 is plotted in the lower right of Figure 12.7. In the upper left of the figure are a series of profiles for the line, calculated for successive values of the product $gf\epsilon$, where ϵ is the abundance. The nature of saturation is clearly seen for the profile with $\log(gf\epsilon) = -6.6$. This line is as black as it can get in the core. Further growth of the line can only occur in the wings. It is fairly straightforward to show that for lines whose strengths are dominated by the wings, the equivalent width grows as $\sqrt{gf\epsilon}$. We will only state the result here, leaving the mathematical details as an exercise.

$$\lim_{gf\epsilon\to\infty} W_\lambda \propto \sqrt{N_{n,t} f_{nm} \gamma_\lambda H/\mu}. \qquad (12.33)$$

It is important to note that the equivalent widths of the strongest lines are, like those of the weakest, independent of the troublesome and often controversial ξ_t parameter. In practice, it is often impossible to use only weak, or very strong lines

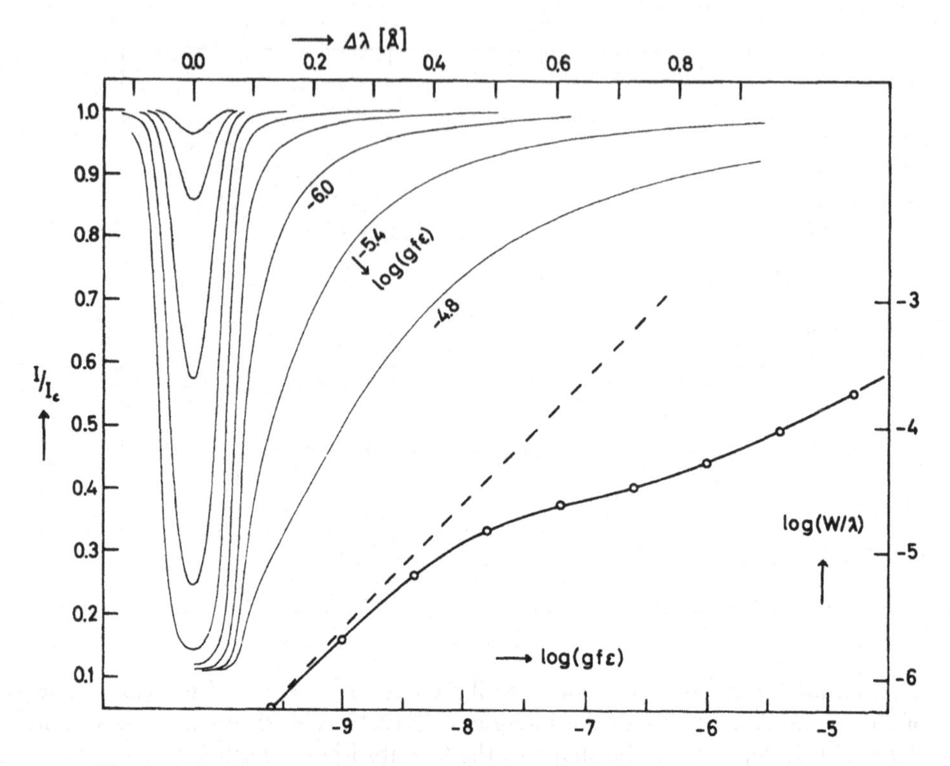

Figure 12.7. Curve of Growth and Corresponding Line Profiles for an Fe I Line in a Solar Model (Courtesy of G. E. Elste). Each line profile is labeled by the value of $\log(gf\epsilon)$, and points corresponding to these values appear along the curve of growth, whose abscissa is $\log(gf\epsilon)$. The profiles are given for steps of 0.6 in $\log(gf\epsilon)$. The weakest profile shown has $\log(gf\epsilon) = -9.6$.

for abundances. This can be the case for numerous reasons. One may be working on faint objects with modest spectral dispersion, which precludes the use of truly weak lines. In the analysis of some of the spectra from the satellite ultraviolet, the noise levels and blending problems are so severe, that only intermediate and strong lines may be considered safely identified. It therefore appears that stellar cosmochemists will need to deal with the curve of growth for some time into the future. We therefore discuss the empirical determination of its shape.

If we make the assumption that all of the lines of a given species, say Fe I, follow the same curve of growth, we may then proceed empirically, following techniques used by Minnaert and Mulders (1931) in a classic analysis of the solar spectrum.

The curve of growth may be considered a plot of $\log(W_\lambda/\lambda)$ *vs*, $\log(W'_\lambda/\lambda)$. If we look at Equation 12.30, it is clear that we may obtain the *shape* of the curve of growth by plotting $\log(W_\lambda/\lambda)$ vs. $\log(gf\lambda) - \theta\chi$. The proper value of θ is found by trial and error, by minimizing the scatter between the lines with different excitation potentials. In many of the early studies, one made plots for narrow ranges of excitation potential on semitransparent paper, and fitted the resulting curves together with the help of a light table. The various horizontal displacements that were required gave the mean value of θ, which was then said to be characteristic of the atomic excitation. Thus $5040/\theta$ gave the "excitation temperature" for the atomic species used to construct the curve of growth.

Note that the absolute value of the abscissa of the curve of growth, the constant in 12.30, is unimportant. What one wants to know is the correction to apply to a measured equivalent width in order to be able to use the formula 12.28. The corrections are obtained *directly from a prolongation of the straight-line portion of the curve of growth.*

Given an empirical curve one may superimpose a theoretical curve such as that of Figure 12.6, and obtain the corresponding value of $r_0 \Delta\lambda_D$. If r_0 has been determined, the Doppler width will furnish ξ_t via Equation 12.23 once an appropriate T is known. Often, one uses an excitation temperature, or a value of T in some numerical model for τ(continuum) ≈ 0.2. In most cases, virtually any reasonable value for T will yield a finite value of ξ_t, or zero.

For stellar work, one may prefer to use curves of growth obtained for the flux (Equation 12.3), rather than for the specific intensity that has been used in our analysis thus far. Such curves have been worked out for the slab model in the flux, but their *shapes*, which are the important thing for us, are so close to those for the specific intensity that it is not worth presenting them separately. The appropriate calculations were carried out by Hunger (1956).

Modern work is likely to proceed in an automated fashion, and unfortunately without the insight that is sometimes gained by the old procedures. Many abundance workers use codes in a black-box fashion, and are not attuned to the fact that the numerical procedures have many of the same difficulties that are present with the slab model. For the determination of ξ_t, it has become standard procedure to plot abundances vs. the equivalent widths, and to vary the ξ_t until no correlation is found. Workers who use this procedure are often unaware that the r_0 (or R_0 for the flux) is important. Virtually all calculations of this kind are done in LTE, where R_0 depends on the value of the Planck function at the "top" of the atmosphere, that is, at the smallest optical depths for which the model is tabulated. Our current theoretical ideas are totally inadequate to fix the conditions in these layers, whether in LTE or non-LTE. The numerical models simply converge toward a constancy of the radiative flux. This is guaranteed, at these highest layers: small τ means that little radiation is intercepted, and the flux must be constant.

A second difficulty with the process of randomizing an abundance vs. W_λ plot is that it makes the ξ_t parameter depend on the weakest lines at least as much as on those of intermediate strength. The slab model shows clearly that in the flat part of the curve of growth, the equivalent widths are most sensitive to the ξ_t and R_0 parameters, and they are relatively insensitive to the abundance. Consequently in a determination of ξ_t we should weight the intermediate-strength lines most heavily. The older methods do this, the newer ones do not. In the latter, the ξ_t-values can be seriously distorted by systematic errors in either the equivalent width measurements of the weakest lines, or the system of oscillator strengths. In the late 1960s a series of nearly supersonic ξ_t-values were determined for A stars. It eventually became clear that the high ξ_t-values were the result of errors in the oscillator strengths.

It is sometimes thought that the modern technique of spectral synthesis (see below), in which entire spectral regions are calculated point by point, are free from the ξ_t and R_0 difficulties of the curve-of-growth method. As we shall see, the problems are only palliated.

12.9 The Method of Spectral Synthesis

It is only in an ideal situation that an equivalent width may be attributed to a unique atomic or molecular transition. In general, spectral lines blend together and make the methods based on equivalent widths impractical. Early attempts to deal with more realistic cases treated the problem of "blends" of two or more lines (cf. Unsöld 1955, Section 122, Schmalberger 1963). It was obvious, however, that many problems required the point-by-point reconstruction of broad regions of stellar spectra.

One of the earliest spectrum synthesis calculations was made at The University of Michigan, by Climenhaga (1960), who worked on the $^{12}C/^{13}C$ ratio in late-type giant stars. Climenhaga's calculations were carried out with a desk calculator, and were based on a slab model. This method could only come into its own in the era of the electronic computer. It is then not difficult in principle to calculate the line absorption coefficient for many line absorbers, add them together, and calculate the emergent intensity (for the sun) or flux (for stars). Figure 12.8, taken from Bolton (1971), illustrates the procedure. At classification dispersion, certain anomalous features could be detected in a class of chemically peculiar stars known as Am's (§13.4.3). Were these features due to atmospheric effects, or abundances? Bolton calculated synthetic spectra using both assumptions in order to see which best fit the observations.

Before synthetic stellar spectra can be calculated, it is necessary to know the relevant absorption lines to be put into the calculations. Two rather different approaches have been taken. Bell, Gustafsson, and their coworkers (see Gustafsson and Bell 1978) made use of the of solar wavelengths (Moore, Minnaert, and Houtgast 1966) as the primary source of atomic lines.

The basic lists have been supplemented and modified in a variety of ways depending on the nature of the star whose spectrum is to be calculated. Especially in the calculation of molecular spectra, it is often necessary to calculate the positions of the individual lines from the theory, since their lines are often unresolved on the solar spectrograms. *Most* of the weaker solar features are unidentified, and it is necessary to make some plausible assumptions about what they are before they can be included in a synthesis.

Kurucz and Peytremann (1975) undertook the task of providing a list of wavelengths generated from atomic data tables. Their work, "A Table of Semiempirical gf Values," contained data for some 250 000 atomic lines. The aim was to create an atomic data base sufficient for the calculation of any atomic feature in a stellar spectrum. The project was one of the most ambitious undertakings in the history of astronomical research. Perhaps as astonishing as the audacity of the project itself

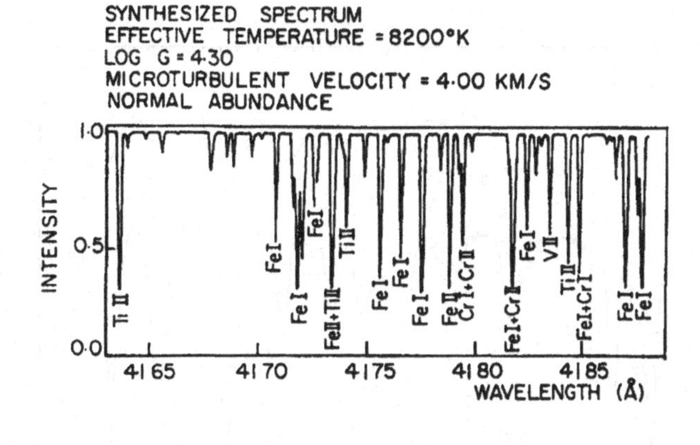

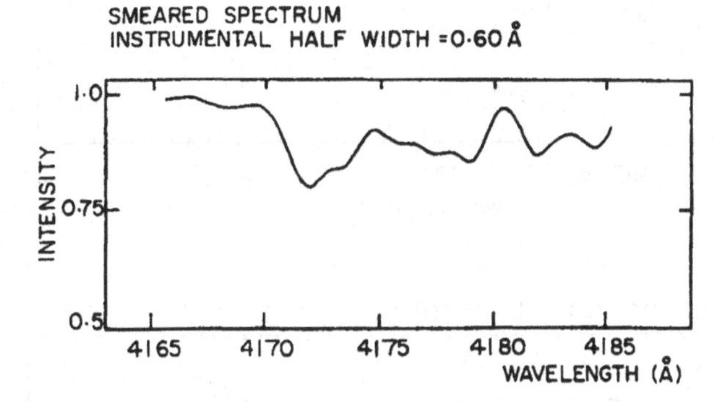

Figure 12.8. An Early Application of Spectral Synthesis. Both spectra are theoretical. The upper one is at essentially infinite spectral resolution. In the lower, an attempt has been made to simulate spectra as they would appear at the much lower resolution, commonly used in classification work. The low-resolution features are clearly the result of blending of numerous lines.

is the fact that it was brought to fruition in some five years time. It has now been largely superseded (Kurucz 1991a). If this work did not stand as *fait accompli*, many would still claim it impossible.

There is no doubt that both the empirical and theoretical approaches are needed to provide data for the interpretation of stellar spectra. In addition to this, there is a very real need for new atomic (and molecular) data from spectroscopic laboratories.

Many computer programs for the analysis of stellar spectra are available through electronic mail. One source of a wide variety of programs and atomic and molecular data is the "Collaborative Computational Project No. 7," or CCP7. This organization publishes a biannual *Newsletter on Analysis of Astronomical Spectra* (cf. Jeffery 1989, 1993), containing a description of archived material as well as instructions for obtaining it electronically. This information is also available by e-mail to csj@st-andrews.ac.uk.

Figure 12.9 shows new work based on observations made with the Hubble Space

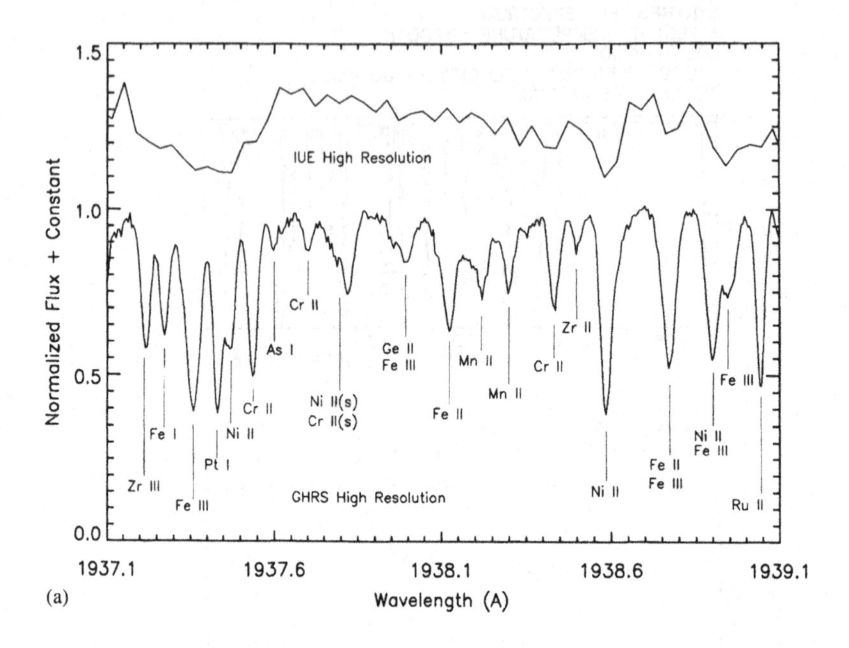

(a)

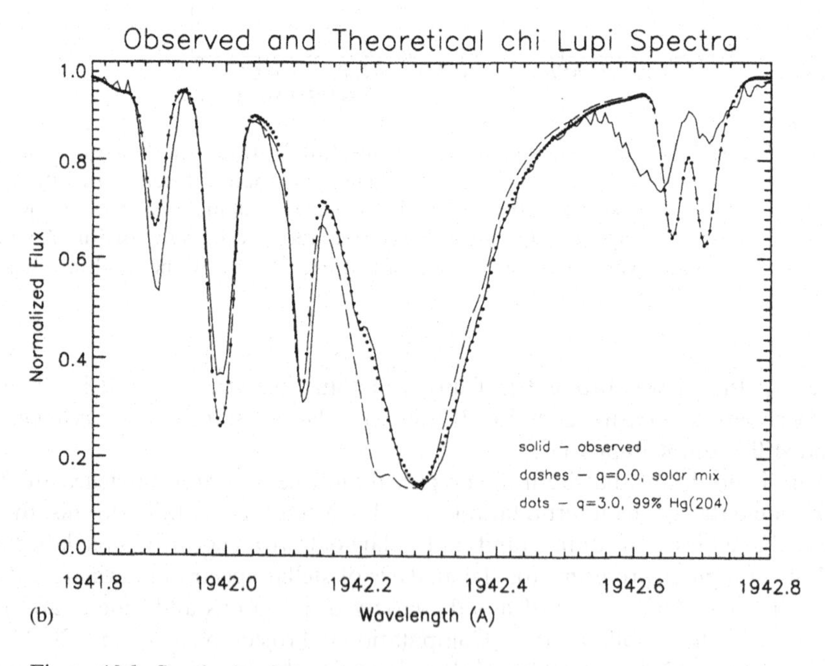

(b)

Figure 12.9. Synthesis of ultraviolet spectra. (a) Observations with IUE (above) and HST (below). (b) Synthetic and observed spectra for χ Lup near Hg II λ1942. Courtesy of D. S. Leckrone.

Telescope (Leckrone, Johansson and Wahlgren 1991). The spectrum of χ Lup is shown in the region of Hg II λ1942. This particular line shows broad hyperfine structure due to nuclear volume effects (§11.8). Figure 12.9(a) merely gives the contrast in resolution between older observations made with the International Ultraviolet Explorer (IUE, above) and the Hubble Space Telescope (HST, below). In Figure 12.9(b), we see the fit by Leckrone and his coworkers to the strong Hg II feature. What is most remarkable about this particular spectrum is that the observations (solid) require the assumption that the mercury be assumed to be almost entirely ^{204}Hg. In the solar mixture, this isotope is only about 7% of the total. Calculations based on the solar mixture are shown with a dashed line; they clearly do not fit.

The next logical step to the synthesis of a stellar spectrum is to compute the emergent spectrum of an unresolved binary star. While examples of such stars are legion, this particular example has been largely bypassed. Intensive efforts have been directed to the synthesis of stellar systems – globular clusters and galaxies. Multiple star systems and clusters are discussed in §13.1.

It is straightforward to compute the spectrum of a stellar system, given certain assumptions about the content of the system. This is perhaps the ultimate application of spectral synthesis, at least within the broad domain of the visible universe.

12.10 Problems

1. Compare the intensities of the Fe I lines in Multiplet 4 with the *equivalent widths* from Moore, Minnaert, and Houtgast (1966). Do you get a good quantitative correspondence? Is there a qualitative correspondence? Which would you expect?
2. Table 12.1 contains sufficient data to illustrate some of the major features of curve-of-growth work. The 17 Cr I lines here will outline the straight-(weak-)line portion and the first saturation. The lines are not strong enough to cover the square-root portion. Plot $\log(F)$ vs. $\log(gf) - \theta\chi$ for various guesses of θ until you obtain the best fit of the two groups of lines with different excitation potential. What θ gives the best excitation temperature for this material? (The $\log(F)$ data are from Cowley and Cowley (1964) for the sun; gf's from Martin, Fuhr and Wiese (1988) – don't expect perfectly smooth curves.)
3. Assume that λ5344.77 is *weak*, and apply Equation 12.28 to obtain a number density of absorbers. Take $r_0 = 0.85$, $\mu = 1.0$, and $H = 1/\kappa(\text{cm}^{-1})$. Use the conditions in the solar model of Gray (1976) for $\log(\tau_0) = 0.0$, $T = 6429$ K, $\log(P_g) = 5.11$, $\log(P_e) = 1.78$. Note: Gray's $\kappa_0\rho$ is our $\kappa(\text{cm}^{-1})$, for λ5000. Assume the opacity at this wavelength for all of the Cr I lines in Table 12.1.
4. You can obtain the total number of Cr atoms per cm^3 from N_n by applying the Boltzmann and Saha relations of Chapter 4 (Equations 4.60 and 4.77). You will need the *partition functions* for Cr I and II. Use values for 6000 K: 12.67 and 8.26 respectively. You will find that Cr is mostly first ionized so to a good approximation $N(\text{Cr}) = N(\text{Cr}^+)$. Take unity for a mean molecular weight, and assume $N(\text{H}) = P_g/kT$. Thence find the chromium-to-hydrogen ratio. Compare your result with the SAD (Table A4). You should be within a factor of 2.
5. Draw a smooth curve through your points from Problem 2. Use the curves of Figures 12.6 and 12.7 as a guide. Now *extend* the straight-line portion and obtain

Table 12.1. *Data for a Rudimentary Curve of Growth for Cr I.* $\log(F) = \log(W/\lambda) + 6.0$ *from Center of Solar Disk* ($\mu = 1.0$).

Cr I Wavelengths	$\log(F)$	$\log(gf)$	χ(eV)
4646.15	1.30	−0.70	1.03
4652.16	1.29	−1.03	1.00
4600.75	1.26	−1.26	1.00
4591.41	1.17	−1.74	0.97
4613.36	1.21	−1.68	0.96
4351.06	1.32	−1.45	0.97
4412.25	0.79	−2.70	1.03
4373.26	1.00	−2.35	0.98
5072.93	0.79	−2.73	0.94
4964.92	0.84	−2.53	0.94
5272.01	0.71	−0.42	3.45
5287.19	0.28	−0.91	3.44
5304.19	0.46	−0.69	3.46
5312.88	0.48	−0.56	3.45
5318.79	0.42	−0.69	3.44
5344.77	0.15	−1.06	3.45
5340.46	0.47	−0.73	3.44

a correction for λ4646.15. Use the corrected, W' to obtain a Cr/H ratio. Your answer should be close to that of Problem 4, but will depend on how you draw the curve through the observed points. The line at λ4351 falls too high. Use it to estimate the uncertainty of this technique.

13
The Chemistry of Stars and Stellar Systems

13.1 The General Framework

In this chapter we consider the results of the analytical methods discussed in Chapter 11 insofar as they apply to stars, and to the integrated starlight of some star clusters. Many of these results were obtained with methods that are directly applicable to the analysis of extragalactic systems. We shall postpone a general discussion of the chemical evolution of our own and external galaxies until Chapter 16.

The main pillars of any description of stars and stellar systems are *spectral classification*, and the *Hertzsprung–Russell (H–R) diagram* or some variation of it. Classification itself provides mostly information about the temperatures and pressures in the photospheres of stars. Since the majority of stars have rather similar compositions, classification mainly discriminates between the broad categories of normal and peculiar chemistry. The chemically peculiar stars are in many ways the most interesting. The position of a star on the Hertzsprung–Russell diagram (§13.3) indicates its state of evolution. Usually, this tells us something of the chemistry of the star's interior, but occasionally, what has happened in the interior of a star is manifested on its surface. Only a very brief summary of these concepts can be given here.

Most stars belong to *double* or *multiple* systems. Double stars were discovered by Sir William Herschel in the late eighteenth century, and a century later, their study was well developed. The marvelous Father Angelo Secchi (1878, p. 228) suggested that perhaps half of the visible stars had physically bound companions. Batten (1973) and Heintz (1978) estimate that more than 70% of *main-sequence* stars are double. Abt, Gomez, and Levy (1990) discuss higher percentages.

Visual binaries are physical pairs that may be seen at the telescope as separate stars. Modern techniques, whose details we must leave to the references (McAlister 1991), make it possible to resolve systems down to 0.02 seconds of arc. *Spectroscopic binaries* are systems whose binarity (duplicity) is revealed from spectra, through radial velocities that vary as a function of time. Sometimes, lines from both stars may be seen in the spectra, in which case one speaks of a *double-lined binary*. When only one system of lines have variable radial velocities the system is said to be single lined. Binaries provide the fundamental determinations of *stellar masses*.

Multiple star systems, triples, quadruples, etc., merge into star clusters (Hesser 1980). *Open or galactic* clusters in our own Galaxy are made up of stars similar to the sun. These clusters are typically young and rarely as old as the sun. Most galactic clusters are disrupted in less than 10^9 years by various forms of gravitational interactions. The *Catalogue of Star Clusters and Associations* (Alter, Balázs, and Ruprecht 1970) lists open clusters with fewer than a score of stars. The very largest open clusters have a thousand or more members. Two well-known examples are the *Pleiades* and the *Hyades*, both in the constellation of Taurus.

A large number of astronomical catalogues are now available in digital form from data centers. We have already mentioned the Collaborative Computation Project No. 7 (§12.9). Data sources through June 1990 are reviewed by Wilkins (1991) for IAU Commission 5 on Documentation and Astronomical Data. The book *Data in Astronomy* by Jaschek (1989) reviews sources of astronomical data, and has a valuable chapter on international organizations. Carlos Jaschek was for many years the director of the *Centre des Donnes Stellaires* in Strasbourg, France, and has made enormous contributions to the compilation and dissemination of data. This field is moving so rapidly that significant new sources are likely to be added to any that can be referenced here. *On-line* sources of information are becoming available. For example, the Astronomical Data Center at the Goddard Space Flight Center publishes an electronic newsletter edited by Lee E. Brotzman (1992, leb@hypatia.gsfc.nasa.gov). As of this writing, automatic subscription is available by sending e-mail to listserv@hypatia.gsfc.nasa.gov. The message need only contain the words "SUBSCRIBE ADCNEWS Full Name", where the writer should make the appropriate substitution for "Full Name."

The Galaxy is surrounded by a spherical system of some 150 *globular clusters*. These systems contain as many as 10^5 stars within a radius of a few tens of parsecs. They have ages typically greater than 10^{10} years, and belong to a different *population* from the sun. Globular clusters are well-known components of external galaxies (Grindlay and Philip 1986).

Stellar Populations were introduced by Walter Baade (1944) in his classic discussion of the nucleus of the galaxy M31, and related questions. The meaning of the term has altered somewhat since that time. It is perhaps most useful to think of the term population as a means of describing the age of a subsystem of stars, along with a variety of related phenomena that are correlated with the age of stars. For example, the older stars in the solar neighborhood have systematically higher space motions than the normal ones – called Population I by Baade. The former objects, he called Population II. In addition to the higher space motions, Population II stars have statistically more eccentric and highly inclined orbits about the center of the Galaxy.

Galactic clusters are typical Population I objects, while globular clusters belong to Population II. Both galactic and globular clusters have played important roles in modern astronomy. Mihalas and Binney (1981) give a beautiful summary of this work.

13.2 Spectral Classification

The earliest spectral surveys were made with visual spectroscopes in the nineteenth century. Secchi (1878, see Section 11.4) distinguished four main types. Type I stars had spectra resembling that of Sirius, Type II that of the sun. Types III and IV we now recognize as belonging to cool stars whose spectra show molecular bands. TiO is prominent in Type III, while C_2 and CN bands are seen in the rare Type IV. Secchi's work was extended and applied to the spectra of some 225 000 stars by Annie J. Cannon at Harvard. The notation employed by these workers is still in use. The hottest stars were designated O, with successively cooler stars being classified as B, A, F, G, K, M. The mnemonic – Oh, be a fine girl, kiss me – has delighted aficionados of astronomy for decades. Alas, it is sexist.

It is now known that these classes describe a temperature sequence of stars with similar chemical compositions. In the earliest work, it was not clear that classes O and B described hotter stars than A, which is the reason for the inverted order. Some of the earlier classes were dropped, and we are left with the sequence O–M. Three other classes used by the Harvard workers belong to stars with a surface chemistry that is distinct from that series. Stars of types R and N show strong CN and C_2 bands (Secchi's Type IV), while in type S, ZrO bands are strong.

The O–M sequence is essentially one-dimensional, and the early Harvard workers made little attempt to distinguish among what we now know to be high- and low-luminosity stars. An exception to this was the "c" category of stars noted by Antonia Maury, which turned out to be high-luminosity objects. Evidence gradually accumulated that a two-dimensional classification scheme was possible. We must pass over the remarkably high-quality work on the spectroscopic determinations of stellar luminosities by the early workers, especially at the Mount Wilson Observatory. and come to the modern system established by W. W. Morgan and P. C. Keenan (see Morgan and Keenan 1973). Classifications on this system are now known as *MK types.*

In addition to the spectral types, MK classification adds a *luminosity class*, indicative of the intrinsic brightness of the star. The brightest giant stars are designated I's. For a given spectral type, the intrinsic luminosity decreases from the I's through the V's (see Figure 13.2 below).

The two-dimensional MK system has had an enormous impact on modern astronomy – in ways that might never have been suspected when the work was undertaken. At first glance, their major accomplishment was merely a refinement of the existing schemes. The great strength of their method was the major role played by direct comparison with *standard stars*. They insisted that the spectra of unknown and standard stars be obtained with the same instrumentation – spectrographic configuration, photographic emulsions, development techniques, etc. – and directly compared. Earlier classification methods had been based on a verbal description of the appearance of certain lines and line ratios. The classifications were carried out with a variety of instruments. Now the appearance of spectral features, seen at low resolution, can change in subtle ways from one spectrogram to another if the spectral purity varies. This is simply because of the way that features blend together. The result "looks" different for different resolutions.

Most workers were unaware of the importance of resolution on the visual appearance of spectra, nor was the remarkable power of the eye as a "null photometer" clear. Classifiers can make the judgment that an unknown is "like" one of their standards with an accuracy that is superior to many of the older quantitative spectrophotometric methods.

The Harvard classification of stellar spectra has been repeated at the University of Michigan, largely through the efforts of N. M. Houk (see Houk 1982). The two-dimensional types are closely integrated with the MK system. Because of the different instrumentation, they cannot be identical to it. The spectra were all made with uniform methods, and MK *standards* were employed. They therefore have the advantage of the eye as a null photometer.

A major achievement of modern (MK) classification has been the clear delineation of *normal* and *peculiar* types. The latter consist simply of stars that cannot be classified within the MK scheme.

We can get a rough estimate of the fraction of stars that cannot be accommodated within the MK scheme by looking at the statistics of the classifications of the first four volumes of the *Michigan Spectral Catalogue* (Houk 1992) , where some 124 000 stars were classified. Some of these stars were assigned various peculiarity types. Of these "peculiar" groups, a sizable fraction are spectroscopically peculiar in a way that does not necessarily imply chemical peculiarity. For example, a large category of peculiar spectra are described as *composite*, that is, the spectrum belongs to two stars with about the same brightness, but with temperatures sufficiently different that the characteristics of the two are detectable in the combined or composite spectrogram. In such cases, there is usually no indication that either star is chemically peculiar.

If we exclude composites, rapid rotators, and some other objects in categories whose chemical peculiarity is not established or controversial, the percentage of types that appear to be chemically peculiar drops to about 3%. Of these stars with apparent chemical peculiarities, some 65% are upper main-sequence objects whose unusual chemistry is now thought to be related to *in situ* differentiation. This probably means that the bulk chemistry of the stars is not peculiar. (We shall return to these objects – now called CP stars – in §13.4.3).

Many of the remaining objects, now about 1% of the stars classified, are *evolved* stars whose peculiar surface chemistry could reflect the results of nucleosynthesis in the interiors followed by convective or other mixing processes that bring the chemically altered material to the surface. The population of unevolved, *lower main-sequence* objects whose *low-resolution* spectra indicate an aberrant surface chemistry is very small – less than 0.1%.

We shall see that this classification, when combined with quantitative abundance work, confirms the general notion now more than half a century old that the majority of stars have very similar compositions. While it is now fashionable to stress departures from the solar composition in stellar populations in conjunction with various chemical (nuclear) evolution scenarios for our Galaxy and other stellar systems, *the dominant characteristic of stellar chemistry is its apparent uniformity.*

13.3 The Hertzsprung–Russell (H–R) Diagram

Our knowledge of the intrinsic luminosities of stars began with Friedrich Wilhelm Bessel's announcement of the parallax (distance) to the star 61 Cygni in 1838.

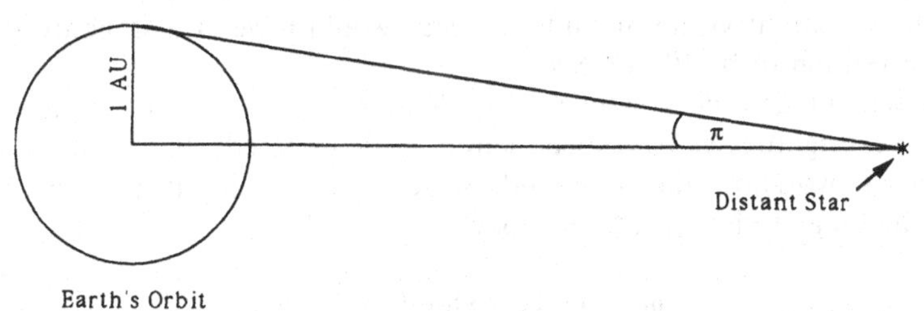

Figure 13.1. Stellar Parallax Defined. The parallax of a star, traditionally called π, is defined as the angle subtended at the star by the radius of the earth's orbit, or 1AU.

William Herschel (1738–1822), whose method of star gauges represented the first attempt to explore the three-dimensional structure of our stellar system, had made the assumption that the intrinsic brightnesses of all stars were similar. It soon became apparent that this was not the case.

Distances to nearby stars are found by the method of parallax, illustrated in Figure 13.1. The parallax of a star is defined as the angle subtended by the average radius of the earth's orbit at the star. This average radius is called the *astronomical unit*, or AU. The parallax angle is traditionally designated by a Greek pi (π). Stellar parallax is measured in units known as "parsecs." A star whose distance is 1 parsec (1 pc) has a parallax of 1 second of arc. There are 3.26 light years in a parsec. In practice, parallaxes are measured by the apparent shift in the position of the π-star with respect to more distant, background stars as the earth moves around the sun. Since the parallaxes are less than a second of arc, it is necessary to measure the positions of stars very accurately. In practice, this method is good only to some 100 pc, which is less than 1% of the radius of our Galaxy.

As the nineteenth century progressed, "positional" astronomy became highly developed. Numerous stellar parallaxes were measured, as well as accurate positions and angular velocities of stars. Astronomers call the latter *proper motions*. Line-of-sight velocities were also measured by the Doppler effect. The Danish astronomer Ejnar Hertzsprung combined this information to show that there were some stars with intrinsic luminosities greatly exceeding that of the sun. He was the first to use the terms *giant* and *dwarf* to describe stars. The basis for this description is best seen in a diagram first made by the American astronomer Henry Norris Russell in 1913, but now known as an H–R diagram. We shall not present an original diagram, but describe in general terms the kinds of H–R diagrams used today.

The ordinate of an H–R diagram is some measure of the intrinsic luminosity of the star. This luminosity may refer to the entire spectrum, in which case it is known as the bolometric luminosity, or it may refer to some limited spectral region, such as that visible to the eye of a ground-based astronomer – a visual luminosity. Theoreticians often plot $\log(L/L_\odot)$, where $L_\odot$ is the luminosity of the sun ($1.98 \cdot 10^{33}$ erg/sec). Astronomers still use the system of logarithmic brightness known as magnitudes, which originated with the ancients. Thus an observer's H–R diagram might use as an ordinate the absolute visual magnitude M_V, which is defined as the apparent visual magnitude a star would have at a standard distance, taken by convention to be 10 parsecs.

The apparent magnitude system is now defined by a system of standard stars, whose relative brightnesses have been carefully determined. If b_1 and b_2 are the brightnesses of two stars, the magnitude scale is defined in such a way that the corresponding magnitudes m_1 and m_2 obey

$$m_2 - m_1 = 2.5 \log \left(\frac{b_1}{b_2} \right) . \tag{13.1}$$

Thus a large number for a magnitude means a faint star. The definition was made in order to preserve the rough system of brightnesses used by the ancient astronomers, whose only photometric instrument was the eye. It was found that a difference of 5 of the old magnitudes corresponded to a brightness ratio of 100.

The factor 2.5 in 13.1 is 5 divided by the logarithm of 100. The brightness ratio between magnitudes n and $n + 1$ is the fifth root of 100, or 2.5119. Sirius (α Canis Majoris), the brightest star in the sky, has an apparent visual magnitude of -1.46, while the planet Venus reaches about -4.6. The faintest stars that can be seen with the unaided eye are about $+6$ in the most favorable circumstances, but for the average city dweller, it is often difficult to see third magnitude stars. Some of the fainter stars measured by standard modern techniques have 24th magnitude; the Faint Object Camera of the *repaired* Hubble Space Telescope may (just) be able to reach the projected (Hall 1982, see p. 5) 28th visual magnitude.

Figure 13.2 is an H–R diagram prepared by Houk (1992) and her collaborators, in which the relative numbers of stars at a given type and absolute magnitude are indicated by the area of the symbol plotted. It is based on Michigan spectral types (Houk 1982). The luminosity classes of the MK system are indicated. The *main sequence* of unevolved stars, burning hydrogen in their cores, have luminosity class V. These stars are also called dwarfs. The *white dwarfs* are not seen on this plot. They form a roughly parallel sequence some 8–10 magnitudes below the main sequence. Stellar masses vary monotonically along the main sequence, from nearly 20 solar masses for a B0 to roughly half a solar mass at M0.

There are many intriguing features about this plot. Many are merely manifestations of the way the figure was constructed. The absolute magnitudes for the stars were obtained by applying a formula to the *discrete* luminosity classifications (Schmidt-Kaler 1982). This accounts for the apparent paths leading from the main sequence to the giants. These paths must not be confused with evolutionary tracks.

The dearth of intrinsically faint stars is an observational selection effect. In any volume of space, there should be more K and M stars than other types. They do not appear on the diagram because they are very faint. The scarcity of very hot stars is a reflection of star-formation and evolutionary processes. Relatively few very hot (and massive) stars form, and they evolve rapidly into the giant region. We can account for a large number of giant K stars in terms of the theory of stellar evolution, which funnels stars from various positions along the main sequence into this region.

The representation of the main sequence narrows near A6, for reasons not completely understood. Certainly one reason is that a number of CP stars which

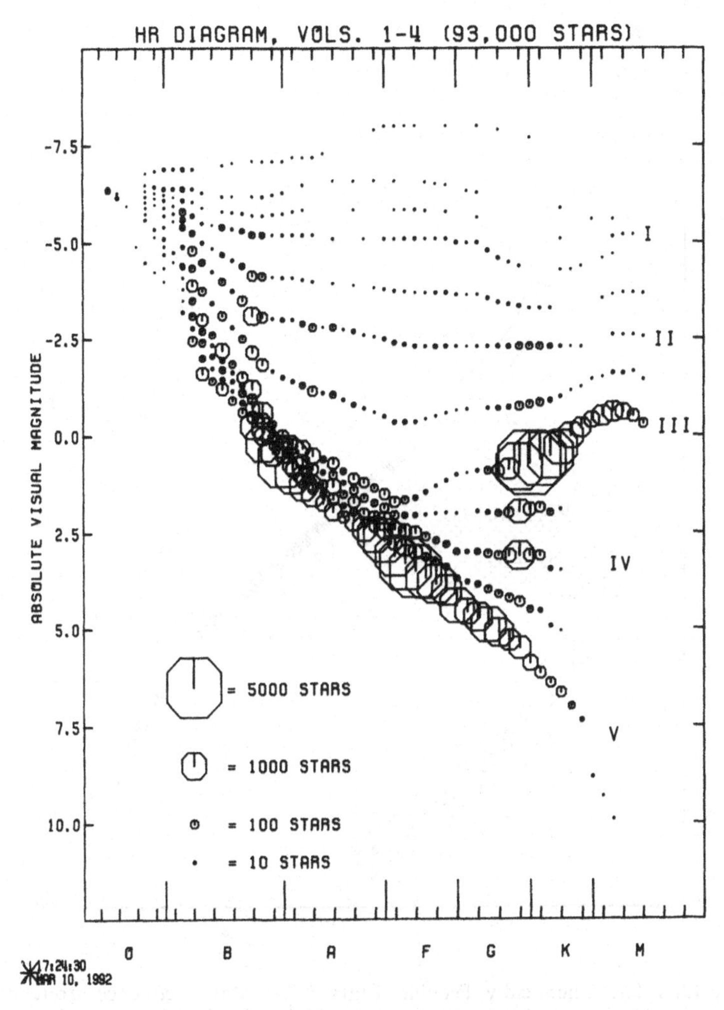

Figure 13.2. Hertzsprung–Russell Diagram from Spectral Types. Note that the spectral types along the horizontal axis mark the center of the ranges for the spectral types, and the longer vertical markings are on the boundaries between types. Thus, A0 starts just under the 'S' in the key for the symbol used for 10 stars. (Courtesy N. Houk 1992.)

occur in the region have not been plotted. Also, the luminosities are difficult to assign, and some fraction of the stars classified IV and IV–V may belong with the V's. However, it is roughly in this region that the CP stars begin to be found. It is also near the domain where stars are unstable to small pulsations. More work needs to be done on stars in this part of the H-R diagram.

Figure 13.3 has the same structure as Figure 13.2, which contains mostly "normal" stars, where the notion of normality as used here means that chemical peculiarities are not noticeable at the resolution of classification spectra. In Figure 13.3, we have

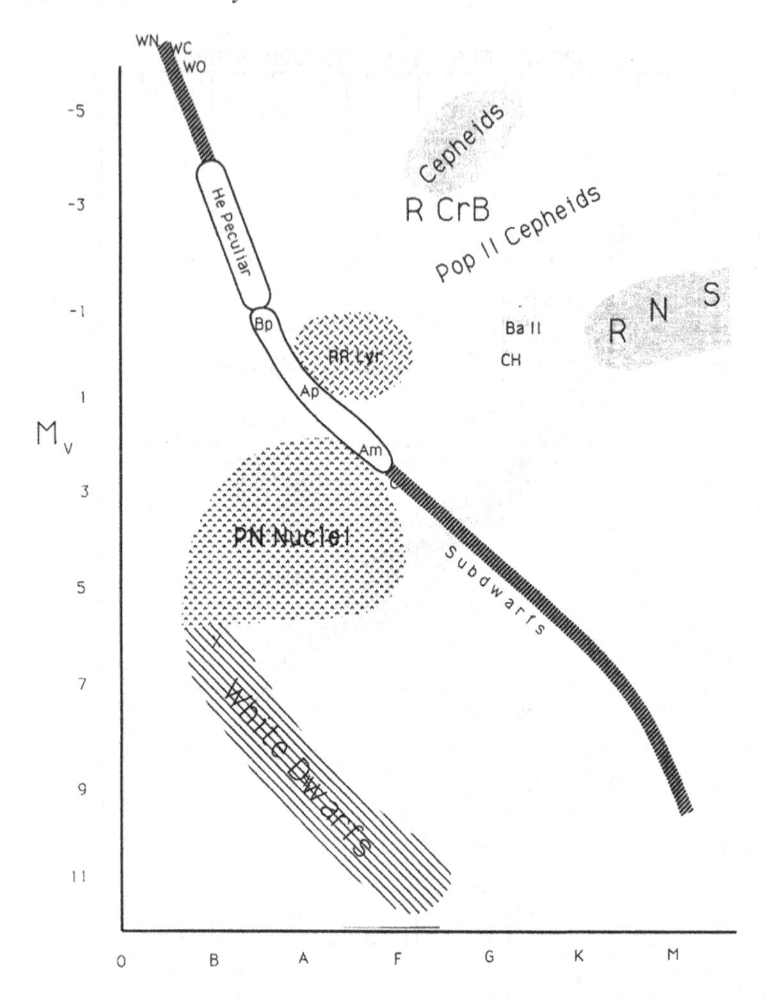

Figure 13.3. The Chemically Peculiar Stars. The Main Sequence stretches from upper left to lower right. Some chemically peculiar stars fall along the main sequence, but most are rather highly evolved objects. WN, WC, and WO are varieties of Wolf–Rayet stars. PN Nuclei refer to nuclei of planetary nebulae (§15.2) and related stars. The cool peculiar types (e.g., R, N, and S) are placed on this diagram using their M_V's and temperatures, which overlap with the late K's and M's. Many of these types are described in detail by Jaschek and Jaschek (1987).

located, in an approximate way, a few categories of stars with palpably abnormal surface chemistry. While all of these stars are believed to be chemically peculiar, it has become customary in recent years to designate those chemically peculiar stars *that lie on or near the upper main sequence* as CP stars (Preston 1974).

Most of the peculiarity types listed in Figure 13.3 will be discussed individually in the following sections. Briefly, they may be divided into two categories, one associated with the notion of stellar Populations, and the remainder belonging to miscellaneous categories that may arise as a result of *in situ* differentiation, mass loss, the mixing of material from processed regions of the stellar interior.

The subdwarfs are the prototype of the unevolved (or mildly evolved) Population II dwarfs. CH stars and the Population II Cepheids are evolved members of the older Population. Among the latter are the shorter-period W Vir stars and the longer-period RV Tauris. Considerable attention has recently been focused on late-type variables, especially carbon stars (R and N) and an intermediate type which perhaps has a C/O ratio near unity known as S, because of the identification of these objects in nearby dwarf galaxies, including the Magellanic Clouds (Haynes and Milne 1991). The S stars have chemical peculiarities similar to those of the Ba II stars.

13.4 Stellar Abundances

The nature of abundance variations is quite different for the different categories of stars shown in Figure 13.3. We shall discuss them under three main groupings. First, we shall treat abundance variations that are thought to be related to Populations. Next, we shall discuss some of the abundance peculiarities that arise from the mixing of processed materials to stellar surfaces. Finally, we treat chemical peculiarities in main-sequence stars which are best accounted for in terms of an *in situ* chemical differentiation.

13.4.1 Population-Related Abundance Patterns

We have indicated a group of stars on Figure 13.3 known as subdwarfs. Since the time of Hertzsprung, what we now call main-sequence stars have been known as dwarfs. However, a few of the stars in the solar neighborhood appeared to fall well below the main sequence, and these were called subdwarfs. We now know that the element-to-hydrogen ratios for all elements with usable spectral lines are below the solar values. A good description of these stars is "weak lined," relative to normal objects at the same temperature. However, the early workers were unaware of the surface temperatures of these stars, and interpreted the weakness of spectral lines as an indication of a high temperature, or an early spectral type. Consequently, they were plotted on the H–R diagrams much too far to the left, which placed them below the main sequence: hence the epithet subdwarf. The most careful modern work shows that these objects still (perhaps arguably) lie slightly below the main sequence of normal, or Population I stars. This may be explained in terms of the theory of stellar structure, given a deficiency of elements heavier than helium.

Ironically, the colors of these stars did not reveal that they were much cooler than their spectral types indicated. This is because of a phenomenon known as *line blocking*. The concept plays a major role in modern work, especially on faint stars, so we shall explain it in some detail.

If one considers the distribution of atomic wavelengths, as revealed for example in the major references such as the Multiplet or MIT Wavelength Tables, it is easily seen that the average strengths as well as the density of lines decrease markedly from the ultraviolet to visible wavelengths.

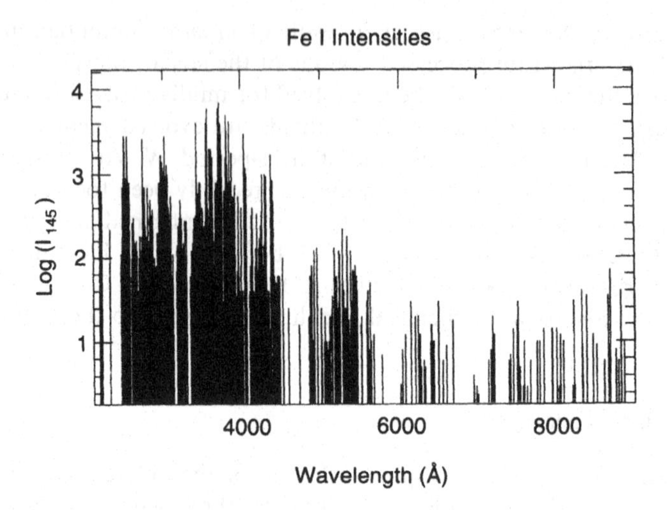

Figure 13.4. Distribution of Intensities in the Fe I Spectrum. Note that the vertical scale is logarithmic.

Figure 13.4 shows the distribution of intensities of lines of Fe I from NBS Monograph **145** (Meggers, Corliss, and Scribner 1975). One can see a steep decline in the intensities from roughly $\lambda 3700$, which is a practical limit for many astronomical spectrographs, toward longer wavelengths. In stellar absorption spectra, this emission reverses; it removes energy from the continuous spectrum. Because of the distribution of wavelengths and intensities, more energy will be removed from the violet than from the blue, and more will be removed from the blue than longer wavelengths. In actual stars, the presence of other absorbers tends to smooth out the irregularities of the Fe I spectrum, and make the overall effect of "selective" blocking by lines marked.

This is shown in Figure 13.5, where the fraction of energy ϵ_λ removed from the continuous spectra of several stars is plotted as a function of wavelength. These plots, from a classic study by Wildey *et al.* (1962), resemble Figure 13.4. In fact, the overall absorption continues to increase toward shorter wavelengths, but this effect is hidden in Figure 13.5 by observational selection effects. In any case, it is clear, first that the lines remove more light from the stellar ultraviolet than the visible, and second, that this removal is much more noticeable in the case of the sun than for the extreme subdwarf HD 19445. The net effect of line blocking is therefore to change the color of a star in a systematic way. The *lack* of effective blocking in the ultraviolet makes the subdwarfs appear to be bluer, and this fact, in addition to the weakness of the atomic lines, made the early workers on subdwarfs overestimate the temperatures of their atmospheres.

Workers in stellar atmospheres distinguish between the terms "blocking by lines" and "blanketing." The former is merely a phenomenological description of the function ϵ_λ of Figure 13.5. This may be obtained by simply adding up equivalent widths in some wavelength band; the latter refers to the overall effect of line

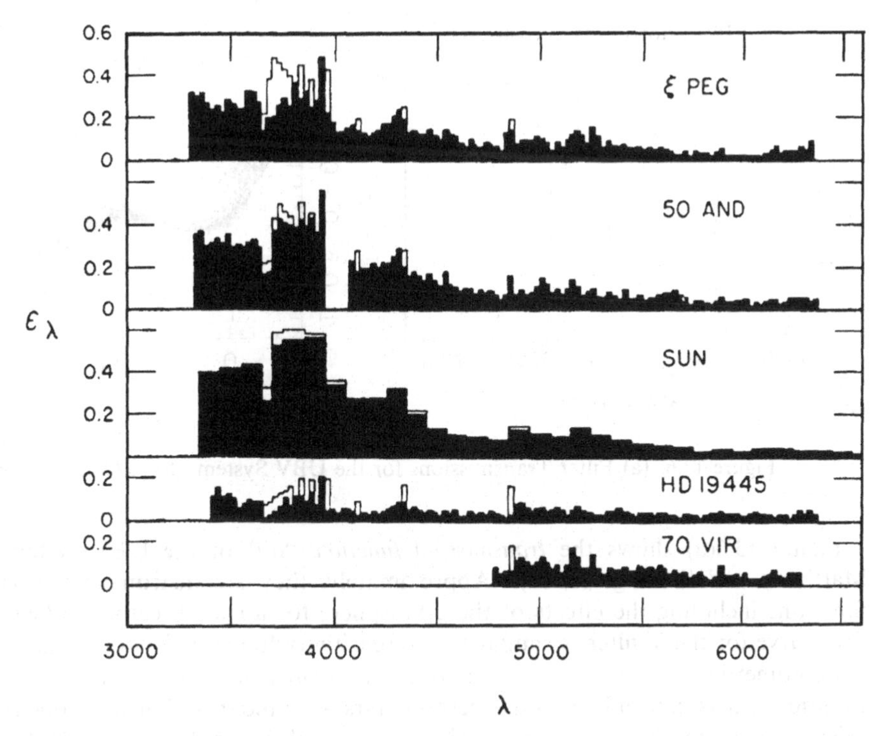

Figure 13.5. Blocking Coefficients for Stellar Spectra. The fraction ϵ_λ of energy removed from stellar continua is plotted against the wavelength in angstrom units.

absorption of the atmospheric structure and emission properties. The absorbed radiation in some cases emerges in unexpected spectral domains. However, for the present purposes, we assume this effect to be small, and a straightforward interpretation of stellar colors in terms of line blocking is then valid: more lines make the star appear redder.

The overall result is that the presence of atomic absorption alters the colors of stars in a predictable way, and because the effect is variable with wavelength, it is possible to distinguish between the effects of line blocking and the temperature of stellar photospheres. The simplest way to see this is to imagine measurements of the stellar continua at wavelengths where the line absorption is virtually negligible. If the indices from such measurements indicated the temperature of the sun, while indices from wavelengths including the ultraviolet were compatible with a much higher temperature, we might conclude that our star had weak lines.

In fact, it is unnecessary to go to wavelengths that avoid all, or virtually all, lines, because the effect of line blocking is systematic. For this reason, astronomers have made extensive use of photometric measurements in the near ultraviolet and visible. We shall discuss the most widely used system, called the UBV (ultraviolet, blue, visual) system, where the symbols roughly indicate the position of the maxima of a system of filters through which stellar brightnesses are measured.

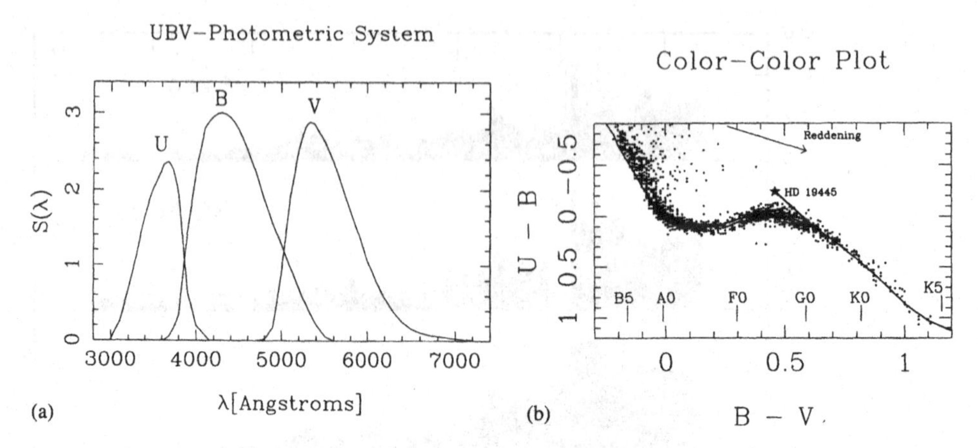

Figure 13.6. (a) Filter Transmissions for the UBV System. (b) A Color–Color Plot.

Figure 13.6(a) shows the *transmission functions* $S(\lambda)$ of the UBV system from Matthews and Sandage (1963). Approximately, they are instrumental sensitivity functions including the effects of the atmosphere for a star directly overhead. The $S(\lambda)$ curve for the V filter is similar to the sensitivity function for the human eye.

Photometric measurements are almost all *relative to some standard.* For astronomers, it is convenient to use certain stars as standards. Let us suppose there exists a single standard star, whose brightnesses as observed through the U, B, and V filters have the values $b(\mathrm{U}_0)$, $b(\mathrm{B}_0)$, and $b(\mathrm{V}_0)$. The system of visual or V *magnitudes* is then defined such that if $b(\mathrm{V})$ is the brightness of an arbitrary star measured with the V-filter (cf. 13.1)

$$\mathrm{V} - \mathrm{V}_0 = 2.5 \cdot \log \left[\frac{b(\mathrm{V}_0)}{b(\mathrm{V})} \right], \tag{13.2}$$

provided appropriate corrections are made for variations in the transparency of the atmosphere. We must refer to works on photometry for such details (see Golay 1974).

In Equation 13.2, V is said to be the "V-magnitude" of the star and V_0 the V-magnitude of the standard. Similar relations hold for U, and B. Two *color indices* are defined for this the UBV system: (U − B) and (B − V). For an arbitrary star, (B − V) is defined by

$$\mathrm{B} - \mathrm{V} = 2.5 \cdot \log \left[\frac{b(\mathrm{V})}{b(\mathrm{V}_0)} \frac{b(\mathrm{B}_0)}{b(\mathrm{B})} \right]. \tag{13.3}$$

A similar relation defines (U − B).

Johnson and Morgan (1953) set up the system in such a way that the color indices (i.e. U − B and B − V) of six main-sequence A0 stars were *defined* to average 0.0 for both indices.

In Figure 13.6(b) we show a two-color or *color–color* plot for main-sequence stars. The points were taken from Hoffleit (1982), with corrections kindly made available by the author and Wayne H. Warren. The numbers displayed as ordinates and abscissae represent measurements made on the UBV system for main-sequence stars. On such a plot, cool stars lie to the right, and down. Note that (U − B) is traditionally plotted increasing downward. The (U−B) positions of a few luminosity class V spectral types are indicated.

The solid line through most of the points is the locus of the so-called *zero-age main sequence*, or ZAMS, according to Schmidt-Kaler (1982). Most of the observed points for the early stars (B stars) fall to the right of the ZAMS because of effects of reddening and also stellar evolution.

The arrow pointing downward and to the right is a *reddening trajectory*. It shows how interstellar reddening changes the UBV colors of stars. The components of this arrow are called the color *excesses*, E(B − V) for the horizontal component, and E(U − B) for the vertical one. Reddening acts nearly the same way as line blanketing. In both instances, stars have redder colors. However, the trajectories are not identical in slope.

The position of one of the classical subdwarfs, HD 19445, is indicated, from the measured colors: B − V = +0.46, U − B = −0.24. A line connects this position (the star) with the position of normal main-sequence dwarfs with the same effective temperature. We have adopted figures from the paper by Bell and Oke (1986) for HD 19445. Early work on subdwarfs showed that their position above the normal main sequence could be reasonably accounted for by the lack of metal lines in their ultraviolet spectra.

The vertical distance of the star above the main sequence is defined as the *ultraviolet excess*, or $\delta(U - B)$. It is a measure of the overall affect of the lines on the star's color, and we shall see that in many cases it is a good predictor of the results of abundance determinations. The amount of the excess, however, cannot be expected to be the same for stars of all temperatures. Sandage (1969) investigated the ultraviolet excess of stars with a *given* iron abundance, [Fe/H], as a function of $(B - V)_0$ (see Equation 13.4). As a result of his work, modern calibrations are usually of [Fe/H] vs. $\delta(U - B)_{0.6}$, where the latter means the ultraviolet excess of stars with $(B - V)_0$ fixed at 0.6 (see Cameron 1985).

While the "reddening trajectory" is not (anti-)parallel to the blanketing vector, it is easy to see that careful photometric work is necessary to separate the two effects. It may be done with confidence only in special instances, for example, when the parallax of the star is so large that one can be sure that the reddening is very small. We shall have more to say about reddening, and the nature of interstellar dust, in Chapter 14.

We may now state briefly the results of the analysis of abundances that are thought to be related to populations. The vast majority of stars have abundances that are within a factor of two or three of the solar values. Since the accuracy with which many abundances have been determined may not be much better than this, it is difficult to know how sharply peaked the distribution of abundances really is.

Abundances that have been determined from modern, low-noise spectra should have *much* better accuracies than a factor of two or three. However, we still do not have a large sample of stars with very accurate abundances. Moreover, the way in which the stars have been selected for study precludes the use of the resulting abundances for many statistical purposes – the results would be biased.

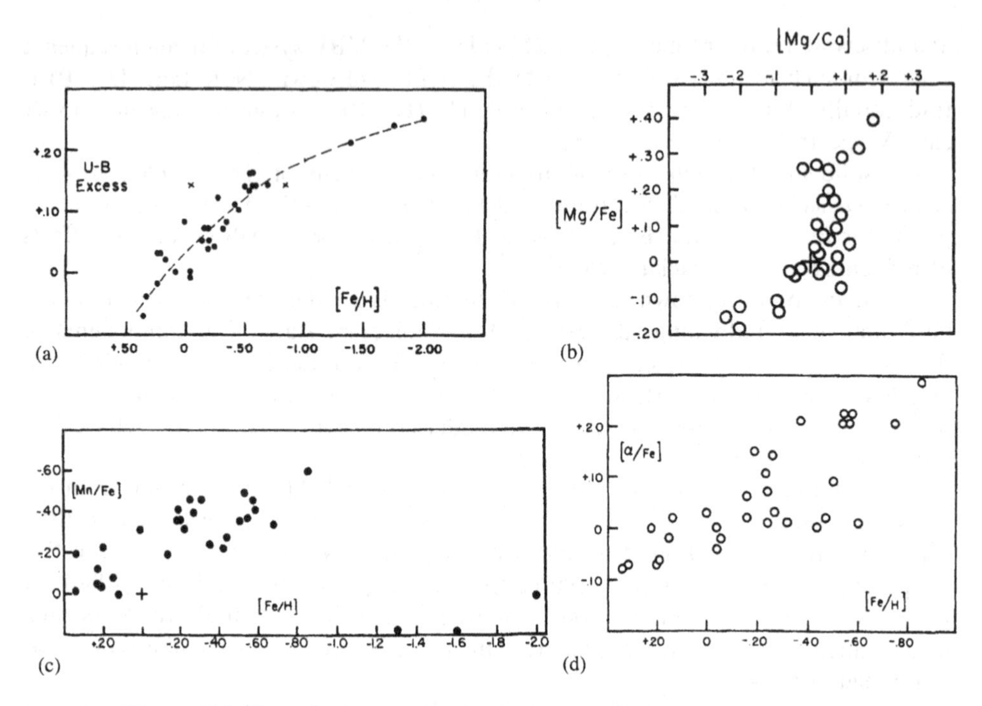

Figure 13.7. Some Systematic Results From Early Abundance Surveys of Late-Type Dwarfs. (a) The Ultraviolet Excess, $\delta(U-B)$ vs. [Fe/H]. (b) [Mg/Fe] vs. [Mg/Ca]. (c) [Mn/Fe] vs. [Fe/H]. Note that the [Mn/Fe] ratio appears to increase more rapidly than [Fe/H]. (d) [α/Fe] vs. [Fe/H]. The α-elements are those with isotopes that can be thought of as composed of α-particles, e.g., ^{24}Mg,^{28} Si, and ^{40}Ca.

The other main result from population-related abundance work is that the vast majority of elemental abundances vary together, or *in lockstep*. Nevertheless, surveys of abundances in field dwarfs, for example those summarized by Wallerstein (1962), *did* show some rather interesting systematic effects – *departures* from lockstep. We illustrate a few results in Figure 13.7(a–d) from Wallerstein's survey of G dwarfs.

Figure 13.7 uses the "bracket" notation of abundance workers. If x represents some measured abundance in a program star and a standard star, which is often taken to be the sun, then

$$[x] \equiv \log(x)_{\text{program star}} - \log(x)_{\text{standard}}. \tag{13.4}$$

When the bracket is written without any special subscript, it is almost always the sun (SAD) which is taken to be the standard. Occasionally, a subscript is given, e.g., $[\text{Fe/H}]_\epsilon$, which would mean that the star ϵ Virginis (Cayrel and Cayrel 1963) is taken to be the standard.

Astronomers frequently speak of abundance differences in units of "decimal exponentials," or *dex*. For example, if two stars differ by 1 dex in their iron abundance, it means the abundances differ by a factor of ten.

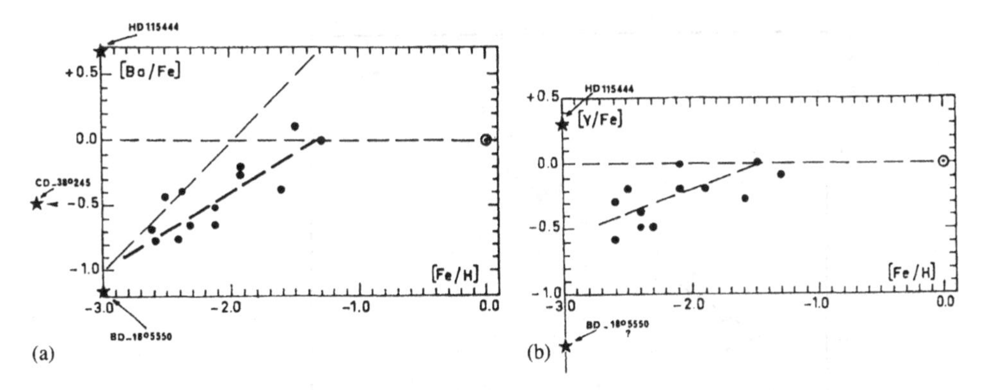

Figure 13.8. Abundance Work Showing Violations of Lockstep. Abundances of barium and yttrium are low, relative to iron, in the metal-poor stars. The SAD abundances of both these elements have larger *s*- than *r*-process contributions. There is now good evidence that the *s*-process operated less efficiently in the past than the *r*-process. Courtesy of F. Spite and the International Astronomical Union.

Of the systematic effects shown in Figure 13.7, the only one that has become universally accepted is (a). The rest were regarded with skepticism. The points scatter roughly a factor of two (0.3 dex) from the mean, and this is about the level of accuracy of the determinations themselves.

As recently as 1976, A. Unsöld, the doyen of stellar atmospheres, argued that "The relative abundance [*sic*] of the heavier elements are the same as in the sun except for elements and stars which obviously show signs of individual nuclear evolution which brought partially burnt matter into their atmospheres" (see Unsöld 1976).

Without doubt, the most obvious population-related abundance variations are of element-to-hydrogen ratios. Helium should probably be excluded from this generalization because it is not possible to obtain its abundance reliably in late-type stars. These late stars are just the ones whose surface abundances are thought to reflect general "population" trends. A number of modern results have confirmed that the lockstep assumption is violated, especially in *halo* objects, that is those with low values of [Fe/H], traditionally considered to belong to Population II. A few examples are shown in Figure 13.8, from the review article by Spite (1983, see also reviews by Pagel and Edmunds 1981, and Lambert 1988).

Most of the stars analyzed here are giants rather than dwarfs, and with giants, there is always the possibility that surface abundances have been altered by the products of interior nucleosynthesis. The star HD 115444, indicated in Figure 13.8, may be such a case. Its abundances resemble those of the so-called CH stars, a well-explored group of objects thought to be the Population II analogues of Ba II stars, to be discussed below.

The abundance variations shown here are more convincing than those of Figure 13.7, primarily because they represent much larger effects.

Spite makes the critical point that the approximate validity of lockstep variations

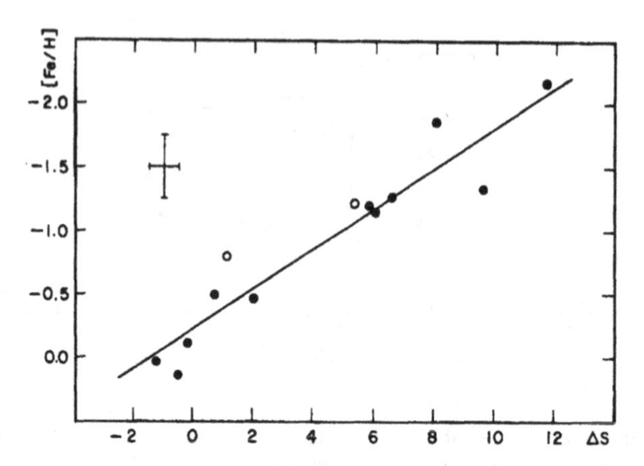

Figure 13.9. Butler's (1975) Calibration of the ΔS Parameter.

makes possible a number of analytical tools that would otherwise be useless. We have already seen one example, the calibration of the ultraviolet excess to yield values of [Fe/H]. A number of other photometric systems are commonly used to measure the "metal-to-hydrogen ratio," in stars. Two of the better-known systems are those of Strömgren (1966) and the David Dunlop Observatory DDO (McClure 1976). A technique that has achieved a remarkable success is based on the difference in the spectral class, ΔS, using the K line of Ca II and the hydrogen lines (Preston 1959, Smith 1984).

One might think the ΔS technique had no chance of working at all! In the first place, it was introduced for the highly evolved RR Lyrae variable stars, whose surface chemistry has had ample opportunity to be changed in ways that would violate the lockstep assumption. Second, the method is based on the Ca II K line in relatively hot stars, which as we shall see shortly are notorious for showing a capricious variation in their calcium abundances.

Figure 13.9 shows a remarkable calibration of the ΔS parameter by Butler (1975). The [Fe/H] values were determined by Butler from standard curve-of-growth procedures, and it would not be difficult to attribute virtually all of the scatter in Figure 13.9 to analytical errors rather than to an intrinsic variation of [Ca/Fe] vs. [Fe/H].

There is certainly some evidence that [Ca/Fe] is variable in these stars (see the references in Smith 1984), but most of the variations are within a factor of three, and could be dismissed by all but those responsible for them (see Butler, Dickens, and Epps 1978). Butler has pointed out the similarity of his results to those of Wallerstein and his coworkers, which we discussed above, saying that the effects were too small to be generally credible. Smith applies a mean relation, $[Ca/H] = 0.8 \cdot [Fe/H]$, in order to be able to infer [Fe/H] from the ΔS measurements. The idea that a mean relation might be valid is surprising to anyone familiar with the chemistry of upper main-sequence stars. As we shall see, [Ca/H] is highly variable in these objects, and it would appear to be hopeless to try to infer the [Fe/H] value in an upper main-sequence star from its [Ca/H].

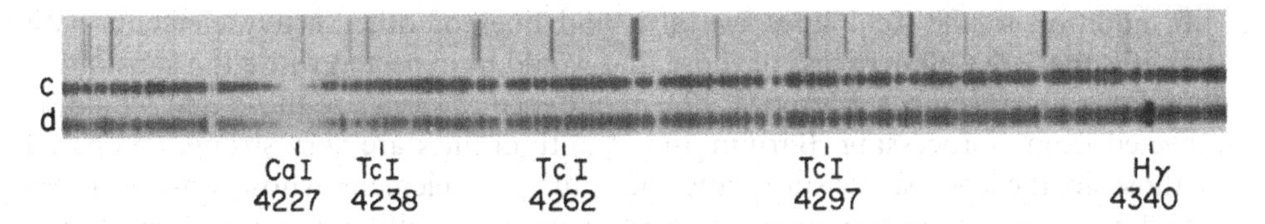

Figure 13.10. Spectrum of Tc I in the Evolved Giant R And.

These reservations aside, there is no doubt that a large body of apparently consistent results have been assembled by workers who have made the lockstep assumption, for $\delta(U - B)$, for ΔS, or for some comparable measurement, such as calibration of the m_1 index of the Strömgren (1966) system of photometry as an indication of [Fe/H]. Perhaps as the accuracy of determinations improves, it will still be possible to apply corrections so that [Fe/H] can be inferred, at least statistically, from ΔS or $\delta(U - B)$.

The abundance systematics of those stars whose abundances reflect populations is essentially the subject of the chemical evolution of galaxies. We shall therefore continue our general discussion of stellar abundances, and return somewhat later to the matter of overall chemical evolution in stars and stellar systems – clusters and galaxies (see Chapter 16).

13.4.2 Abundance Variations Attributed to In Situ *Nucleosynthesis*

Paul Merrill (1952) discussed the identification of the element technetium ($Z = 43$) in the spectrum of the long-period variable star R Andromedae. Since the longest-lived isotope of technetium, ^{98}Tc, has a half-life of some $4.2 \cdot 10^6$ years, the presence of this element has been widely accepted as palpable (*prima facie*) evidence for ongoing nucleosynthesis. At the time of the discovery it had been widely believed that the heavy elements were created in a primordial event, according to ideas associated with G. Gamow and his colleagues.

A small amount of technetium occurs in terrestrial rocks as a product of the fission of ^{238}U and ^{235}U. However, the amounts are minuscule. In pitchblende, an ore mineral of uranium, primarily UO_2, Kuroda (1982) found some 2 to 3 times 10^{-13} gm of ^{99}Tc per gram of material. Compare this tiny fraction of Tc with the mass fractions of niobium and molybdenum present in the solar (SAD) mixture: 10^{-8}–10^{-9} gm/gm. However, spectra of R And reveal that the amount of Tc present is *comparable* to that of any of the heavy elements that can be detected in the stellar spectrum. This is illustrated in Figure 13.10.

The illustrations are from the celebrated paper by B^{2}FH. The top spectrum is of a mid-M giant, 56 Leonis. The lower spectrum is of R Andromedae, an S-type long-period variable star. Several Tc I lines are marked.

Long-period variable stars have very complicated atmospheres, so that quantitative abundance estimates must be very crude. But the spectrum shows that the strengths of the Tc I lines are comparable to those of surrounding metallic lines, and this indicates a surprisingly high abundance for this unstable element.

In addition to the Tc I lines, we also find lines of other heavy elements with strengths that are well in excess of those that would be expected without a substantial abundance enhancement. Moreover, the abundance *patterns* are those that would be expected from *s*-processing. Barium and strontium lines are very strong, compared to those in the coolest M stars, and europium, an element whose only isotopes have large neutron capture cross sections, is not markedly enhanced. Cowley and Hensberge (1981) have reviewed the classical identifications in R And with the help of wavelength coincidence statistics, and found excellent agreement with them.

The traditional interpretation of Tc and the enhancements of lines from other heavy elements in certain long-period variables such as R And is that the products of nucleosynthesis have been transported to the surface by deep convective currents that stir the envelopes of such cool stars.

It has been known for more than a century that the spectral sequence divides, for the coolest stars, into three principal branches. The M's are an oxygen-rich sequence with abundances similar to the SAD, and the R's and N's a carbon-rich sequence, sometimes with enhancements of heavy elements. The S stars form a final branch, for which oxygen is probably about equal to carbon in abundance, again with enhancements of the heavy elements. R And is an S star.

Accurate abundance work is nearly impossible for the coolest giant atmospheres. Slightly hotter giants known as Ba II stars and their Population II analogues the CH stars are more tractable. These stars are related chemically to the later types, and have been analyzed by traditional abundance methods. Lambert and his collaborators have analyzed the isotopic abundances of C and Mg in some of these objects as well as elemental abundances for numerous elements (see Smith and Lambert 1990 and references therein).

Predictions of elemental abundances from *s*-process calculations (Cowley and Downs 1980, Malaney and Lambert 1988) fit the abundances of heavy elements obtained for the barium stars, and presumably also their Population II congeners. However, one has a certain freedom in the kinds of assumptions that one can make in this fitting of theory and observation. It is difficult to be unbiased about these fits.

The discovery by McClure and his coworkers (see McClure 1984) that a very large percentage of the barium stars are binaries has cast doubt on the mixing scenario. An alternate hypothesis has been made possible by this ubiquitous binarity. It is very likely that material from highly evolved companions has been transferred to the objects now seen as Ba II stars.

Scalo (1976) placed the late-type peculiar stars on an H–R diagram (Figure 13.11). The ordinate is the *bolometric* magnitude. This is a logarithmic measure of the total energy output of the star, integrated over all wavelengths (see §13.3). The MS stars represent a class between the M's and the classical S stars.

The very large range in the luminosity of these stars should be noted, especially in view of the discovery that dwarf and irregular members of the Local Group (see §16.1) of galaxies are anomalously rich in (highly luminous) carbon stars. We know little of possible analogues of the Ba II stars in these systems.

Scalo concluded that "the positions of these stars cannot be reconciled with current stellar evolution calculations." It seems to the present writer that the imprint

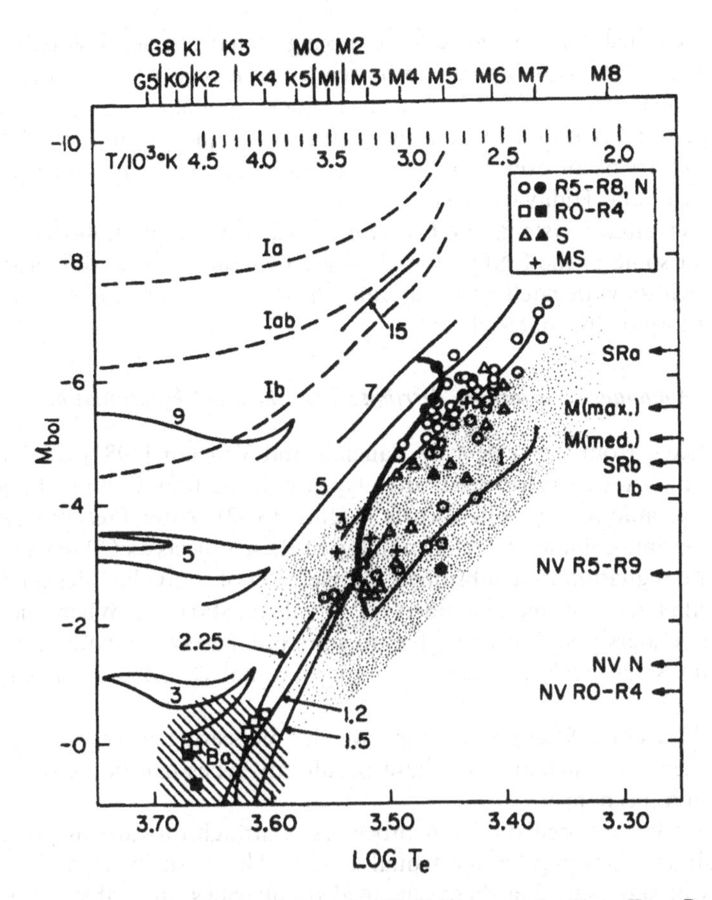

Figure 13.11. Scalo's Composite H–R Diagram for Late-Type Peculiar Stars. Note that the ordinate is absolute bolometric magnitude rather than absolute visual magnitude

of nuclear processes is rather clear, but the history of the matter has yet to be clarified.

Before we leave this section on stars whose abundances were modified by *in situ* evolution, we mention briefly the so-called CNO stars, at the opposite temperature extreme to the objects we have been discussing. These stars are usually too hot to show lines of the heavier metals ($Z > 26$), and discussion usually focuses on the elements carbon, nitrogen, and oxygen themselves. Early stars with helium anomalies are also known, but these objects are considered unrelated to the CNO stars.

In situ evolution could cause nearly all of the CNO anomalies of early stars. Most of these objects are supergiants, for which there is the possibility of mixing of nucleosynthetic products. In particular, the nitrogen-rich members of this group could arise as a result of the mixing to the surface of the enhanced ^{14}N expected from hydrogen burning on the CNO cycle. On the other hand, nitrogen-deficient stars have also been identified, and this makes the mixing scenario much less palatable. At least some of the CNO stars appear to be related to a class of hot young stars known as Wolf–Rayet stars (see §10.2), whose chemical anomalies arise from mass loss. These divide roughly into a carbon and a nitrogen sequence, and the ratio of the two may be a part of an overall pattern of abundance variations on a galactic scale (van der Hucht and Hidayat 1991).

Mass transfer in binary systems, and *in situ chemical* differentiation, have also been suggested for some of the CNO anomalies. We turn now to a consideration of a group of stars whose variegated surficial abundances are thought to be unrelated to the nuclear history of the material.

13.4.3 Abundance Variations Attributed to Chemical Fractionations

Figure 13.12 is taken from the review by Vauclair and Vauclair (1982) of element segregation in stars. A variety of peculiarity types may be found along the main sequence as well as among the white dwarfs (Figure 13.12). After Preston's (1974) review paper, it became common to refer to many of these objects as CP (chemically peculiar) stars, a designation that subsumed the plethora of individual designations such as Ap (peculiar A), Am (metallic-lined A), Si (silicon star), etc. When one uses CP, it is generally understood that one refers to stars on or near the main sequence, so that the barium stars, while chemically peculiar, are not CP stars in the current sense.

Vauclair and Vauclair (1982) show, following the pioneering work of Michaud (1970), that the general systematics of these peculiarity types can be explained in terms of a diffusion mechanism.

The physics of the proposed diffusion processes is straightforward in principle, but very difficult to work out in individual cases. The basic postulate is that the atmospheres of stars showing these chemical peculiarities are stable to mixing currents or convection to a very high degree. This crucial point, dealing with questions of stellar hydrodynamics, seems, from this point in time, impossible to establish with any kind of rigorous theory. We must consider the proposed diffusion currents, which move at speeds generally under 10^{-3} cm/sec, compared with the other relevant velocities in the atmosphere, such as the sound speed, the microturbulent velocity, or the rotational velocity. The latter are all some 10^5 cm/sec or higher, so the stability requirements amount to eight or more orders of magnitude. In view of the lack of any kind of proof that such stability really obtains, plausibility arguments have been advanced. They have been found rather convincing.

The arguments generally begin by noting that the temperature domain in which the chemical peculiarities occur is that in which no *strong* mixing currents are expected. The stars named in Figure 13.5 have envelopes that are predominantly in radiative equilibrium. While shallow convection zones do exist in some of these stars, their net effect is to maintain a uniform composition throughout those specific zones, while permitting diffusion to occur in regions either above or below the mixed zones.

A second overall argument that diffusion in stable zones is relevant is that the

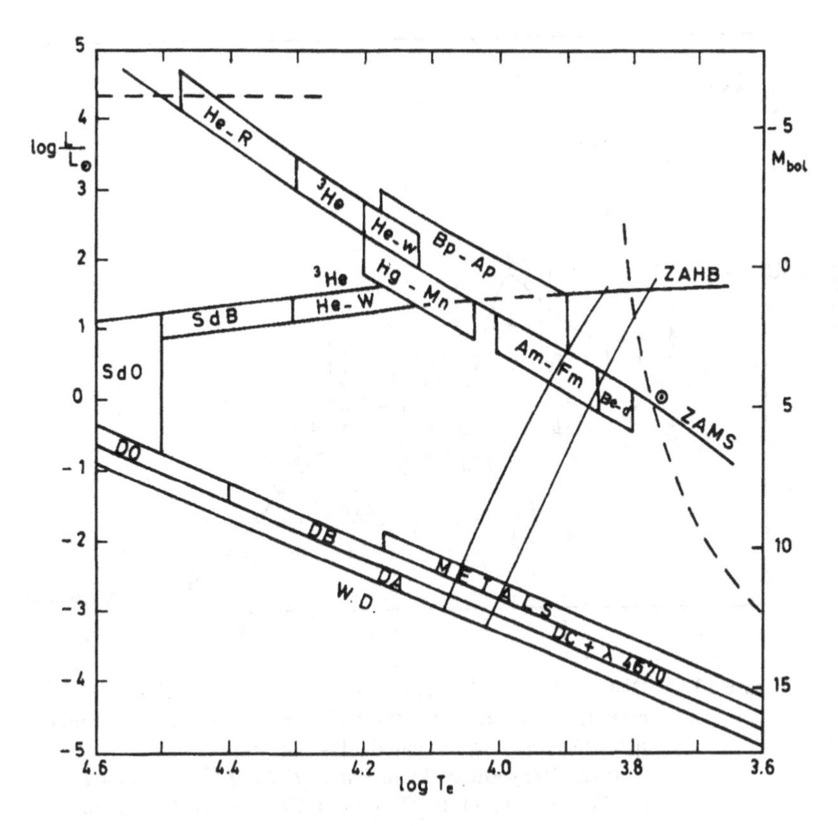

Figure 13.12. Location of the CP Stars on the H–R Diagram. See text for description. Reproduced with permission from *Annual Review of Astronomy and Astrophysics*, Vol. 20, ©1982 by Annual Reviews Inc. Courtesy of S. and G. Vauclair.

majority of CP stars are slow rotators. Certainly, one would expect the typical rotations of A and B stars (100–200 km/sec) to stir currents that could preclude the effectiveness of 0.001 cm/sec diffusion velocities. Figure 13.13 shows the distribution of rotational velocities projected on the line of sight ($v \sin(i)$) for "normal" stars of luminosity class I, III, IV, and V (dwarfs) for spectral types O–G, along with those for the CP stars (Ap and Am's). Figure 13.13 is from a review by Fukuda (1982) on stellar rotational velocities.

The arguments noting slow rotation and radiative envelopes throughout the CP stars are plausibility arguments: if the phenomenon (effective diffusion) can happen anywhere, it can happen here; it *is* in just this domain where we observe the chemical peculiarities; therefore it is reasonable to assume the necessary stability as a working hypothesis, and explore the possibilities. Such explorations have been made by Michaud, the Vauclairs, and their coworkers in an impressive array of papers cited in the Vauclairs' review. An excellent overall introduction to this entire subject may be found in the book by Wolff (1983): *The A-Type Stars: Problems*

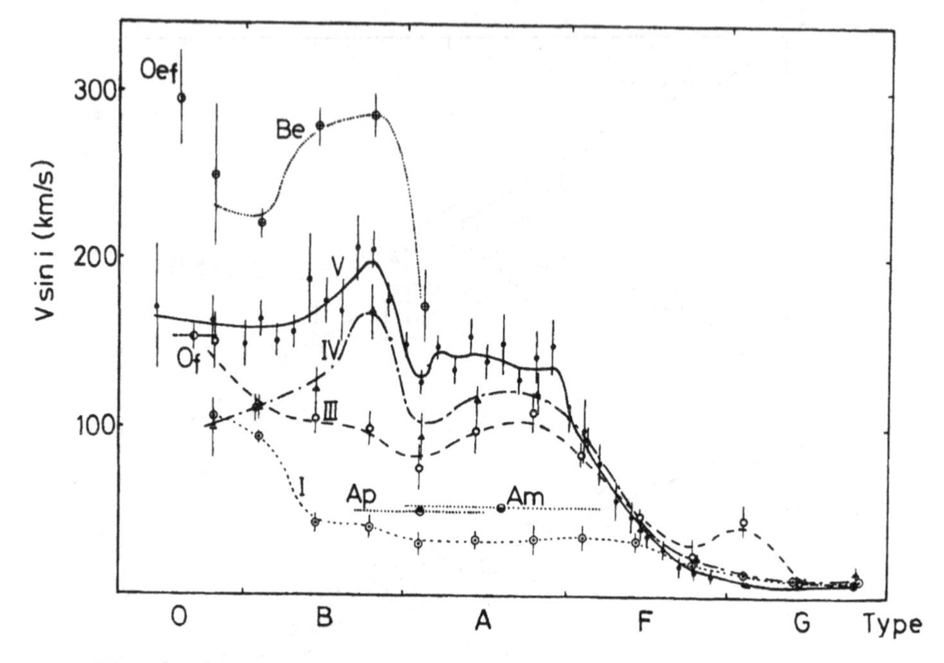

Figure 13.13. Distribution of Projected Stellar Rotational Velocities. The Roman numerals indicate the different luminosity classes. Some peculiar spectral types are indicated. The Of stars are O stars with emission features. Reproduced from *Publications of the Astronomical Society of the Pacific*, Vol. 94, p. 271 (1982). Courtesy of I. Fukuda.

and Perspectives. Recent work is summarized in proceedings edited by Michaud and Tutukov (1991) and Dworetsky, Castelli, and Faraggiana (1993).

In an absolutely stable atmosphere, all particles would attempt to achieve their own scale height, subject to the restriction that electric fields prevent all but a small separation of the positive ions and electrons. This electric field, sometimes called a Pannekoek–Rosseland field by workers in solar wind theory, can be simply derived in the case of a two-component isothermal atmosphere by requiring no net separation between ions and electrons. The result, for a proton–electron gas, is a scale height of $\mathscr{R}T/g\bar{\mu}$ where $\bar{\mu}$ represents the mean molecular weight of a proton and an electron. The electrical force is generally small relative to the gravitational and radiation forces, and will be neglected in the discussion below.

Let us calculate the upward radiative force in a stellar atmosphere on an atom or ion with cross section $\alpha(\nu)$, due to a single absorption line. Let the net upward flux in the line be designated πF_ν, with dimensions ergs/cm^2/sec per unit frequency interval. We can think of the number of photons per cm^3 in the frequency interval ν to $\nu+d\nu$ as given by $\pi F_\nu d\nu/ch\nu$. Dividing by $h\nu$ gives the flux of photons (number/cm^2/sec) and dividing again by their velocity, c, gives the number density of photons (number/cm^3) in ν to $\nu + d\nu$. We now formulate the net upward force as a collision problem, with the collision frequency being of the form (number density $\times$ cross section $\times$ velocity), and the momentum transferred per collision being $h\nu/c$. We thus have for the upward, radiative force (or momentum flux) F_{RAD},

$$F_{\mathrm{RAD}} = \int_0^{\infty} \frac{h\nu}{c} \frac{\pi F_\nu d\nu}{ch\nu} \cdot \alpha(\nu)c. \tag{13.5}$$

If we only consider this one (weak) line, and take the flux to be approximately constant across it, then all of the variables are effectively constant for the small region of frequency where the absorption cross section is constant, and the integral over $\alpha(\nu)$ is $\pi e^2/mc$ multiplied by the oscillator strength f_{nm}, as discussed in §12.5.

Take $f_{nm} \approx 1$ (compensating somewhat for the neglect of other lines). Then

$$F_{\mathrm{RAD}} \simeq \pi F_\nu \left(\frac{\pi e^2}{mc^2}\right). \tag{13.6}$$

Let us now make a crude estimate of how this force would compare, for a cerium atom (atomic weight 140) in the atmosphere of an A star with an effective temperature of 10 000 K. We may obtain the upward flux, for example at $\lambda4000$, from calculations on stellar atmospheres (e.g., Kurucz, Peytremann, and Avrett 1974), or for the present purpose, we may simply compute the Planck function for 10 000 K, and multiply it by π to account for all 2π (outward) steradians (see Problem 3 below). The Planck function for $\lambda4000$ times π is $5.5 \cdot 10^{-4}$, about a factor of 2 larger than the value that we would get from the reference cited for a $\log(g) = 4.0$ model atmosphere at $\lambda4000$. In cgs units, $\pi e^2/mc^2$ is $8.85 \cdot 10^{-13}$. The net upward radiative force is thus

$$F_{\mathrm{RAD}} \simeq (5.5 \cdot 10^{-4}) \cdot (8.85 \cdot 10^{-13}) \simeq 4.9 \cdot 10^{-16} \text{ dynes}. \tag{13.7}$$

This upward force should be compared with the downward force due to gravity, which in a $\log(g) = 4$ atmosphere, for a cerium atom is $10^4 \times 140 \times 1.7 \times 10^{-24}$ or about $2.4 \cdot 10^{-18}$ dynes.

The result of this rough calculation is fairly typical. The upward radiative force on certain atoms can exceed the downward gravitational force by several orders of magnitude. The atmosphere as a whole does not blow off because certain of the assumptions that we made in obtaining the numerical values just presented do not hold for abundant species. In particular, the radiative flux through an absorption line can be reduced by many orders of magnitude if the line absorption coefficient is larger than the continuous absorption coefficient *at* the relevant frequencies where the atoms are absorbing photons.

This brings us to a cogent argument that this overall diffusion mechanism is relevant to the problem of CP stars. By and large, the elements that are enhanced with respect to the SAD are the trace species, whose line absorption coefficients would not be so large that the net upward flux is prohibitively reduced. On the other hand, abundant species, especially the light elements such as helium, oxygen and calcium, are typically depleted with respect to the SAD.

In order to calculate the net upward force on some elemental species, it is necessary to take account of all relevant stages of excitation, ionization, and external factors in addition to the upward radiative flux. Possible additional factors are magnetic fields and temperature gradients, both of which can be important. Upward radiative momentum can derive from transitions into the continuum as well as from lines. In special circumstances a curious phenomenon known as *thermal diffusion* (see Hirshfelder, Curtiss, and Bird 1954) can be important.

We must leave details of the calculations of diffusion rates to the review by Vauclair and Vauclair (1982) and the references cited therein. It is necessary, however, to emphasize the complexity of the problem of making these calculations. Not only must theoreticians make a global leap of faith concerning the nature of slow mixing currents in the star, but they must take accurate account of an extraordinary number of microscopic processes at the atomic level. In many cases the data necessary for the calculations are unavailable. Often, the critical processes for separation take place at the top of stellar atmospheres where the useful LTE assumption cannot be made. A diffusion calculation then takes on the full complexity of a non-LTE abundance study, with all of the additional gas kinetic details that must be included to treat the diffusion. The final indignity is the responsibility of the observers themselves, who have been unable yet to provide a sound account of what the observations really mean in terms of abundances.

While there is no doubt that the photospheres of the CP stars are very different in composition from the SAD, there remain very grave uncertainties in the quantitative abundance work. Needless to say, reliable abundances are absolutely essential as a guide to the theoretical work. We shall illustrate the problem with a discussion of perhaps the most bizarre stellar spectrum in the known universe – an object referred to by astronomers as Przybylski's star.

The Polish-Australian astronomer A. Przybylski (1961) called attention to "HD 101065 – a G0 Star with High Metal Content." The truly unusual nature of this spectrum did not emerge until a subsequent paper (Przybylski and Kennedy 1963), in which it became clear that the majority of the identifiable spectral lines belonged to lanthanide rare earths. Of the fourteen lanthanides, La–Lu, Przybylski and Kennedy claimed to have found all but the unstable promethium, and the two heaviest, ytterbium and lutetium. An undue emphasis was placed on the identification of the intermediate lanthanide holmium, partly as a result of confusion that arose because of the misclassification by atomic spectroscopists of two strong lines of Ho II, $\lambda\lambda$3796.75 and 3810.73. They had been incorrectly assigned to the first spectrum. Przybylski's star then became known as the "holmium star," as though holmium were the most abundant or dominant element.

A few elements other than lanthanides were identified by Przybylski and Kennedy, but they concluded that "of the first 53 elements in the periodic system, only hydrogen and calcium could be said to be present beyond any doubt." To any stellar spectroscopist, the absence of strong lines of Fe I in a star with the color of HD 101065 is an extraordinary thing. It was certainly unique in the experience of the present writer, who had spent more than a decade studying the spectra of CP and related stars.

Relative abundances in Przybylski's star have been the subject of some debate. The star is *so* peculiar, that traditional methods of abundance determinations break down. The reader interested in more details should consult the references in Wolff (1983).

The currently favored interpretation of the surface chemistry of the CP stars brings with it a rather new aspect of sidereal astronomy. For many years it was thought that cosmical abundances could be interpreted within the framework of nuclear astrophysics. In particular, it was thought that any influence of *chemical* factors could be ignored, or perhaps avoided.

It now appears that a very large portion of the upper main sequence shows important abundance deviations from those of the SAD. For virtually every element that is observable, anomalies have been found in some upper main-sequence star, and the fraction of objects that may be considered to have abundances which deviate by less than a factor of two from the SAD may be less than 50%.

As we saw in §7.3, chemical differentiation is thought to take place between the gas and solid phases of the interstellar medium (see also §§14.5 and 14.6). It was more or less an act of faith that such chemical separations happen on a scale that leads to chemical uniformity for the large volumes of gas that eventually form stars. While accretion processes have been discussed in connection with the CP stars (see Havnes 1974), they usually have not focused on the possibility of sweeping up material that is already highly differentiated chemically, such as interstellar dust. This possibility was considered, but not in connection with CP stars (see Alcock and Illarionov 1980). In §7.3, we mentioned the suggestion by Venn and Lambert (1990) that chemical fractionations might be responsible for the abundances in the λ Boötis stars. This interesting suggestion has received additional attention by stellar spectroscopists. A number of papers in the proceedings edited by Dworetsky, Castelli, and Faraggiana (1993) deal with this question. Holweger and Stürenburg (1993) suggest the fractionation takes place at the time of star formation.

We must conclude that the chemistry of upper main-sequence stars and related objects such as horizontal branch stars is not well understood. While the general mechanism of elemental separation under the influence of gravitational and radiative forces is well conceived, it is not clear that this mechanism alone can account for the wide variety of abundance patterns that can be seen in these stars. This writer has only echoed the opinions of knowledgeable spectroscopists before him that many of the abundance patterns in the CP stars suggest nuclear processes. For example, we find – insofar as the low accuracy of stellar abundance work permits – a persistence of the odd–even alternation in abundances that are many orders of magnitude away from those of the SAD. There are also instances of extreme fractionations – 10^5 to 10^6 – in the case of mercury and platinum, in stars whose other abundances look rather commonplace. It is almost as though the heavy elements had a different chemical history from the intermediate ones.

The main shortcoming of hypotheses involving nuclear processes is that plausible "scenarios" have not been recently constructed to account for the emplacement of matter with an exotic composition in the atmosphere of a main-sequence star. Some years ago, van den Heuvel (1967, 1968) proposed mass transfer in binary systems as a reasonable mechanism for the origin of CP stars. Related ideas were also discussed by Renson (1967). These seem as attractive to the present writer as ever, in view of numerous interesting developments in the field of binary evolution, nucleosynthesis on white dwarfs, the discovery of CP stars among blue stragglers, etc. However, it is clear that if such mass transfer is relevant to the Ap stars, processes beyond those carefully considered by Proffitt and Michaud (1989) must be at work. The latter authors concluded that only a few percent of the CP-star abundances might be accounted for in terms of mass transfer in binaries.

13.5 Problems

1. A very useful equation in astronomy relates the absolute magnitude M, the apparent magnitude m, and the distance of a star, d, when the latter is measured in parsecs. Use the definition of the absolute magnitude (§13.3) to derive

$$M = m + 5 - 5 \cdot \log(d). \tag{13.8}$$

2. If a color excess, $\delta(\mathrm{U}-\mathrm{B})$, of 0.25 *magnitudes* could be attributed *entirely* to line blanketing in the U filter (and none in the B), what fraction of the radiative *flux* in the B band would be removed by the spectral lines of normal stars?
3. Explain why one multiplies the Planck function by π and not 2π to get the upward radiative flux from a hemisphere as used in Equation 13.7.

14
Cold, Non-stellar Material in Galaxies

14.1 Introduction

In 1910, the British astronomer Arthur Eddington published an influential monograph with the impressive title of *Stellar Movements and the Structure of the Universe.* Eddington, who became "The most distinguished astrophysicist of his time" (Chandrasekhar 1983), was only 28 when *Stellar Movements* was published, but his clarity of exposition and physical insight are readily seen in this small volume. Nevertheless, our present view of the Galaxy in which we live, and the universe around us, is completely different from that limned by Eddington at the end of the century's first decade. Not only were the astronomers of that time uncertain of the nature of the spiral nebulae we now call galaxies (Chapter 16), but they thought the solar system was at the center of our own system of stars.

It had been known since the time of the star gauges (counts) of William Herschel (1738–1822) that faint stars did not increase in number as one would expect, but indicated an "end" of the entire system. Today, at visual wavelengths we can in some sense detect the end of our Galaxy if we look out of the plane, toward its poles. Within the plane, starlight is significantly dimmed by interstellar material – by 1 to 2 magnitudes per kiloparsec (kpc) at visual wavelengths (§§13.3, 14.4).

If we merely count stars as a function of brightness, there is no way to distinguish between the effect of dust and an "end" of the stellar system. But the difference between the effects of dust and simple exhaustion of stars can be determined in other ways.

Long before the radio observations had detected material at the center of our Galaxy, astronomers knew it was located in the direction of the constellation Sagittarius. Shapley's work on the distribution of globular clusters (§13.1) showed this (cf. Mihalas and Binney 1981, §1.1). While Eddington did not have these results in 1910 when he wrote *Stellar Movements*, this writer has always found it fascinating that the frontispiece of his book shows the famous *star clouds* of Sagittarius (Figure 14.1). Did Eddington suspect there was something special about this direction?

Direct photographs of star fields clearly show abrupt changes in the density of stars (number per square degree), but it had been traditional to interpret them as voids in the stellar population rather than obscuration by intervening material. Two illustrations of interstellar material are given in Figure 14.2. Both bright and dark

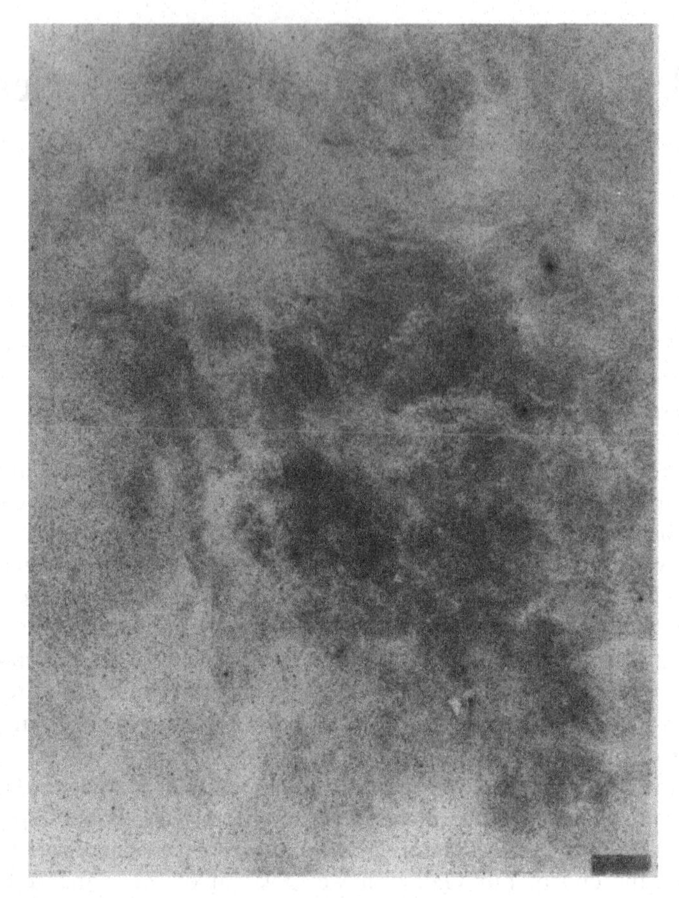

Figure 14.1. The Star Clouds in Sagittarius Toward the Galactic Center. The star images are dark on this negative reproduction. Courtesy California Institute of Technology.

interstellar matter can be seen. The former results when ultraviolet photons from young stars heat the interstellar gas. We shall discuss these regions in the next chapter. Lynds's (1962) "Catalogue of Dark Nebulae" lists some 1800 of the latter objects.

Non-stellar objects are most commonly known to astronomers by their *Messier* or *NGC* numbers. Messier's original work appeared at the end of the eighteenth century. A modern version of his catalogue was published by Mallas and Kreimer (1978). Dryer's "New General Catalogue" (NGC) and two additions, the Index Catalogues or IC's, were published in the late nineteenth century. They also have modern versions (e.g., Sulentic and Tifft 1973).

The study of interstellar material has advanced enormously in the second half of the present century. New instrumental techniques have revealed a variety of forms and physical conditions. The pioneers of radio astronomy had already begun

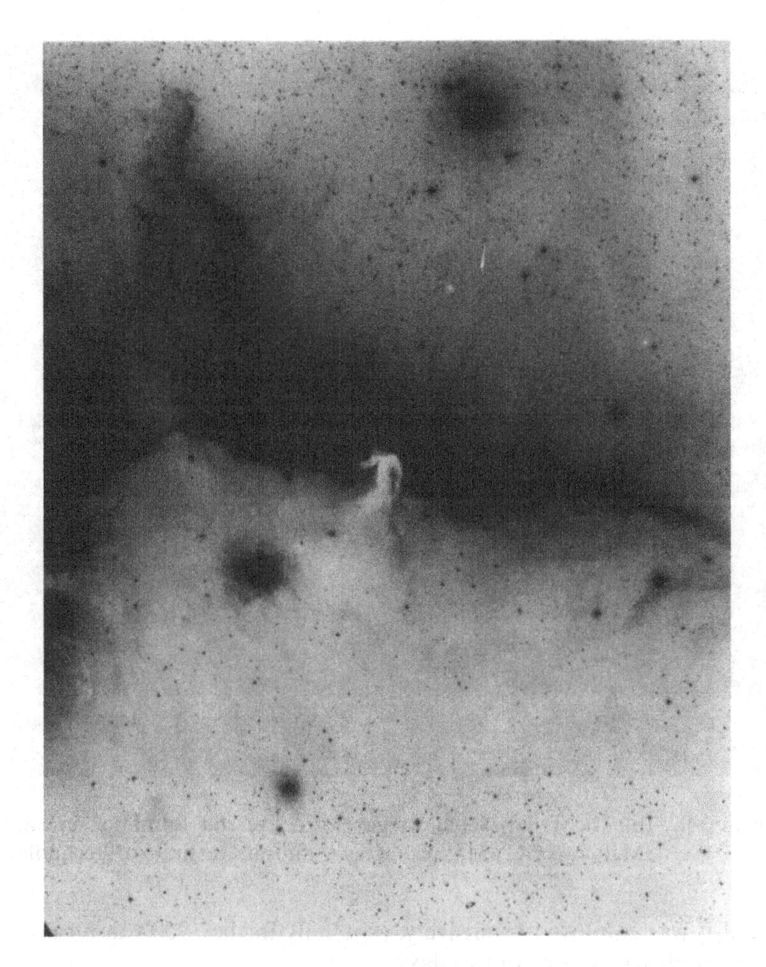

Figure 14.2. Illustrations of Interstellar Material in the Galaxy. (a) A negative photograph of the Horsehead Nebula in the constellation Orion.

their observations in the 1930's. The new astronomical window bore rapid fruit following World War II, when the highly developed radar techniques began to be applied in astronomical research. The exploration of space, first by rocket flights and later by manned and unmanned satellites, opened most of the remainder of the electromagnetic spectrum to observational science.

One of the principal contributions of radio astronomical techniques has been in the field of interstellar matter. The early receivers, beginning with those of Jansky and Reber (see Shklovsky 1960), picked up continuous emission from hot, diffuse gas, in both our own and external galaxies. The first spectral line to be observed at radio frequencies was the 21 cm line of neutral hydrogen. It is emitted when the spin of the electron "flips" from a parallel orientation with respect to the proton spin to an antiparallel orientation. The 21 cm line can be observed in both absorption and

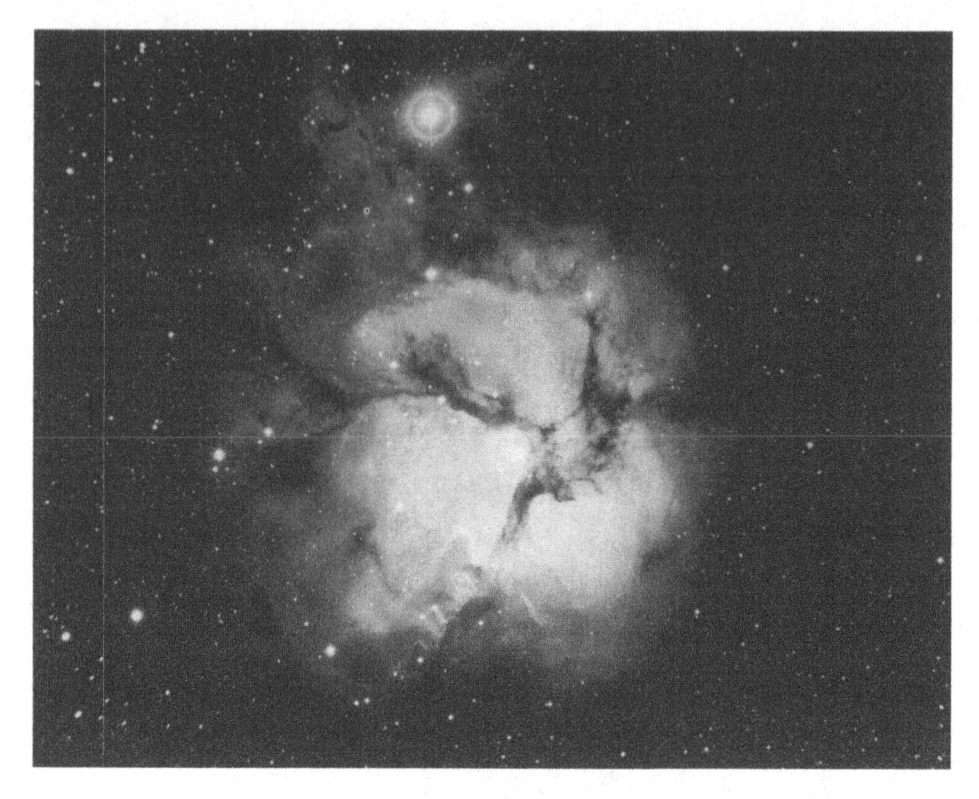

Figure 14.2. (b) The Trifid Nebula, region of active star formation in Sagittarius (M20 = NGC 6514). Courtesy California Institute of Technology.

emission, and it was employed extensively, especially in the 1950's in an attempt to map out the spiral structure of our Galaxy.

It is straightforward to obtain the temperature of the clouds of neutral hydrogen observed in the 21 cm line, so long as the cloud is optically thick and may be assumed to be more or less uniform in temperature. We need only put $\tau_B = \infty$ in Equation 12.9, to see that the observed specific intensity gives the source function directly. Radio astronomers frequently employ the Rayleigh–Jeans approximation to the Planck or black body function,

$$B_\nu = \frac{2h\nu^3}{c^2}\frac{1}{\exp(h\nu/kT)-1} \sim \frac{2k\nu^2 T}{c^2}. \qquad (14.1)$$

Since the specific intensity is directly proportional to the temperature, it is common in radio astronomy to speak of a "brightness temperature," meaning the temperature of a black body source that would give the observed signal (brightness temperature is also discussed in §14.2).

In addition to the requirement that the cloud be optically thick in the 21 cm line, it is also necessary that the size of the cloud be sufficient to fill the *beam* of the antenna. To put the matter another way, the cloud must appear as an

Table 14.1. *Interstellar Molecules.*

Simple hydrides, oxides, sulfides, amides, and related molecules			
H_2	CO	$NaCl^*$	CC
HCl	OCS	$AlCl^*$	CS
PN	SO_2	KCl^*	SiS
H_2O	SiO	AlF^*	SiH_4^*
H_2S	NH_3	CH_4	HNO
Nitriles, acetylene derivatives, and related molecules			
CCC^*	HNC	H_3CNC	$HN{=}C{=}O$
$CCCCC^*$	HCN	H_3CCN	$H_3C{-}CH_2{-}CN$
	$HC{\equiv}C{-}CN$	$H_3C{-}C{\equiv}C{-}CN$	
	$H_2C{=}CH{-}CN$		
$CCCO$	$H(C{\equiv}C)_2{-}CN$	$H_3C{-}C{\equiv}CH$	
	$HN{=}C{=}S$		
$CCCS$	$H(C{\equiv}C)_3{-}CN$	$H_3C{-}(C{\equiv}C)_2{-}H$	C_4Si^*
$HC{\equiv}CH$	$H(C{\equiv}C)_4{-}CN$	$H_3C{-}(C{\equiv}C)_2CN$?	
$HC{\equiv}CCHO$	$H(C{\equiv}C)_5{-}CN$	$H_2C{=}CH_2^*$	
$H_2C{=}C{=}C$	$H_2C{=}C{=}C{=}C$	$HCCNC$	$CCCNH$
Aldehydes, alcohols, ethers, ketones, amides, and related molecules			
$H_2C{=}O$	H_3COH	$HO{-}CH{=}O$	H_2CNH
$H_2C{=}S$	$H_3C{-}CH_2{-}OH$	$H_3C{-}O{-}CH{=}O$	H_3CNH_2
$H_3C{-}CH{=}O$	H_3CSH	$H_3C{-}O{-}CH_3$	H_2NCN
$NH_2{-}CH{=}O$	$H_2C{=}C{=}O$	$(CH_3)_2CO$?	
Cyclic molecules			
C_3H_2	C_3H(cyclic)	SiC_2^*	
Ions			
CH^+	HCO^+	$HCNH^+$	H_3O^+
HN_2^+	$HOCO^+$	SO^+?	HOC^+?
	HCS^+		H_2D^+?
Radicals			
OH	C_3H(linear)	CN	HCO
CH	C_4H	C_3N	NO
C_2H	C_5H	H_2CCN	SO
CH_2?	C_6H	C_2S	NS
SiC^*	$HCCN^*$	CP^*	NH
SiN^*	C_2O		

Source: After Irvine (1991, 1992).

extended object when observed with the radio receiver. At the centers of the densest aggregations of interstellar matter, called giant molecular clouds (GMC), a number of complex organic molecules have been identified.

Table 14.1 is a listing of interstellar molecules by Irvine (1991, 1992). Question marks indicate species that required confirmation at the time of Irvine's review. An asterisk indicates molecules identified *only* in envelopes around evolved stars. They

Table 14.2. *Diatomic Molecules of the Most Abundant Chemical Elements. Logarithmic Abundances and Dissociation Constants (eV).*

	H	C	N	O	Mg	Si	S	Fe
Abund	12.00	8.56	8.05	8.93	7.58	7.55	7.27	7.51
H	4.48	3.46	3.47	4.39	1.34	3.06		
C		6.21	7.76	11.09		4.64	7.36	
N			9.76	6.50			4.8	
O					3.5	8.26	5.36	4.20
Mg							2.4	
Si							6.42	3.04
S							4.37	3.31
Fe								1.06

are perhaps more properly called *circumstellar* than interstellar molecules. Stars with circumstellar envelopes may be found with the help of bibliographies by the NASA workers (e.g., Gezari, *et al.* 1993). The more complicated molecules are found only in dense clouds where they are shielded from the ultraviolet photons from hot stars which would dissociate them. Some of the simpler diatomic molecules such as CH and CN have been known and studied for decades. Their absorption lines are readily seen in the spectra of early-type stars whose uncomplicated spectra show absorption due to the interstellar material. Presumably the same absorption lines are present in the spectra of late-type stars. However, the intrinsically more complicated spectra of the latter makes the detection of weak absorption features difficult.

The molecules of Table 14.1 are dominated by H, C, N, and O, as one would expect from the cosmic abundance of these species. Some of the additional elements present in the molecules listed in Table 14.2, Si and S, are also abundant. In Table 14.2 we give the logarithmic (SAD) abundances in the first row from Anders and Grevesse (1989, see also Table A5, Appendix). Dissociation constants for diatomic species (Huber and Herzberg 1979) are given at the appropriate intersections, e.g., the dissociation constant for CO is 11.09 eV. We have omitted the inert gases helium and neon.

It is straightforward to account for the atomic constituents of the molecules in Table 14.1 with the help of the entries of Table 14.2. The elements that are present are the cosmically abundant ones. We have not listed the isotopic species in Table 14.2. Lovas (1992) lists a variety of deuterated molecular species including C≡CD, DC≡C–CN, and DCO^+. According to our present estimates, deuterium is roughly as abundant cosmically as silicon, so its presence among the interstellar molecules is not immediately surprising. We shall return to the question of isotopic species in interstellar molecules in a moment.

Of the cosmically abundant elements, apart from the noble gases, the only one not represented in Table 14.1 is magnesium. This is explained, at least in part, by the very low binding energies of the possible diatomic species that could be formed of magnesium with other cosmically abundant elements. The low binding energy of MgH is noteworthy. Missing entries for diatomic molecules involving Mg mean that the species are not easily prepared in the laboratory, and are probably fragile.

While the chemical elements listed in Table 14.1 are familiar to students of introductory organic chemistry, the molecules themselves are clearly unusual. The familiar aliphatic compounds such as the alkanes methane (CH_4), ethane ($H_3C{-}CH_3$), propane ($CH_3CH_2CH_3$), butane, etc., are nowhere to be seen. These compounds are called saturated because the carbon atoms are all connected by single bonds. The unsaturated alkenes are analogues of the alkanes, but with double bonds: ethylene ($H_2C{=}CH_2$), propylene, butylene, etc. The unsaturated, triple-bond alkynes start with acetylene ($HC{\equiv}CH$).

In §2.4 we mentioned the chemists' notation sp^3 to describe *each* of the four electrons in single carbon bonding. The notation is confusing because in atomic physics, this same notation could mean *four* electrons, three of which were p's and one an s. In this connection, however, the chemist means *each* electron is described by a mixture one fourth s and three fourths p wave functions. When carbon forms double bonds, as in ethylene $H_2C{=}CH_2$, the electrons in the carbon atom are said to be described by *one p* and *three* sp^2 orbitals. In this connection, sp^2 means a wave function that is a mixture one part s and two parts p wave functions. The four-electron configuration might be called $p(0.33s0.67p)^3$ – but it isn't! Now consider the four electrons surrounding a carbon atom with a triple bond. The triple bond is sometimes said to involve sp orbitals. We might write the four electrons thus: $p^2(0.5s0.5p)^2$.

For the most part, the compounds that show up in the microwave region of the spectrum through rotational transitions are not pure hydrocarbons but *derivatives*. They are hydrocarbons where one or more of the hydrogens or (CH_n)-groups has been replaced by an element or another group such as CN or NH. These molecules may be detected because of their (lack of) symmetry. Those species possessing a *center of symmetry*, such as methane or ethane, have no permanent electric dipole moments, and therefore no rotational spectra. The same thing is true of the common molecule, CO_2, which is *linear*. These molecules can have transitions among asymmetrical *vibrational* modes which appear in the infrared (see Herzberg 1945, and below).

Quadrupole vibration–rotation lines of the homonuclear H_2 molecule *have* been observed in a number of sources (Shull and Beckwith 1982, Black and van Dishoeck 1987). They were first observed in the dense interstellar clouds associated with the Orion Nebula (Gautier *et al.* 1976). Schwartz, Cohen, and Williams (1987) give a useful energy level diagram for H_2 showing a number of the observed transitions.

Acetylene has been observed in circumstellar gas as has the planar, symmetrical molecule ethylene ($H_2C{=}CH_2$). In both cases, infrared rotation–vibration transitions are involved (see the review by Betz 1987). Similarly, an infrared 15.2 micron feature was first identified as CO_2 in several point-like infrared, IRAS sources (d'Hendecourt and Jourdain de Muizon 1989).

The detection and identification of interstellar molecules are surely favored when they have permanent electric dipole moments. This readily accounts for the large number of hydrocarbon derivatives rather than the pure hydrocarbons themselves in Table 14.1. The table also shows a preference for *unsaturated* species, which may reasonably be understood in terms of the low-pressure environments of interstellar molecules.

Since the earliest studies (see the review by Watson 1976), it has been known that the lifetimes of interstellar molecules subject to photodissociation in the general galactic radiation field are only of the order of 100–1000 years. Consequently, it is clear that shielding by interstellar dust must be important for their survival. Lang (1980, p. 168) has a useful table of photodissociation rates for various molecules.

The presence of interstellar dust was discussed by Russell, Dugan, and Stuart (1938) in the main portion of their classic textbook which was written in 1927. At that time, this material was generally regarded as a nuisance! It prevented the determination of the number of stars per unit volume. Even in this role, the significance of general interstellar absorption was underestimated (see their §896). While the bright, or emission-line, nebulae have been intensively studied since the 1930's, serious work on dark nebulae awaited various technical developments that have emerged in the second half of the twentieth century.

Burgeoning developments in space astronomy have made it possible to demonstrate that the dark, solid material of the interstellar gas may resemble the kinds of condensations we have studied in connection with the formation of the planets. The composition of the dust is difficult to determine. Some information is available from the interaction of this material with starlight, which causes reddening, polarization, extinction, and some selective absorptions (Savage and Mathis 1979). Additional sources of information are the inclusions in meteorites (cf. Kerridge and Matthews 1988) and perhaps cosmic dust (Brownlee 1985).

In §7.3 we discussed *depletions* of heavy elements in the gaseous phase of the interstellar medium. These depletions are best understood in terms of the formation of refractory dust grains. From the relatively large depletion of calcium (see Figure 7.5), we might infer that the refractory oxide perovskite ($CaTiO_3$), or the silicate melilite ($Ca_2MgSi_2O_7$) were among the minerals forming the dust grains.

We shall return to the subject of dust later in the present chapter.

14.2 Molecular Clouds

The radio astronomers speak of a *ring* of giant molecular clouds (GMC's) in our Galaxy, between roughly 4 and 6 kiloparsecs from the galactic center (Thaddeus 1991). Within this ring, some 90% of the interstellar matter is in the form of these giant clouds, whose total mass is comparable to that of the stars in the same volume of space. There are several thousand of these clouds, with masses that can be as large as 10^5–10^6 solar masses. The clouds themselves are some tens of parsecs across (see Solomon and Edmunds 1980). It is within such clouds that the polyatomic molecules listed in Table 14.1 are found.

The principal constituent of these clouds is undoubtedly molecular hydrogen, with particle densities of the order of 10 to 100 cm^{-3} (Spitzer 1982). Temperatures, which may be inferred from the rotational excitation levels of the H_2 molecules, are of the order of 80 K. Toward the centers of these clouds the densities must be considerably higher, since star formation almost certainly takes place within them. Whether or not the temperature is very much lower than the typical 80 K at the heart of some of these regions depends upon the stage of star formation there. Hot, young stars will ionize a region of space surrounding them, and a young cluster (§13.1) can disperse the remnant gas from which it was formed. This is undoubtedly what has happened in the region of the Orion Nebula, where young stars are found in close proximity to dense molecular gas (Figure 14.3). The Orion cloud complex is discussed in a review article by Genzel and Stutzki (1989).

In clouds that are not too dense, it is possible to observe ultraviolet lines from electronic transitions in molecular hydrogen in absorption against the continua of hot, early stars. It is then possible to construct a curve of growth for these molecular lines in a manner that is essentially the same as that used for absorption lines in stars. Figure 14.4 shows such a curve of growth taken from Spitzer's book. The rotational quantum number J measures the kinetic energy of rotation of the H_2 molecules. We assume that the distribution of rotational levels follows a Boltzmann distribution

$$N(E_n) \propto (2J+1)\exp(-E_n/kT) \tag{14.2}$$

where E_n is $J(J+1)\hbar^2/(2I)$, and I is the moment of inertia of the molecule (see Equation 11.1). Since the transition probabilities f_{nm} for the rotational transitions are known, an empirical curve of growth may be plotted. A standard method would be to plot $\log(W)$ (or $\log(W/\lambda)$) vs. $\log(gf)$ for transitions from the same vibrational level. The horizontal displacement necessary to superimpose the curves for all J yields the *excitation temperature* for the molecules (Problem 2 of Chapter 12). With the theory developed in §12.7, the numerical values of the abscissae given in Figure 14.4 may be derived from the weak lines.

While the GMC's consist primarily of H_2, they are found principally by microwave (radio) techniques which do not detect the molecular hydrogen directly because it lacks a permanent dipole moment. Thus, most of our information on GMC's has come from observations of the carbon monoxide molecule. A transition from the first excited rotational state of CO produces an emission line at 0.26 cm (cf. Table 11.3).

Analysis of these emission lines may be made with the help of the slab-model theory of §12.4. In a great many instances, the emission from the ^{12}CO molecule is not directly proportional to the amount of emitting gas. This situation is analogous to saturation in absorption lines, which the laboratory spectroscopist often calls self absorption. Let us work out the case for an emission line that can be seen *above* a background continuum with intensity I^0 (see Figure 14.5).

Instead of Equation 12.11, we have

$$r_\nu \equiv \frac{I_\nu - I^0}{I^0} = \frac{S_\nu - I^0}{I^0}\left[1.0 - \exp\left(-\frac{\tau_\nu}{\mu}\right)\right]. \tag{14.3}$$

Here, τ_ν is the same as τ_B, the optical depth at the edge of the cloud away from the observer, and we now assume $S_\nu > I^0$ at frequencies near the emission line. Because the geometry of interstellar clouds is uncertain, we usually take $\mu = 1$ in these relations.

We may define an r_0, in a manner analogous to the definition given by 12.15, but

Figure 14.3. The Orion Nebula Complex. Courtesy California Institute of Technology.

again introducing a minus sign, so Equation 14.3 becomes

$$r_\nu = r_0 \left[1.0 - \exp\left(-\tau_\nu/\mu\right)\right]. \tag{14.4}$$

For the case of small optical depth τ_ν in the cloud, the exponential may be expanded as in §12.7. If only the first term is retained, the integral is trivial, and we

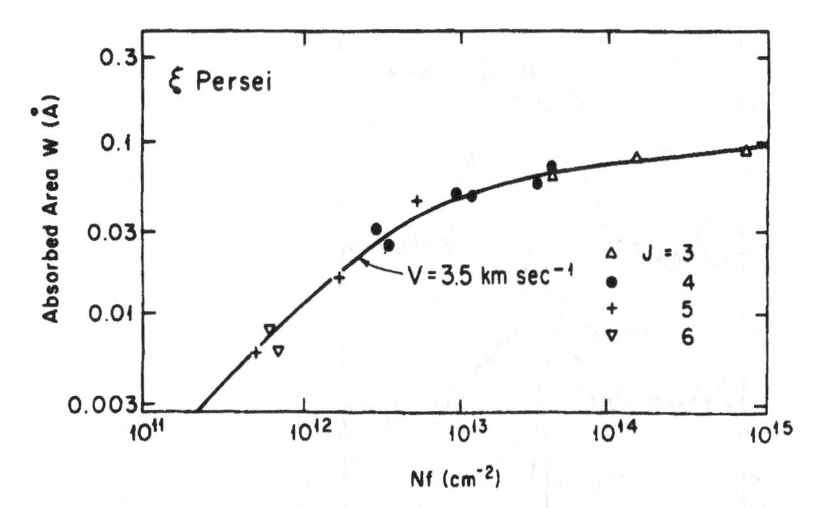

Figure 14.4. A Curve of Growth for Interstellar Molecular Hydrogen. From *Searching Between the Stars* by Lyman Spitzer, Jr., Yale University Press, ©1982, with permission.

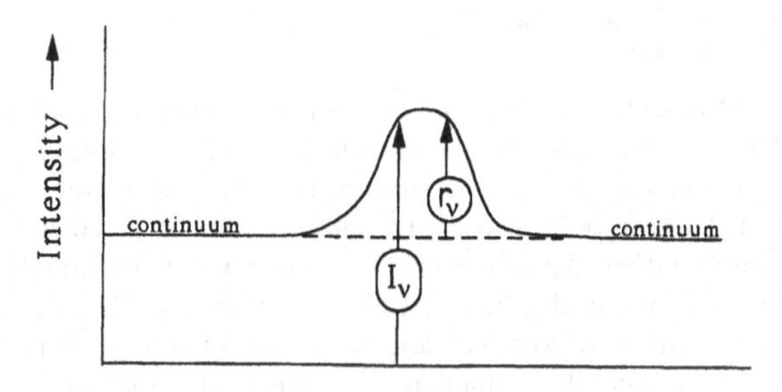

Figure 14.5. Definition of Terms for the Analysis of Emission Lines. The quantity r_ν is defined in an analogous manner to the line depth of §12.4. It is actually the negative of the quantity we called the line depth; this change is made so that the new r_ν, which we may call the "relative intensity," will be positive for an emission line.

have

$$W_\nu \equiv \int r_\nu d\nu \approx \frac{\pi e^2}{mc} r_0 N_n f_{nm} H/\mu. \tag{14.5}$$

Note that r_0 is the limit of r_ν as $\tau_\nu \to \infty$. Therefore, this parameter may be determined empirically for those clouds whose emission is "flat topped," or for which there is reason to think that the maximum observed intensity approaches that which would be observed for $\tau_\nu \to \infty$. We shall discuss this point in more detail below. Once the value of r_0 is fixed, we may equate the weak-emission equivalent width to physical constants times the effective column density, $N_n H/\mu$, which has

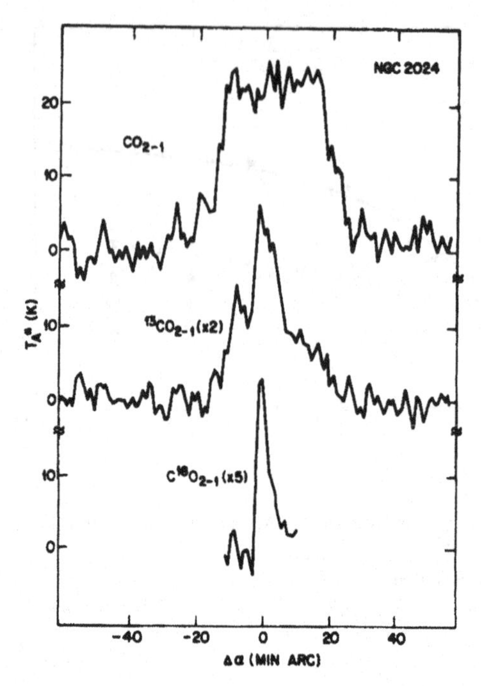

Figure 14.6. Carbon Monoxide Observations of the Dark Cloud Cloud Associated with NGC 2024. Curves are shown for three isotopically distinct species of CO molecules. The subscripts (2-1) refer to the rotational transition from $J = 2$ to $J = 1$. The ordinate is given as antenna temperature rather than intensity in accordance with Equation 14.1 and the discussion in the text. The abscissa is an angular coordinate on the sky, in minutes of arc. The angular extent of this cloud is between 40 and 50 arc seconds in diameter, as seen in the dominant molecule $^{12}C^{16}O$. The two less abundant isotopic species peak only at the center of the cloud where the density is (presumably) highest. From Phillips *et al.* (1979).

units of number per cm^2. One cannot determine the number per unit volume, without knowing something about the cloud geometry.

For clouds whose angular size may be delineated, a typical ploy is to assume a spherical geometry. The physical size of the cloud is then known if it is possible to estimate its distance. This may often be done with the help of nearby stars that are thought to be closely associated with the complex of which the GMC is a part. With an estimate of the physical diameter of the cloud, one may obtain a mean (NB) number density from the column density.

In the case of the giant molecular clouds, a typical situation is that the background continuum, I^0, cannot be distinguished from zero, as in Figure 14.6. In this case, the analysis proceeds much as in Equations 14.2 to 14.4. One simply puts $I^0 = 0.0$ in Equation 12.9, and integrates over frequency to obtain a relation for I_ν of the same form as our Equation 14.4, but with S_ν playing the part of r_0. The quantity S_ν may

Table 14.3. *Molecular Ratios Used for* $^{12}C/^{13}C$. *If the isotope is not indicated for an element, the most common isotope should be understood.*

$\frac{H_2CO}{H_2\,^{13}CO}$	$\frac{C^{18}O}{^{13}C\,^{18}O}$	$\frac{HCO^+}{H\,^{13}CO^+}$
$\frac{HC_3N}{H\,^{13}CC_2N}$	$\frac{NH_2CHO}{NH_2\,^{13}CHO}$	$\frac{OCS}{O\,^{13}CS}$
$\frac{C\,^{34}S}{^{13}CS}$	$\frac{C\,^{18}O}{^{13}CO}$	$\frac{H_2C\,^{18}O}{H_2\,^{13}CO}$
$\frac{HC\,^{18}O^+}{H\,^{13}CO^+}$	$\frac{HC\,^{15}N}{H\,^{13}CN}$	

be obtained empirically for an optically thick cloud, after making allowance for the angular size of the cloud relative to the "beam" of the radio antenna.

We mentioned (see Equation 14.1) that it is common in radio astronomy to make use of the Rayleigh–Jeans approximation to the Planck formula, and to use temperatures rather than intensities. If the geometrical correction has been made for a partially filled radio beam, then one may use a "brightness temperature," T_B, which may be converted to a specific intensity on multiplication by $2k\nu^2/c^2$. If this geometrical correction is unknown or uncertain, the radio astronomer often speaks of an "antenna temperature," T_A. Both temperatures are a measure of the energy that would be received per unit solid angle and per unit frequency interval if the radio telescope were inside a cavity (black body) whose walls had reached an equilibrium temperature T_B or T_A.

There are several clues to the fact that the optical depth in an emission line is large. An obvious clue is when the profile is flat topped, as a function of frequency. Another clear indication of optical thickness may come from observations of two (or more) emission lines whose relative intensities are known under the conditions of optically thin emission. If the intensities of such lines are the same at the center of the profiles, we may assume both profiles have reached their maximum ($\tau \to \infty$) intensity. In the case of the GMC's the frequency profiles are rarely flat topped, and it is typical to infer optical thickness by comparing the maximum intensity of some trace isotopic molecular constituent, or from the equality of transitions from the $J = 2 \to J = 1$ and $J = 1 \to J = 0$ profiles at their centers.

A very common situation is that the ^{12}CO profile is strongly self absorbed or optically thick. It is then useful for the determination of the source function, while the ^{13}CO profile is from an optically thin source, and suitable for a determination of number density.

One may then obtain the number density, or column density, of the dominant species ^{12}CO after assuming a value for the $^{12}C/^{13}C$ ratio. This ratio is given directly by the ratios of the observed intensities (or brightness temperatures) in those regions where the emission is weak ($\tau_\nu \ll 1.0$).

In order to connect the CO column densities to H_2 column densities, one must combine observations of H_2 made in the satellite ultraviolet with the radio observations of CO. These observations are typically made at the edges of clouds, where the optical depths in the relevant lines are small. Typical values for $N(^{13}CO)/N(H_2)$ given by Edmunds and Solomon (1980) are 0.5 to 2.0 times 10^{-6}. In principle, the infrared H_2 quadrupole transitions might be used in dense clouds to investigate their C/O ratio, at least for the Orion region.

Many total H_2 masses are inferred from CO brightnesses that are integrated both over the profiles and over areas of the sky containing the emission. The technique can have only statistical accuracy because the observed intentities are often from optically thick emission. A number of the methods are reviewed by Wolfendale (1991), and in an extragalactic context by Booth (1991).

The $^{12}C/^{13}C$ ratio in the GMC's is of considerable cosmochemical significance, since we would assume that this ratio would decrease during the nuclear evolution of the galaxy due to CNO processing (see §10.2). Wannier (1980) gives a value of about 60 for the $^{12}C/^{13}C$ ratio in the GMC's of the "galactic ring" (4–6 kpc from the center).

The GMC's avoid the region of the Galaxy from 2–4 kpc inward, until a region some 750 pc from the center is reached. The innermost regions of the Galaxy contain the largest molecular complexes of all (Oort 1977). These clouds are peculiar, not only in their masses, which are an order of magnitude or so greater than those of the typical clouds in the galactic ring, but also in the $^{12}C/^{13}C$ ratio. It is variable, but significantly lower than in the ring. For two giant complexes in the central region of our Galaxy, Wannier (1980) has given a number of estimates of the $^{12}C/^{13}C$ ratio, based on observations of several molecules. These vary from less than 12 to 60, with typical values between 20 and 30.

An amazing variety of isotopic species may be observed in the interstellar gas. In Table 14.3, we assemble the ratios upon which the $^{12}C/^{13}C$ value is estimated.

Note that for five entries of Table 14.3, the numerators and denominators contain different isotopic species in addition to the ^{13}C and ^{12}C. Some assumptions are therefore necessary before the $^{12}C/^{13}C$ ratio may be extracted. The standard assumption is that the other isotopic ratios are in "cosmic" proportions. The latter may be read from an ordinary chart of the nuclides, if one's favorite tabulation of the SAD is not at hand. Wannier points out that when these isotopic corrections are made, the results are not always consistent. For example, for a cloud known as W51, the $^{12}C^{34}S/^{13}C^{32}S$ ratio $\times 23$ indicates a $^{12}C/^{13}C$ ratio of 74 ± 5. The factor 23 is the SAD value of $^{32}S/^{34}S$. The $^{12}C^{18}O/^{13}C^{16}O$ ratio $\times 500$ gives 62 ± 1, where the factor of 500 is the SAD $^{16}O/^{18}O$ ratio. In the cloud DR21, the corresponding numbers are 115 ± 12 and 34 ± 9. Thus these ratios are inconsistent with one another, and in a way that is not yet understood. Whether the question will be resolved in terms of further isotopic variations from the SAD or simply in analytical techniques remains to be seen.

Our knowledge of the chemistry of interstellar processes of all varieties, including molecular formation, is severely handicapped by the fact that the relevant conditions cannot be easily explored in terrestrial laboratories.

One of the more interesting interstellar processes is the fractionation of deuterium, because of the importance of the overall D/H ratio as a constraint on models of the universe. McCullough (1992) gives $1.5(\pm 0.2) \cdot 10^{-5}$ for the interstellar D/H ratio. If this abundance is primordial, the corresponding cosmological inference is that the density of matter in the universe is about one tenth that necessary for closure. On the other hand, it is possible that the observed deuterium may have been reduced from its primordial (big bang) value by the nuclear reactions that can take place even in stellar envelopes. This would mean that the margin by which the universe is "open" is still wider – assuming no dark matter (§16.10).

The DCN/HCN and DCO^+/HCO^+ ratios observed in GMC's indicate severe partitioning of the deuterium into the molecules, as would be expected on the basis of the arguments presented in §8.5. The deuterium enhancements are very large; Watson (1976) quotes values ranging up to five orders of magnitude over the value $\approx 10^{-5}$ then accepted for the solar neighborhood.

14.3 The Theory of Interstellar Chemistry

In this section we will present an introduction to the theory of chemical rate processes with an emphasis on the specific reactions thought to be taking place in interstellar space. We begin with a general discussion of the rates of chemical reactions, such as might be found in a textbook on physical chemistry. More details and references can be found in the article by Viala (1990).

It is well known that in interstellar space, the physical conditions are far from those of thermal equilibrium, where the relations expressing the populations of atomic and molecular energy states, ionization, molecular dissociation, and the distribution of particle velocities all involve a single temperature. Under these non-equilibrium conditions, there is no alternative to a kinetic approach. In many instances, it is possible to postulate a stationary state, and attempt to derive the relevant populations by setting the time derivatives equal to zero.

The basic approach is then much the same as that used in the non-LTE treatment of stellar atmospheres, although the relevant processes are quite different. We shall focus on those processes relevant to the diffuse and dense *molecular* clouds, rather than traditional, emission regions ($T \approx 10\,000$ K) or the hotter interstellar gas responsible for the emission of O VI. These regions will be discussed later in the next chapter.

Consider a general chemical reaction such as

$$\mathrm{A} + \mathrm{B} \rightarrow \mathrm{C} + \mathrm{D}. \tag{14.6}$$

We *define* the rate of this reaction in terms of the time derivative of the number density of one of these species – whichever is convenient. In chemistry, the reaction rate is often given in terms of the rate of change of the *concentration* (e.g., moles/liter) of some species, but such units are inconvenient and unnecessary in the present context.

It is common to distinguish between first-order and second-order reactions, depending upon whether the rate depends on the first or second power of the number density (or concentration) of a species. A paradigm first-order reaction would be radioactive decay, where

$$\frac{dN}{dt} = -\lambda \cdot N. \tag{14.7}$$

Here, λ is a decay constant, equal to 0.693 divided by the half-life.

In discussing nuclear reactions, it is common to speak of projectiles and targets, a nomenclature that clearly arose from laboratory experiments with particle beams. When the reacting particles are simply together in a "soup," the distinction between target and projectile is not an obvious one. Nevertheless, we may arbitrarily adopt one species, and focus our attention on it. Then, for strictly two-body interactions, what we have called the *collision frequency*, ν_{coll} (for projectiles, say), is the product $N_{\text{T}} < \sigma v >$, where N_{T} is the number density of targets, and the angular brackets indicate an average of the cross section times the velocity. The frequency of collisions per cm^3 per sec is then the product $N_{\text{P}} N_{\text{T}} < \sigma v >$. Since the dependence is on the product of the number densities, the reaction is second-order. A chemical reaction such as 14.6, could be first-order *if* the number density of one of the species were essentially constant. This would happen if a trace species were reacting with some much more abundant constituent. The nuclear reactions of the CNO cycle are all *initially* first-order, since the hydrogen abundance does not change appreciably, at least until hydrogen is substantially exhausted.

Three-body reactions would in general have *third-order* reaction rates. Such reactions are well known in the laboratory, and are not unknown in astronomical contexts. The nuclear 3α process (§10.2) is one example. We shall meet another when we discuss the formation of the H_2 molecule. Here, the third body is macroscopic!

The general approaches to the rates of chemical reactions that we shall introduce have close parallels in the theory of nuclear reactions (§9.5). Indeed, in both cases we can introduce a cross section σ for the reaction, and write the number of reactions per second as the product of the number densities of targets and projectiles, the cross section, and the relative velocity. The dimensions of such a rate are those of number/cm^3/sec. In chemistry, one defines a rate constant, k, such that the number of reactions per cm^3 per sec is given by the product of the reacting species (for second-order reactions) with k. This rate constant must not be confused with Boltzmann's constant, for which we also use the symbol k.

In the astronomical literature, the rate constants are traditionally designated by a Greek α, a symbol that is also used to designate the photoabsorption cross section per particle. Like many of our notational tribulations, there is little to be done about it!

The rate "constants" often show an exponential temperature dependence, which we shall discuss below. We shall show the temperature dependence explicitly, writing $\alpha(T)$.

The approach to reaction rates via cross sections is usually called "collision theory" in chemistry. There is an alternative approach that is conceptually valuable, which also has an analogue in nuclear reaction theory. This approach is called "transition-state theory," and the nuclear analogue is called Hauser–Feshbach, or compound-nucleus theory. The rationale of the transition-state theory may be traced to the great Swedish chemist Svante Arrhenius (1859–1927). He pictured chemical equilibrium in terms of the equality of the forward and backward reaction rates for a chemical equation such as 14.6.

Let us suppose that the reaction 14.6 proceeds by way of an intermediate state, which we shall write $(AB)^*$. This complex is the analogue of the compound nucleus in nuclear reaction theory. Equation 14.6 then proceeds by the intermediate stage

$$\mathrm{A} + \mathrm{B} = (\mathrm{AB})^*. \tag{14.8}$$

Under equilibrium conditions, the theory developed in §§4.6–4.8 applies (see also §5.4). We may thus write

$$\frac{N(\mathrm{A})N(\mathrm{B})}{N(\mathrm{AB})^*} = K_+(T), \tag{14.9}$$

where we have introduced number densities rather than pressures, dividing each pressure by kT. The resulting equilibrium "constant" is the "pressure" equilibrium constant, divided by kT for the reaction 14.8. We write this quantity $K_+(T)$ in order to distinguish it from an analogous $K_-(T)$ for the inverse reaction

$$\mathrm{C} + \mathrm{D} \rightarrow (\mathrm{AB})^*. \tag{14.10}$$

According to the theory of Chapter 4, these equilibrium constants contain a term $\exp(\Delta G/kT)$, where ΔG is the Gibbs energy change for the relevant reactions, Equation 14.8 or 14.10. Under conditions of constant volume, the ΔG would be replaced by ΔH, the "heat of formation" of the compound. In transition-state theory, this heat of formation of the *intermediate* species $(AB)^*$ is identified with what is known as the "energy of activation" of a chemical reaction.

It has been found empirically that, even in the case of exothermic reactions, there is a certain energy *threshold* that the reactants must have before the reaction will proceed. One may visualize this concept with the help of a plot of potential energy versus a *reaction coordinate* (r). This reaction coordinate is not precisely specified but we picture it as a *measure* of the separation of the centers of the two species A and B of Equation 14.6 or Figure 14.7.

When the reaction coordinate is large, we have the reactants, A and B, at an energy that is higher than that of the product C + D, which occurs at the minimum of the energy plot, at small values of the reaction coordinate. The maximum of the curve corresponds to the activation energy, which we may identify with the ΔH or ΔG of 14.8.

Physically, the activation must consist of that energy necessary to break the orbital bonds of the constituent species A and B, prior to the formation of the new ones in the complex $(AB)^*$. If A and/or B are themselves molecules, they may have internal degrees of freedom such as vibration or rotation. Formation of the complex will impose constraints on such internal motions in the reacting molecules before the products can form.

The activation energy introduces a threshold for chemical reactions even when the (overall) ΔH or ΔG's of 14.6 are negative. We may write a formal expression for

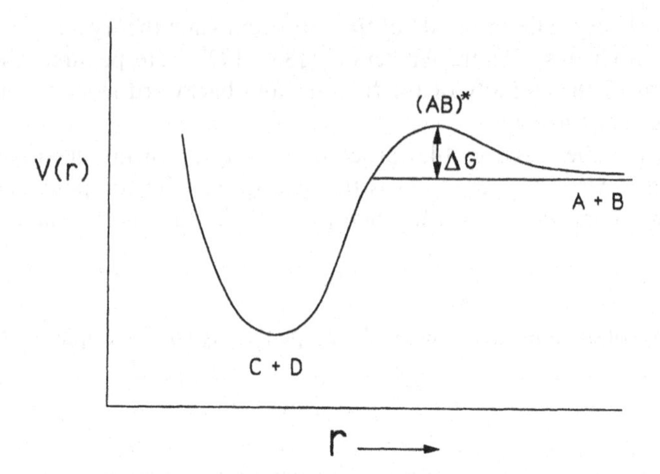

Figure 14.7. The Concept of Activation Energy. In this figure, it has been assumed that the reaction is exothermic, since the potential energy curve is lower for the complex after the reaction (C + D).

the reaction rate as

$$N(\mathrm{A})N(\mathrm{B})\alpha(T) = N(\mathrm{A})N(\mathrm{B}) \int_{\epsilon_0}^{\infty} \sigma(\epsilon)\phi(\epsilon)d\epsilon, \tag{14.11}$$

where we have taken account of the fact that the reaction cross section will generally be a function of the relative energy of the reacting species

$$\epsilon = \frac{1}{2}\mu v^2. \tag{14.12}$$

Here μ is the reduced mass, and v is the relative velocity. The function $\phi(\epsilon)$ is given by Equation 9.15.

We shall write the expression for the reaction rates with two different schematic assumptions about the functional form of the cross section $\sigma(\epsilon)$. First, let us assume that this quantity is a simple constant, $\sigma(\epsilon) = \sigma_0$. Then a straightforward integration of Equation 14.11 yields

$$\alpha(T) = (\sigma_0/\pi) \left[\frac{8\pi kT}{\mu}\right]^{1/2} (1 + \epsilon_0/kT) \exp(-\epsilon_0/kT). \tag{14.13}$$

If we assume a somewhat smoother threshold, by postulating the functional form

$$\sigma(\epsilon) = \sigma_0 (1 - \epsilon_0/\epsilon), \tag{14.14}$$

the resulting expression for the rate constant is

$$\alpha(T) = (\sigma_0/\pi) \left[\frac{8\pi kT}{\mu}\right]^{1/2} \exp(-\epsilon_0/kT). \tag{14.15}$$

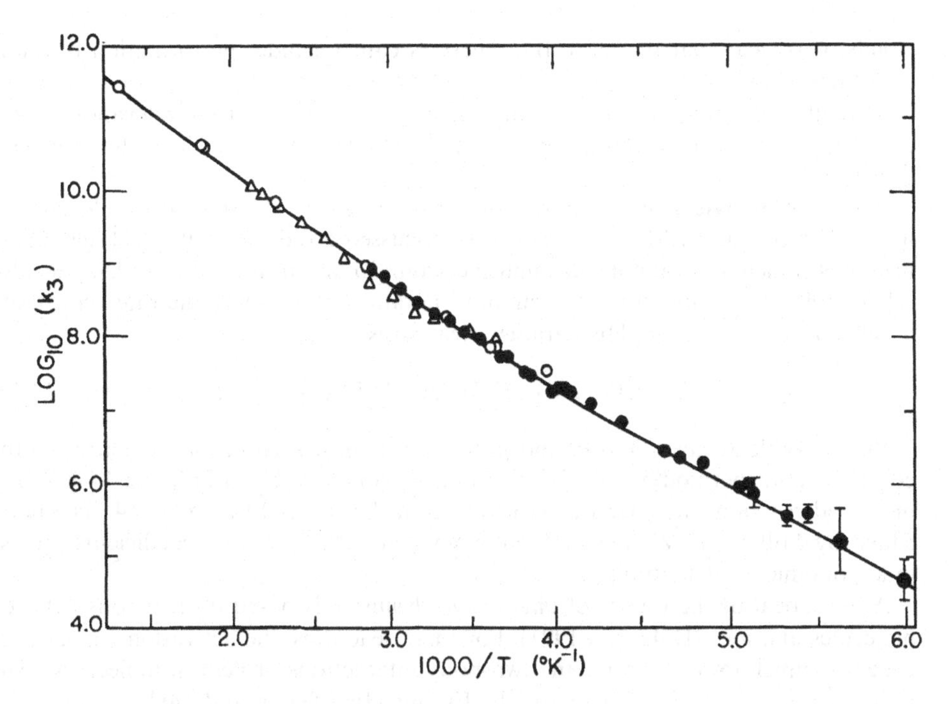

Figure 14.8. Measured Rate Constants for the Reaction $D + H_2 \rightarrow HD + H$. Courtesy American Institute of Physics.

In both of these instances, the rate has a strong, exponential dependence on the reciprocal of the temperature, with a much weaker, square-root, temperature dependence.

Ab initio calculations of chemical reaction rates are of limited value at the present time. Progress depends, as in the case of much chemical thermodynamics, on using the theory as a guide to laboratory measurements. Both Equations 14.13 and 14.15 suggest that if we can measure the rate of the disappearance of a species such as A or B in the laboratory at a series of temperatures, the logarithm of these rates will plot very nearly as a straight line against $(1/T)$. An example is shown in Figure 14.8, taken from Mitchell and Le Roy (1973).

It is common to parameterize such chemical reactions with an expression of the form

$$\alpha(T) = A \exp(-E/kT), \tag{14.16}$$

where E is the activation energy.

While the activation energies, E, are typically only a few tenths of an electron volt, at the temperatures of relevance to interstellar chemistry, they can represent formidable barriers. Within the densest parts of molecular clouds, the temperatures are thought to be of the order of 10 K. In the so-called diffuse clouds, where both CO and H_2 can be observed, the temperatures are 100 K or lower. In terms of the variable $\theta = 5040/T$, these temperatures give $\theta = 50.4$ and 5.04 respectively, and the corresponding exponentials in 14.16 are $10^{-50.4E}$ and $10^{-5.04E}$. In the cool clouds, it is clear that the activation energies can preclude the formation of many molecules.

Virtually all *endothermic* reactions must be precluded from relevance to the chemistry of molecular clouds by the kinds of two-body collisions that we have discussed so far.

Viala (1986) assembled a *library* of 1074 reactions of 80 species relevant for interstellar chemistry. Millar *et al.* (1991) discussed a data file with 2880 gas-phase reactions which is available through electronic mail from CCP7 (§12.9, see also Jeffery 1993). We adopt Viala's (common) notation here, where the rate "constant" is called k rather than 'α'. His parameterization is

$$k = (AT + B) \cdot T^{-\alpha} \exp(-\beta/T) \text{ cm}^6 \text{ sec}^{-1}. \tag{14.17}$$

Viala's Table A gives A, B, α, and β, as well as information on the source of the constants and thermodynamic information. A sample is given in Table 14.4. Many measured reaction rates depart markedly from the forms 14.13 or 14.14. The book edited by Millar and Williams (1988) is devoted entirely to rate coefficients for use in astronomical calculations.

A great deal of the theory of *interstellar chemistry* is based on reactions between molecules and *ions* (Dalgarno 1991). For these reactions, the activation energies are generally much lower than for the two-body interactions of neutral molecules . The idea apparently goes back to a paper by Eyring, Hirshfelder, and Taylor (1936), who pointed out that in the reaction between the H_2 and H_2^+ molecules, the attraction due to polarization forces will persist to very small separations. This attraction can be enough to cancel the activation energy that could arise from rotation of one or both of the molecules.

A formalism often used for *ion–molecule reactions* is called the Langevin (1905) model. The model has been criticized (e.g., Henchman 1972), and qualified (Herbst and Millar 1991). Nevertheless, it is often in agreement with experiment (cf. Clary 1990), and we review it briefly here.

Suppose a charge q_1 approaches a polarizable molecule. Let the molecule have an induced dipole moment represented by a charge q_2 separated by a distance p. Its *polarizability* α, the dipole moment, and the inducing field are then related by

$$q_2 p = \alpha E. \tag{14.18}$$

Here, E is the electric field at the center of the dipole. In the present treatment, we shall not need to consider the vector nature of the field E, nor possible tensorial nature of the polarizability α. Indeed, it is sufficient to assume the dipole always points at the incoming charge.

Since the electric field derives from the charge q_1, the separation of the dipole is a function of the distance r of the (center of the) molecule from the ion. Indeed, it follows from 14.18 that $p = \alpha q_1/(q_2 r^2)$. We may now write the expression for the attractive force on the charge q_1 due to the dipole. If we retain only the first-order term, this force is

$$F_r = -\frac{2q_1 q_2 p}{r^3} = -\frac{2q_1^2 \alpha}{r^5}. \tag{14.19}$$

We now calculate the work necessary to remove the charge q_1 from the position r to infinity. This work will be positive, because the force is attractive. It will also be the negative of the potential energy of the configuration of charges. Thus

$$-V(r) = \int_r^\infty \left(\frac{2q_1^2\alpha}{r^5}\right) dr = \frac{q_1^2\alpha}{2r^4}. \tag{14.20}$$

Orbits in this potential have interesting properties foreign to the astronomer who is accustomed to the inverse-square potential. Unlike the latter situation, the attractive force is not always overwhelmed by the centrifugal potential, and orbits are possible that spiral in to the center of force.

Use the angular momentum $l = \mu r\dot{\theta}$ to eliminate $\dot{\theta}$ from the equation of energy, as outlined for example in Problem 6 below. This leads to an effective potential for the one-dimensional problem:

$$V_{\text{eff}} = -\frac{q_1^2\alpha}{2r^4} + \frac{l^2}{2\mu r^2}. \tag{14.21}$$

Here, μ is the reduced mass of the incoming ion. Unlike the situation with the $1/r$ potential, this function has a *maximum* at a radius $r = R$, such that

$$R^2 = 2q_1^2\alpha\mu/l^2. \tag{14.22}$$

Consider a series of trajectories with *impact parameters b*, increasing in size. The trajectories belonging to very small b will "feel" the strong, short-range force, and spiral in to the center. For some limiting value of b, the trajectory will become tangent to a circle around the center of force, and "just" move away to infinity. We seek this limiting b, which obviously depends on velocity at infinity. When the trajectory moves tangent to the circle about the force center, $\dot{r} = 0$, and if this is also the boundary between capture and escape, we must have $r = R$ for this b as well – the potential *gradient* must vanish. These two conditions allow us to eliminate R from the energy equation and solve for b. We now make the further assumption that a chemical reaction will *surely happen for those trajectories leading to classical capture*. The *Langevin* cross section Q is then

$$Q = \pi b^2 = \pi q_1 \left(\frac{4\alpha}{\mu}\right)^{1/2} \frac{1}{V_\infty}. \tag{14.23}$$

Details are left as a problem (Problem 6, below). The collision frequency, or alternatively, the *rate constant*, will be *independent of the energy of the incoming ion.*

While it is well known that there *is* an energy dependence of the rates of ion–molecule reactions (see e.g., Henchman 1972), such dependences are thought to be generally small for the reactions of astrophysical interest, and tabulations typically do not indicate a temperature variation. Huntress (1977) in a review article on ion–molecule reactions of astrophysical interest, states that "the rate constants are most likely temperature insensitive over the range 80–300 K." Additional information on ion–molecule reactions may be found in the volume edited by Millar and Williams (1988) and the tables of Ikezoe (1987).

Table 14.4. *Some Reaction Rate Constants.*

Reactants				Products			B	α
H	+	HCO	$\rightarrow$	H_2	+	CO	$1.15 \cdot 10^{-11}$	−0.50
C	+	CH	$\rightarrow$	C_2	+	H	$4.20 \cdot 10^{-12}$	−0.50
C	+	OH	$\rightarrow$	CO	+	H	$4.20 \cdot 10^{-12}$	−0.50
H^+	+	CH	$\rightarrow$	CH^+	+	H	$1.90 \cdot 10^{-09}$	0.0
H^+	+	C_2	$\rightarrow$	C_2^+	+	H	$3.10 \cdot 10^{-09}$	0.0
N^+	+	H_2	$\rightarrow$	NH^+	+	H	$4.80 \cdot 10^{-10}$	0.0
H^+	+	e^-	$\rightarrow$	H	+	$h\nu$	$1.37 \cdot 10^{-10}$	0.60
C^+	+	e^-	$\rightarrow$	C	+	$h\nu$	$1.35 \cdot 10^{-10}$	0.60
O^+	+	e^-	$\rightarrow$	O	+	$h\nu$	$1.22 \cdot 10^{-10}$	0.60
CO^+	+	N	$\rightarrow$	NO^+	+	C	$2.00 \cdot 10^{-11}$	0.0
O_2^+	+	NO	$\rightarrow$	NO^+	+	O_2	$4.40 \cdot 10^{-10}$	0.0

Source: Viala (1986).

Table 14.4 is from Viala's Table A and gives rate constants for a few ion–molecule and neutral–neutral reactions, as well as electronic recombination and charge-exchange rates, to be discussed below. For purposes of illustration, we have chosen examples where the exponential in 14.17 is zero. The neutral–neutral reactions all have a $\sqrt{T}$ dependence, while the ion–neutrals do not. Since the relevant temperatures are typically between 10 and 100 K, the $(BT^{-\alpha})$-term will *increase* the neutral–neutral rates over that with $\alpha = 0$, but by one order of magnitude or less. The rates for the reactions involving ions are therefore an order of magnitude or more greater than those involving only neutral species.

Throughout the vast domains of interstellar space, the photons of starlight or collisions among shock-heated particles cause most of the ionization. Phenomena of this kind are likely to occur around the periphery of the dense clouds, but not deep within them. Near the centers of the cold, dense clouds, a residual ionization is thought to be due to the cosmic radiation. Indeed, the interiors of these clouds are well shielded from the ultraviolet photons of starlight, which would preclude much of the complex chemistry that is now of such interest.

Molecular and atomic ions may lose their charge by radiative association or by charge-transfer. A few rate constants from Viala's tabulation are included in Table 14.4.

The formation of the H_2 molecule itself is thought to occur primarily on the surfaces of grains. Since H_2 is a homonuclear molecule, it has no electric dipole moment, and ordinary vibration–rotation transitions must involve electric quadrupole terms. Such transitions go very slowly, with the result that two-body radiative combination of H_2 molecules is less likely than combination with the help of interstellar grains. The process can take place through a sequence that is reasonably well understood in principle. Our understanding is hampered by the fact that we are unsure of the nature of the grains. This certainly complicates the interpretation of any laboratory investigation.

The first stage in H_2 formation is the adsorption of atomic hydrogen to the grain surface. Heats of adsorption of molecules to various solids have been measured. From these measurements it is possible to estimate the sticking probability for atomic hydrogen onto a grain, with, let us say, an ice mantle. Once an atom of hydrogen is captured, it will migrate across the surface by thermal *hopping* and quantum tunneling through the potential due to the lattice structure of the grain onto which it has been captured. This migration has been estimated to be very efficient. It is thought that any two hydrogen atoms on an interstellar grain will find each other and combine chemically, before one of them hops off the grain, or is otherwise detached.

The production rate of H_2 thus depends on the number density and size distribution of solid particles. Dalgarno (1991) outlines the subsequent buildup of complex species by ion–molecule reactions, starting with the formation of H_2^+.

14.4 Interstellar Grains: Optical Properties

Much of what we know about interstellar grains comes from the direct interaction of such particles with the radiation from stars and bright nebulae. Some of the earliest studies of interstellar dust were made by counting the number of stars within given brightness limits. These counts were compared with numbers expected on the assumption that the stars were distributed uniformly in space. By such counts one can get an idea of the distance and thickness of a dust cloud, but virtually no information on the chemistry of the particles that make up the cloud.

Information about the size and composition of interstellar dust grains may be obtained from the fact that these dust grains act differently on light with different wavelengths. By convention, the dimming of starlight by dust is called extinction.

We can make straightforward measurements that will give the dependence of interstellar extinction on wavelength. The energy distributions of stars may, for the present purposes, be assumed to be identical for all stars with a given spectral type. Consider, then, the intensities received from two stars (1 and 2) with the *same* spectral types, as a function of wavelength. If there were no extinction of the stars' light, the flux received per unit wavelength interval would be the same for all wavelengths, and would reflect the difference in the distances (d_1 and d_2) of the two stars:

$$\frac{F_\lambda(1)}{F_\lambda(2)} = \frac{d_2^2}{d_1^2}. \tag{14.24}$$

Let us choose a *standard wavelength*, usually $\lambda 5550$, and take the ratio of received fluxes at λ to that at $\lambda 5550$.

$$f_\lambda \equiv \frac{F_\lambda}{F_{\lambda 5550}}. \tag{14.25}$$

If we now take the ratio of the f's for two (identical) stars at a variety of wavelengths, the unknown distances will cancel. If interstellar extinction either were nonexistent or had no color dependence, the ratios $f_\lambda(\text{star}_1)/f_\lambda(\text{star}_2)$ should be unity. Departures from unity are thus a measure of the *color dependence* of interstellar extinction. In astronomy, brightnesses, or quantities proportional to brightnesses, are often converted to logarithmic quantities called *magnitudes* (Equation 13.1). Consequently, we write the difference in extinction in magnitudes ΔA_λ at wavelength λ:

$$\Delta A_\lambda = 2.5 \cdot \log\left[f_\lambda(1)/f_\lambda(2)\right]. \qquad (14.26)$$

Since the f's become unity at the standard wavelength, it is clear that $\Delta A_{\lambda 5550}$ will be zero for $\lambda = 5550$ Å. Moreover, if the properties of the material responsible for the extinction are the same throughout the Galaxy, the ΔA_λ vs. λ curve should be a universal one. Detailed observational results (Mathis 1986) show evidence for a standard extinction curve for $\lambda > 3000$ Å. In the satellite ultraviolet, especially for $\lambda < 1600$ Å, there are "certainly real variations."

Assume that $F_{\lambda 5500}$ is the same as the quantity we called $b(\mathrm{V}_0)$ in Equation 13.3. Then ΔA_λ becomes a *color index* $\mathrm{B}-\mathrm{V}$ defined in §13.4.1, when λ is the wavelength of the B-band (see Figure 13.6). For an arbitrary wavelength, λ, we might write the color index as $(m_\lambda - \mathrm{V})$, where m_λ is the apparent magnitude measured with a filter centered at λ. The sources to which we shall refer simplify this notation. They write their color indices for an arbitrary wavelength as $(\lambda - \mathrm{V})$, thus subtly changing the meaning of the symbol 'λ.' Again, there is nothing to be done about this. The reader must simply recognize when λ refers to a wavelength, and when it means a magnitude measured with a filter centered near the wavelength λ.

If the color of a star has been changed (reddened) as a result of intervening dust, we may define the color excess, by

$$E(\lambda - \mathrm{V}) = (\lambda - \mathrm{V}) - (\lambda - \mathrm{V})_0, \qquad (14.27)$$

where $(\lambda - \mathrm{V})_0$ is the color index of an unreddened star.

Figure 14.9 is taken from Witt, Bohlin, and Stecher (1984). It shows a variety of stellar extinction curves along with the "average" curve of Savage and Mathis (1979). The authors plot $E(\lambda - \mathrm{V})/E(\mathrm{B} - \mathrm{V})$ vs. $1/\lambda$, where the wavelength is measured in micrometers (known to most people as microns). The average is apparently well defined, but there are clearly differences in a few cases (cf. HD 38087 and 204827).

Surely the most striking feature of this curve is the hump near $\lambda 2200$, $(1/\lambda = 4.55)$. It has been attributed, in part, to graphite as will be discussed below. Figure 14.10(a) shows theoretical fits to a mean galactic interstellar extinction curve. It is taken from Spitzer (1978, see also Gilra 1971). Figure 14.10(b) is adapted from Fitzpatrick (1989). It illustrates differences in the interstellar extinction curves in our own Galaxy and the large (LMC) and small (SMC) Magellanic Clouds. The average galactic extinction curve according to Seaton (1979) is shown, along with curves for stars in the region of 30 Doradus, a giant gas–dust complex in the Large Magellanic Cloud. Note that the $\lambda 2200$ feature is only marginally present in 30 Dor, and in the Small Cloud, it may be absent altogether.

Absorption in the interstellar medium has also been observed in the *infrared*. Sandford *et al.* (1991) discuss features between 2.8 and 3.7 μm which they attribute to C–H and O–H *functional groups*. Features of this nature will be discussed in more detail below (§14.5). Sandford *et al.* find the strengths of the absorptions

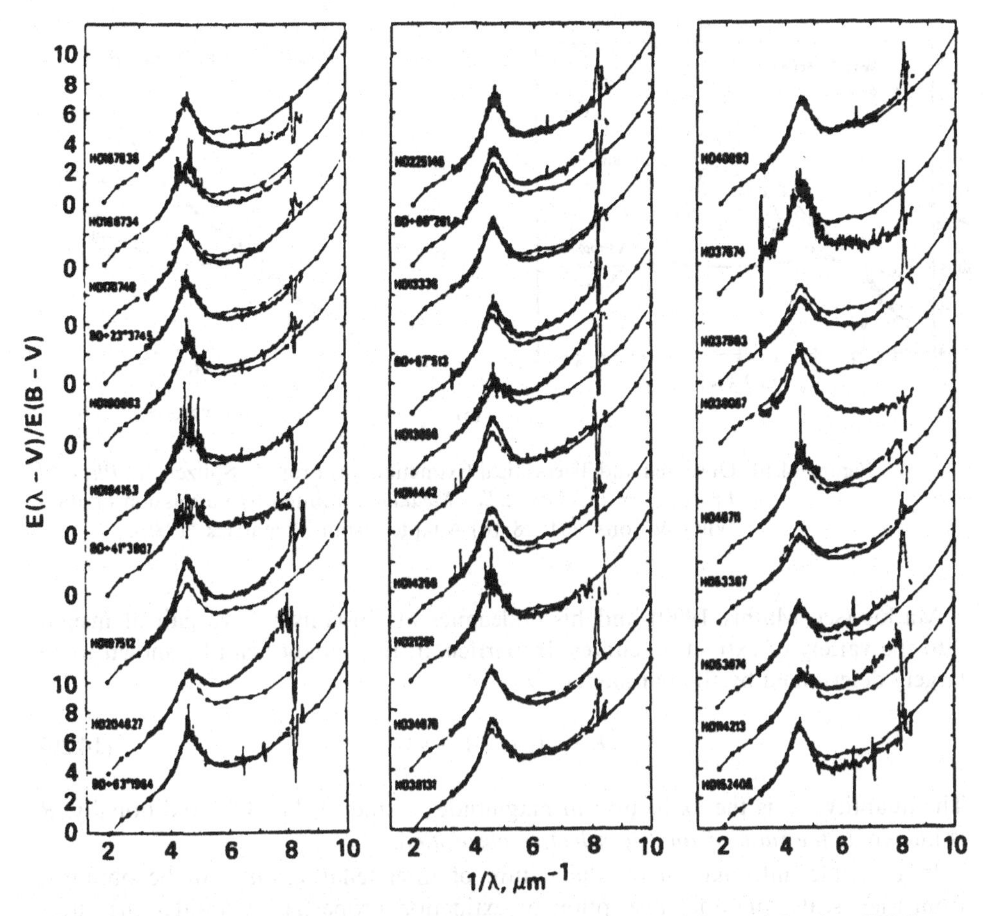

Figure 14.9. Observed Interstellar Extinction. Witt, Bohlin, and Stecher (1984).

attributed to the two groups are uncorrelated, and conclude they belong to different compounds.

Fitzpatrick and Massa (1986) show that a *very close* fit to the ultraviolet extinction curves, between 3.3 and 5.9 micron^{-1}, can be achieved with a five-parameter fit consisting of a "Drude profile with a linear background:"

$$\frac{E(\lambda - \mathrm{V})}{E(\mathrm{B} - \mathrm{V})} = a_1 + a_2/\lambda + \frac{a_3}{\left[(1/\lambda) - (1/\lambda_0)\right]^2 + \gamma^2}. \tag{14.28}$$

They tabulate these parameters for some 45 stars. It is straightforward to see from their tabulations that the central peak is remarkably stable in position, while the half width γ can vary by some 50%.

Figure 14.10(b) was constructed with the help of analytical formulae given by Fitzpatrick and Massa (1988). The relations in the 1988 paper are somewhat more elaborate than 14.28, so they may cover a larger wavelength region. The data points shown for the SMC come from Prévot *et al.* (1984).

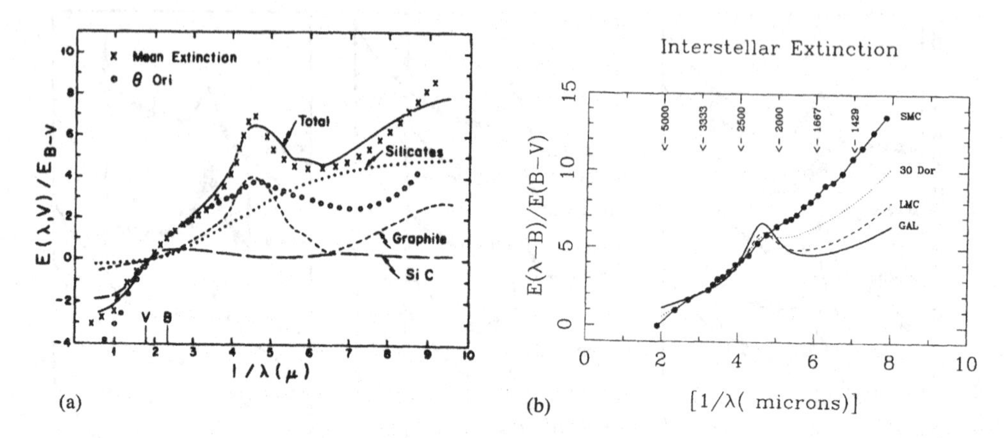

Figure 14.10. Observed and Theoretical Extinction. (a) From L. Spitzer, Jr., *Physical Processes in the Interstellar Medium*, reprinted by permission of John Wiley & Sons, ©1978. (b) Adapted from Fitzpatrick (1989).

Mathis (see Mathis 1990) and his colleagues at Wisconsin have gained insight into the variety of extinction curves. In particular, their shapes can be shown to be largely determined by the parameter

$$R = A_{\rm V}/E({\rm B}-{\rm V}). \tag{14.29}$$

The quantity $A_{\rm V}$ is the extinction in magnitudes at the "V-band." Traditionally, R is known as *the ratio of total to selective absorption.*

In principle, information on the nature of interstellar grains can be obtained from their scattering and absorption or extinction properties. Consider the effect of an incident plane electromagnetic wave on a dielectric particle. Some of the radiation is reflected (or scattered) from the surface while some is transmitted, and passes into the solid. After entering the solid, the radiation will in general have a complicated history that depends on the composition, shape, and size of the grain. Eventually some fraction of the radiation will emerge, in a direction that depends on the orientation of the surface. It is easily appreciated that this problem will be a very complex one in general.

The problem was formally solved many years ago, for the case of scattering by homogeneous spheres with isotropic (complex) indices of refraction. While real interstellar grains are almost certainly *not* homogeneous spheres, the solution for this case may nevertheless retain some validity, perhaps as a representative *average*. Mie (1908) solved this complicated problem for the uniform sphere. Closely related work was done by the remarkable Peter Debye (1909), but in astronomy, this formalism is simply called *Mie theory.* It is the theoretical basis for much of the work that has been done on interstellar dust. Greenberg (1991) discusses more elaborate grain shapes and models.

The real and imaginary parts of the index of refraction of the idealized spheres depend on their composition. It is therefore possible, in principle, to determine something about the chemistry of interstellar dust grains from their observed scattering properties.

Since the discovery by Hall (1949) and Hiltner (1949) of the interstellar polarization of starlight, it has been generally known that interstellar dust particles are *not* spherical. Spherical particles do not cause polarization. We cannot discuss here the mechanisms by which elongated particles are preferentially aligned in weak interstellar magnetic fields. We refer to the book by Spitzer (1978), as well as symposia on interstellar matter (e.g., Allamandola and Tielens 1989).

Purcell and Pennypacker (1973) proposed the *discrete dipole approximation* approach to light scattering by dust particles. This method was discussed in some detail by Draine (1989), who concluded that it supported many of the results of Mie theory.

Before we consider the theory of Mie, let us discuss a related problem discussed in most texts in electricity and magnetism or optics (e.g., Hecht 1989, Jackson 1975). Consider the normal incidence of a plane-polarized wave at the boundary of two semi-infinite media, where the indices of refraction are respectively n_1 (above) and n_2 (below). In general, these refractive indices will be complex numbers.

The reflection and transmission coefficients are derived by requiring continuity of the electric and magnetic fields at the boundary. Let us use the subscripts 'I' for incident, 'R' for reflected, and 'T' for transmitted waves. The boundary condition leads to two equations for the amplitudes of the electric and magnetic vectors. The magnetic vector may be eliminated with the help of Maxwell's relations, and one finds

$$E_I + E_R = E_T \tag{14.30}$$

and

$$n_1 E_I - n_1 E_R = n_2 E_T. \tag{14.31}$$

The only subtlety here is the minus sign in 14.31, which arises because of the 180° change in direction of the reflected wave. The Poynting flux $[(c/4\pi)\boldsymbol{E} \times \boldsymbol{H}]$ for this plane-polarized wave must reverse direction on reflection. This means that the direction of either $\boldsymbol{E}$ or $\boldsymbol{H}$ must reverse for the reflected wave. Here, we assume it was the magnetic field that reversed direction at the boundary, or as is often said, had a phase change of 180°. These geometrical relationships are shown in Figure 14.11.

We can eliminate successively E_R and E_T to obtain the transmission coefficient $T = (E_T/E_I)^2$ or that of reflection $R = (E_R/E_I)^2$. If we can assume that the magnetic permeabilities of both media are unity, there result

$$T = \frac{4n_1^2}{(n_1 + n_2)^2} \tag{14.32}$$

and

$$R = \frac{(n_1 - n_2)^2}{(n_1 + n_2)^2}. \tag{14.33}$$

The next step in complication is to consider a plane wave incident on a uniform

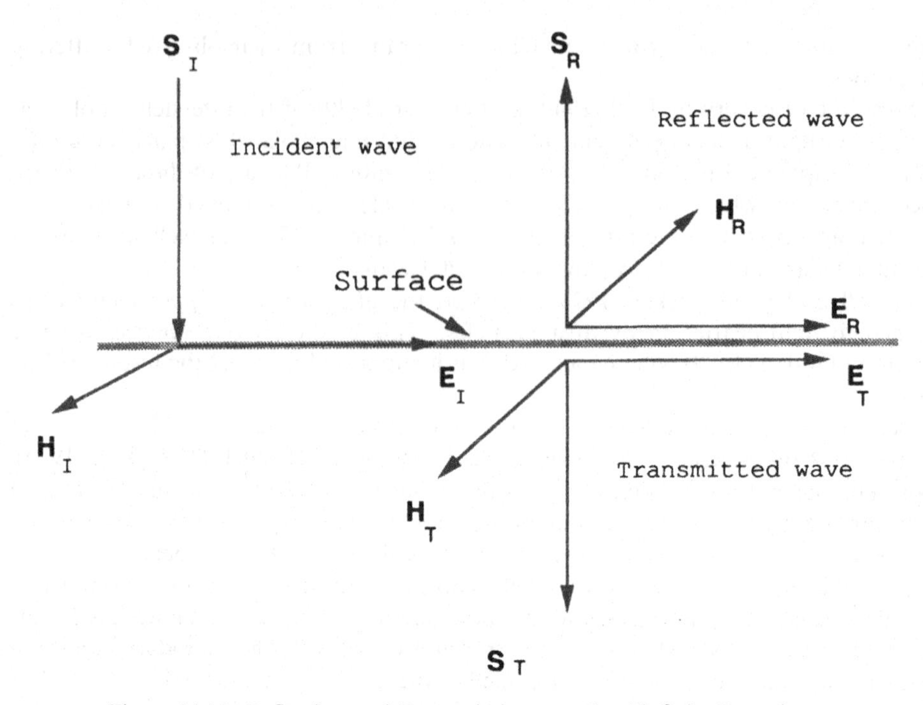

Figure 14.11. Reflection and Transmission at a Semi-infinite Boundary.

slab of finite thickness d. In this case, one calculates the electric and magnetic fields at the second boundary, taking into account that the amplitude of the waves in the slab may decline due to absorption (complex index of refraction). Boundary conditions are again set up, but the situation is now more difficult because of the possibility of (multiple) internal reflections within the slab. Stratton (1941) works out the reflection and transmission assuming only single reflections at the two boundaries.

Stratton's treatment of this simple case is already rather complicated. The situation is much more tedious when one considers the scattering of plane waves by dielectric spheres. In the first place, it is necessary to consider *spherical* boundary conditions. In addition, we have the possibility of *scattering*, where the direction of the incoming wave is modified, by the boundary of the sphere. This complication does not enter when we have plane boundaries extending to infinity.

For this book, we therefore abandon attempts at a fundamental derivation of the relevant absorption and scattering coefficients, and proceed phenomenologically with the help of the concept of cross sections. We refer readers to Stratton or the remarkable book by van de Hulst (1957) for details of the derivations of these cross sections. More recent references may be found in the work by Greenberg (1991) cited above.

Consider radiation of intensity I_λ incident on a volume containing N dielectric spheres per cm^3. Then, just as in §12.4, we may develop and solve the equation of transfer. For an optical depth element $d\tau$, let

$$d\tau = -N(\pi a^2)\, Q_{\rm ext}\, dz. \tag{14.34}$$

We use an established notation in this field where the cross section is written as the geometrical cross section, πa^2, multiplied by an *efficiency factor*, designated by a 'Q.' Let us distinguish two kinds of extinction. In one case, the oncoming light has its direction changed by interaction with the sphere. We say this light is scattered. In the other, the light is absorbed within the sphere. The total efficiency for extinction is the sum of an absorption and a *scattering* term:

$$Q_{\rm ext} = Q_{\rm abs} + Q_{\rm scat}. \tag{14.35}$$

Mie theory provides series solutions for these Q's. In addition, it provides information on the angular distribution of the scattered light. This is described by a so-called phase function, $S(\psi)$, defined as the ratio of the incoming flux to the flux scattered into unit solid angle about the direction (ψ, ϕ). For a spherical particle, there is no ϕ dependence, so that the phase function depends on the single variable ψ. Witt (1988, 1989) reviews attempts to probe the nature of interstellar grains by analyzing the distribution of *diffuse* or scattered interstellar light.

The Mie parameters are rather straightforward to evaluate if one takes advantage of the ability, for example, of FORTRAN to do complex arithmetic. We have found the book by Wickramasinghe (1973) very useful. A computer program to evaluate the Q's requires as input the following parameters:

1. A dimensionless length $x = 2\pi a/\lambda$, where a is the radius of the sphere, and λ the wavelength of the incoming light.
2. The real and imaginary parts of the index of refraction of the material. Note that some authors write the complex index of refraction as $n - i\kappa$, while others write $n + i\kappa$.

In general, dielectrics are birefringent, that is, they have different indices of refraction, depending on the orientation of the solid to the oncoming light ray. This is particularly true for graphite, one of the most commonly discussed possibilities for interstellar extinction. The real and imaginary parts of the indices of refraction are shown in Figure 14.12 for light with the electric vector perpendicular to the graphite planes. Corresponding coefficients for light with the electric vector parallel to the planes are not shown. These constants were computed from tables and relations published by Draine (1985).

If one looks at the prominent feature of the curves in Figures 14.9 and 14.10, it is apparent that the "hump" is located at $1/\lambda$ between 4 and 5 reciprocal microns. This is close to the location where the optical constants of graphite are changing rapidly. It is difficult to locate "a" (reciprocal) wavelength of greatest activity. A common practice for birefringent materials is to use an average of the properties computed for perpendicular and parallel light. For graphite, the weight (2/3) is given to the case where the electric vector is perpendicular to the plane of the sheets. Results for parallel light are weighted by (1/3).

Figure 14.13 shows the extinction efficiencies, Q, for graphite particles with radii of 0.01 and 0.04 microns. The plots are for extinction of light with the E-vector

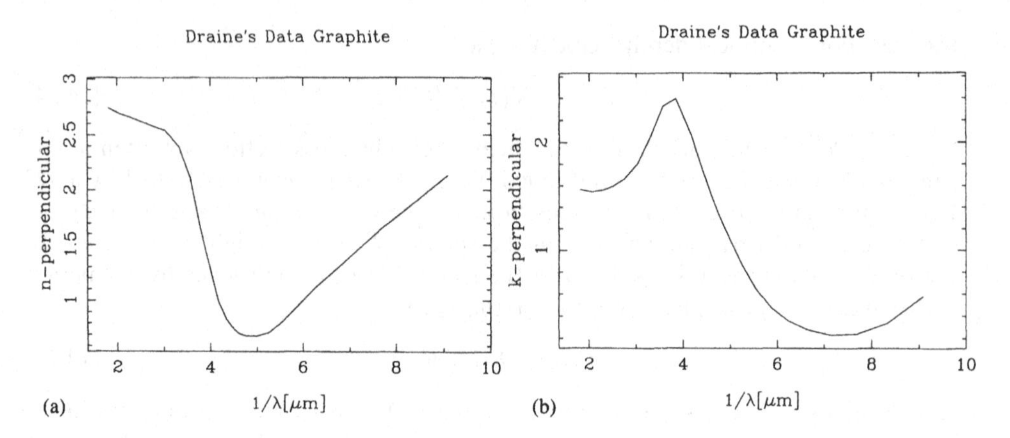

Figure 14.12. Optical Constants for Graphite. Real and imaginary parts of the index of refraction for light incident perpendicular to the graphite planes.

perpendicular. The prominent maximum for $a = 0.01$ microns is obviously suggestive of the $\lambda 2200$ peak shown in the interstellar extinction curve. As the particle size increases this maximum shifts slightly toward smaller values of $1/\lambda$, largely due to the growth of a secondary maximum whose presence is already evident at $a = 0.04$ microns.

Modern work uses an average over an assumed particle size distribution as well as over orientations. Characteristic grain sizes are of the order of 1 to 10×10^{-6} cm. Mathis, Rumpl, and Nordsieck (1977) were able to fit the observed interstellar extinction curve using a mixture of olivine and graphite grains with a distribution of radii a,

$$n(a) = A \cdot a^{-3.5}, \tag{14.36}$$

for $0.005 \leq a \leq 0.25$ microns. For this particular distribution, the average value of the radius is only slightly larger than the lower limit (cf. also Problem 1 below). Other materials are incorporated into grain models in order to match the overall observed curves shown in Figure 14.9.

Draine (1989) discusses graphite as well as a number of other possible explanations of the interstellar $\lambda 2200$ absorption peak. Other possible materials include nongraphitic carbonaceous solids, OH^- on small silicate grains, polycyclic aromatic hydrocarbons (§14.5), and several miscellaneous mechanisms including the desiccated microorganisms discussed by Hoyle and Wickramasinghe (1991). Draine finds difficulties with all of the mechanisms, including graphite, and concludes that "the $\lambda 2175$ feature remains a puzzle...".

14.5 Interstellar and Circumstellar Features

A series of diffuse, interstellar absorption bands have been recognized in the spectra of early stars for many decades. The strongest of these absorptions are at $\lambda\lambda 4430$, 4882, 6177, and 6283. Some one hundred of these bands are now known.

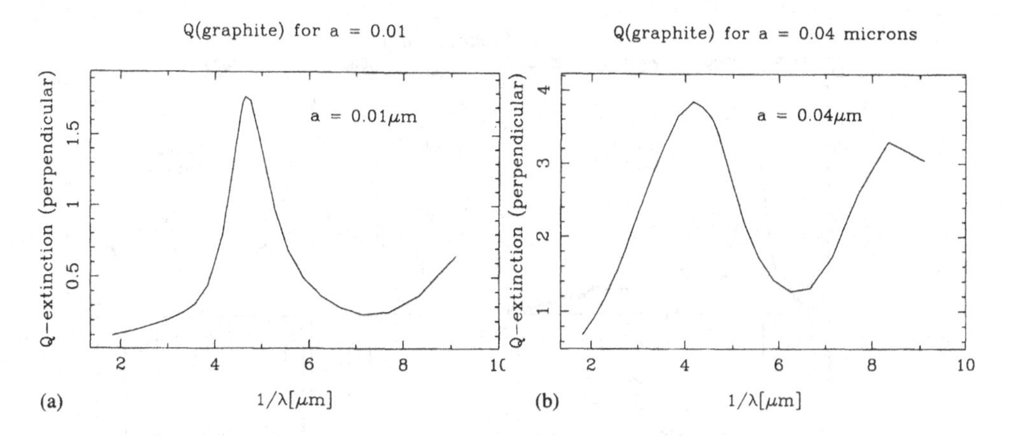

Figure 14.13. Extinction Coefficients Q for Graphite.

Their strengths are closely correlated with the general interstellar absorption or color excess: $E(B-V)$. Herbig's (1975) article summarized what was known about these features. Unlike most of the absorptions that appear in stellar spectra in this wavelength region, these diffuse bands remain unidentified (Herbig and Leka 1991). It is possible, however, that these features may arise from relatively large organic molecules possibly responsible for both emission and absorption features that appear in the infrared (see Puget and Léger 1989).

The infrared region of the spectrum stretches in wavelength from the limits of detection of red light ($\approx \lambda7500$) by the human eye to 1 or perhaps a few millimeters. Infrared astronomy, like its space and radio counterparts, became a mature discipline in the decades following World War II. The development of sensitive detectors, and their use from ground-based observatories, balloons, and aircraft, made this a rapidly developing area of research even before the results from the *Infrared Astronomical Satellite* (IRAS) "forever changed our view of the sky" (Beichman 1987). Soifer and Pipher (1978) surveyed the impressive observational tools available more than a decade ago. This spectral domain has given important new insights into the nature of interstellar dust.

Infrared features are now well known, in both emission and absorption, in spectra of a wide variety of objects. Much recent excitement has been generated by the possibility of explaining most of the emission features as due to polycyclic aromatic hydrocarbons, or PAH's (Puget and Léger 1989).

Some PAH's are illustrated in Figure 14.14 taken from Allamandola, Tielens, and Barker (1989). "Schematic" absorption spectra are indicated by the vertical lines. These are calculated spectra based on laboratory measurements obtained with the molecules suspended in KBr. The environment distorts the spectra relative to those of free molecules. The three molecules illustrated appear in the 1991–92 *HCP's* section on rules for nomenclature of organic chemistry, p. **2**-54. They are Examples (18), (19), and (30). Allamandola, Tielens, and Barker discuss two size ranges of PAH's. They think the sharper infrared features may be due to molecules with 10 to 40 atoms. More diffuse features could be caused by what they call PAH clusters

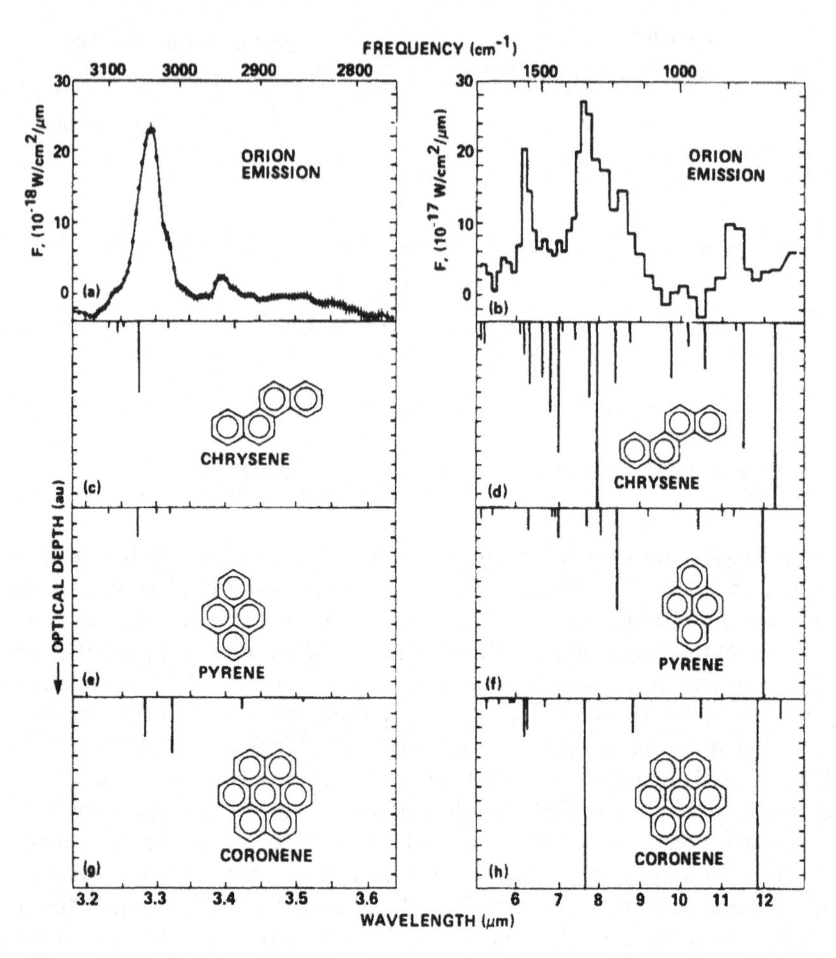

Figure 14.14. Spectra of PAH's and Interstellar Features. Allamandola, Tielens, and Barker (1989).

and amorphous carbon, with 100 to 500 atoms. The PAH's are large molecules, but not macromolecules such as proteins or polymers.

Pure spectra of these molecules in the gaseous phase have not yet been obtained over the domain of Figure 14.14. Cherchneff and Barker (1989) measured the C–H stretch feature near 3.29 μm (μm = microns) in the vapor-phase spectrum of azulene ($C_{10}H_8$, Example (4), op. cit.). They found a peak near 3050 cm^{-1}, as expected for the stretching of the C–H bond when the carbon is a part of an aromatic ring (see Table 14.5). Brenner and Barker (1992) present more recent laboratory spectra relevant to the PAH hypothesis.

Infrared spectra are widely used in analytical chemistry. The photon energies are commensurate with molecular vibrational energies, roughly, multiples of 10^3 cm^{-1} (see Equation 11.7 and following text). Brief introductions to these techniques are given in books on physical and organic chemistry (Castellan 1983, Eğe 1989). The

Table 14.5. *Some Characteristic Infrared Signatures*

Bond	Compound type	Wavenumber (cm^{-1})	Wavelength (μm = microns)
≡C–H	alkynes	3300	3.03
C–H	aromatic	3000–3100	3.23–3.33
–C–H	alkanes	2850–2960	3.38–3.51
C≡N	nitrile	2220–2260	4.42–4.50
C=O	aldehydes	1736	5.76
C=O	aliphatic ketones	1724	5.80
C=O	aromatic ketones	1680–1645	5.95–6.08
C=C	alkenes	1650	6.07
Si–CH_3		1210–1270	7.86–8.26
Si–O–Si		1000–1100	10.00–9.09
Si–C		690–860	11.6–14.5

Handbook of Chemistry and Physics contains numerous charts for the identification of many functional *groups* – usually pairs of atoms in a given bond environment, whose vibrational frequencies are useful analytical signatures. Detailed information may be found in the book by Lin-Vien *et al.* (1991), and many other sources.

Table 14.5 lists a few of the characteristic features from the sources cited. Generally, the vibrational modes that stretch the bonds have features at larger wavenumbers than bending or "rocking" modes. The fingerprint region lies roughly from 200 to perhaps 4000 cm^{-1} (2.5 to 50 μm). Infrared spectra in this domain tend to be characteristic of the molecule as a whole, rather than of specific bond vibrations. Shimanouchi's (1972, 1977) tables identify the vibrations of specific molecules and also describe the nature of the vibrations: stretching, rocking, etc.

Figure 14.15 shows three tracings from the *Aldrich Library of FT-IR Spectra* (Pouchert 1989). The molecules were in the *vapor phase.* Figure 14.15(a) is of 1-pentyne, and illustrates features related to both single and triple carbon bonds. The C–H stretch absorption appears at both 3.0 and 3.4 μm. The tougher, triple bond has a signature at the shorter wavelength absorption. The C–H stretch associated with double and aromatic carbon bonds falls between these positions. The absorptions between about 6.7 and 8.2 μm are due to C–H bending. The center of gravity of the complex is tantalizingly near the 7.7 μm features of Figure 14.16 (below), but this molecule is surely *not* responsible for the astronomical features.

Figure 14.15(b) is of an aliphatic hydrocarbon, with a C≡N at one end – a nitrile. The feature most characteristic of C≡N is near 4.5 μm, and is quite weak in the spectrum of this particular molecule. Features due to C–H stretching and bending are also present in butyronitrile. Figure 14.15(c) illustrates features due to Si–O–Si, Si–C, and Si–CH_3, which appear in the silane octamethyltrisiloxane. The silicon–oxygen bond may play a role in the 9.7 μm feature discussed below.

d'Hendecourt, and Allamandola (1986) prepared a small atlas of laboratory infrared spectra for use in astronomical applications. Their spectra are from thin,

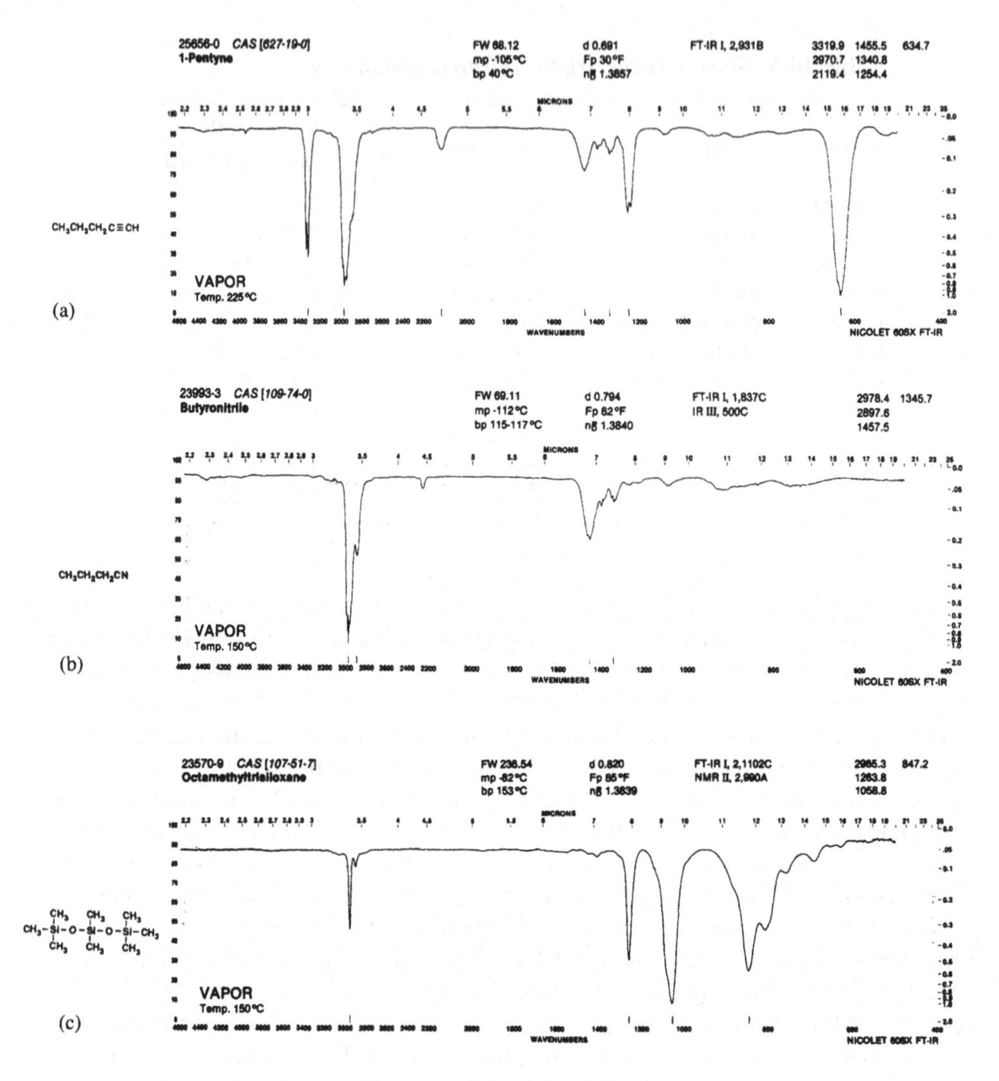

Figure 14.15. Infrared Spectra of Organic Molecules.

solid materials at a temperature of 10 K. Even though the observed features may come from different materials than the ones chosen, they provide useful illustrations and descriptions of vibrations giving rise to similar features. Finally, they give data from which the number molecules in the line of sight may be estimated.

We now turn to astronomical infrared spectra. Several unidentified emission features are indicated in the spectrum of the hot planetary nebula NGC 7027 seen in Figure 14.16 from Russell, Soifer, and Willner (1977). The emission peak near 3.3 μm is obvious. Similar features can be seen in the spectra of NGC 7023 and 2023 by Sellgren *et al.* (1985).

The IRAS *Atlas of Low Resolution Spectra* (IRAS Science Team 1986) illustrates

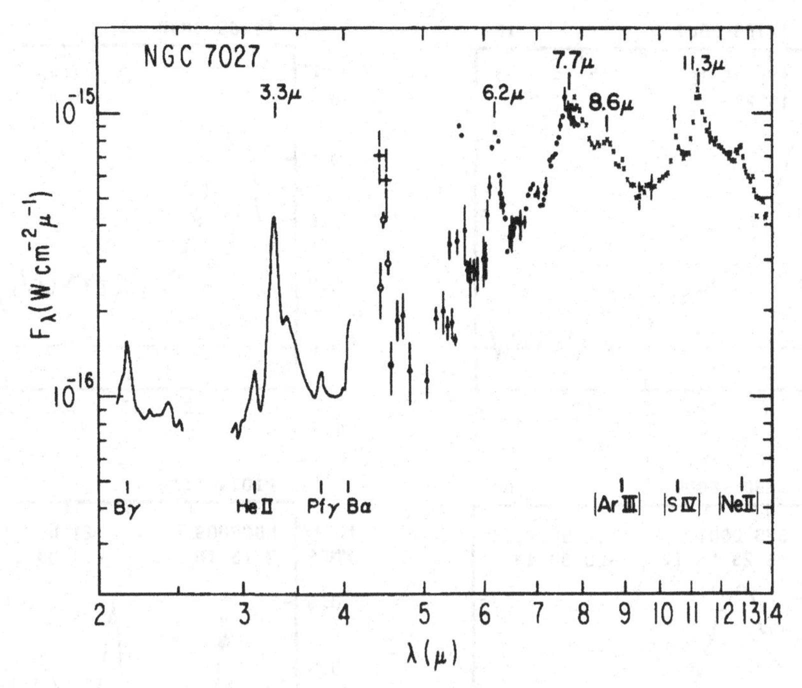

Figure 14.16. Infrared spectra of NGC 7027, from Russell, Soifer, and Willner (1977).

a variety of astronomical objects including stars, emission regions, and molecular clouds. Perhaps the most prominent feature in the *Atlas* is centered at 9.7 μm. The width is typically 2–3 μm. It is in emission in the spectra of cool giants and supergiants, and can be due to circumstellar rather than interstellar dust. Because the feature is not seen in the spectra of carbon stars, it is plausible to attribute it to silicate dust. Early references to the identification with vibrations of the Si–O bond may be found in the reviews by Merrill and Ridgway (1979), or Savage and Mathis (1979).

According to Aitken (1981), the wavelength of the 9.7 μm feature is not quite correct for ordered silicate minerals (e.g., olivine), but could be explained by a "disordered" silicate. On the other hand, Aitken *et al.* (1988) discuss a narrow feature at 11.2 μm that could be due to "a more structured silicate, similar to olivine."

In Figure 14.17(a–d), we show several examples from the IRAS *Atlas*. Figure 14.17(d) is of the planetary NGC 7009, the "Saturn" Nebula.

Parts (e) and (f) of the figure show laboratory spectra of solid material collected with the help of high-altitude aircraft (cf. Sandford and Walker 1985). The dark dots in part (e) are emission spectra from Comet Kohoutek that have been inverted for purposes of comparison. Unfortunately there is a gap in the Kohoutek spectra near the maximum just beyond 11 μm that is characteristic of crystalline olivine. This feature is present in the Halley spectrum (open squares) shown in part (f).

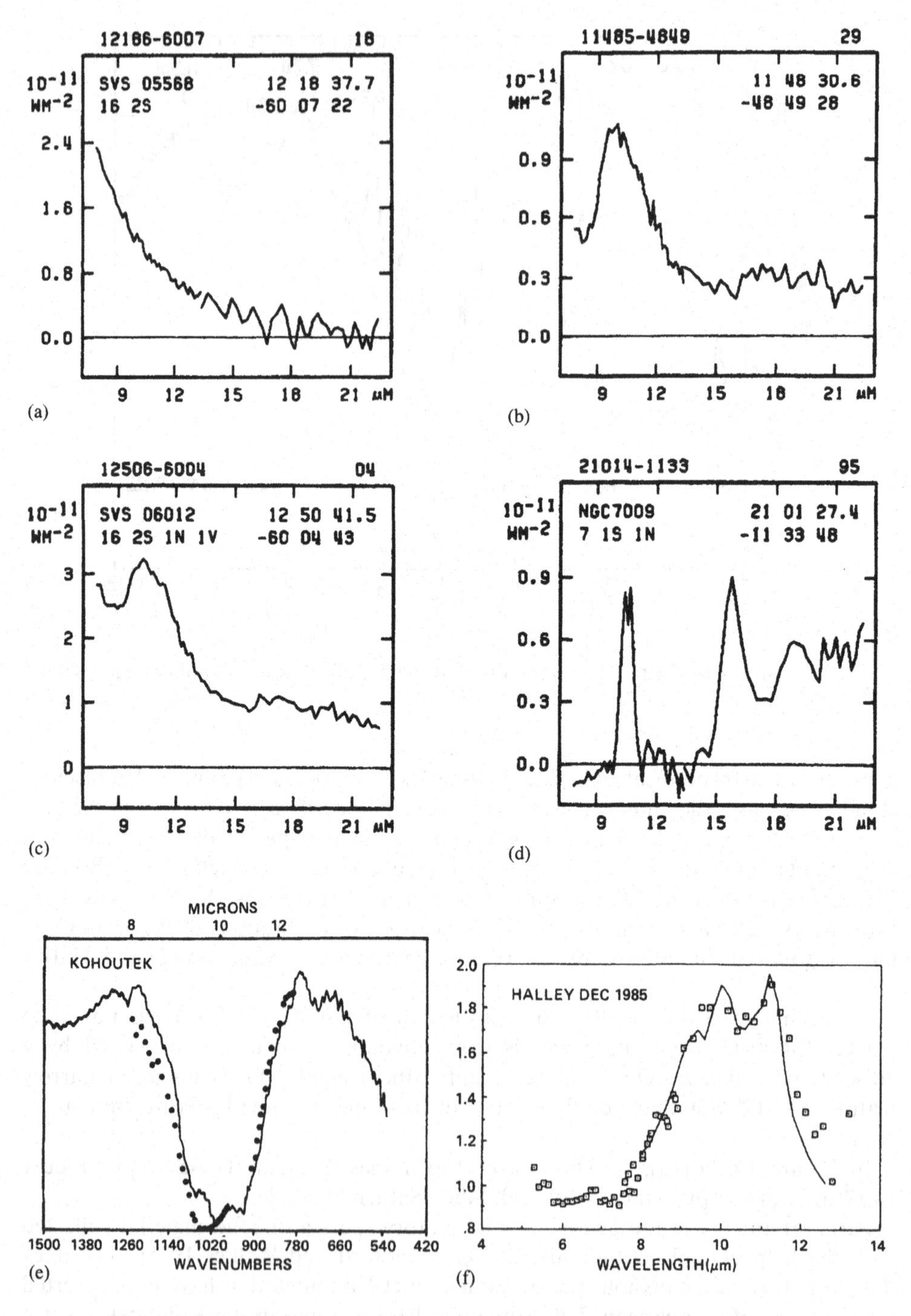

Figure 14.17. IRAS Low Resolution Spectra (a–d). Courtesy of IRAS Science Team. Comet Spectra, courtesy of S. A. Sandford (e,f).

The laboratory spectrum (solid) is a mixture of 35% pyroxene and 55% olivine taken from IDP's or *interplanetary dust particles* described by Sandford and Walker (1985).

Sandford and Walker's IDPs' samples and olivine have, in addition to the broad 10 μm absorption, additional absorption near 16–20 μm. Both features are due to SiO; the shorter wavelength absorption results from Si–O stretching, the longer to Si–O–Si bending. If these *two* features could be established in circumstellar spectra of stars of the oxygen sequence, it could strengthen the identification with silicates.

A broad feature at 11 μm is associated with carbon stars and carbon-rich planetary nebulae. Higher-resolution spectra than those of Figure 14.17 are required to distinguish it from that at 9.7μm. The proposed identification is with SiC.

A feature at 3.07 μm has been seen in sources located deep in molecular clouds. An identification has been made with ice on the basis of Mie theoretical calculations that used optical constants for H_2O ice. The O–H stretch of water is well illustrated in Figure 1a of d'Hendecourt and Allamandola (1986).

The various infrared features cannot be regarded as securely identified, at least not in the same sense that many atomic and resolved molecular features are. In the case of atomic features, we can make use of many wavelength coincidences. We can also consider expected relative intensity behavior. It is obvious near 10 μm, where both emission and absorption occur, that the identification of the responsible substance is a much more formidable task. Workers in this field feel much more confident of the identifications of the *functional groups or bonds* (cf. Table 14.5) than individual species.

It is probable that only in the domain of atomic spectroscopy can we begin to claim anything like necessary completeness of data for reliable cosmochemistry. A paper by Donn (1986) lists some useful references for optical properties of solid substances as well as molecules.

In spite of the vast amount of chemistry thought to go on in star formation regions, there is little evidence for an influence on the bulk chemistry of the stars. Should we not expect some stars to have more than an average share of grains relative to the residual gas? Is there no chemical separation of the various molecular species prior to star formation? Models of the overall chemistry of galaxies have largely ignored such possibilities. This has been because the evidence for chemical differentiation is dubious among the lower main-sequence dwarfs, whose surfaces are not subject to *in situ* differentiation.

Nevertheless, some speculations concerning the relevance of this chemistry to stars and stellar evolution have been made. The ideas fall into two categories. First, there has been some thought given to chemical separation prior to star formation, either by gravitational sedimentation or through the influence of electromagnetic forces. Some references are given in a paper by Lattanzio (1984). We have mentioned recent ideas concerning grain–gas segregation and stellar abundances in §7.3.

A second approach to the influence of interstellar processes on stellar cosmochemistry has been to consider the accretion of material after star formation. Such processes were mentioned in §13.4.3 as a possible explanation of the variegated surface chemistry of upper main-sequence stars. See the references to Alcock and Illarionov (1980) as well as Talbot and Newman (1977).

14.6 The Formation of Dust

Using the methods outlined in the earlier sections of this chapter, we may derive cloud column densities ($N \cdot H$, number/cm^2) for species giving rise to absorption or emission lines. These column densities can be turned into number densities (N, number/cm^3) if the path length is known. Path lengths can be estimated if we assume the clouds are roughly spherical, and we know their approximate distance and angular extent.

While estimates of number densities made in this way can claim only modest accuracy, there is little reason to think that published estimates are in error by more than an order of magnitude. They are in broad agreement with the densities required for the molecular reactions to produce the observed species. Goldsmith (1987) summarizes the properties of interstellar clouds. According to his summary, in the densest regions, which he calls clumps, the particle densities reach $> 10^6\,\mathrm{cm}^{-3}$.

The pundits have traditionally said that the densities of interstellar clouds are insufficient for the formation of solid particles. Draine (1989) suggested this might not be so, but we must refer to his paper for additional details. The traditional sources of solid particles are *stellar atmospheres.* In the solar photosphere, the particle density is roughly 10^{17}. In cool red giants, typical values are only two orders of magnitude or so lower. Consequently, the stellar particle densities are many orders of magnitude higher than the value of 10^6 just given for dense clouds.

Since *stars themselves* are thought to form in these clouds, the argument that the densities in them are not high enough for dust formation seems flawed. Nevertheless, it is observationally clear that many late-type stars are surrounded by *circumstellar* material, revealed by both atomic absorption lines and the kinds of infrared features discussed in the preceding section. In addition, planetary nebulae are known to contain both molecular features, and dust, and these objects are thought to be ejecta of red giants (§15.2). Accordingly, it is conventional wisdom that interstellar dust is formed in late-type *field* giants and supergiants, and that it is later *concentrated* into the regions we now know as dark clouds. We must refer to the literature (e.g., Hollenbach and Thronson 1987) for details of possible mechanisms, and turn now to a discussion of the mechanism of dust formation.

The technical term for the formation of solids or liquids from the vapor phase is *condensation.* It is well known from the study of rain clouds that this process is one that proceeds most readily in the presence of *condensation centers.* In the case of rain clouds, the initial condensates form on dust grains. Formation on previously existing centers is *inhomogeneous condensation.*

We consider here the question of *homogeneous* condensation or nucleation, in which a pure substance changes from the vapor to a solid or liquid. We begin with some elementary considerations that may be found in the introductory text of Barrow (1988). Other references used in the present section are the books of Adamson (1990), and a general work on chemical rate processes, Gardiner (1969). Nucleation has been discussed in an astronomical context by Draine (1979), Draine and Salpeter (1977), and Salpeter (1977).

There is a conceptual problem with homogeneous nucleation. With no initial condensation center, there is no target for a molecule in the vapor to strike to become a part of the condensed phase. Nature obviously has a way around this difficulty, and man has devised one as well. It starts with a consideration of small droplets in a condensed phase.

Consider the condensation of n moles of vapor at pressure P to a liquid. We shall write this as a chemical equation:

$$\text{A}(n \text{ moles, vapor}) \rightarrow \text{A}(n \text{ moles, liquid}). \tag{14.37}$$

Let P^0 be the pressure of a vapor that is in equilibrium with the liquid phase. Then the ΔG for 14.37 is the same as that to take the vapor from the pressure P to the pressure P^0, because if the vapor is in equilibrium with the liquid at pressure P^0, the ΔG to go from the vapor at P^0 to the liquid is zero. Thus for the temperature T, we have (e.g., from Equation 4.27)

$$G(\text{liquid}) - G(\text{vapor}) \equiv \Delta G = n\mathcal{R}T \ln(P^0/P). \tag{14.38}$$

Now consider the further transformation

$$A(n \text{ moles, liquid}) \rightarrow$$
$$A(n \text{ moles of liquid in } n_1 \text{ droplets with radii } r_\text{d}). \tag{14.39}$$

Strictly speaking, we should consider the surface energies for the configurations on both sides of 14.39. However, we suppose the surface energy is negligible for the A on the left-hand side (relative to the internal energy) and only consider it for the small droplets of A, which we take to be spherical with surface area σ_1. This is because the ratio of the number of particles in a surface layer of thickness t to those in a volume of radius r becomes very small for finite r. We therefore neglect entirely the surface energy associated with the liquid on the left of 14.39, and assume that the conjugate Gibbs energy change is entirely due to the formation of the surface area $n_1\sigma_1$ of the n_1 droplets on the right.

Work must be done on the system to make droplets. The reason for this is illustrated in Figure 14.18. Molecules within the body of a liquid feel attractive forces on all 4π steradians, and are thus more tightly bound than those at the surface, which are held by forces from (roughly) only a hemisphere. Molecules at the surface of droplets feel these attractive forces from a relatively smaller fraction of the sphere. The smaller the radius of curvature of the sphere, the smaller the attractive surface holding any particular molecule to the rest of the liquid.

We see that the free energy change to make droplets must therefore be positive: the droplets would not form spontaneously from a macroscopic liquid. Since there is no volume change in creating the surface, we write simply

$$\Delta G = n_1\sigma_1\gamma = n_1(4\pi r_\text{d}^2)\gamma. \tag{14.40}$$

We have made the simplification that the surface energy depends only on the surface area and not on the size of the droplet. This complication has been discussed, for example, by Draine and Salpeter in the papers cited above. However, it seems likely that other uncertainties in nucleation are larger than this particular effect, and we work with a single "mean" surface energy.

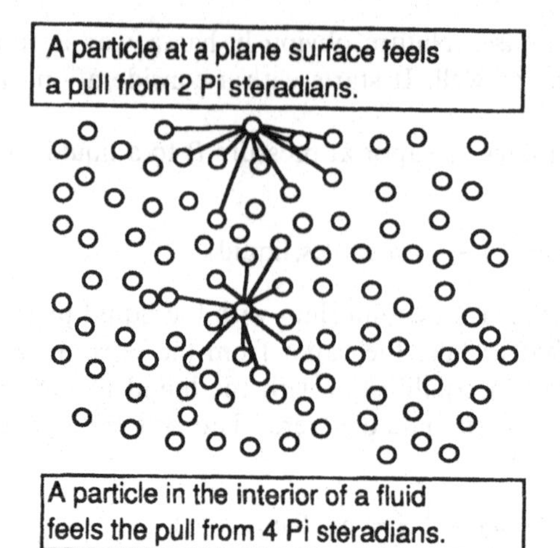

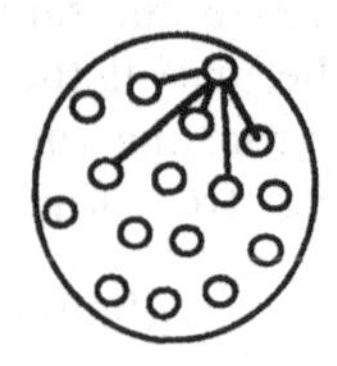

Figure 14.18. Surface Energy and Surface Tension.

We now combine the ΔG's for Equations 14.37 and 14.39, and obtain the free energy change for the formation of droplets within the scope of our specified assumptions. In order to emphasize that the combined ΔG involves droplets rather than bulk liquid, we introduce a new symbol $\Delta\Gamma$. Then

$$\Delta\Gamma = n\mathscr{R}T \ln\left(P^0/P\right) + \gamma n_1 \sigma. \tag{14.41}$$

For a fixed P^0 and P, the two terms on the right of 14.41 have different dependences on r_d. The total number of moles n may be written

$$n = \left[\left(\frac{4}{3}\right)\pi r_d^3 \left(\frac{\rho}{\mu}\right)\right] \cdot n_1, \tag{14.42}$$

where μ is the molecular weight. The term in the square brackets gives the number of moles in each droplet. Both terms on the right of 14.41 depend linearly on n_1, but the first term depends on the cube of r_d, while the second goes as r_d squared ($\sigma = \pi r_d^2$). Therefore, as r_d increases, the first term becomes relatively more important. Likewise, as r_d decreases, the importance of the surface term increases.

Now consider the relative signs of the two terms. The surface term is intrinsically positive, as we have argued above, since work must be done against surface tension to increase the surface area. The first term could thus far be positive or negative, since P has not yet been specified. The quantity P^0 is fixed, of course, since it is the pressure of the vapor in equilibrium with the macroscopic ($r_d \rightarrow \infty$) liquid. If we stipulate at this point that the vapor is *supercooled*, then P must be larger than P^0.

If the droplets are in equilibrium with the vapor, we must also have $P > P^0$, since a higher vapor pressure is required to keep the less tightly bound molecules at the surface of the droplets than the "flat" liquid. This means that for this critical P (equilibrium with the droplets) and above, the logarithmic term on the right of 14.41

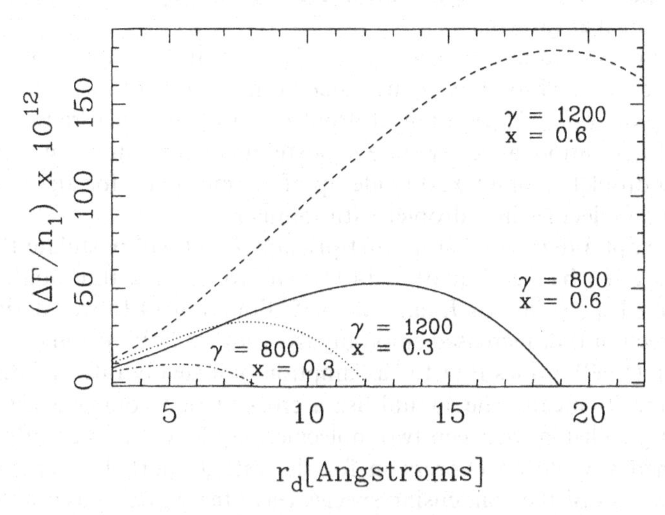

Figure 14.19. $\Delta\Gamma$ as a Function of Drop Radius r_d.

is less than zero. We shall use $\Delta\Gamma$ to describe *that* ΔG that holds when P is the pressure in equilibrium with the droplets.

If $P^0/P < 1.0$, we can always find a value of r_d for which $\Delta\Gamma$ reaches a maximum. Since the first term has the higher r_d dependence, for larger r_d, $\Delta\Gamma$ will *decrease*. Indeed (see Figure 14.19), put

$$\left(\frac{d\Delta\Gamma}{dr_d}\right) = n_1 4\pi r_d^2 \left(\frac{\rho}{\mu}\right) \cdot \mathscr{R}T \cdot \ln\left(P^0/P\right) + n_1 \cdot \gamma \cdot 8\pi r_1 = 0. \tag{14.43}$$

We can now solve for a critical value of r_d, say, r_c at the maximum of $\Delta\Gamma$:

$$r_c = \left(\frac{2\gamma\mu}{\rho\mathscr{R}T \cdot \ln\left(P/P^0\right)}\right). \tag{14.44}$$

We can use 14.44 to eliminate the pressure term from 14.43. At the maximum, we have, for $n_1 = 1$,

$$(\Delta\Gamma)_{\max} = 4\pi\gamma r_c^2/3. \tag{14.45}$$

This is a classical result, attributed to Gibbs. The right-hand side of 14.45 is one third the surface free energy of a droplet with the critical radius.

Classical, homogeneous nucleation considers a variety of droplet sizes, made up of $\mathcal{N}$ molecules. Each drop is called a complex, and one may ask what the free energy change will be if $\mathcal{N} \rightarrow \mathcal{N} + 1$, that is, the drop increases by one molecule.

Clearly $\Delta\Gamma$ will *increase* with $\mathcal{N}$ for droplets that are smaller than the critical size (r_c). The formation of such droplets or complexes is thus thermodynamically unfavorable, and would require pressure or temperature fluctuations to make them possible. However, once the critical droplet size is reached, addition of another molecule reduces the Gibbs free energy, and there is a steadily decreasing thermodynamic barrier to further growth.

Droplets less than r_c in radius are expected to "dissolve" in the vapor, while those with larger radii are assumed to grow. Thus, calculating the *rate* of nucleation is the same as calculating the rate of formation of droplets with the critical radius. In the classical picture of nucleation, a steady state is postulated with a uniform "flow" of molecules through droplets (complexes) made up of $\mathcal{N}$ molecules for all $\mathcal{N} < \mathcal{N}_c$, where there are $\mathcal{N}_c$ molecules in a droplet with radius r_c.

We shall not attempt a derivation from first principles, but will postulate that the rate ($cm^3\ sec^{-1}$) is given by our Equation 14.15. The role of the threshold energy ϵ_0 is played by the $\Delta\Gamma_{max}$ from our Equation 14.45. Equation 14.15 was derived by assuming a cross section that increased smoothly from a threshold to some constant value σ_0. Adamson (1990) argues that the assumption of a steady rate of formation of droplets with *any* $\mathcal{N}$ means one should use a cross section commensurate with the most elementary collision: between two molecules. We leave σ_0 as an adjustable parameter, and merely write an expression for the rate J (particles per cm^3 per sec), at which molecules of the condensing species leave the vapor phase (and enter droplets). If there are N molecules per cm^3 of vapor,

$$J = N^2\sigma_0(kT/m)^{1/2}\exp\left(-\Delta\Gamma_{max}/kT\right). \tag{14.46}$$

A possible factor of (1/2) because of collisions between similar particles, along with other constants, has been absorbed into σ_0.

We may obtain a characteristic time from Equation 14.45 for the disappearance of the species from the vapor phase. Since J may be interpreted as dN/dt, we have for this time

$$t_c = N/J. \tag{14.47}$$

Put $P/P^0 = x$ into Equation 14.44. Then 14.45 becomes

$$\Delta\Gamma_{max} = \frac{4\pi\gamma}{3}\left(\frac{2\gamma\mu}{\rho\mathcal{R}T\ln(x)}\right)^2. \tag{14.48}$$

It is important to ask whether condensible material in an expanding stellar chromosphere (wind) will have time to nucleate before the density drops prohibitively low. Thus, we need to compare the t_c of Equation 14.47 with H/w, where w is a characteristic, chromospheric wind speed, and H the chromospheric density scale height. First, it is necessary to comment that the theory of nucleation presented here is very crude at best. It is not at all clear that the thermodynamic concepts that have been employed will be valid for real nucleation. Second, the nature of the dust particles that may form in stellar atmospheres is not known. In a carbon-rich star, the condensate might be graphite, soot (microcrystalline graphite), or amorphous carbon (imperfect crystals). Therefore the surface energies cannot be accurately specified, and they appear as the third power in $\Delta\Gamma_{max}$!

With the large uncertainties duly confessed, we discuss briefly the results of an exploratory calculation outlined in Problem 5 below for the nucleation of graphite.

The nucleation time N/J is extraordinarily sensitive to the degree of supersaturation of the vapor, which enters in the term $\ln(P/P^0)$. P^0 depends exponentially on temperature (or $1/T$) (§5.5). A small drop in T causes a large change in P^0.

Since at least some estimates lead to short condensation times in cool stars, it seems plausible that nucleation does indeed take place in stellar atmospheres. Certainly the large uncertainties of the current theory need to be reduced.

14.7 Problems

1. Assume the distribution of dust grain radii given by Equation 14.48. Give an argument that a reasonable estimate for a characteristic grain radius would come from averaging a^2, and taking the square root.
2. For purposes of the present problem only, let us assume a general extinction of 2 magnitudes per kiloparsec. Assume also an extinction cross section equal to the geometrical cross section for a grain radius of $2 \cdot 10^{-6}$ cm. How many grains per cm^3 are then implied by this extinction? If the grains are graphite, calculate the number of carbon atoms per cm^3 tied up in the grains. Finally, assume a general hydrogen atom density of 1 per cm^3, and calculate the "smoothed out" C/H ratio. Compare this with the C/H ratio of the SAD. Suggest refinements and applications of this kind of calculation.
3. Consult your favorite reference books for values of the surface tensions of a variety of substances. What sort of errors might you make if you assumed $\gamma = 10^3$ erg/cm^2 for graphite condensation from a stellar atmosphere?
4. Calculate the equilibrium line for graphite for temperatures between 2500 and 5000 K using the methods of Chapter 5 (see §5.5). If the vapor pressure of carbon is $5 \cdot 10^{-5}$ atmospheres, what is the condensation temperature?
5. A very crude estimate of the condensation time N/J may be made for graphite. We shall "cook" the following figures a bit to get a plausible result. The reader should vary some of these values to get a feel for the uncertainties. For σ_0 take a near geometrical cross section for the carbon atom with a radius of 1 Å. Take $\log(P_g) = 3.6$ and $T = 2000$ K. For this temperature, the equilibrium calculation above shows that graphite will condense if its partial pressure is roughly 3.3×10^{-5} dynes/cm^2. Assume a carbon-rich composition ($\epsilon_C = 8.5 \cdot 10^9, \epsilon_O = 3.6 \times 10^8$). Equilibrium calculations with 15 molecules, including acetylene (C_2H_2), give a partial pressure of atomic carbon of $1.02 \cdot 10^{-4}$ cgs, so the vapor would be supersaturated ($x = 3.09$) if the carbon had not condensed. For the critical surface tension, take $\gamma = 500$ cgs. Show that these figures lead to a condensation time N/J that is about a day. Compare this value with prudent guesses for H/w. For H, you might estimate a few isothermal scale heights, $H = \mathcal{R}T/g\mu$, with g the surface gravity for a giant or supergiant K or M star. The value of w should not be too far from the sound speed $\approx \sqrt{\mathcal{R}T/\mu}$. Data for stars are given by Allen (1973) and Lang (1992). Does condensation of graphite in giant atmospheres seem possible?
6. Use the angular momentum $l = \mu r^2 \dot{\theta}$ to eliminate $\dot{\theta}$ from the equation for the

conservation of energy for a general central force problem:

$$\frac{1}{2}\mu(\dot{r}^2 + r^2\dot{\theta}^2) + V(r) = E. \tag{14.49}$$

Thence write an equivalent energy equation for the resulting *one-dimensional problem* with an *effective potential*, V_{eff}. This is a standard problem in classical mechanics (Goldstein 1980, §3-3, Eq. 3-22′). Obtain the specific form 14.21 using the potential 14.20.

7. Write down the equation for the conservation of energy of a particle in the field of the potential 14.20. Use a coordinate system with origin at the center of force. Let the energy be $(1/2)\mu V_\infty^2$, where V_∞ is the velocity at infinite separation. Then the angular momentum $l = \mu V_\infty b$, where b is the impact parameter. Set $\dot{r} = 0$ for $r = R$, and use 14.22 to eliminate R. Solve the resulting relation for b^2, and thus obtain the Langevin cross section.

15
Emission-Line Regions and their Chemical Abundances

15.1 Emission Regions

The dark clouds seen in the plane of the Galaxy are often accompanied by bright, *diffuse nebulae* such as those shown in Figure 14.2. A catalogue by Lynds gives some 1100 such objects. Roughly as many *planetary nebulae* (see the following sections) are given in a very useful new catalogue of Acker *et al.* which includes finding charts and keys to the naming conventions. These works are referenced in Table 15.1 along with additional galactic and extragalactic emission-line sources. In the optical region of the spectrum, the *emission nebulae* radiate primarily in atomic lines. Optical continuous emission is present mostly from the recombination of hydrogen ions. The Balmer continuum of hydrogen ($\lambda < 3647$ Å) is readily detected, even on photographic spectra, while the Paschen continuum ($\lambda < 8204$ Å) is much weaker. Continuous emission is also provided by helium, and from the decay of the $2s^2S$ level of neutral hydrogen by *two-photon* decays (Osterbrock 1989, §4.3).

Among the diffuse nebulae, there are some for which continuous emission is observed due to the reflection of starlight from dust particles. Well-known reflection nebulosity is associated with stars of the Pleiades.

Supernova remnants are an important class of emission-line objects of keen interest in modern astronomy. Direct optical photographs of the brightest of these objects were published by van den Bergh, Marsher, and Terzian (1973). Nova

Table 15.1. *References for Emission-Line Regions*

Object	Reference	Approx. No.
Bright Galactic Nebulae	Lynds (1965)	1100
Galactic Planetary Nebulae	Acker *et al.* (1992)	1143
Supernovae Remnants	van den Bergh *et al.* (1973)	24
Extragalactic H II Regions	Hodge and Kennicutt (1983)	125
Seyfert Galaxies, etc.	Hewitt and Burbidge (1991)	935
Quasars	Hewitt and Burbidge (1987, 1989)	4400
Quasars and AGN's	Véron-Cetty and Véron (1991)	7927

remnants may often be seen as a nebulosity surrounding the stellar image, after a period of some years. If the nova outburst is not seen, it may be difficult to decide upon the nature of the remnant (de Muizon *et al.* 1988).

The continuous emission from various bright nebulae has been studied in the *radio region* for many years. With the improvement in the sensitivity of receivers, recombination line radiation was also observed, in both the radio and microwave regions. This occurs as electrons cascade from orbits with principal quantum numbers of the order of sixty (microwave) or a hundred or more (radio). Lines of a few elements beyond hydrogen and helium have been observed. Silvergate (1984), for example, discusses observations of C I 166α, that is, the emission line produced when a transition in neutral carbon takes place from principal quantum number $n = 167$ to $n = 166$. Gordon (1988) has reviewed radio recombination lines within the context of these galactic emission or *H II regions.*

Many emission regions may be detected in other galaxies. Hodge and Kennicutt (1983) have published *An Atlas of H II Regions in 125 Galaxies.* Additional references may be found, for example, in Kennicutt (1988). The abundance analyses of extragalactic H II regions have played a key role in our understanding of the chemical evolution of galaxies. Much interest in modern astronomy centers around *active galactic nuclei* or *AGN's* and the related *quasars.* The spectra of these objects are similar in many ways to those of galactic H II regions, and their abundance analyses involve similar techniques. We shall return to these objects at the end of the present chapter, and in the following one.

The early visual observers said that gaseous nebulae were green. This was because of the great strength of lines near λ5000 (green!) that were unknown in laboratory spectra. At one time it was supposed that they were due to an unrecognized, *ad hoc* element, *nebulium.* The lines, which we now know to be due to forbidden transitions of doubly ionized oxygen, are still sometimes called the nebulium lines. The identifications with forbidden transitions in doubly ionized oxygen, $\lambda\lambda$4959 and 5007, were made by the American astronomer I. S. Bowen (1928), who recognized that these lines could be very strong under the conditions of extremely low density that prevailed in emission nebulae (see §15.3 below). The forbidden-line spectrum of an element or ion is indicated by the presence of brackets about the usual designation. Thus, $\lambda\lambda$4959 and 5007 are said to be due to [O III] (cf. §11.2).

In the last section of this chapter, we have assembled representative "simulated" spectra of a variety of emission sources. Details of how these spectra were produced are given below. They show the variety of the spectra of nebulae within the domain $\lambda\lambda$3500–7500. The material is summarized in Table 15.2. We shall refer to it in the sections that follow.

15.2 Planetary and Diffuse Nebulae: The Hydrogen Lines

The planetary nebulae may be the simplest of the emission regions to understand and analyze, so we shall treat them first. Figure 15.1 shows the most famous of the planetaries, the *Ring Nebula* in Lyra.

It is currently believed that planetaries are ejected gaseous shells from red giants,

Table 15.2. *Index of False Spectra*[a] *of Emission-Line Regions*

Designation	Object	Remarks
M1-44	Planetary nebula	Low temperature
NGC 6833	Planetary nebula	High temperature
NGC 6720	Planetary nebula	M 57, Ring Nebula
K648	Planetary nebula	Population II planetary
M1	Supernovae remnant	Crab Nebula, helium-rich filament
Cygnus loop	Supernova remnant	High temperature
Cygnus loop	Supernova remnant	Lower temperature
DQ Her	Nova ejecta	Nova Herculis 1934
Orion Nebula	Diffuse nebula	Associated with GMC

[a]See Figures 15.9–15.17

which have since evolved into hot, subluminous stars. Figure 15.2, taken from Osterbrock (1989), shows a schematic evolutionary track of a star similar to the sun. Prior to the ejection of the nebular shell, the progenitor stars have been red giants twice – once with hydrogen-burning shells, and a second time with both hydrogen- and helium-burning shells. Stars in the latter stage are said to be on the asymptotic red giant branch.

The nature of the central stars of planetary nebulae – surface temperatures and luminosities – is reasonably well known. The analysis, to be outlined below, tells us that the ejected gas is typically some tenths of a solar mass. Spectroscopic observations indicate that it is expanding at roughly the speed of sound ($\approx$ 20 km/sec). From this expansion, it is straightforward to calculate that these objects will drop below typical detection limits in a time of the order of a few tens of thousands of years.

Ultraviolet photons from the central stars supply the energy to the nebulae. Only photons with frequencies beyond the Lyman limit ($\nu_0 = 3.3 \times 10^{15}$ sec^{-1}) are energetic enough to ionize the hydrogen in the ground state. With only a few exceptions, other photons from the central star pass through the nebular gas unimpeded. The standard analyses of such objects are based on the work of Herman Zanstra, and Donald Menzel and his coworkers. References may be found in the book by Osterbrock cited above.

Planetary nebulae are by no means simple, but they are surely less complicated than the heterogeneous array of objects known under the general label of diffuse nebulae. The latter may be associated with the star-forming regions which also contain the giant molecular clouds discussed in the previous chapter. In this case, the emission regions are powered by the ultraviolet photons from massive young stars rather than the highly evolved single object at the center of planetaries.

Many smaller diffuse emission regions are undoubtedly the remnants of giant clouds that were disrupted by radiation and gas pressures involved in star-formation processes. Massive stars formed within these regions may run through their lifetimes

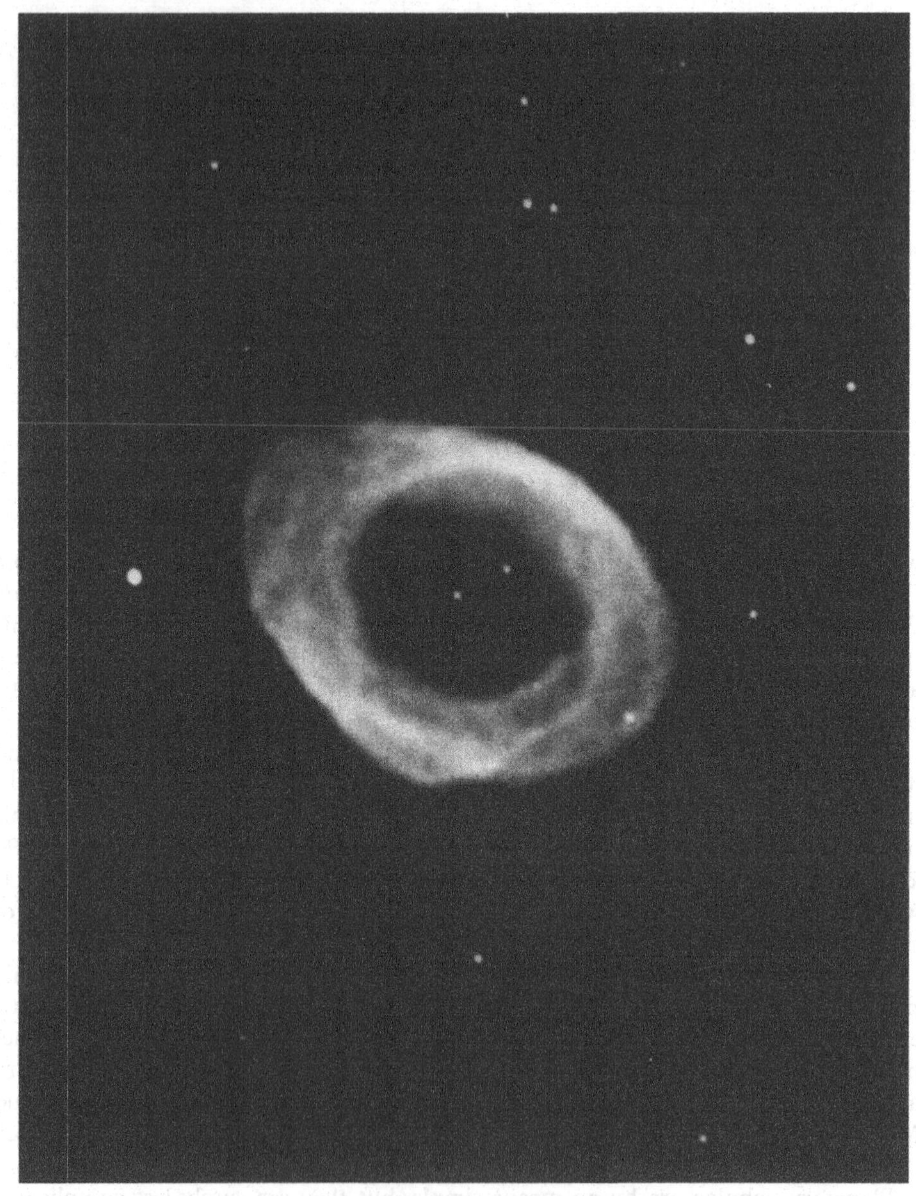

Figure 15.1. The Ring Nebula in Lyra. Courtesy California Institute of Technology.

and explode as supernovae, further disrupting the main complex and leaving smaller clouds to capture the ultraviolet photons of young stars.

The spectra and physical conditions within emission regions that are powered by ultraviolet photons are generally similar. The planetaries are somewhat hotter and denser than typical emission nebulae. Other bright ionized regions are powered

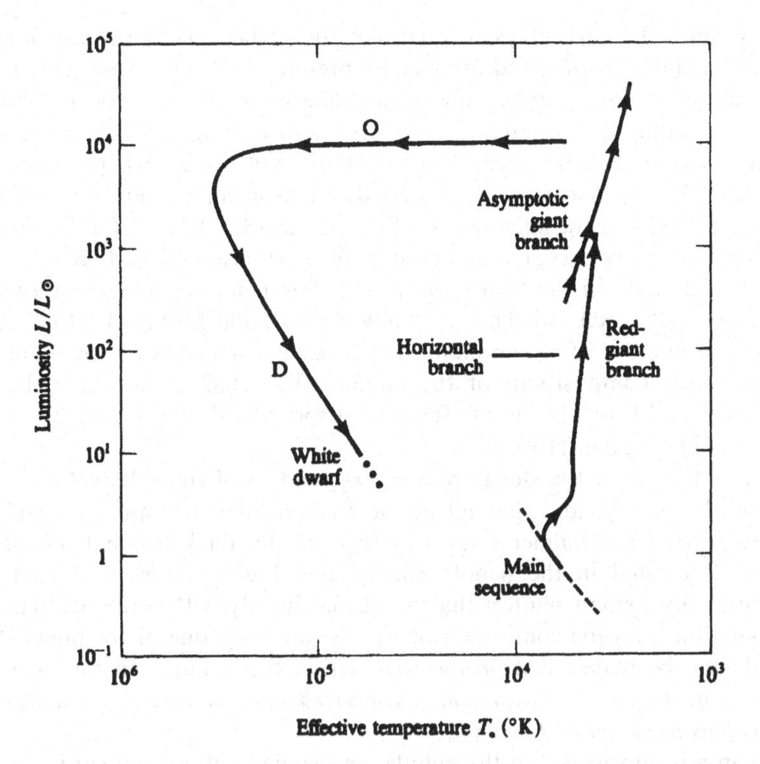

Figure 15.2. Central Stars of Planetary Nebulae. Courtesy of D. E. Osterbrock. Reprinted with permission from University Science Books.

by shock waves generated by a supernova explosion. One of the best examples of such *supernova remnants* is the Cygnus Loop. Another well-known supernova remnant, the Crab Nebula, is morphologically very different from the Cygnus Loop. The electron temperature within the Cygnus Loop is considerably higher than that of planetary and H II regions associated with young stars. Abundances in these remnants can depart markedly from the solar pattern.

Consider a planetary nebula and its central star. In a stationary state, the number of ultraviolet photons caught by the nebula per second must equal the number of recombinations per second. Let us make the simplifying assumption that the nebula is pure hydrogen, and that *all* of the ultraviolet photons are absorbed by it. Then if we write the luminosity of the star as L_ν [ergs/sec/Hz], we have

$$\int_{\nu_0}^{\infty} (L_\nu/h\nu)\, d\nu = \int_V N_i N_e \langle \sigma v \rangle d^3x, \tag{15.1}$$

where N_i and N_e are the ion and electron number densities. The integral on the right is over the emitting volume, V. The integrand contains the velocity averaged product of the *recombination cross section*, σ, and the relative velocity v of the proton and electron. This average, of course, is carried out over a Maxwellian velocity distribution (see 9.14). The recombination cross sections for hydrogen have been known since the early days of quantum mechanics. They do not, however, have simple functional forms, and are usually presented in tables (see below).

The nomenclature for hydrogen line transitions is as follows. The Lyman lines are transitions ending (or beginning) on the ground level $n = 1$. The lines within the Lyman series are distinguished by the symbols α, β, γ, etc., where α means the transition from $2 \to 1$, β from $3 \to 1$, and so on. Still higher series members are distinguished by the n of the *upper* level. Thus, we have Lα, Lβ, $\cdots$, Ln. Balmer lines are designated similarly except that the symbol 'H' is usually used. So Hα is the transition $3 \to 2$, and Hn the transition $n \to 2$. Paschen lines are transitions with lower level $n = 3$ (Pα, etc.), Brackett, with lower $n = 4$, and Pfund with lower $n = 5$.

In detail, the number of recombinations per second depends on the state (temperature, pressure, composition) of the nebula. We shall outline here the most important case, and leave deviations from it to specialized works such as those of Osterbrock (1989) or Aller (1984).

We assume that *all* of the stellar photons beyond the Lyman limit ($\nu > \nu_0$) are absorbed by the nebula, and that *all* of the Lyman lines are optically thick. We further assume that the Balmer lines are *not* optically thick, so that any Balmer photon that is created in the nebula can escape directly from it. Under these circumstances, any Lyman photon that is created locally within the nebula, either by recombination from the continuum or by cascade from one of the higher levels, will immediately be reabsorbed *on the spot* within the nebula. In this way, *each stellar photon in the Lyman continuum is converted into Balmer, Paschen, Brackett, etc. photons plus a photon of Lyman* α.

The Lyman α is imprisoned in the nebula, and slowly diffuses outward. These Lα photons are not directly observable in planetaries, because the general interstellar medium is optically thick at this wavelength. However, Zanstra realized that we can count the number of these photons created by the Lyman continuum of the central star. *Every emitted Lα photon must have been preceded by the emission of a photon from the Balmer series or Balmer continuum.* We assume the latter can escape the nebula freely, and be observed. In this way, we can count the number of Lα photons that leave the central star by counting the number of Balmer plus Balmer continuum photons leaving the nebula.

The situation described here is known to astronomers as Baker and Menzel's (1938) "Case B". Their paper, along with a number of the classical works on gaseous nebulae and related physical processes, was reprinted in a Dover publication edited by Menzel (1962). Case A occurs when the Lyman lines and continuum are also optically thin. It is of far less practical importance, though as many as half of the planetaries may not be completely optically thick to photons in the higher Lyman lines.

For Case B, we need to *add* the recombinations to all levels of the hydrogen atom *except* the ground state. The net result of a rather sophisticated calculation may be written in such a way that the number of recombinations per cm^3 per sec is given by the product $N_i N_e \cdot \alpha_B$. The *recombination coefficient* α_B is equal to the appropriately weighted average of σv that appears in Equation 15.1, accounting for recombinations on all levels *except* $n = 1$. Table 15.3 gives a few numerical values for hydrogen, and neutral and first ionized helium.

Table 15.3. *Recombination Coefficients for Hydrogen and Helium*

	α_B ($\times 10^{13}$ cm^3 sec^{-1})		
ion	5 000 K	10 000 K	20 000 K
H^+	4.54	2.59	2.52
He^+	4.34	2.73	1.55
He^{2+}	25.6	15.4	9.08

The populations of the individual levels of atoms in a gaseous nebula cannot be obtained from the Boltzmann formula (Equation 12.29), since nebulae are typically far from local thermodynamic equilibrium (LTE). Instead, it is necessary to work out the relative level populations on the basis of an assumed statistically steady state. For each level, we assume that the *sum* of all rates that populate the level is balanced by the *sum* of all rates that depopulate the level. In strict *thermodynamic equilibrium* the *individual* rates are assumed to balance – this is called detailed balancing.

Emission nebulae were the first astrophysical situations in which the concept of level population balancing or kinetic equilibrium routinely replaced the earlier statistical mechanical methods which we reviewed briefly in Chapter 4. In stellar atmospheres, the techniques are often referred to as *non-LTE* methods. We shall illustrate the procedure in a simplified situation in a moment. For the present, we assert that the postulate of a statistically steady state will allow us to calculate the relative populations of the hydrogen atom. The electrons may be safely assumed to obey a Maxwell–Boltzmann distribution with an *electron temperature* T_e. Departures from LTE are traditionally expressed in terms of parameters called b_n's ("b sub n's") introduced by Menzel and his coworkers. With their aid, one may write the combined Boltzmann–Saha equation for hydrogen in the form

$$\frac{N_n}{N_i N_e} = b_n \cdot \frac{\varpi_n}{2} \cdot \left(\frac{h^2}{2\pi mkT_e}\right)^{3/2} \cdot \exp\left[(\chi_i - \chi_n)/kT_e\right]. \tag{15.2}$$

Use of the symbol ϖ_n for the statistical weight of an atomic level 'n' is typical by workers in the field of emission regions, although Osterbrock (1989) uses ω_n. The symbol 'g' is used throughout most of stellar and atomic spectroscopy (cf. Cowan 1981) for the statistical weight. For most atomic levels, g, or ϖ, is $2J+1$ (see §11.2).

The b_n's, sometimes called departure coefficients, allow one to use the Boltzmann and Saha formulae in their familiar forms. It might have been more useful to define corrections to the temperature, since as we have seen in Chapter 4, the most probable distribution of atomic states with the two classical side conditions is an exponential one (see §4.9). However, that was not done, and the b_n-notation is firmly entrenched.

Menzel and his coworkers did not discriminate among the different l-states of hydrogen for a given n. Modern computing power makes such simplifications unnecessary, so that it is now proper to calculate and use b_{nl}'s (e.g., Brocklehurst,

1971). The departure coefficients themselves are rarely tabulated in modern work, although one can frequently find plots in the literature, especially in discussions of stellar atmospheres and chromospheres, where sophisticated versions of the methods developed in the study of gaseous nebulae are often applied (non-LTE, cf. Mihalas 1978, §7-5).

For nebulae, it is more useful to tabulate quantities related to observations. In the Case B approximation that we use here for planetaries, the Lyman lines are assumed to be unobservable. Lyman α itself is absorbed in interstellar space, and the higher Lyman members are converted to members of other series, primarily the Balmer series, plus Lα. Thus, in this case, the one of the most practical importance, it is the ratio of the Balmer and higher series lines that are of interest.

It is rather typical to normalize intensities to Hβ. At any given point in a planetary nebula, the ratio of the emission in Hn to Hβ is simply given by the ratios of the level populations, the appropriate Einstein A's, and the photon energies, $h\nu$:

$$\frac{j(n)}{j(\beta)} = \frac{N_n A_{n2} \cdot h\nu_{n2}}{N_3 A_{32} \cdot h\nu_{32}}. \tag{15.3}$$

Moreover, if we assume uniform physical conditions, this should represent the ratio of Hn to Hβ for the entire nebula.

A similar formula will be valid for *any other atomic line for which we can make the assumption that the optical thickness is negligible.* Therefore, if we have a way of connecting the number of atoms in the level n to the total number of atoms of a given element, we can obtain information on the relative abundances of the elements to which the lines belong. We therefore turn to a discussion of the physical conditions in emission regions that will enable us to relate the population of a single atomic level to that of an elemental species. We turn first to temperature and density indicators of emission regions.

15.3 Electron Temperatures and Densities in Emission Regions

Menzel, Aller, and Hebb (1941) used the relative intensities of three forbidden oxygen ([O III]) lines to determine electron temperatures in some 30 planetary nebulae. The transitions and the relevant structure of O III are shown in Figure 15.3. The figure also shows the analogous N II structure and transitions that can be used in T_e determinations. The three low levels all belong to the p^2 configuration, and transitions among them are forbidden by the Laporte parity rule (§11.2). Certain of the lines violate other electric dipole selection rules as well (Problem 3 below).

Fortunately, sufficient atomic data are available to allow calculation of all of the important collisional as well as radiative rates that contribute to the population of these low levels in O III.

Let us take the simplest situation, which occurs in the limit of low electron densities. Then the number of downward transitions from the highest level, the ^1S to the ^1D, is given by the rate at which the ^1S level is populated by electron collisions from the (ground) ^3P levels. All other rates *into* the ^1S are much smaller, and rates *out* of the ^1S *other* than the radiative decays are also small.

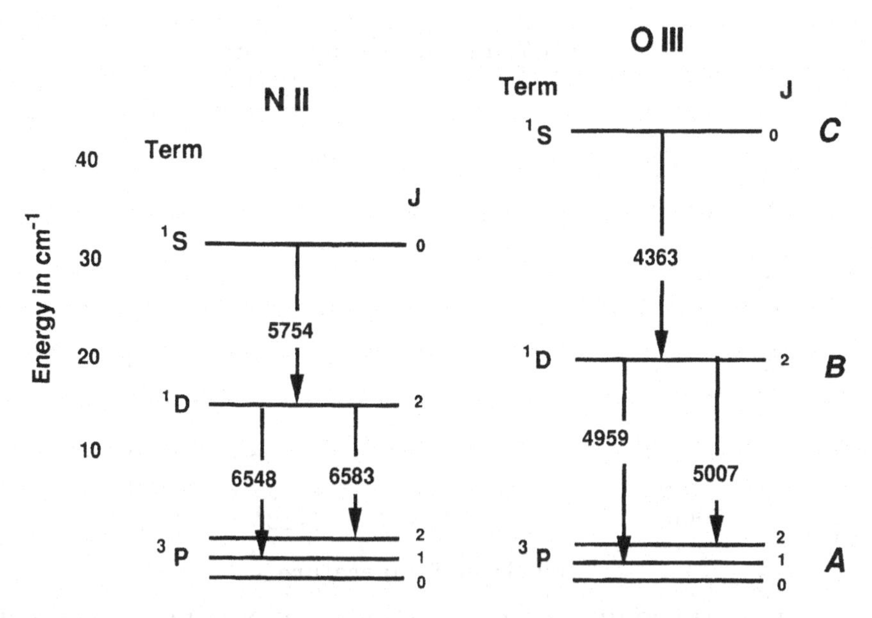

Figure 15.3. The [N II] and [O III] Lines Used for T_e Determinations.

This rate, per O^{2+} ion in the ground term, may be written $N_e\langle\sigma v\rangle$. Here, σ is the appropriate cross section for the excitation, v the relative velocity, and the brackets indicate an average over the electron's Maxwellian velocity distribution – a function of the electron temperature, T_e.

These averages, $\langle\sigma v\rangle$, are usually designated q_{AB}, where the subscripts stand for the appropriate levels. In the present case we group the three levels of the ^3P term together, and designate them A (see Figure 15.3). The q's are a function of temperature, and atomic constants, including the so-called *collision strengths* Ω, which conceal all of the sophisticated quantum mechanics of the cross section calculations. They are tabulated, for example, by Mendoza (1983; see also Aller 1984, or Osterbrock 1989) for most transitions of interest. If we lump all of the numerical and physical constants into one, $q_0 = 8.63 \cdot 10^{-6}$ (cgs), we have (see Problem 2 below)

$$\langle\sigma v\rangle \equiv q_{AB} = q_0 \frac{\Omega(A,B)}{\varpi_A} \frac{1}{\sqrt{T_e}} \cdot \exp\left(-\chi_{AB}/kT_e\right), \tag{15.4}$$

for collisional excitations $A \rightarrow B$. The rate for collisional de-excitations is similar, except that ϖ_A is replaced by the statistical weight of the *upper* level, ϖ_B, and the exponential term does not appear.

For the ^1D level, again in the simplest approximation, the important rates are collisional excitation from the ^3P and radiative decay downward in the $\lambda\lambda$4959 and 5007 lines. The former are given by an equation precisely analogous to 15.4 above.

In order to get the ratio of the energy in the two [O III] lines to λ4363, we multiply each transition rate by the energy in the photons. The latter are *inversely*

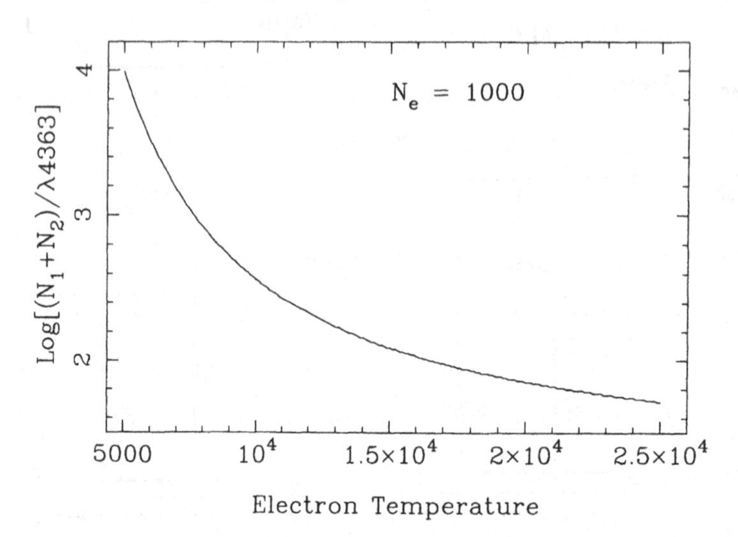

Figure 15.4. [O III] Line ratios as a function of T_e. N_1 and N_2 are the "nebulium" lines of [O III]. The relevant transitions are shown in Figure 15.3. A constant electron density of 10^3 cm^{-3} was assumed in this calculation.

proportional to the relevant wavelengths. Thus, we have in this approximation,

$$\frac{I\,(N_1+N_2)}{I(\lambda 4363)} = \frac{\Omega(A,B)}{\Omega(A,C)} \cdot \exp\left[(-\chi_{AB}+\chi_{AC})/kT_e\right] \frac{\lambda 4363}{\lambda\,(N_1+N_2)}, \tag{15.5}$$

where by $\lambda(N_1+N_2)$ we mean an appropriately weighted wavelength for $\lambda\lambda$4959 and 5007.

It is easy to see that the temperature of the emitting region may be obtained from Equation 15.5, given a measurement of the intensity ratio of N_1+N_2 and λ4363.

In a more sophisticated treatment, easily done even on a modest personal computer, one calculates the ratio of the populations of the upper levels N_B and N_C, taking into account various processes that we have neglected in the above heuristic treatment. This ratio again depends on the electron temperature, although in a more complicated way than illustrated by Equation 15.5. For higher electron densities, the (N_1+N_2)-to-λ4363 ratio depends on N_e as well as T_e. This dependence is taken into account by the more sophisticated treatment, but the interpretation of the line ratio is somewhat more complicated when N_e is important.

Figure 15.4 shows the dependence of the (N_1+N_2)-to-λ4363 ratio as a function of temperature for an assumed electron density of 10^3 cm^{-3}.

In addition to O III and N II, Ne V, S III, and Ar V have s^2p^2 configurations, and therefore have similar forbidden lines that could in principle be used as temperature indicators. In some cases, the relevant lines lie in the satellite ultraviolet or infrared, and their study is a challenge to the modern astronomer whose observations are no longer restricted to the traditional optical domain.

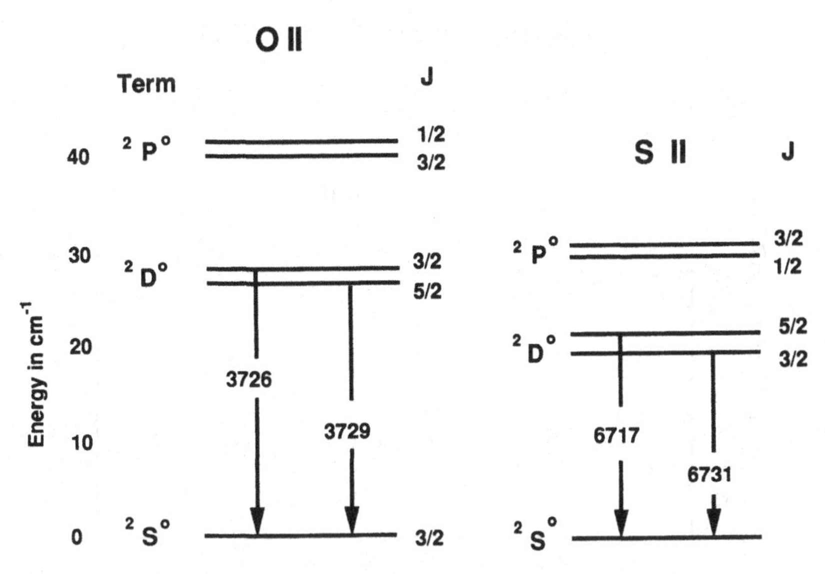

Figure 15.5. Levels and Transitions Used to Determine N_e.

Electron temperatures for planetary (and other) nebulae may also be determined from measurements of the radio continuum. The principles are the same as those discussed in §14.2, and may be based on the slab solution (Equation 12.9) in the limit at $\tau_B \to \infty$. Then, the observed intensity is the same as the source function in (the surface of) the slab. This may be translated into the language of the radio astronomer with the help of the Rayleigh–Jeans approximation, according to which the Planck function $B_\nu \propto T$. Then, assuming that the antenna beam is filled by the source, and instrumental losses can be neglected, the observed or *antenna temperature* is equal to the electron temperature in the observed, optically thick source.

Radio recombination lines (e.g., H109α) can also be used to determine electron temperatures. Since the technique involves use of the theory of continuous emission, and corrections for departures in the level populations from LTE, we must refer to Gordon (1988) and the references therein for details.

Electron *densities* may be determined by observing the ratios of lines of [O II] or [S II]. [Cl III] and [Ar IV] may also be useful. The relevant structures in O II and S II are shown in Figure 15.5. Note the order of the J values in the $^2D^0$ and $^2P^0$ terms of O II and S II. The intensity ratios as a function of $\log(N_e)$ are given in Figure 15.6. These ratios depend somewhat on T_e as well as N_e. The S II ratios are illustrated for 10^4 K.

Consider the [S II] lines. In the limit as the density becomes very high, the level populations will approach that given by the Boltzmann distribution 12.29. Because the upper levels are close to one another, the *difference* in the Boltzmann factors $\exp(-\chi_n/kT)$ will be small. The population of the upper levels giving rise to the $\lambda\lambda$6716 and 6731 lines will be proportional to the statistical weight factors (called g_n in Chapter 12, but ϖ_n by emission-line workers). The expected ratios of the two

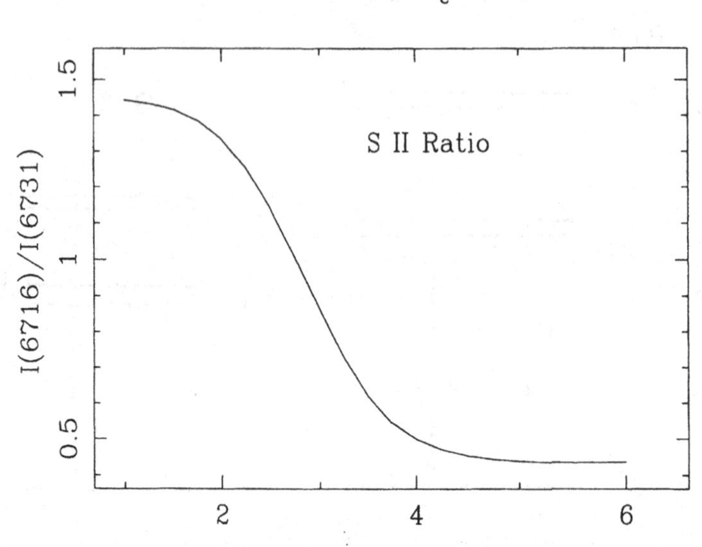

Figure 15.6. Density-Sensitive Line Ratios as a Function of N_e for 10^4K.

lines are therefore

$$\frac{I(6716)}{I(6731)} = \frac{6}{4} \cdot \frac{A(6716)}{A(6731)} \cdot \frac{6731}{6716}, \tag{15.6}$$

The last factor accounts for the ratio of the energies in the photons, which is proportional to the frequency ratio, and thus inversely proportional to the wavelength ratio.

The ratio of the Einstein A's according to Mendoza and Zeippen (1982, cf. Osterbrock 1989) is 0.000 26/0.000 88, so that the expected ratio is 0.44, as is shown in Figure 15.6. In the low-density limits, the intensity ratio is essentially set by the rate of upward collisions; this is the situation we considered in our simplified treatment of the $(N_1 + N_2)/\lambda 4363$ ratio of [O III]. The relevant upward rates are calculated with the help of equations of the form 15.4 above. Because of the similarity in the energies of the two ^2D levels, the temperatures very nearly cancel, and the upward rate ratio is just the ratio of the respective collision strengths, Ω. In this case it is precisely the ratio of the statistical weights of the two upper levels. So at the low-density limit,

$$\frac{I(6716)}{I(6731)} = \frac{6}{4} \cdot \frac{6731}{6716} \simeq 1.5, \tag{15.7}$$

as shown in Figure 15.6. This figure was actually made with a more sophisticated calculation that includes all relevant rates among the five levels shown in Figure 15.5.

Electron densities for emission regions may be estimated in a variety of additional ways. Transitions in C III between the $2s^2\,{}^1S$ and $2s2p\,{}^3P^o$ at $\lambda\lambda 1907$ and 1909 are useful when high-resolution spectra are available from satellites. Intersystem ($\Delta S \neq 0$) transitions are sometimes designated by a single bracket, e.g., C III] to indicate a "semi- or spin-forbidden" transition (see §11.2). Of the two lines, $\lambda 1907$ is highly forbidden, since it involves a $\Delta J = 2$ jump in addition to $\Delta S \neq 0$. The $\lambda 1909$ feature is useful in setting an upper limit to the electron densities in the so-called broad region of active galactic nuclei (see below). The absolute intensities of optically thin lines or radio continua may also be used to estimate the densities in emission regions whose size and distance are known. The latter are typically rather poorly known so that densities that depend on them are used as consistency checks on values derived from line ratios.

The early analyses of emission-line regions were made under the assumption of constant and uniform T_e and N_e. Modern work allows for both temperature and density fluctuations, but the determination of these fluctuations is a matter of considerable uncertainty and debate.

The most sophisticated treatments of emission regions now use the method of modeling, which is similar in principle to the technique used to study stellar atmospheres. A model includes an energy input mechanism, from exciting stars, accretion disks surrounding black holes, or fast particles such as cosmic rays. There may be density and temperature fluctuations and molecular and dust inclusions, etc. On the basis of these assumed properties, the emitted spectrum is calculated (hopefully) with all relevant physical processes accounted for.

Early modeling of this kind was done by MacAlpine (1971) and Davidson (1972) for active galactic nuclei. More recent efforts are incorporated in a 16000 FORTRAN statement code by Ferland (1991). As of the present writing, the code is in a state of continuous revision.

While modeling has the advantage of being able to treat such matters as density or temperature fluctuations explicitly, it suffers from the standard disadvantage of this procedure; the question of uniqueness of the solution is unresolved. How do we know some other model won't fit the observations equally well? Throughout much of the history of the analysis of emission regions, it has been assumed that the derived abundances are not sensitive to the model uncertainties. Such questions can now be probed with the help, for example, of the Ferland code.

15.4 Determination of Abundances

When the modeling technique is used, the determination of abundances is straightforward, at least in principle: when the model satisfactorily fits the observations, the abundances are those of the model. Much of the analysis of emission-line regions is still done with the help of older techniques, which we shall now describe. Many workers in the field now think that these methods should only be regarded as provisional, with definitive results coming from detailed models.

Let us again use the simplest approximations, and assume the emitting region is uniform in temperature, density, and chemical composition. Single values of the electron temperature and density may then be determined using the methods of the preceding section.

Consider first the determination of the critically important helium-to-hydrogen ratio. Lines of both of these elements arise from *recombinations* whose theory is generally well understood. The $H\beta$ line is usually used as a reference, and the intensities of other lines given as ratios to $H\beta$. The recombinations per second which lead to the emission of an $H\beta$ photon may be written $N_p N_e \alpha^{\mathrm{eff}}_{\mathrm{H}\beta}$. This $\alpha^{\mathrm{eff}}_{\mathrm{H}\beta}$, which is tabulated, for example, by Osterbrock (1989), is very nearly equal to the recombination coefficient for the $n = 4$ level ($H\beta$ is $4 \to 2$). The small fraction of Paschen-α transitions ($4 \to 3$) is nearly compensated by downward cascades from $n > 4$ to $n = 4$.

We use N_p for the number of ionized hydrogen atoms per cm^3 rather than the usual N_i (cf. Equation 15.2), because we shall need to discuss several species of ions. Typically, N_i is only slightly larger than N_p because hydrogen is so much more abundant than the other elements. The volume emission coefficient j_β ergs cm^{-3} steradians^{-1} for the entire $H\beta$ line may now be written

$$j_\beta = \frac{1}{4\pi} N_p N_e \alpha^{\mathrm{eff}}_{\mathrm{H}\beta} \cdot h\nu_\beta. \tag{15.8}$$

The $\alpha^{\mathrm{eff}}_{\mathrm{H}\beta}$'s have only a slight dependence on electron density, which may be taken into account.

For He, recombination theory leads to an analogous recombination coefficient for the $\lambda 5876$ line. Thus the *ratio* of the intensities of He I $\lambda 5876$ to $H\beta$ is

$$\frac{I(\lambda 5876)}{I(\mathrm{H}\beta)} = \frac{j(\lambda 5876)}{j(\mathrm{H}\beta)} = \frac{N_{\mathrm{He}^+}}{N_p} \cdot \frac{\alpha^{\mathrm{eff}}_{\lambda 5876}}{\alpha^{\mathrm{eff}}_{\mathrm{H}\beta}} \cdot \frac{4861}{5876}. \tag{15.9}$$

We can extract the $\mathrm{He}^+/\mathrm{H}^+$ ratio from the observed intensity ratio. In most emission regions, hydrogen is virtually completely ionized, so we have the He^+-to-hydrogen ratio. The He I intensity ratios can be modified from those of pure recombination theory by collisional excitations, especially from the metastable 2^3S level. This is discussed by Clegg (1987) and Osterbrock (1989).

Recombination theory for He II leads in a similar way to the He^{2+} to hydrogen ratio.

It is now important to recognize that for recombination lines, observations of a given stage of ionization yield information on the number density for the next higher stage of ionization. In the case of helium, observations of the He I and He II lines provide information on He^+ and He^{2+}, but not on neutral helium.

A standard practice in gaseous nebulae is to correct for an unobserved ionization stage with the help of observations of lines from ions with similar ionization potentials. For example, the ionization energy of neutral helium (He^{0+}) is 24.6 eV, while that of singly ionized sulfur is 23.3. As a very rough approximation, then, we might assume

$$\frac{\mathrm{He}^{0+}}{\mathrm{He}^+} \simeq \frac{\mathrm{S}^+}{\mathrm{S}^{2+}}, \tag{15.10}$$

and thus a determination of the number ratio of $\mathrm{S}^+/\mathrm{S}^{2+}$ could allow us to correct for the fraction of neutral helium.

We can now understand why *high-excitation* emission regions are useful for the primordial helium abundance (§10.7). When most of the helium is ionized, the uncertain correction for the neutral species will be small.

The emission coefficient (per steradian) for a forbidden transition such as [S II] $\lambda6716$ is given by $N(^2D_{5/2})A(\lambda6716)h\nu(\lambda6716)/(4\pi)$. In the low-density limit the number of downward ($^2D_{5/2} \rightarrow {}^4S$) transitions is the same as the number of collisional excitations ($^4S \rightarrow {}^2D_{5/2}$). Thus

$$j(\lambda6716) = \left(\frac{1}{4\pi}\right) N_e N\left({}^4S\right) q_{^4S\rightarrow{}^2D_{5/2}} h\nu(\lambda6716). \tag{15.11}$$

In this case,

$$\frac{I(\lambda6716)}{I(\mathrm{H}\beta)} = \frac{N\left({}^4S\right) q_{^4S\rightarrow{}^2D_{5/2}} h\nu(\lambda6716)}{N_p \alpha^{\mathrm{eff}}_{H\beta} h\nu(\lambda4861)}, \tag{15.12}$$

and we may readily extract the ratio of $N(^4S)/N_p$, which is essentially the $N(\mathrm{S}^+)$-to-hydrogen ratio. For higher electron densities, a solution of the (non-LTE) equations of statistical equilibrium provides the number densities of any of the S II levels in terms of any of the others. For example, if $N(^2D_{5/2})/N(^4S) = x(T_e, N_e)$, then

$$\frac{I(\lambda6716)}{I(\mathrm{H}\beta)} = \frac{x\left(T_e, N_e\right) N\left({}^4S\right) A(\lambda6716)h\nu(\lambda6716)}{N_p \alpha^{\mathrm{eff}}_{H\beta} h\nu(\lambda4861)}, \tag{15.13}$$

and we can again get $N(^4S)/N_p \approx N(S^+)/N(\mathrm{Hydrogen})$.

These general principles illustrate the methods by which relative abundances, element-to-hydrogen ratios, are extracted from emission lines.

Even the earliest workers realized planetary nebulae could not be of uniform temperature and density. The most energetic ultraviolet photons, capable of removing the second electron from helium, are exhausted relatively near the source of ionizing radiation, e.g., the central star of a planetary nebula. Images of planetary nebulae obtained with *slitless spectrographs* or suitable optical filters show that the emission regions vary inversely in spatial extent with the relevant ionization energies. Modern optical and radio imaging of planetaries reveals complicated structures that vary from line to line (see Bignell 1983, Icke, Preston, and Balick 1989).

The nearby emission regions (see Figure 15.1) may be resolved spatially, and *average* values (along the line of sight) of T_e and N_e may be determined as a function of position on the sky. Similarly, *average* abundances may be determined as a function of position. Variations in the conditions along the line of sight cannot, of course, be directly observed, and assumptions are necessary if any allowance for them is to be made. Similar remarks apply to variations in the conditions perpendicular to the line of sight, but on a physical scale that is *unresolved* by the observations.

Emission-line regions in extragalactic sources have been the subject of large research efforts in the last decades. For these sources, it is typically the case that the regions are spatially unresolved, and weak lines, such as the critical $\lambda4363$ of [O III], are not available. For diffuse nebulae and unresolved planetaries, the temperature of the exciting stars cannot be obtained by the well-established procedures used in the study of planetary nebulae. Workers have devised a variety of semiempirical techniques to deal with such difficulties. Some details may be found in the article by Pagel (1990).

Inhomogeneities may be taken into account explicitly in detailed models. The problem is that one does not know the model! Unlike stellar atmospheres, in which the pressure–temperature structure is basically set by a one-dimensional equation of hydrostatic equilibrium, even the simpler planetary nebulae contain cold inclusions (including dust and molecules) whose structure is neither well understood nor readily predicted. Against these serious drawbacks is the single very great advantage that in most respects the emitting gas may be considered optically thin, and the very difficult problem of radiative transfer may be treated in its simplest approximation.

15.5 Abundances in Emission Regions

In what follows, we shall summarize the analytical abundance results for emission regions in our own and external galaxies. The interpretations of these abundances will be discussed in the following chapter. Aller (1989) has summarized abundance results for ejecta from highly evolved stars. His very useful tables give estimates of the rate of mass returned to the interstellar medium from Wolf–Rayet stars, planetary nebulae, and novae.

By and large, the abundances in gaseous nebulae resemble those of the sun and stars. This means most abundances resemble those of the sun itself. With a few exceptions, it is only when we look at relatively small variations – of a factor of two or three – that interesting and systematic variations occur. However, in view of the relatively short lifetimes of most emission regions, abundance variations related to populations are less marked than for stars. But population-related variations for nebulae are observed, as Figure 15.7 shows (Kaler 1985).

Figure 15.7 shows that the N/O ratio is correlated, although weakly, with the He/H ratio. The latter may be explained in terms of *in situ* stellar evolution rather than general chemical evolution of the interstellar gas. The solid, dotted, and dashed lines indicate predictions based on various evolutionary model calculations in which nuclear-processed material is *dredged up* by convection currents in pre-nebula phases of the stellar lifetimes.

Aller and Keyes (1987) surveyed 51 planetary nebulae for abundances. We show two of their histograms for the abundances of nitrogen and oxygen in Figure 15.8(a,b). A more recent discussion of the systematics of planetaries is by Amnuel (1993).

The abundance of carbon is particularly difficult to determine in emission regions. As can be seen from the figures in §15.6, there are no strong forbidden transitions in the traditional optical region. The permitted C II doublet λ4267 (Multiplet 6) has been used for abundances, but according to Osterbrock (1989, p. 154), their mode of excitation – recombination or resonance fluorescence – is uncertain. This makes them less useful than, for example, the lines that are known to arise primarily from electron collisional excitation. Carbon abundances are often determined from lines in the satellite ultraviolet (see e.g., Aller and Keyes 1987).

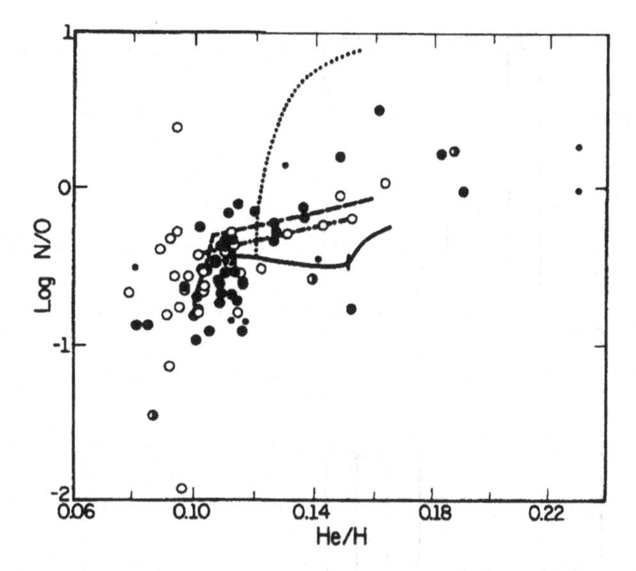

Figure 15.7. Population and Evolutionary Effects in Planetary Nebulae. Open circles on the diagram are planetaries that can be generally associated with (old) Population II stars. One can see that *statistically* the latter have smaller He/H ratios. Reproduced with permission from the *Annual Review of Astronomy and Astrophysics*, Vol. 23, ©1985 by Annual Reviews Inc. Courtesy J. B. Kaler.

The analytical results thus far show an enhancement of the carbon abundance in planetaries over that of the SAD by about a factor of two, while oxygen is depleted by roughly the same factor. In addition, nitrogen is nearly normal (SAD), although in some planetaries the N/O *ratio* can be much larger than solar, and an interpretation in terms of CNO burning is possible, though it is not easy (cf. Kaler and Jacoby 1990). Some workers have discussed the ejection of a carbon-rich shell (see below).

We can apply standard statistical tests to histograms such as those in Figure 15.8 to test against the *null* hypotheses that the distributions are normal (Gaussian) (and possibly arise from errors). We can use, for example, the procedure outlined by Burington and May (1970, see §14.30) to find estimates of the mean and standard deviations from the histograms themselves, and then do a χ^2-test of the null hypothesis. A variety of other tests could, of course, be used.

Only in the case of the oxygen abundances can the null hypothesis be rejected at higher than the 99.9% confidence level, and this is clearly because of the skewness of the data. The formal mean oxygen abundance for Figure 15.8(b) is 8.6, *below* the solar value, 8.9. The reason for this systematic displacement is not certain, although it may be due to temperature fluctuations in the sources themselves. It turns out that the oxygen abundance in *most* galactic and extragalactic emission regions is below the SAD.

A conservative conclusion from the material on planetaries presented thus far

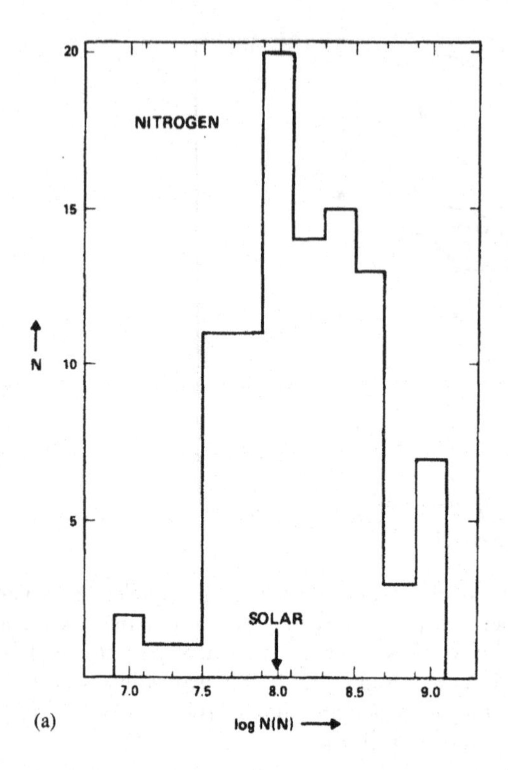

(a)

SPECTROSCOPIC SURVEY OF PLANETARY NEBULAE

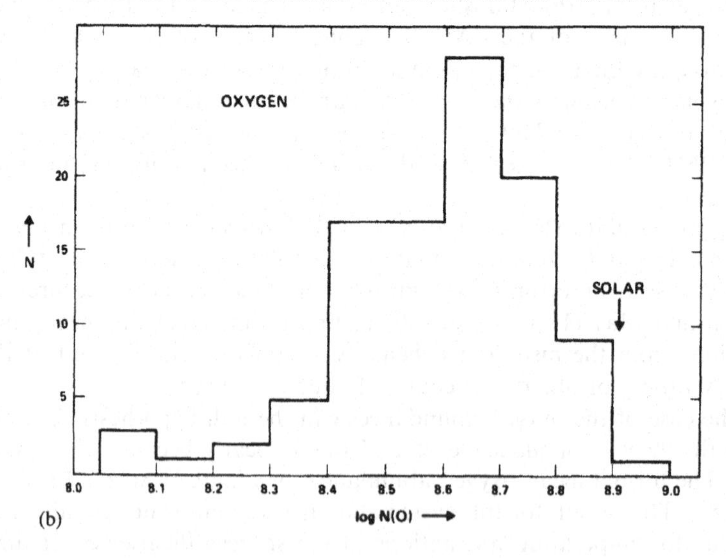

(b)

Figure 15.8. Abundance Histograms for Planetary Nebulae. Aller and Keyes (1987).

Table 15.4. *Abundances in Novae Ejecta*

Nova	Year	Mass Fraction					
		H	He	C	N	O	Ne
RR Pic	1925	0.53	0.43	0.0039	0.022	0.0058	0.011
HR Del	1967	0.45	0.48		0.027	0.047	0.0030
T Aur	1891	0.47	0.40		0.079	0.051	
V1500 Cyg	1975	0.49	0.21	0.070	0.075	0.13	0.023
V1668 Cyg	1978	0.45	0.23	0.047	0.14	0.13	0.0068
V693 CrA	1981	0.29	0.32	0.0046	0.080	0.12	0.17
DQ Her	1934	0.34	0.095	0.045	0.23	0.29	
V1370 Aql	1982	0.053	0.085	0.031	0.095	0.061	0.47

would be that the analytical errors are of the order of 0.3 dex (a factor of two), and that *most* of the scatter in the results (e.g., in Figure 15.8) may be attributed to various errors and uncertainties. Certainly, more elaborate interpretations of these and similar results have been given (see the references in Osterbrock 1989). The statistical test that we performed gives us no basis for assuming that the spread in the data is due to errors rather than a true *cosmic* abundance scatter. Indeed, there must be some cosmic scatter.

The situation is not unlike that which obtained some decades ago regarding stellar abundances. Abundance variations of a factor of two or less were not *generally* taken seriously, and only the much larger differences, associated with extreme Population II, were widely accepted.

Planetary nebulae are very rare in the quintessential Population II objects, the globular clusters (§13.1). Until recently, only one was known, K648, whose spectrum is illustrated in Figure 15.12 below. The designation is from a tabulation of direct image positions in M15, Küstner (1921). Pease (1928) discovered the true nature of the object. In a brief note to the *Publications of the Astronomical Society of the Pacific* he described two four-hour spectrograms taken with the Mount Wilson 100-inch reflector. "The continuous spectrum is O type; the bright lines, in order of their intensities, are N_1, Hβ, N_2, Hδ, Hϵ, 4471, 4363, 3888, and 3868" (see our Figure 15.12). He concluded that the object "...is therefore a bright planetary nebula with a relatively weak nucleus." The *stellar* metal abundances in the globular cluster are less than 1% of those of the sun (Sneden *et al.* 1991).

Adams, Seaton, and Howarth (1984) found nitrogen and oxygen abundances (element-to-hydrogen ratios) in K648 to be respectively 1/30 and 1/20 the solar (or SAD) values, while the carbon abundance was essentially solar. These abundances are larger than those of the stellar population, and the authors suggested the carbon abundance could be accounted for by the ejection of nuclear-processed material from the central star, in which case the planetary and the "typical" stars of M15 would not be expected to have similar abundances.

By definition, the young Population I stellar objects are found associated with

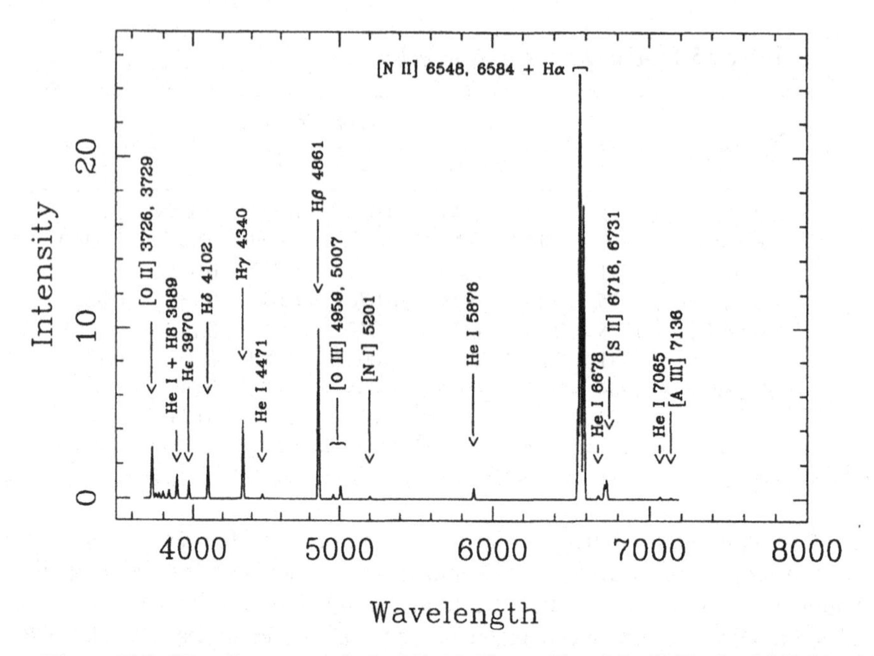

Figure 15.9. The planetary nebula M1-44 (Entry No. 4 in Table 1 of Minkowski 1946). Data from Aller and Keyes (1987). This is an unusual planetary, in that the nature of its central star is unclear. The Perek–Kohoutek (1967) finding chart shows a *nearly* stellar image. There is a G or K star within about 2 seconds of arc of the emission-line source, which Lutz and Kaler (1983) suggest may not be physically related to the nebula. We include the spectrum here because it is an example of a low-excitation source that makes a good contrast with the high-excitation planetary NGC 6833 illustrated in Figure 15.10. Note the relative weakness of the [O III] $\lambda\lambda 4959 + 5007$ ("nebulium") to [O II] $\lambda 3727$ as well as to Hα and Hβ. It is unusual in an emission-line region for [S II] $\lambda\lambda 6716, 6731$ to be comparable in strength to the [O III] lines.

the cold and hot (H II regions) gaseous material associated with star formation. It is therefore not surprising that abundance analyses of emission regions in our own galaxy generally indicate modest variations from solar abundances. Shaver *et al.* (1983) made an extensive study of abundances in nearly 70 galactic H II regions. They give both raw and analyzed materials in tabular form. Optical and radio observations were analyzed. This work is reported as an investigation of galactic *abundance gradients*, which will be discussed in the following chapter. This and similar work has been reviewed by Meyer (1989), and Pagel (1990).

Let us for the moment ignore the galactic gradients in the Shaver *et al.* oxygen and nitrogen abundances. If we then take the 22 sources for which total abundances were reported (without upper or lower limits), we find standard deviations of 0.3 dex (a factor of two) for both elements. This is very similar to the scatter found for planetary nebulae. The mean oxygen abundance is 0.2 dex below SAD, in general

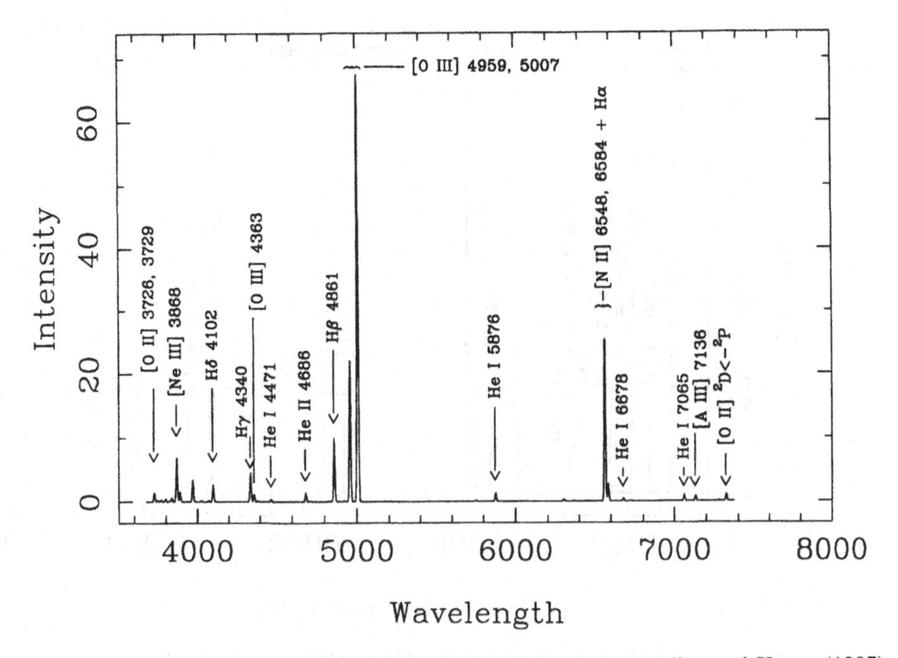

Figure 15.10. The planetary nebula NGC 6833. Data from Aller and Keyes (1987) who give an angular diameter of 2 seconds of arc. This is a *hot* planetary nebula. Notice that the [O II] λ3727 pair is *much* weaker than the [O III] lines $\lambda\lambda$4959, 5007. The auroral line λ4363 of [O III] is present, and in about the same strength as the [O II] λ3727 doublet. Weakness of the [S II] $\lambda\lambda$6716, 6731 doublet is characteristic of hot H II regions where the energy source is ultraviolet radiation from hot stars. The [S II] doublet can be strong in hot sources such as supernovae remnants (cf. the Cygnus Loop and Crab Nebula spectra below). He II λ4686 has been added from Vorontsov-Vel'yaminov *et al.* (1965).

agreement with the planetaries, but the *nitrogen mean is 0.4 dex below SAD*, a result which is not easy to understand, since any material that has been through the CNO cycle (massive stars) would have been enhanced in nitrogen (§10.2).

Shaver *et al.* (1983) show that the electron temperatures of the H II regions increase markedly with distance from the center of the Galaxy. This clearly defined trend is shown in their Figure 16. It is readily understood in terms of the thermal balance within emission regions. The temperature of these regions is determined by a balance between the radiative heating from the hot stars and cooling by radiation in the infrared. The latter is a strong function of the abundances of C, N, O, and heavier elements. When the abundances drop, other things being equal, we expect higher temperatures in H II regions. Shaver *et al.*'s result may be interpreted in terms of a decline in the abundances of H II regions with galactocentric distance. The effect is likely to be enhanced by a hardening of the ultraviolet spectrum of the

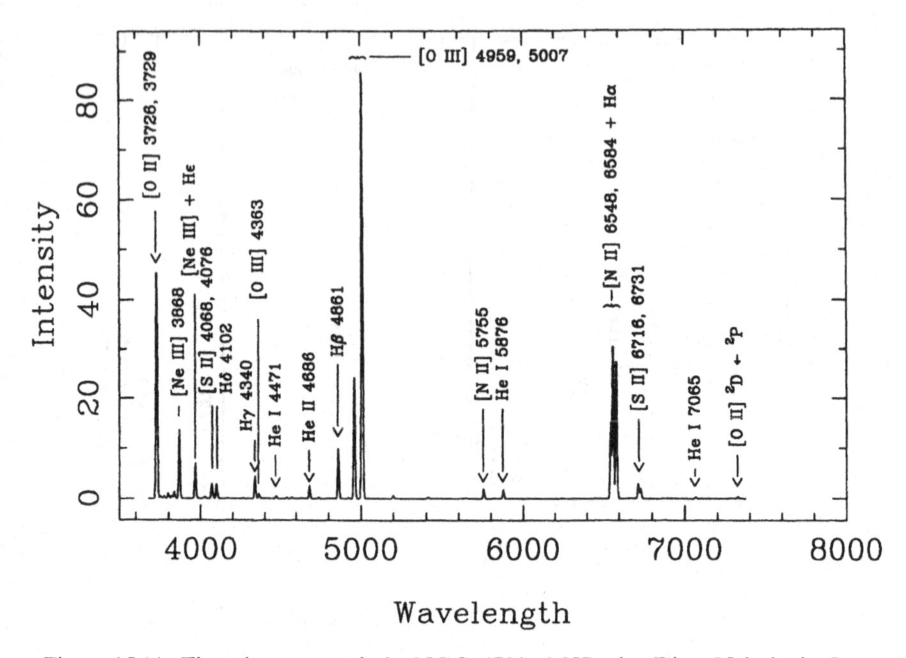

Figure 15.11. The planetary nebula NGC 6720, M57, the Ring Nebula in Lyra (Figure 15.1). The spectra shown are mostly *averages* from Kaler (1976) for this paradigm planetary. The [O II] pair $\lambda 3727$ shows substantial spatial variations over the nebula, and Kaler does not give an average. We have arbitrarily chosen the value from his Region 1. The [S II] pair is also variable, and not rigorously on the $H\beta = 100$ scale.

exciting stars with galactocentric distance. This occurs as a result of decreased line blanketing.

In external galaxies, H II regions generally reflect the underlying stellar abundances. These abundances, which will be discussed in more detail in Chapter 16, fall into two categories. First, the galaxies as a whole may be metal-poor with respect to our own ("the" Galaxy). Second, abundance *gradients* are observed, with the central regions, especially of spirals, being more metal-rich than those further out.

Emission regions have been used to estimate the *primordial* helium abundance. Kunth and Sargent (1983) observed high-temperature emission-line galaxies, for which the correction for the unmeasurable neutral helium would be small. They obtained helium-to-hydrogen ratios ranging down to 0.075 (mass fraction, $Y = 0.23$), which should be compared with 0.098 ($Y = 0.28$) in the SAD. In fact, no helium abundances lower than this have been observed anywhere, with the exception of a few early-type stars where helium could have diffused out of the atmosphere (§13.4.3). The ubiquitous presence of this "floor" of the helium abundance in all astronomical objects is the primary reason for the assumption of cosmological nucleosynthesis of the lightest nuclei (§10.7).

Let us pass now from the planetaries and large H II regions to discuss briefly

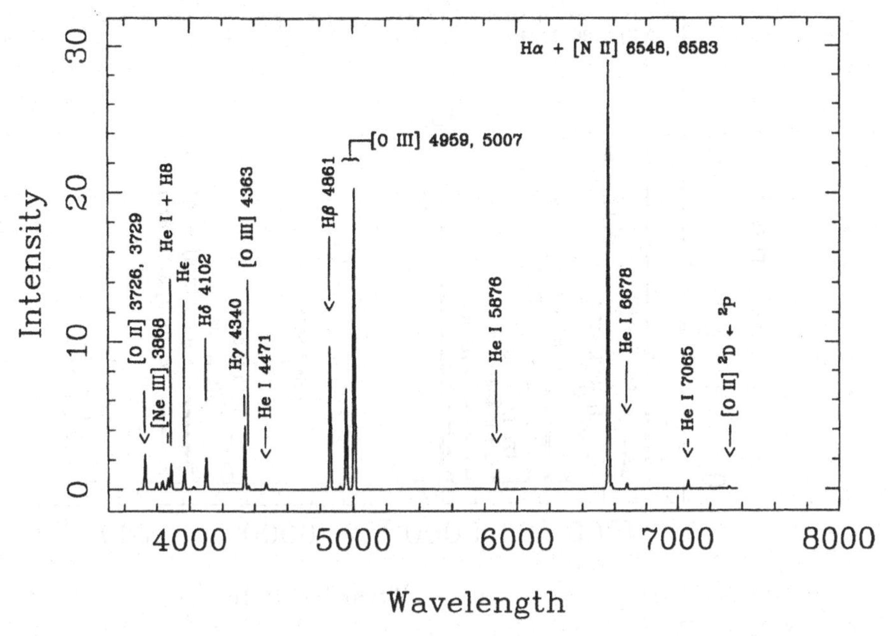

Figure 15.12. The planetary nebula K648. This object is in the globular cluster M15, and is thus a sure case of a Population II planetary nebula. The observations are from Adams *et al.* (1984). The reduced abundances in this source cause a lowered cooling rate that results in an overall temperature of about 10 600 K – somewhat hotter than that of typical planetaries. The high temperature results in higher collisional rates, with the result that the [O III] nebular lines are still strong.

supernovae remnants (SNR) and other specialized objects. Among these sources, we find evidence of large departures from the solar abundance pattern.

Some of the supernovae remnants are very oxygen-rich as revealed by analyses, for example, of Kirshner and Chevalier (1977). Fesen and his coworkers (see Fesen *et al.* 1988, Hamilton and Fesen 1988) discuss the ultraviolet absorption spectrum of the SNR associated with the supernova of AD 1006. Lines of Si II, S II, O I, and Fe II are identified and discussed as ejecta from a supernova of type Ia. The strongest lines are those of Fe II. Fesen, Hamilton, and Saken (1989) took direct images in the region of the supernova S Andromedae in the Galaxy M31. They used optical filters that isolated the Fe I resonance line at $\lambda3859$, and found an apparently iron-rich cloud. Danziger (1989) discusses abundances in SNR's, including SN 1987A, which exploded in the Large Magellanic Cloud.

Anomalous and heterogeneous abundances in the supernova remnant known as the Crab Nebula have been studied by MacAlpine and his coworkers (see MacAlpine *et al.* 1989 and references therein). Helium is generally enriched and highly variable across the face of the nebula. Some of the filaments have exceptionally strong lines of [Ni II] and [Fe II] suggesting abundance enhancements of an order of magnitude

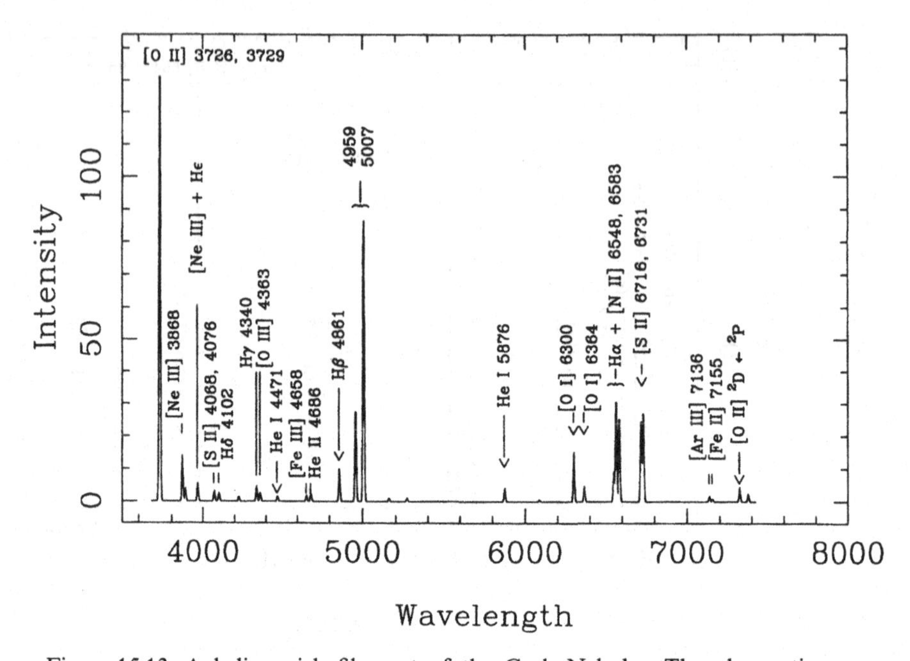

Figure 15.13. A helium-rich filament of the Crab Nebula. The observations are from Fesen and Kirshner (1982, Position 8) and MacAlpine *et al.* (1989). Note the relative weakness of Hα, and strength of the nearby [S II] doublet. The [S II] as well as the presence of [O I] λ6300 is characteristic of heating from radiation with a power-law spectrum (or shocks). The region chosen here is not *the* most helium-rich of the Crab filaments.

or more over those of the SAD. A variety of complications make the analysis of this remnant difficult.

The "game" of the observers of SNR's is to connect the observed abundances with predictions that can be based on models of exploding stars (supernovae) of various types. As discussed by Danziger (1989) or Aller (1989) these efforts have been only partially successful.

Aller (1989) recalls how in the early 1940's he and Payne-Gaposchkin applied the standard H II region analytical techniques to the chemical analysis of novae ejecta, finding definite abundance differences from the sun. They were then intimidated by critical remarks from a "distinguished spectroscopist," and their findings were not published. At that time, the notion of *cosmical abundances* was firmly entrenched in the minds even of distinguished astronomers; Aller and Payne-Gaposchkin's experience with apparent violations of this concept can be easily understood.

Time has surely vindicated Aller and Payne-Gaposchkin, as may be seen in Table 15.4 taken from Truran and Livio (1986). These authors are generally able to account for the observed abundances in terms of models of the white-dwarf binary systems responsible for the outbursts. We may only mention briefly the *explosive* burning of hydrogen that may take place on the surface of a white dwarf or neutron

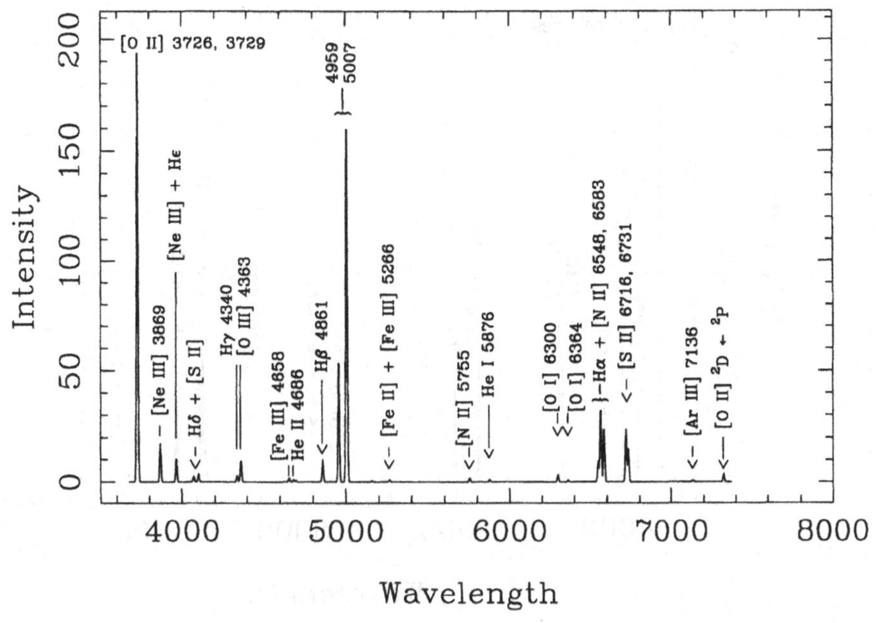

Figure 15.14. A high-temperature filament of the Cygnus Loop. Data from Fesen, Blair and Kirshner (1982, Position C). The [S II] is strong in spite of the high temperature because of shock heating. Note that [O III] λ4363 is stronger than Hβ.

star (see Champagne and Wiescher 1992). The resulting rapid addition of protons, or *rp*-process, may account for some of the lighter, proton-rich nuclides in the SAD.

Before we leave the general area of abundances in emission-line regions, we must consider the possible influence of dust. We have already mentioned in §§7.3 and 14.1 that the general interstellar gas is depleted in heavy elements, and that the most logical explanation of this result is that these elements are locked up in grains. At one time it was perhaps possible to believe that dust grains would be so rapidly vaporized *in H II regions* that their influence on abundances would be negligible. However, this is no longer the case. There is overwhelming evidence for dust within H II regions. A rough estimate by Osterbrock (1989) puts the gas-to-dust mass ratio *in* H II regions at about 100.

Now the mass of *all* of the elements heavier than hydrogen and helium in the SAD is only about 2 percent of the total mass. Therefore, if we acknowledge that Osterbrock's estimate is good at best to an order of magnitude, we see that we should not be surprised to find most of the heavy elements in H II regions to be locked in grains! The palpable fact that this is *not* so in most H II regions hardly means that we can neglect the possible influence of grain formation on abundances in the gaseous phase! Even in the relatively simple planetary nebulae there is evidence not only for dust, but for cool inclusions containing molecules. The chemical elements will partition themselves among these various phases – ionized and neutral gas,

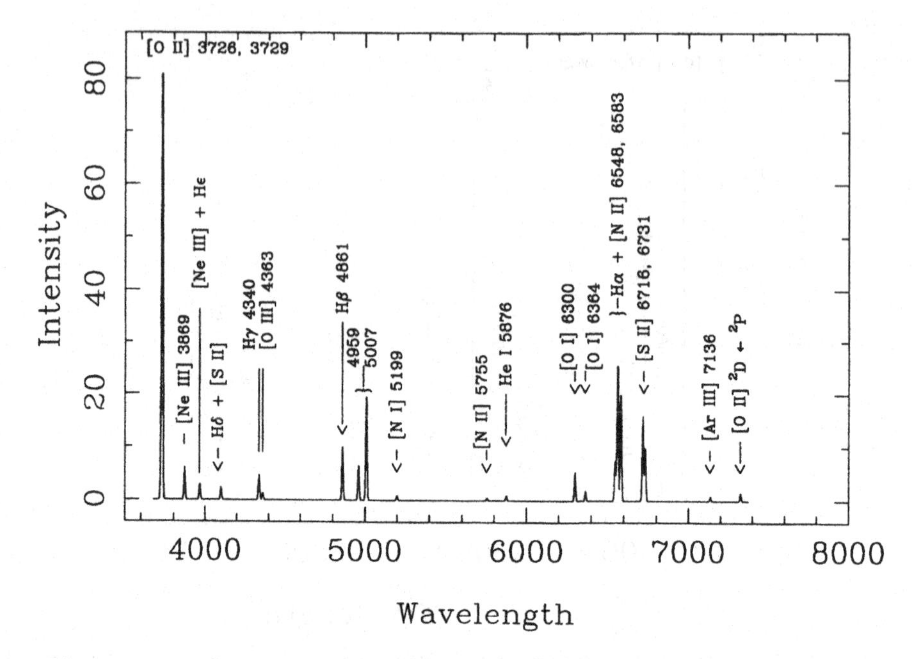

Figure 15.15. A low-temperature filament of the Cygnus Loop. Data from Fesen, Blair, and Kirshner (1982). Note that [O III] λ4363 is now weaker than Hβ, unlike the situation for the high-temperature filament. Also note the presence of the neutral oxygen lines $\lambda\lambda$6300, 6364.

vapor and solid – according to the laws of chemistry and physics. Large abundance fractionations are known to occur in the cool gas. In the hotter, H II regions, the analytical techniques are not yet sufficiently precise for the detection of chemical fractionations.

15.6 Atlas of Simulated Emission Spectra

Figures 15.9–15.17 were generated from *tabulated* intensities taken from sources referenced. Intensities are corrected for reddening, and with a few exceptions, referred to Hβ = 100. The intensities were converted to simulated spectra with the help of a computer program which ensured that the areas under the individual emission lines equaled the tabulated intensities. Simple Gaussian profiles were used for all lines. We caution the reader that *these are not true spectra.* In one or two cases we have added to the tabulated data. For example, we may have inserted [N II] λ6548 with an intensity equal to one third (the theoretical value) of the *tabulated* [N II] λ6584 line. Often authors will not include weak features in their tables. Such lines cannot appear in our spectra.

These simulations have the didactic advantage of uniform visual presentations. The student who has studied them along with the comments in the text should, for

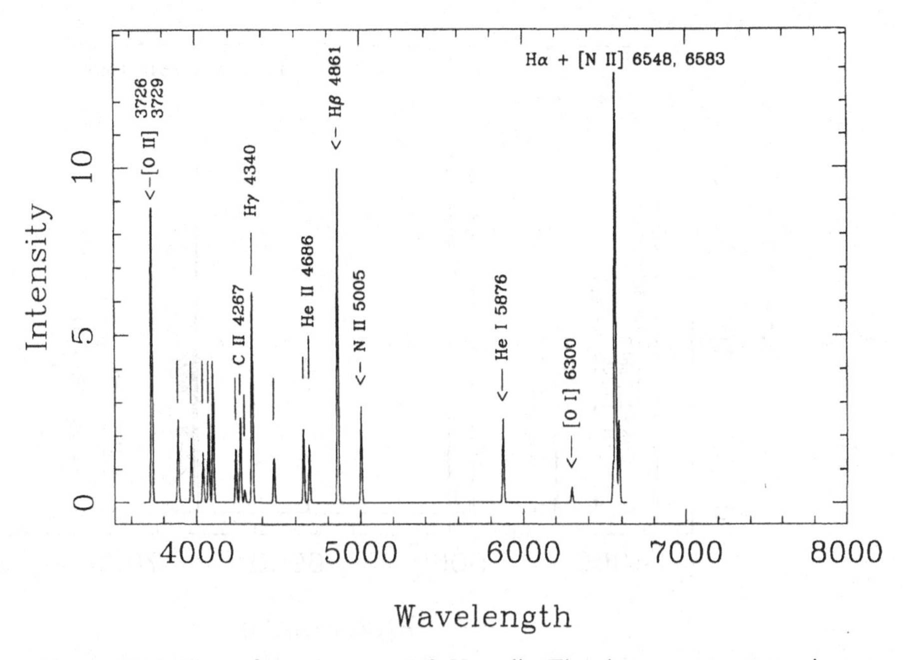

Figure 15.16. Ejecta from the nova DQ Herculis. The electron temperature is very low ($\approx$ 500 K), as a result of cooling due to enhanced abundances of CNO. Data from Ferland *et al.* (1984). We have added the following lines with the help of Figure 2 of Williams *et al.* (1978): $H\zeta$ + He I λ3889, $H\epsilon$ λ3970, N II λ4041, O II + C II λ4075, N II λ4239, O II λ4294, He I λ4471. These features do not represent photometric measurements (NB), but should approximate the relative line strengths for Region A. The permitted lines and probably a substantial part of the forbidden emission are due to recombination.

example, be able to distinguish between the spectrum of a typical planetary and a supernova remnant or a dense H II region from a low-density one.

The spectra of active galactic nuclei and quasars are not included here (see §16.1). Many of these objects are a composite of two regions, one emitting narrow lines with spectra rather similar to those below. There is superimposed emission from a region with much broader lines. Spectra of emission-line galaxies will be discussed in Chapter 16.

All wavelengths are in angstrom units.

15.7 Problems

1. The text asserts that it might have been better to define the departure coefficients b_n as corrections to the temperature than as in Equation 15.2. Given the considerations of §4.9, explain why a *single* departure coefficient would *not* be adequate to describe the atomic populations of hydrogen in gaseous nebulae. In what way might corrections to T_e be better behaved than the conventional b_n's?

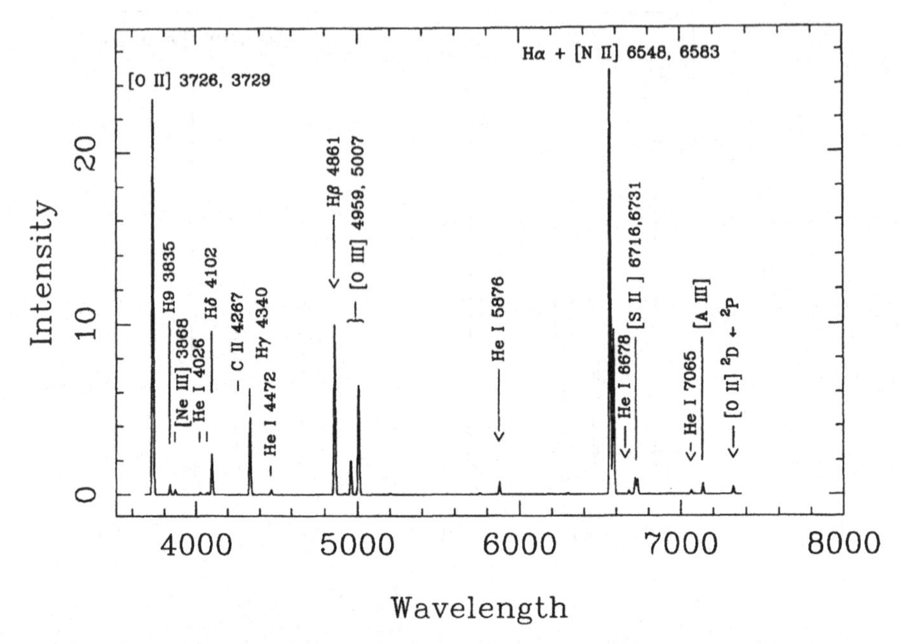

Figure 15.17. The Orion Nebula, also known as M42, or NGC 1976. The observations are from Torres-Peimbert, Peimbert, and Daltabuit (1980), for the region designated Orion 7b, a bright, nearly central part of the nebula. The spectrum is similar to that of a planetary nebula. It resembles the cool planetary M1-44 (Figure 15.9), but has stronger [O II] λ3727 and [O III] $\lambda\lambda$4959 and 5007. Diffuse emission nebulae are characteristically somewhat cooler than planetaries.

2. Assume that the cross section for excitation from level A to B may be written as follows:

$$\sigma_{AB} = \sigma_{\mathrm{H}} \frac{v_{\mathrm{H}}^2 \Omega(A,B)}{v^2 \varpi_A}. \tag{15.14}$$

Here, σ_{H} is the cross section of the first Bohr orbit of hydrogen, and v_{H} is the velocity of the electron in the first Bohr orbit. Perform the average of this cross section times a relative velocity over a Maxwellian (Equation 9.14) distribution, with the assumption that the cross section is zero for $\frac{1}{2}m_e v^2 \leq \chi_{AB}$. You should obtain Equation 15.4, and $q_0 = kh^2(2\pi m_e k)^{-3/2}$. The Ω's may be regarded as a correction to an intuitive, dimensionally correct expression. They are typically (though not always) of order unity.

3. The forbidden lines of [O III], [O II], [S II], etc., often violate more selection rules than the Laporte parity rule. Give several examples of LS-coupling selection rules that are violated by these transitions. The $J = 0$ to $J = 0$ jump is forbidden for electric dipole transitions in all coupling schemes (§11.2). Are any such transitions observed in gaseous nebulae?

16
Abundances of the Elements in Galaxies

16.1 An Introduction to Galactic and Extragalactic Research

In the first half of the twentieth century it became clear that our own stellar system was but one of a very large number of galaxies. To be sure, philosophers such as Immanuel Kant had speculated upon the notion of "island universes," but it was really the work of Edwin Hubble in the mid 1920's that established the great distance of the Andromeda Nebula (galaxy), and clarified the nature of the extragalactic domain as we know it today. His marvelous book *The Realm of the Nebulae* (Hubble 1936) is now primarily of historical interest.

Hubble made use of certain highly luminous variable stars known as Cepheids. The Harvard astronomer Henrietta Leavitt had shown that the intrinsic brightnesses of these stars could be obtained from their periods of variation. With this relationship in hand, it was only necessary for Hubble to locate such stars in the Andromeda galaxy, and measure their periods and apparent brightnesses. Their distances followed immediately.

An interesting historical sidelight concerns the errors in the calibration of the absolute brightnesses of the Cepheid variables. By a curious combination of errors, Hubble underestimated the distance to the Andromeda galaxy by a factor between two and three. Harlo Shapley, using similar methods to determine the size of our own system, obtained results that were much more nearly correct because of a cancellation of effects of which he was unaware. These matters were sorted out in the 1950's, largely as a result of the observational work of Walter Baade at the Mount Wilson Observatory during World War II, where he had been "confined" as a German citizen.

Baade's (1963) *Evolution of Stars and Galaxies*, like Hubble's book, is a classic. Extragalactic astronomy has grown so rapidly in the last several decades that modern works cannot hope to have similar lasting value. We must, however, cite the volume of Sandage, Sandage, and Kristian (1975), which remains useful. For modern references on extragalactic astronomy, the student is advised to consult the reports of the International Astronomical Union, especially those of Commissions 28 (Galaxies), 40 (Radio Astronomy), and 47 (Cosmology). We will mention a few secondary sources below, such as the texts by Mihalas and Binney (1981), and Binney and Tremaine (1987).

Galaxies come in a wide variety of forms and associations. We shall describe

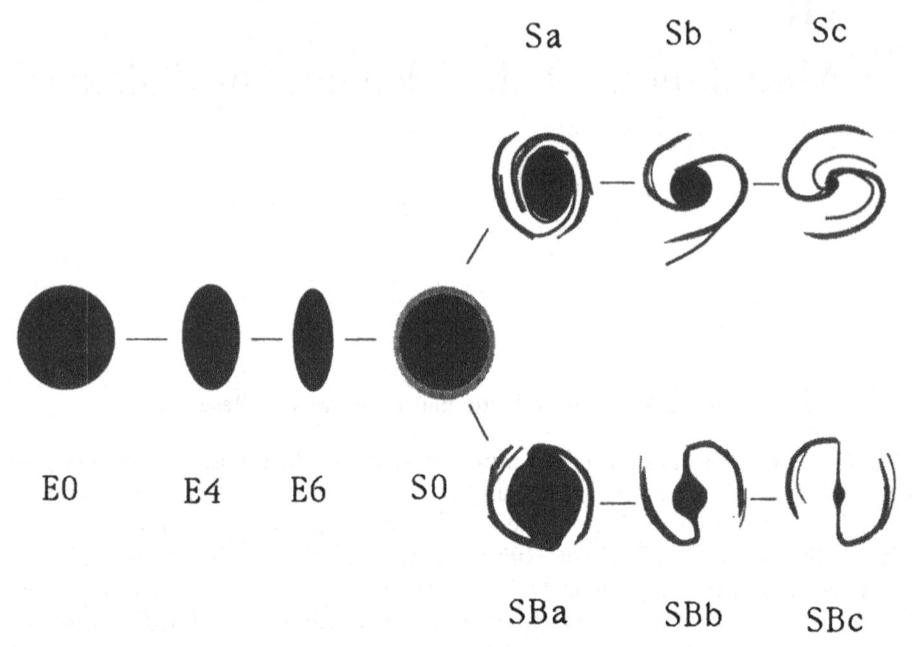

Figure 16.1. Hubble's Classification of Galaxies.

them briefly, and then move on to a discussion of abundances in our own Galaxy and external systems.

"The" Galaxy is a flattened system some 30 kiloparsecs in diameter surrounded by a diffuse halo of stars and globular clusters (§13.1). There could be substantial amounts of nonluminous material in this halo (§16.10).

Edwin Hubble introduced a *tuning fork classification* that is still useful for *normal* galaxies (see Figure 16.1). Astronomers refer to galaxies at the left of the tuning fork as *early*, and those on the right as *late*. The origin of these terms has only historical interest.

Spiral galaxies range in size from some 6 to 30 kiloparsecs in diameter. Their masses are a part of a general problem that is currently under intensive investigation (see §16.10). However, the portion of spiral galaxies delineated by starlight has been found to range from some 10^9 to $4 \cdot 10^{11}$ solar masses ($M_\odot$). Elliptical galaxies are more variable in size than spirals. Within our Local Group of some 20 galaxies, there are a number of "dwarf ellipticals" with diameters of only a few hundred parsecs. This is not a great deal larger than some of the galactic globular clusters (§13.1). Indeed, some of these aggregations are found beyond the distance to the Magellanic Clouds, and the distinction between these objects and dwarf elliptical or spheroidal systems is perhaps not a sharp one. The largest ellipticals are extraordinarily bright, and massive objects. Their intrinsic luminosities range up to an order of magnitude greater than that of the Andromeda spiral galaxy.

In addition to the classical Hubble types, spirals and ellipticals, one may distinguish a few percent as "irregulars." Perhaps the best-known examples of irregulars are the Magellanic Clouds, the two closest members of our Local Group of galaxies, located between one and two galactic diameters from us.

Perhaps one percent of galaxies are classified as "active." Active galaxies have compact central regions or nuclei, a few hundred parsecs across. Some of the best-studied objects of this kind are known as Seyfert galaxies, after the American astronomer who studied them in the 1940's.

The nuclei of Seyfert galaxies have strong emission-line spectra, often with both broad and narrow profiles. Seyferts with broad profiles are called Seyfert 1's while those with only narrow profiles are called Seyfert 2's. The Doppler motions that can be inferred from the broad lines range from 10^3 to 10^4 km/sec. The narrow lines have breadths corresponding to motions of 500 to 10^3 km/sec. Figure 16.2 and Figure 16.3, from Kollatschny (1989), are representative of several categories of active galaxies.

It is generally accepted that quasars or quasi-stellar objects (QSO's) are Seyfert galaxies seen at distances sufficiently great that only the quasi-stellar nuclei are apparent. The spectra of Seyferts and quasars are essentially the same. LINERs (Figure 16.3(a)) are L(ow) I(onization) N(uclear) E(mission) R(egions). Their main characteristic is the presence of low-ionization species such as [O I] and the weakness of [O III] relative to [O II]. Osterbrock (1989) suggests they may represent the low-energy end of the spectrum of active galactic nuclei.

An enormous amount of current astronomical research is now devoted to the study of active galaxies. Many books and symposia have been devoted to them. Osterbrock's (1989) text provides a general background with many references. The active radio galaxies exhibit bizarre jets and beams of relativistic particles (Hughes 1991). Barry Parker (1990) has written a popular introduction to active and colliding galaxies.

We now turn to a résumé of what is known of abundances in *active galactic nuclei* – the Seyfert Galaxies and their more distant congeners, the quasars. Early studies by Gaskell, Shields, and Wampler (1981), Uomoto (1984) and others indicated essentially solar (SAD) abundances. Abundances in quasars are briefly discussed in §16.9.

Research efforts have been largely devoted to the construction of increasingly elaborate models with energy input from the "central engine," the latter being a euphemism for the "B-word." The central engine is thought to be an accreting black hole, but astronomers in their conservatism prefer a vaguer term. These models typically have three components: the central engine, the *broad-line region*, and the *narrow-line region.* The models that predict the observed spectra most closely are assumed to have the right compositions. Current models are usually based on (nearly) solar abundances. The work is described in some detail by Osterbrock (1989).

From the point of view of cosmochemistry, one of the important classifications of galaxies is that of the *starburst galaxy* (Tuan, Montmerle, and Van 1987). These are galaxies in which the star-formation rate is much higher than normal. Of course, this rate cannot be measured directly; it is inferred from strong emission, especially in the infrared. Intense sources of infrared radiation are typically attributed to heating of dust by massive young stars. Such heating may also come from the

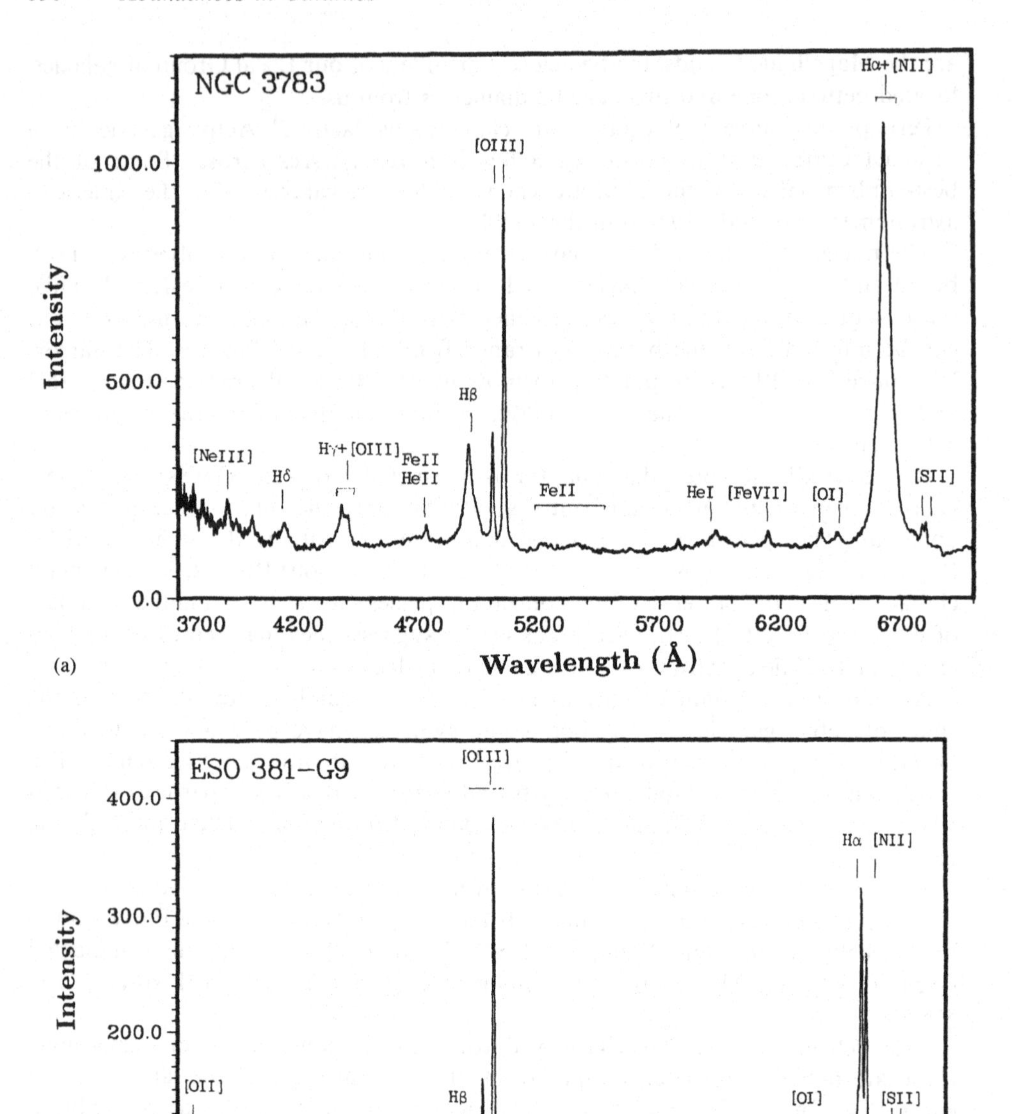

Figure 16.2. Spectra of Two Seyfert Galaxies. (a) shows the Seyfert 1 Galaxy NGC 3783. The broad hydrogen lines distinguish it from (b), the Seyfert 2 called ESO 381-G9. Courtesy of Wolfram Kollatschny and *Sterne und Weltraum.*

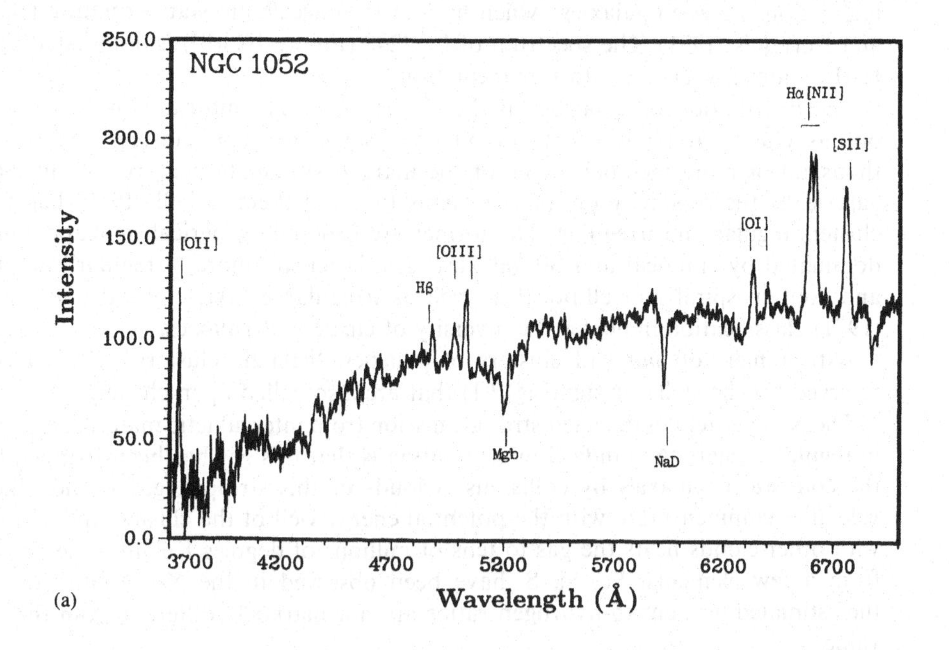

(a)

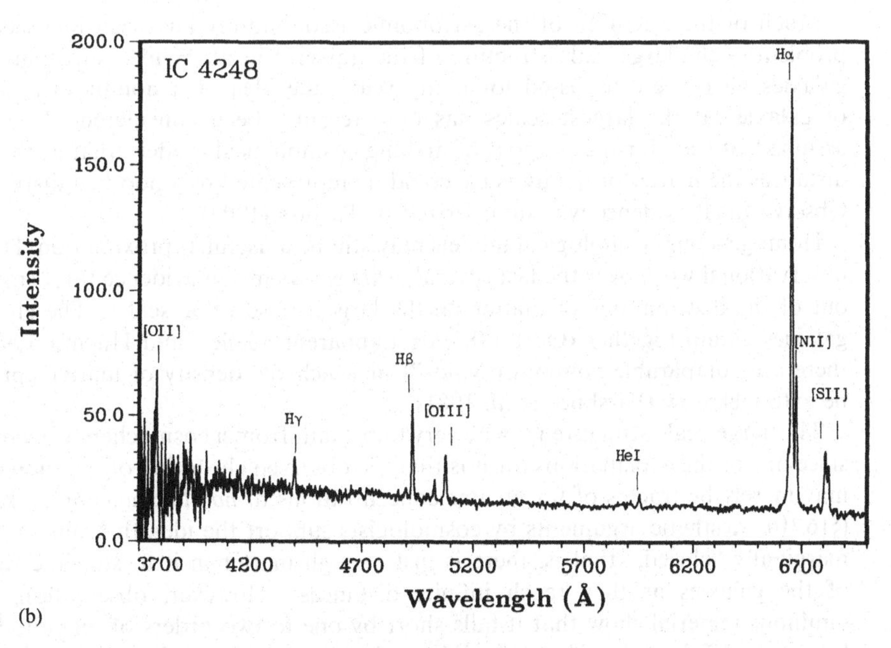

(b)

Figure 16.3. Spectra of a LINER and a Starburst Galaxy. (a) The LINER NGC 1052 and (b) the Starburst galaxy IC 4248. The latter has a spectrum resembling Galactic H II regions (apart from the weakness of [O III]). Courtesy of Wolfram Kollatschny and *Sterne und Weltraum*.

interactions between galaxies, which may in turn accelerate star formation (Barnes and Hernquist 1991). The spectrum of IC 4248 (Figure 16.3(b)) is dominated by the H II regions associated with star formation.

The distribution of galaxies on the sky is far from uniform. Like stars, these objects cluster at all levels, from binary systems to giant clusters containing a thousand or more members. One of the marked systematic features of clusters of galaxies is the proclivity of spiral systems to avoid them. Abell (1958) has called clusters regular and irregular. The former are generally spherical in shape and are dominated by elliptical and S0 galaxies. The latter are more "irregular" in shape, and contain spiral and elliptical as well as irregular galaxies. Rood and Sastry's (1971) classification indicates the diversity of clusters of galaxies.

Astronomers do *not* call clusters of galaxies "galactic clusters." This term is reserved for the stellar systems (§13.1) that are also called open clusters.

The X-ray satellites detected strong emission from intergalactic material, especially in regular clusters. A standard interpretation is that this gas has been stripped from the constituent galaxies by collisions. Clouds of this stripped gas would take on velocities commensurate with the potential energy well of the cluster, and collisions with other clouds heats the gas to tens of millions of degrees Kelvin. Line emission from a few elements, Fe, Si, S, have been observed in the X-ray emission, and the estimated element-to-hydrogen ratios are not markedly different from the solar value.

Much of the attention of the astronomical community has been focused on the problem of the large-scale structure of the universe itself. While "superclusters" of galaxies have been discussed for many years, the idea of a nonuniform structure of galaxies at the largest scales has only recently been considered. It has been assumed, at least for the purpose of making cosmological models, that at the largest distances the matter of the universe could be represented by smoothed distributions. Observational evidence was summarized by Peebles (1993).

Homogeneous cosmological models may still be a useful approximation. However, observational work over the last several years shows mixed evidence for a smoothing out of the distributions of matter on the largest observable scales. The clusters of galaxies clump together (Oort 1983, de Lapparent, Geller, and Huchra 1991), and there are comparable volumes ("voids") in which the density of matter appears to be extremely low (Kirshner *et al.* 1981).

This large-scale structure may be very important from a cosmochemical viewpoint, since one of the explanations for it is that the observed clustering of luminous matter may merely be tracers of far greater concentrations of *non-luminous* or dark matter (§16.10). Aesthetic arguments by cosmologists support the idea that our universe is marginally "closed," that is, there is just enough matter in it to stop the recession of the galaxies as they reach infinite distances. However, observations of the luminous material show that it falls short by one to two orders of magnitude from being of sufficient density to provide such a marginal universe. Moreover, there *is* observational evidence (see §16.10), for either considerable amounts of unseen matter, or a breakdown of the law of gravitation at scales larger than tens of kiloparsecs.

If the amount of this "missing mass" is sufficient to provide a marginal universe, it has to be greater, again by one to two orders of magnitude, than the ponderable material that has thus far been the object of our cosmochemical study. Because of this, it must be the unseen matter that dominates the large-scale structure of the universe, and gives rise to the existence of superclusters and voids.

16.2 Basic Data for the Chemical Evolution of the Solar Neighborhood

After this digression to the limits of the universe, we return to our own Galaxy and its overall chemical evolution. The basic concepts to be developed here will then be applied, insofar as possible, to the extragalactic domain. Good introductions to the modern theory are given by Pagel and Patchett (1975) and Tinsley (1980).

Within the neighborhood of the sun, we find unevolved, lower main-sequence stars whose surface chemistry *may* be uninfluenced by *in situ* differentiation or accretion. Let us assume the surface chemistry of these stars reflects the composition of the interstellar medium from which they were formed. We may then use them to measure the gradual transformation of primordial hydrogen and helium into heavy elements. We begin with an assembly of basic data, which is simply a tabulation of the relative numbers of stars with given abundances.

Since the pioneering work of van den Bergh (1957) and Schmidt (1963), it has been the practice to make use of photometrically estimated abundances rather than spectroscopic determinations. Only a firm belief in the assumption that stellar element-to-hydrogen ratios vary in lockstep has made this approach credible. But we have seen (§13.4.1) that the lockstep assumption has been remarkably durable, and in spite of abundant evidence that it is violated, it is still useful. On the other hand, spectroscopically determined abundances come from a severely biased sample, since most stars were chosen for analysis because their spectra were unusual.

Schmidt made use of observations of ultraviolet excesses ($\delta(U - B)$'s) of only 56 late G-type dwarfs. He converted these measurements to [Fe/H] with the help of a relation given by Wallerstein (1962), which is shown in our Figure 13.7. A more recent conversion is by Cameron (1985). We reproduce a modern version of Schmidt's compilation in Table 16.1. The figures are adapted from Pagel (1989), who provides an estimate of equivalent oxygen-to-hydrogen abundance ratios. Pagel gives the parameter ϕ, where

$$\log \phi \equiv [O/H] = 0.5[Fe/H], \quad [Fe/H] \geq -1.2, \tag{16.1}$$

and

$$\log \phi \equiv [O/H] = [Fe/H] + 0.6, \quad [Fe/H] \leq -1.2. \tag{16.2}$$

Oxygen abundances are now considered more representative of "the" heavy elements than iron (§16.5). The distinction would be unimportant if the lockstep assumption were completely valid.

Table 16.1 shows that most of the stars have the solar abundance $[Fe/H] \leq 0.0$. There is no evidence of very metal-poor stars. Many workers have wondered about the dearth of metal-poor stars. If the chemical elements are mostly the product

Table 16.1. *UV Excesses*[a] *and Abundances for Stars in the Solar Neighborhood*

$\delta(U-B)_{0.6}$	No. Stars	[Fe/H]	ϕ
0.170	1	−1.2 to −1.1	0.25 to 0.28
0.159	3	−1.1 to −1.0	0.28 to 0.32
0.148	2	−1.0 to −0.9	0.32 to 0.36
0.137	3	−0.9 to −0.8	0.36 to 0.40
0.125	10	−0.8 to −0.7	0.40 to 0.45
0.113	10	−0.7 to −0.6	0.45 to 0.50
0.100	8	−0.6 to −0.5	0.50 to 0.56
0.087	17	−0.5 to −0.4	0.56 to 0.63
0.073	15	−0.4 to −0.3	0.63 to 0.71
0.058	14	−0.3 to −0.2	0.71 to 0.80
0.042	11	−0.2 to −0.1	0.80 to 0.89
0.025	14	−0.1 to +0.0	0.89 to 1.00
0.006	17	+0.0 to +0.1	1.00 to 1.12
−0.015	7	+0.1 to +0.2	1.12 to 1.26

Source: Pagel (1989) and Cameron (1985).
[a]See §13.5 for notation.

of stellar nucleosynthesis, one would expect the oldest stars to show little or no evidence of metals.

We can tell the ages of unevolved stars in several ways. If the stars are in a cluster, we can tell the age of the cluster from its color–magnitude diagram with the help of theories of stellar evolution. We find the "turn-off" point, where stars leave the main sequence, and estimate how long it takes for such stars to exhaust their central hydrogen. It is almost always assumed that all of the stars in a cluster formed at the same time. For the field dwarfs in the solar neighborhood, we can estimate an age with statistical accuracy from their kinematics or motions.

Briefly, those stars that move in the plane of the Galaxy, along with the gas–dust clouds where stars are currently being formed, must be young. The oldest stars hardly partake in the general rotation of the disk. Along with the system of globular clusters, they belong to the so-called galactic halo.

In fact, the oldest unevolved stars that we know are indeed metal-poor. However, we may ask if the *relative number* of these stars can be accounted for in our picture of the chemical evolution of our Galaxy. Many workers have thought the answer to this question was "no." Fewer very metal-poor stars than expected are found. People still speak of a *G-dwarf problem*, meaning there are fewer low-metal G-dwarfs than expected. In the next section we shall outline a very simple mathematical model that allows one to predict the expected number of unevolved stars within specified abundance ranges.

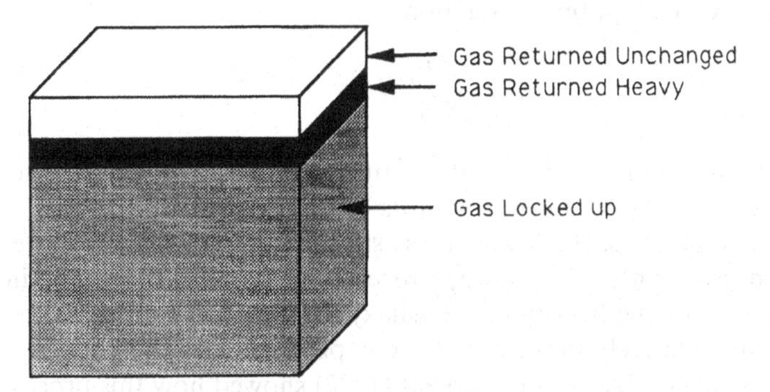

Figure 16.4. Definition of Quantities for the Star-Forming Event.

16.3 Analytical Models of Chemical Evolution: Basic Relations

Let us consider some fixed volume of space which at time $t = 0$ contains a mass M_0 of gas from which stars can form. We shall assume that the stars formed within this volume do not leave it, but we allow for a net gain or loss of *gas* from the volume.

There is a rough correlation between the mass of elliptical galaxies and their metal content. This may just mean that more massive systems can retain the enriched gas expelled by dying stars. The star deaths themselves can cause mass loss from a low-mass galaxy. There is a strong energy input to the interstellar medium from the blast waves from supernovae, which can cause the gas to escape. It is also possible for interstellar gas to be stripped in a close encounter with another galaxy.

In order to take account of such gains or losses of gas, we assume that at every star-formation event, some increment of gas, positive or negative, is added to our volume.

Let us define some quantities relevant to a star-forming event with reference to Figure 16.4. We show the mass of material, ΔM, as a cubical volume, to emphasize the highly schematic nature of this analytical approach.

Let us refer to the overall process as the *Event*. It is perhaps most easily visualized as the formation and dissolution of a cluster of stars, especially since it is thought that most stars are born in clusters. The net result of the Event is assumed to go as follows. A certain amount of mass is returned to the interstellar gas (GR – gas returned), and some remains locked in low-mass stars (GL). Of the gas that is returned, a portion is returned as processed material, or heavies (GRH). It will be assumed that we can treat all of this "heavy" material as a single entity, as though there were just one element beyond hydrogen (and helium) to be considered. Of the gas that is returned to the interstellar medium, some is processed to heavies, and some is simply returned in its original form. If ΔM is the amount of mass in the Event we define

$$\alpha = \frac{\Delta M_{\rm GL}}{\Delta M}. \tag{16.3}$$

We further define the yield, y, by the relation

$$y = \frac{\Delta M_{\mathrm{GRH}}}{\Delta M_{\mathrm{GL}}}. \tag{16.4}$$

An amount $\alpha y \Delta M$ of pure heavy material is returned to the gas after the Event, and an amount $(1 - \alpha - \alpha y)\Delta M$ is returned with its original composition.

In the case of real clusters, the lowest-mass stars undoubtedly form more slowly than the more massive ones. Moreover, processed material will be continuously returned to the gas, over the history of the galaxy. However, the most massive stars will be born and die relatively rapidly, and accomplish most of the nucleosynthesis.

In a fundamental paper, Searle and Sargent (1972) showed how this process could be described by a remarkably simple system of equations. They neglected entirely the relatively slow nuclear processing by low-mass stars, and assumed the entire Event happened over a short time span. This assumption is called *instantaneous recycling*. We shall generalize their work following Pagel and Patchett (1975), but with a simplification due to Hartwick (1976). He showed how the analytical nature of the analysis could be preserved when the total mass changes. One need only assume the amount of mass added to the system was proportional to the mass ΔM of the star-forming Event. If M_t designates the total mass of the system, we assume that as the result of the Event

$$\Delta M_t = \gamma \Delta M. \tag{16.5}$$

Let M_Z denote the mass of the heavy material *in the gas*. If M_G is the mass of gas in the volume, then we define the *heavy element abundance* Z by

$$Z = \frac{M_Z}{M_G}. \tag{16.6}$$

The Event causes a mass, $Z\Delta M_{\mathrm{GL}}$, of the heavy material to leave the gas and become locked up in stars. An amount ΔM_{GRH} of *pure* Z is added to the interstellar gas.

If we assume star formation expels local gas, the composition of this gas will simply be Z. If we assume infall, the composition might take on any value, although it is standard to imagine any infalling material would have a very low heavy element content. Usually, in the former case, the composition of the gas lost would be a *variable* while in the latter, we would assume infall of gas with a *fixed* composition denoted by ζ below.

The fraction of the total mass of gaseous material within our volume at any time will be designated μ. Thus as a result of the Event, we have

$$\Delta \mu = \frac{M_t \Delta M_{\mathrm{G}} - M_{\mathrm{G}} \Delta M_t}{M_t^2}. \tag{16.7}$$

Similarly, the change in the fractional quantity Z is given by

$$\Delta Z = \frac{M_{\mathrm{G}} \Delta M_Z - M_Z \Delta M_G}{M_G^2}. \tag{16.8}$$

We have for the change in the mass of the gas,

$$\Delta M_{\mathrm{G}} = (\gamma - \alpha)\Delta M, \tag{16.9}$$

and for the change in the mass of the heavy material,

$$\Delta M_{\mathrm{Z}} = -Z\Delta M + (1 - \alpha - \alpha y)Z\Delta M + \alpha y\Delta M + \zeta\gamma\Delta M. \tag{16.10}$$

In 16.10, the first term on the right is the Z content of the material that goes into the Event. It represents a loss. The term with the parenthesis represents the mass of heavy material that comes out as it went in, that is Z times ΔM_{GR}. Next we have the ΔM_{GRH}, which is "pure" Z. Finally, there is the heavy content of the infalling material, $\zeta \cdot \gamma\Delta M$.

If we simplify Equation 16.10, and factor out the common ΔM, we obtain

$$\Delta M_{\mathrm{Z}} = [\alpha y + \zeta\gamma - (\alpha + \alpha y)Z]\Delta M. \tag{16.11}$$

Remarkably, this equation gives us a *maximum* value of Z under the present model for which *Events* will increase the mass of heavy material. No more net heavy matter can be added through star formation when the quantity in brackets in 16.11 becomes zero! This critical value is

$$Z_{\mathrm{crit}} = \frac{\alpha y + \zeta\gamma}{\alpha + \alpha y}. \tag{16.12}$$

If we assume no infall, then the critical value of Z is directly related to the yield, y. Indeed

$$Z_{\mathrm{crit}} = \frac{y}{1 + y}. \tag{16.13}$$

Now Z_{crit} is not necessarily the *maximum* value of Z, because Z is the *ratio* of M_{Z} to M_{G}, and Z may still increase if M_{G} is used up. Indeed, that may happen.

Use 16.5, 16.9 and 16.11 in 16.7 and 16.8. Then replace the ratios M_{G}/M_t by μ and $M_{\mathrm{Z}}/M_{\mathrm{G}}$ by Z. It is possible to eliminate the ratio $\Delta M/M_t$ between the two equations relating changes in μ and Z. We shall write the resulting relation with the incremental relations replaced by differentials:

$$\frac{dM}{M_t} = \frac{\mu dZ}{\alpha y + \zeta\gamma - (\alpha y + \gamma)Z} = \frac{d\mu}{-\alpha + \gamma - \mu\gamma}. \tag{16.14}$$

Equation 16.14 is the basic relation with which we shall work. For the simplest case where there is no infall, $\gamma = \zeta = 0$, the exact solution that satisfies the boundary conditions $Z = 0$ at $\mu = 1$ is

$$Z = 1 - \mu^{y}. \tag{16.15}$$

From Equation 16.14, we may obtain a limiting value for Z itself. When the denominator on the left becomes zero, the increment dZ must also go to zero whence we obtain a limiting value

$$Z_{\max} = \frac{\alpha y + \zeta\gamma}{\alpha y + \gamma}. \tag{16.16}$$

If $\gamma = 0$, Z_{max} could be unity! This value would be reached, by Equation 16.15, when $\mu = 0$, that is, when all of the gas is used up.

Equations 16.16 as well as 16.12 contain the four constant parameters of our analytical model: α, γ, ζ, and y. In these, as well as other important relations, we may reduce the number of *independent* parameters by dividing by α. Let us therefore define

$$r = \gamma/\alpha. \tag{16.17}$$

When $r = 1$, infall precisely balances gas lockup in stars. If, in addition, we assume the Z-content of the infalling gas is zero, the Z_{max} of Equation 16.16 is the same as the Z_{crit} of Equation 16.13. This result is often cited in its approximate form for small yield, $Z \rightarrow$ yield. As we shall see in the next section, this would occur when an infinite amount of mass has been accreted!

Observations reveal a number of stellar systems with virtually no gas, but no objects are known with a value of Z even approaching unity. There are several possible explanations of this. A nonzero value of γ, for example, can change Z_{max} substantially. Perhaps as the gas in a system is used up, a point is reached where supernovae blast waves simply remove all of the residual material. Then there would be no chance for the maximum enrichment. Thus far, little serious consideration has been given to the possible existence of ultra-metal-rich stars. Perhaps there are a very small number of such objects that have remained undetected. The bizarre object discovered by Przybylski (§13.4.3) certainly must have a high value of Z. In external systems, we only observe an "average" Z, and so would not be able to detect one or two objects with extreme compositions.

If one makes the assumption that Z is small, for the case of $\gamma = \zeta = 0$, Equation 16.14 simplifies to

$$\frac{dZ}{\alpha y} = -\frac{d\mu}{\mu\alpha}, \tag{16.18}$$

with the approximate solution

$$Z = y \ln(1/\mu). \tag{16.19}$$

Equation 16.19 was first given by Searle and Sargent (1972). Many workers now speak of "the" simple model, to describe the assumptions upon which it is based. Because of the *logarithmic dependence*, a large change in μ means only a small change in Z. This result has enormous appeal because most abundances in stars and galaxies do not depart greatly from those of the sun. Since Equation 16.19 is derived under the assumption of *small* Z, difficulties at $\mu = 0$ do not arise.

16.4 Stars and the Total Mass

If we allow the total mass to change, then

$$\Delta M_t = \gamma \Delta M. \tag{16.20}$$

Clearly,

$$\frac{\Delta M}{M_t} = \frac{\Delta M_t}{\gamma M_t}. \tag{16.21}$$

The ratio $\Delta M/M_t$ is the same as the ratios appearing in Equation 16.14. Using differentials rather than incremental quantities, we write

$$\frac{dM_t}{\gamma M_t} = \frac{d\mu}{\alpha + \gamma - \mu\gamma}. \tag{16.22}$$

If we apply the boundary condition that $\mu = 1$ when $M_t = M_0$, we get the simple solution

$$M_t = \alpha M_0 \left(\frac{1}{\mu\gamma - \gamma + \alpha} \right). \tag{16.23}$$

So long as $\alpha > \gamma$, the gas content of the volume will go to zero, and a final, limiting mass

$$M_{\text{limit}} = M_0 \frac{\alpha}{\alpha - \gamma} \tag{16.24}$$

will be reached.

Let us assume an initial total mass M_0, and define a fractional quantity S by the relation

$$\Delta S = \frac{\Delta M}{M_0}. \tag{16.25}$$

The quantity S thus represents the fraction of material that has formed into stars relative to the *initial* mass. It is not the fractional mass in stars, but if there is zero infall ($\gamma = 0$), then

$$\mu = 1 - \alpha S. \tag{16.26}$$

Then αS represents the fraction of the total mass currently in the form of stars.

We can put $\Delta M/M_0 = (\Delta M/M_t)(M_t/M_0)$ in 16.25, and use 16.23 to eliminate (M_t/M_0). Thus

$$\Delta S = \left(\frac{\alpha}{\mu\gamma - \gamma + \alpha} \right) \frac{\Delta M}{M_t}. \tag{16.27}$$

The fraction $(\Delta M/M_t)$ is the same one that is eliminated in the derivation of 16.14. We can therefore relate ΔS to $\Delta\mu$. In differential form we obtain from the right-hand side of 16.14

$$\frac{d\mu}{-\mu\gamma + \gamma - \alpha} = \frac{\Delta M}{M_t} = \frac{\mu\gamma - \gamma + \alpha}{\alpha} dS. \tag{16.28}$$

It is easy to separate the variables in 16.28 and integrate. With the boundary conditions $S = 0$ at $\mu = 1$, we have

$$S = \frac{1 - \mu}{\mu\gamma - \gamma + \alpha}. \tag{16.29}$$

Solving for μ, we find

$$\mu = \frac{1 + (\gamma - \alpha)S}{1 + \gamma S}. \tag{16.30}$$

If we eliminate $(\Delta M/M_t)$ from 16.28 using 16.27 and the first equality of 16.14, we obtain

$$\frac{\mu dZ}{\alpha y + \zeta\gamma - \alpha y Z - \gamma Z} = \frac{\mu\gamma - \gamma + \alpha}{\alpha} dS. \tag{16.31}$$

Equation 16.30 may now be used to eliminate μ to yield a relation between Z and S:

$$\frac{dZ}{\alpha y + \zeta\gamma - \alpha y Z - Z\gamma} = \frac{dS}{1 + (\gamma - \alpha)S}. \tag{16.32}$$

Equation 16.15 is straightforward to integrate, although a little tedious. In order to simplify the algebra, we define the auxiliary relations

$$A = \frac{\alpha - \gamma}{\alpha y + \gamma} = \frac{1 - r}{y + r}, \tag{16.33}$$

and

$$B = \frac{\alpha y + \gamma}{\alpha y + \zeta\gamma} = \frac{y + r}{y + \zeta r}. \tag{16.34}$$

Clearly B is $1/Z_{\max}$, where $Z_{\max}$ is defined in 16.16. An exact solution of 16.32 which satisfies the boundary condition $S = 0$ at $Z = 0$ is

$$\ln(1 - BZ)^A = \ln[1 + (\gamma - \alpha)S]. \tag{16.35}$$

If the arguments of the logarithms are positive real numbers, we may equate them. The maximum value of Z (Equation 16.16) follows from 16.35 above if we put $BZ = 1$. We could choose combinations of the parameters α, γ, y, and ζ which could make one of these arguments negative. We shall not consider these cases here. We mention also, that we have assumed the parameter ζ, the Z value of the infalling gas, to be a constant. Should we decide to take $\gamma < 0$ (gas loss), it is reasonable to assume $\zeta = Z$, in which case, this parameter is a variable.

With the stipulated reservations, we obtain from 16.35

$$S = \frac{1 - (1 - BZ)^A}{\alpha - \gamma}. \tag{16.36}$$

16.5 The Distribution of Stellar Abundances: Infall

The derivative of 16.36 with respect to Z is proportional to the mass of stars *that have formed* within "the" volume with values of Z between Z and $Z + dZ$ (or in (Z, dZ)). Another way of saying this is that dS/dZ is proportional to the mass of stars formed per unit Z. If we make the appropriate assumptions, we can relate this derivative to the present *number* of stars in (Z, dZ).

First, we must restrict consideration to stars with sufficiently low mass that they have not evolved during the operation of our Events. Let us therefore focus on groups of such stars: G and K stars, for example. We must assume that throughout the history of our model galaxy, the G and K stars always make up the same *fraction* of the mass ΔM of an Event. In the language of the astronomers, we must assume that the stellar *mass function* has been constant.

The mass function, often designated $\xi[\log(m)]$, gives the fraction of stars formed with masses in the interval (m, dm) during an Event. We must refer to the literature for a discussion of this relation, in particular, the monumental review by Scalo (1986, cf. also Rana 1991). An old explanation for the G-dwarf problem of the solar neighborhood has been that a very early generation of massive stars quickly enriched the interstellar medium. These putative massive stars were assumed to have no low-mass counterparts, so evidence of their existence would only be the current, apparent abundance floor – no zero-metal stars are known. This overall situation has been called *prompt initial enrichment* or PIE.

The existence of primordial, massive stars with no low-mass congeners would mean the mass function has *not* been constant over the history of the Galaxy. Stars with masses of tens of solar masses would have run through their lifetimes so quickly that there would be little left now to indicate they ever existed. Synthesized chemical elements *might* point to such stars, but there might be other explanations.

Modern analyses have revealed a number of extremely metal-poor objects (Bessell and Norris 1984, Beers, Preston, and Schechtman 1985, 1988). Possibly enough ([Fe/H] ≤ -4.5 !), from the point of view of our current theory (see below). Thus, for the present, there is no reason to abandon the analytical theory that has been developed thus far. Indeed, it plays a role in most recent discussions of galactic chemical evolution.

Let us construct a histogram, then, of the number of stars in different intervals of "abundance" Z. Traditionally, astronomers have taken the iron-to-hydrogen ratio as a measure of Z, although recently, the oxygen abundance has been preferred (§16.8, Matteucci and François 1992).

If we restrict the stars to low-mass objects that have not evolved over the lifetime of the Galaxy, the empirical distribution should match the theoretical derivative dS/dZ once appropriate normalization has been made.

From Equation 16.36 we have directly

$$\frac{dS}{dZ} \propto p(Z) = \frac{AB(1-BZ)^{A-1}}{1-(1-BZ_1)^A}. \tag{16.37}$$

The *probability* function $p(Z)$ is normalized here so that its integral from zero to Z_1 (the maximum value of Z) is unity.

Stellar abundances are traditionally reported as logarithms, as explained in Chapter 12, so it is useful to transform 16.37 so that the independent variable is a logarithm. Let

$$x = \log(Z/Z_0). \tag{16.38}$$

Here, we may let Z_0 be the solar value of Z, or it could be the abundance of some other standard. The mean abundance of the galactic cluster known as the Hyades is sometimes adopted as a standard. If we set $p(Z)dZ = p(x)dx$, a straightforward transformation gives

$$p(x) = \frac{\ln(10)Z_0 10^x AB\,(1-BZ_0 10^x)^{A-1}}{1-(1-BZ_1)^A}. \tag{16.39}$$

The transformation from Z to x completely changes the shape of the distribution functions 16.37 and 16.39. For small Z, equation 16.37 very closely resembles a *declining* exponential function

$$p(Z) \propto \exp[-B(A-1)Z], \tag{16.40}$$

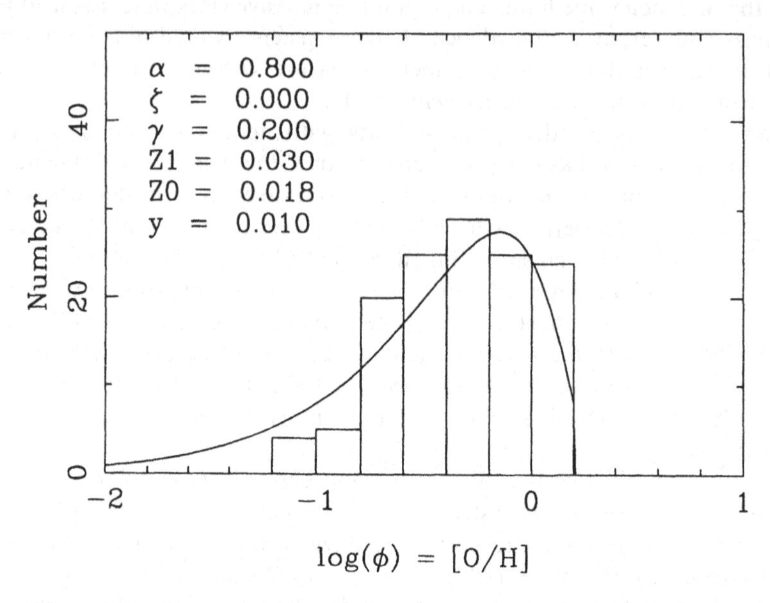

Figure 16.5. Abundance Distribution of Stars in the Solar Neighborhood

a form that can be extremely misleading. Naively, one might interpret 16.40 to mean that there should be more metal-poor than metal-rich stars. Strictly speaking, this is entirely correct. However, that is *not* the G-dwarf problem. The latter is that there are relatively few *very* metal-poor G dwarfs, that is G dwarfs with [Fe/H] ≤ -2.0, say. If we take, roughly, $Z_0 = 0.02$, then [Fe/H] $= -2$ corresponds to $Z = 2 \cdot 10^{-4}$. The *fraction* of stars expected with Z-values less than this amount is the ratio of the *integral* of $p(Z)$ from 0 to $2 \cdot 10^{-4}$ to the integral of $p(Z)$ over its domain. The latter is unity. This ratio is a very small number, so we expect only a few *very* metal-poor stars.

We focus our attention on the logarithmic form 16.39, which is shown in Figure 16.5. We also show a histogram of stellar abundances from Pagel (1989), but with slightly different binning of the data. Following his treatment, as well as the recommendation of Matteucci (1991), we use the logarithmic oxygen-to-hydrogen ratio rather than the more traditional [Fe/H]. At the present time, oxygen is considered more representative of the 'Z' of our treatment than iron. The distinction would not arise if lockstep were strictly followed. Rana (1991) makes corrections to Pagel's data on the basis of the kinematics of stars in different age (or abundance) groups. However, we cannot discuss these details here. The parameters used in the plot are representative, but not definitive.

Figure 16.5 shows the classical G-dwarf problem. At the low-metal, or left, end of the diagram, the solid, calculated curve is higher than the data.

One can ask how serious the departures of the observations from the model are. Granted the simplifications of the theory, one hardly expects a perfect fit. Strictly, one should add up the chi squares, and compare the sum with the expected value *taking errors into account*. These errors are surely larger than the formal square-root of N errors typically applied in goodness-of-fit tests. We leave a more realistic test as an exercise for the student.

We have not found a single set of parameters that will fit the observations for the solar neighborhood, within the confines of the theory we have outlined here. A variety of solutions have been offered. One of the more elegant of these was by Lynden-Bell (1975). His "best accretion model" fit Pagel's (1989, Fig. 4) observations closely. We shall not develop this model here, and must also leave to the references a variety of developments that are both more realistic and more complex than our treatment.

We can simulate the success of some of these treatments by extending our present theory. We simply fit a portion of the data with one set of parameters, the remainder with another. One possible match is shown in Figure 16.6. The low-Z stars are fit with the full theory with infall, with $p(x)$ given by Equation 16.39. High infall rates ($\gamma = 2.5$ here) have the effect of reducing the $p(x)$ for large negative x (low Z), and causing the curve to rise sharply above some x, here above $x = -0.5$. In the figure, we have assumed no infall beyond $x = -0.38$, and *the* simple model from $x = -0.38$ to $x = +0.1$. The yield has the same value for both theoretical distributions. We leave the distribution function and its normalization for the restricted domain of x as an exercise.

This two-model fit lacks the elegance of Lynden-Bell's treatment, but it shares some of its properties. The G-dwarf problem is reduced by infall at the early (low-Z) stages, and the infall is zero at the present time.

The simplistic theory outlined here is rather more successful in fitting other abundance distributions. Hartwick (1976), for example, showed that the galactic globular clusters could be described rather well by a model with infall (see Fig 9-10 of Binney and Tremaine 1987).

Finally, we mention that the function $p(x)$ (Equation 16.39) resembles a Gaussian with a funny tail. Consequently, any distribution that is approximately Gaussian in x will fit one of the families (α, γ, y, etc.) of $p(x)$ – approximately. However, the observed distribution function can be used to gain information about the model parameters. We naturally ask if the observationally determined α, γ, y, etc. make sense in terms of the theory of stellar evolution.

16.6 Fixing the Model Parameters from Observations

There is a simple expression for the maximum of the function 16.39. If we set the derivative of $p(x)$ equal to zero, and solve for x, the result is

$$x_{\text{peak}} = -\log(ABZ_0). \tag{16.41}$$

Since we may determine the maximum of $p(x)$ observationally, this is a fundamental observational constraint on the parameters of our model.

In the simplest case, where $\gamma = 0$, the product AB is the reciprocal of the yield, y.

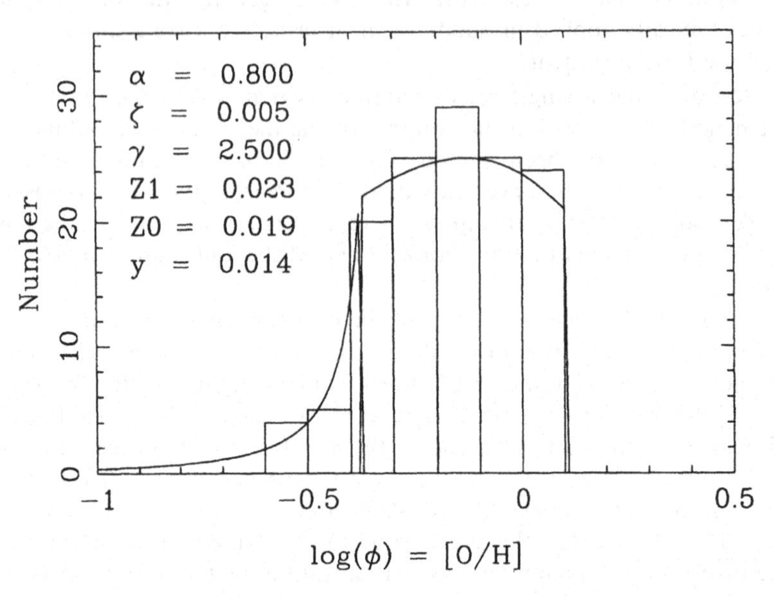

Figure 16.6. Fit of Observed and Theoretical Abundance Distributions for the Solar Neighborhood.

From this, we immediately have $Z_{\text{peak}} = y$, or in terms of the usual variable, x,

$$y = Z_0 10^{x_{\text{peak}}}. \tag{16.42}$$

The numerical value of x_{peak} for the solar neighborhood varies a little from paper to paper. In our Figure 16.5 it is about −0.1, but values as large (in absolute value) as −0.3 may be found. Pagel and Patchett (1975) used the Hyades as a standard, so the maximum in their plots is shifted to values about 0.1 dex *smaller* than it would be if the solar Z were used as a standard. We adopt the value $Z_0 = 0.019$. This is the fraction by mass of all elements other than hydrogen and helium in the abundance compilation of Anders and Grevesse (1989). If we use this figure, *and* adopt a model with no gas infall or loss, then $x_{\text{peak}} = -0.3$ corresponds to a yield of 0.0095, while $x_{\text{peak}} = -0.1$ gives $y = 0.015$.

Laird *et al.* (1988) compare the abundance distributions for high-proper-motion field dwarfs and globular clusters. Both of these distributions resemble predictions of models with $\gamma = 0$, but with very small yields. Note that in their paper, Laird *et al.* report yields *in units of the solar "abundance"*, ($Z_\odot = 0.019$). They also smooth the histogram of observations to simulate uncertainties in the values of the abscissae. Figure 16.7 is from their paper. The compromise model that fits both the globular clusters and the high-velocity field dwarfs has a yield of $y = 0.0006$ – quite small. However, when y is greater than Z_1, the largest Z, Equation 16.41 cannot be used because the distribution 16.39 cuts off before the (formal) mathematical maximum is reached. If one nevertheless uses 16.41, a small yield will inevitably result. The

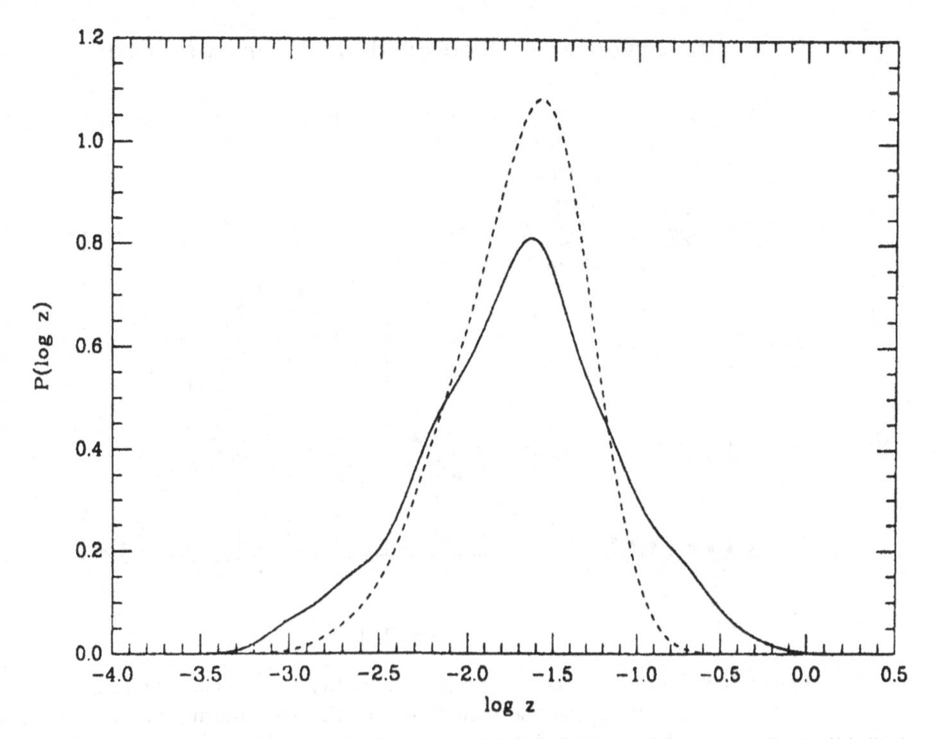

Figure 16.7. Abundance Distribution of High-Velocity Field Dwarfs (solid line) and Globular Clusters (dashed line). Both curves represent observed distributions (see text).

more general form 16.41 might also be used to interpret x_{peak} (Problem 6). Finally, we note that the stars plotted in Figure 16.7 may be drawn from a sample involving multiple populations – possibly resulting from the capture of early satellite systems of our Galaxy. Our theory was not designed to cover such situations.

Rich (1990) has discussed the distribution of metal abundances in the region of the center of our Galaxy. Most stars in this so-called galactic bulge are found to be richer in metals than the sun. Figure 16.8 shows the abundance distribution for bulge giants according to Rich's review paper. The maximum of this distribution is at x_{peak} is $\approx +0.3$, giving a yield of 0.038, assuming $\gamma = 0$.

In the more general case where $\gamma \neq 0$, one cannot get the yield directly from the observed distribution function. Rather, one gets the product

$$AB = \frac{1-r}{y+\zeta r}. \tag{16.43}$$

In this relation, we have used $r = \gamma/\alpha$. When we set the abundance of the infalling material equal to zero, the maximum depends on only *two* parameters. In principle, the yield y (and the "lockup" parameter α) should be determined from the theory of stellar structure and evolution, so our basic model may still be constrained by observations, even if $\gamma \neq 0$.

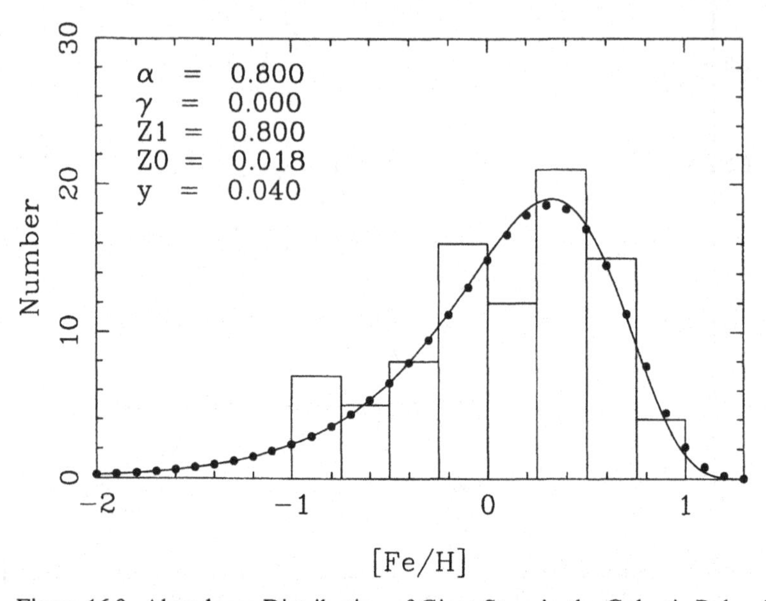

Figure 16.8. Abundance Distribution of Giant Stars in the Galactic Bulge. The dots are for a "simple model" (no infall), with normalization allowing for all values of Z between 0 and infinity. The solid line is for the exact solution 16.39 normalized between $Z = 0$ and Z_1.

Observations provide us with additional constraints on the parameters of our model, namely, the fraction of gas remaining at any time μ, and the maximum value of Z, or Z_1.

We have not yet written down the general expressions for Z as a function of μ, or its inverse. This comes from the integration of Equation 16.14. The solution for no infall or outflow ($\gamma = 0$) is given in 16.15. In the more general case, the solution with $Z = 0$ when $\mu = 1$ is

$$\ln[1 - BZ]^A = \ln\left(\frac{\mu\alpha}{\alpha - \gamma + \gamma\mu}\right). \tag{16.44}$$

The constants A and B are defined by Equations 16.33 and 16.34. For convenience, let us define β such that

$$\frac{1}{\beta} = (1 - BZ)^A. \tag{16.45}$$

Then

$$\mu = \frac{1 - \gamma/\alpha}{\beta - \gamma/\alpha} = \frac{1 - r}{\beta - r}. \tag{16.46}$$

We may use Equation 16.46 along with the *current* values of the gas mass fraction μ_1 and abundance Z_1 to obtain an additional constraint on the parameters of our model. This will be explored in the following section.

16.7 The Distribution of Stellar Abundances: Gas Loss

When gas is expelled from the system, it is reasonable to assume that it has the composition Z of the ambient medium. Consequently, Equation 16.14 simplifies because the $\zeta\gamma$ and γZ terms cancel. The exact solution for gas loss that has $S = 0$ at $Z = 0$ is

$$\ln(1 - Z)^A = \ln[1 + (\gamma - \alpha)S], \tag{16.47}$$

where now

$$A = \frac{1 - r}{y}. \tag{16.48}$$

Note that $\gamma < 0$. If the arguments of the logarithms are real and nonnegative, the solution for S as a function of Z is

$$S = \frac{1 - (1 - Z)^A}{\alpha - \gamma}. \tag{16.49}$$

The properly normalized *probability function* for Z that corresponds to 16.37 is

$$p(Z) = \frac{A(1 - Z)^{A-1}}{1 - (1 - Z_1)^A}. \tag{16.50}$$

A fundamental difference between the cases of infall and gas loss is that in the former there is a natural maximum to the Z-value given by Equation 16.16. With gas loss, our model allows values of Z up to unity.

In terms of the variable $x = \log(Z/Z_0)$, we find

$$p(x) = \frac{\ln(10) Z_0 10^x (1 - Z_0 10^x)^{A-1}}{1 - (1 - Z_1)^A}. \tag{16.51}$$

The maximum of this function is found in the standard way to occur at

$$x_{\text{peak}} = -\log(A Z_0). \tag{16.52}$$

We now integrate Equation 16.14 for the case of $\zeta = Z$. The exact solution with boundary conditions $Z = 0$ at $\mu = 1$ is similar to 16.44, except that A is now defined by the simpler relation 16.48 above. We find

$$\ln(1 - Z)^A = \ln\left(\frac{\mu\alpha}{\mu\gamma - \gamma + \alpha}\right), \tag{16.53}$$

with $\gamma < 0$ by assumption. Let us now define β by

$$\frac{1}{\beta} = (1 - Z)^A. \tag{16.54}$$

Then, for the case of mass loss, the relation between Z and μ is again

$$\mu = \frac{1 - r}{\beta - r}. \tag{16.55}$$

This is the same as Equation 16.46 apart from the change in the sign of γ. In both of these equations, we divide by α, and obtain μ as a function *only* of β and the *ratio* $r = \gamma/\alpha$. The quantity β of 16.45 depends on both the parameters A and B defined by Equations 16.33 and 16.34, but A and B as well as their product can be seen to depend only on the ratio $r = \gamma/\alpha$ and the yield y.

A few applications of these ideas are developed in the Problems at the end of this chapter.

16.8 Recent Developments in Galactic Chemical Evolution

Near the middle of the twentieth century, the American astronomer Jesse Greenstein and his collaborators began a survey of stellar abundances. They used primarily differential techniques based on the principle of the curve of growth (see Chapter 12). Many of his coworkers from those years became leaders in this field of research. We mention only a few: Baschek, G. and R. Cayrel, Helfer, Jugaku, Pagel, Sargent, Searle, and Wallerstein. These efforts were summarized by L. H. Aller (1961, see his Chapter 9), who had also made major contributions in this field.

These early surveys showed tantalizing deviations of individual abundances from lockstep with the reference element iron. In many cases, the deviations occurred in dwarfs, and could not be attributed to the mixing of products of nucleosynthesis to the surface. Some of these results suggested general trends in galactic chemical evolution. An early study by Arnett (1971) appeared to show an increase in the abundances of certain odd-Z elements with time. This was expected from nucleosynthesis. Some bizarre abundance anomalies were eventually attributed to chemical fractionations in stars (§13.4.3).

Unfortunately, the observational uncertainties of this early work were quite large. We pointed out in §13.4.1 that as recently as 1976, the doyen of the theory and analysis of stellar atmospheres argued that most variations from lockstep (in dwarfs) could be accounted for in terms of errors (Unsöld 1976). In addition, many abundance workers were discouraged by blistering criticisms of their use of local thermodynamic equilibrium (§12.3).

Modern work on stellar abundances has benefited enormously from the use of low-noise spectra obtained with electronic devices (see Cayrel de Strobel and Spite 1988). In addition, a number of calculations made relaxing the LTE assumption justified the use of LTE in many applications. First rate spectroscopic analyses of stellar spectra are now performed at a number of institutions in the US, Europe, and Japan. Much of this work would be impossible without the massive efforts of the Harvard-Smithsonian astronomer R. L. Kurucz, who has generously made his programs and data bases freely available (see, e.g., Kurucz 1991b). While many analysts employ the Kurucz codes, a number of independent programs remain in use and provide a "control."

Highlights of the last dozen years or so may be summarized as follows:

1. Radial abundance gradients exist in our own and external galaxies. Typically, the stars at the centers of spiral galaxies are richer in "metals" by factors of 3 to 10 than those at the edge. These results come from analyses of integrated

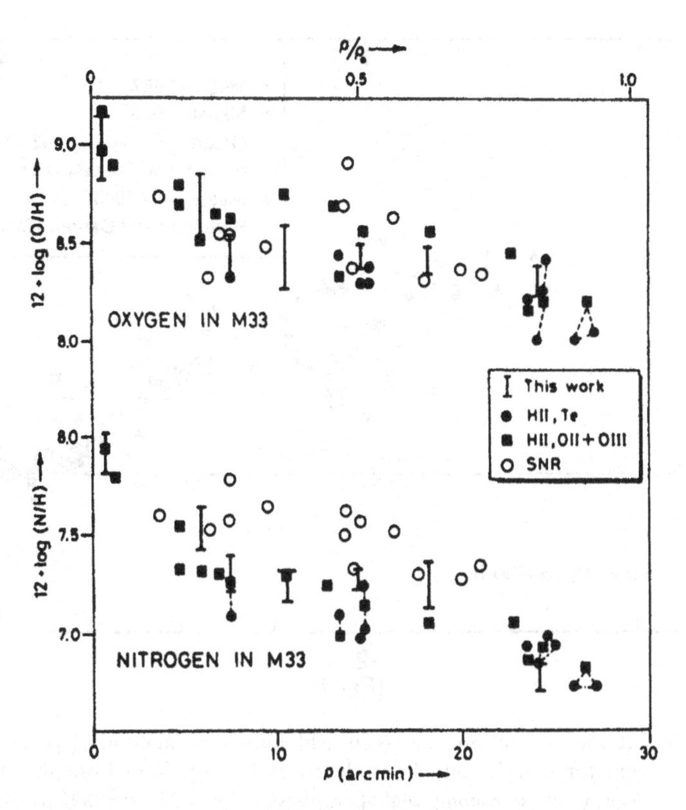

Figure 16.9. Radial Abundance Gradients for oxygen and nitrogen across the Spiral Galaxy M33 (Vílchez *et al.* 1988).

starlight *and* emission regions using the techniques of Chapter 15 (Figure 16.9). Edmunds (1989) summarizes the analyses of these regions and the abundance gradients in external galaxies. Additional information is available from the CO and other molecules (Young and Scoville 1991). Gradients in elliptical galaxies exist, but they are not marked as in spirals (cf. Davies, Sadler and Peletier (1993). Generally, the abundances of elliptical galaxies are well correlated with their overall mass.

2. Credible departures from lockstep can be measured for a number of elements among field stars (dwarfs and giants), as is illustrated in Figure 16.10. We cite reviews by Lambert (1988a), Wheeler, Sneden, and Truran (1989), and Truran (1991).
3. The departures from lockstep as well as the general metallicity gradients may be accounted for in terms of detailed theories of element production and stellar evolution. Figure 16.11 is from the review by Matteucci (1991). It is based on a time-dependent theory of chemical evolution that takes into account detailed calculations of production ratios of individual elements from two principal classes of supernovae. Successful predictions such as those illustrated in Figure 16.11 have been made possible by recent improvements in our understanding of nu-

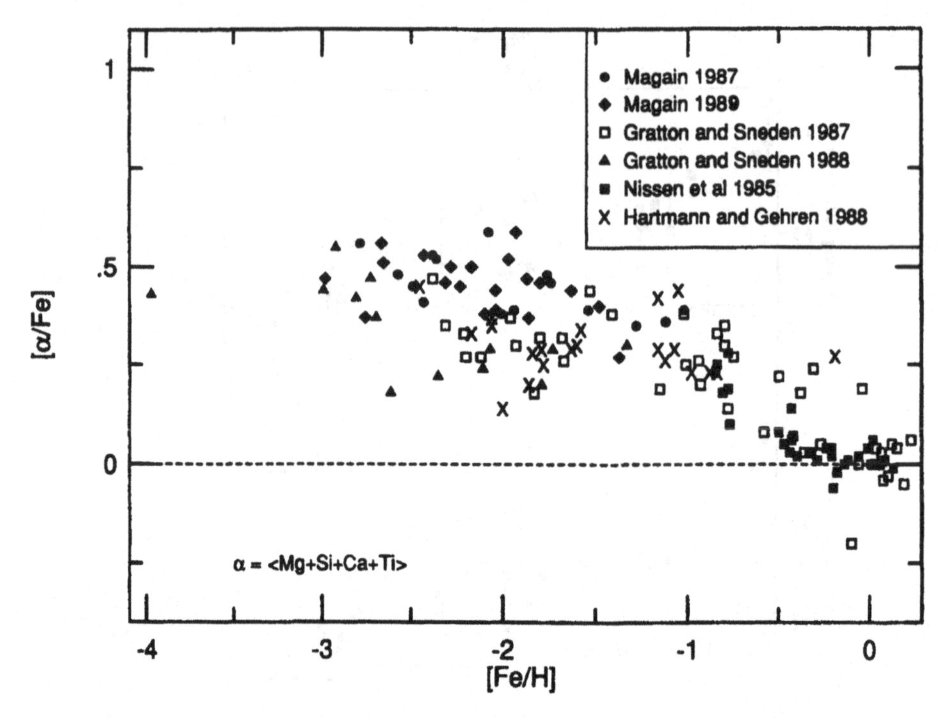

Figure 16.10. Abundances of α-elements in field stars as a function of [Fe/H]. See text for description. Reproduced with permission from the *Annual Review of Astronomy and Astrophysics*, Vol. 27, ©1989, by Annual Reviews Inc. Courtesy of J. C. Wheeler, C. Sneden, and J. W. Truran.

cleosynthesis in massive stars (see Maeder 1991) and supernovae (see Nomoto, Shigeyama, and Tsujimoto 1991). Nitrogen originally is bypassed in the synthesis of α-rich nuclides. After CNO-processed material is returned to the interstellar medium the relative nitrogen abundance increases rapidly. Nitrogen is sometimes called a *secondary* element because other species, C and O, are needed to produce it. Convincing observational evidence that this is in fact so has been slow in coming.

4. Time and production constraints prevent us from more than briefly noting fascinating developments in the subject of *nucleochronology* or *nucleocosmochronology* in a galactic context. These topics concern the age of the Galaxy and the history of nucleosynthesis based on observations of radioactive elements and their decay products. An old, but highly readable treatment is that of Fowler (1972). Many references are given in the review by Cowan, Thielemann, and Truran (1991).

General enrichment of abundances at the centers of spiral galaxies may be simply understood in terms of the analytical theory developed in the early sections of the present chapter. Essentially, the high abundances are explained by the depletion of gas. Oort (1977), for example, gives figures from which one may derive 0.02 for μ within 300 pc of the galactic center. That is certainly lower than the value

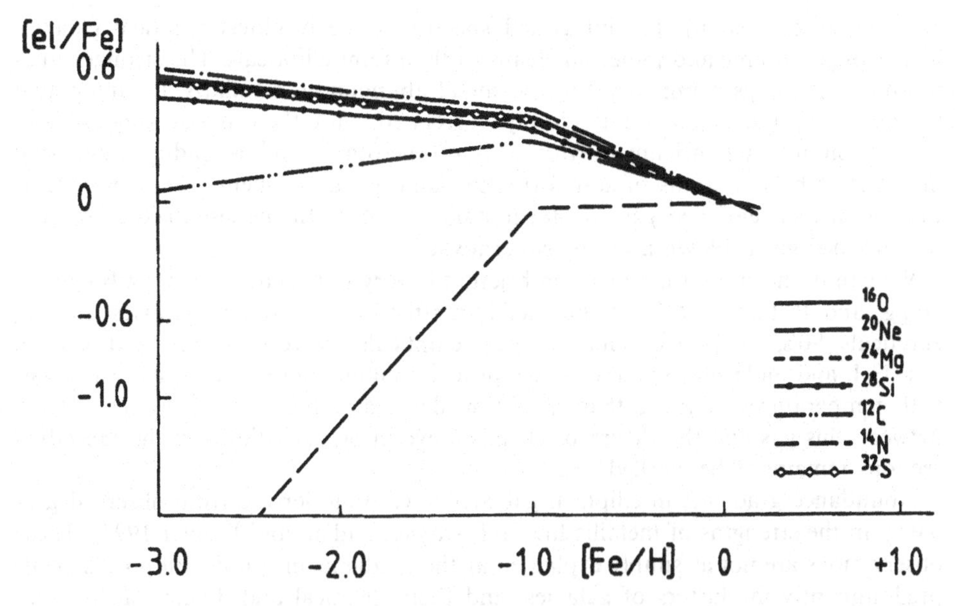

Figure 16.11. Predicted relative abundances for various nuclides. The α-nuclei (^{16}O, ^{20}Ne, ^{24}Mg, and ^{28}Si, show the observed trend (Figure 16.10). The nitrogen (dashes) follows an expected pattern, but the observations thus far show no definitive trend. (Courtesy of F. Matteucci.)

$\mu \approx 0.1$ given, for example, by Rana (1991) for the solar neighborhood. It therefore accounts qualitatively for the enhanced abundances of Rich (1990). Consider the observations of Rich in Figure 16.8. If we take the largest value of [Fe/H] as about unity to imply $Z = 10Z_\odot = 0.2$, then Equation 16.54 with 16.48 (with $\gamma = 0$) gives $1/\beta = 0.8^{25}$. The exponent is $A = 1/y$, and for these giants in the galactic bulge, we found $y = 2Z_\odot = 0.04$ (§16.6). For no infall, $r = \gamma = 0$, and $\mu = 1/\beta = 0.004$.

According to Scheffler (1982), the mass of gas in the central disk ($R < 1.5$ kpc) is about $10^9 M_\odot$, while the stellar mass in the region $R < 2.5$ is $4 \cdot 10^{10}$. The figures are not for the same volumes, so to estimate the current μ, we must scale one figure or the other. We must also decide whether to do it by area or volume: $(2.5/1.5)^2 = 2.8$, while $(2.5/1.5)^3 = 4.6$. Suppose we scale the gas mass within $R < 1.5$ up a factor of three to $3 \cdot 10^9$. Then μ is about 0.08 – rather large. From Oort (1977) we had 0.02 for the galactic center, still larger than the value $\mu \approx 0.004$ we would expect if the maximum Z were really $10Z_\odot$ and $y = 0.04$.

In the solar neighborhood, Rana (1991) gives for the present gas fraction, $\mu_1 =$ 0.10. Since we find lower local abundances than at the galactic center, "the" simple model is at least in qualitative, though perhaps not quantitative, agreement with the observations. This is not surprising since gas gain and loss may be relevant, and the fitting parameters are uncertain (see Problem 4 below).

In the case of elliptical galaxies, the most important parameter in determining the abundances is the total mass (Pickles 1987). It was noted long ago (see Faber 1973, and references therein) that integrated spectra of the brightest elliptical galaxies had stronger atomic and molecular features than faint ellipticals. The stronger lines mean (other things being equal) higher metal abundances. Similarly, the integrated brightness of these galaxies is closely correlated with their mass. The *general* correlation of metal enrichment with mass, on the other hand, is readily understood in terms of basic notions of star formation and gas enrichment. Massive systems can retain their enriched gas and form stars out of it. In the less massive system, the enriched gas is blown away by supernovae.

We can do no more than mention briefly the very important survey by Bregman, Hogg, and Roberts (1992) of interstellar matter in "early-type" galaxies, that is, ellipticals, S0's, Sa spirals. While the pure ellipticals are very poor in cold, neutral gas (H I, and molecular species) they typically contain copious amounts of hot gas, with temperatures of more than a million degrees Kelvin. The possible relation between this gas and the theory of chemical evolution as outlined in the preceding sections remains to be worked out.

Abundance gradients in elliptical galaxies have been derived from observed gradients in the strengths of metallic lines (cf. Davies, Sadler, and Peletier 1993). These observations are not as simply explained as the gradients in spirals. Ellipticals occur predominantly in clusters of galaxies, and their chemical and dynamical histories have been influenced by complicated encounters between systems, including mergers (see Barbuy and Renzini 1992).

16.9 Abundances in Distant Objects

The spectra of the brightest individual stars have been analyzed among a few galaxies of the Local Group. In particular, bright giants in the Magellanic Clouds have been studied (see, e.g., Russell and Bessell 1989). These dwarf, irregular galaxies are only a few tens of kiloparsecs away. In more distant systems, it is necessary to study *integrated spectra.*

Integrated spectra of stellar systems are much more difficult to understand than the spectra of single stars. The problem is that in addition to the usual parameters of stellar atmospheres, $T_e, \log(g)$, and abundances, a host of new variables enter. These new variables specify the *mixture* of stars of various kinds that make up the source in question. This stellar mixture cannot be unfolded from the observations available in the general case. Instead, astronomers have developed *modeling* techniques, based on ever more complicated algorithms (see, e.g., Pickles 1987, Silva 1991).

The efforts to study these systems are often more deeply involved with questions of star formation and galactic evolution than abundances *per se* (Schweizer *et al.* 1990). We cannot pursue the topic here. Instead, we turn to a consideration of emission regions in extragalactic systems.

Bright H II regions have been systematically studied in spiral systems that are still relatively nearby, cosmologically speaking, i.e. usually within 10 megaparsecs (see McCall, Rybski, and Shields 1985, McGaugh 1991). Abundances and spectra of these objects are similar to those of the galactic sources discussed in Chapter 15.

In the most distant objects, astronomers expected to find indications of primordial (metal-poor) material. The quasars would certainly seem to be far enough away to show interesting abundance effects. Indeed, times required for their light to reach us are nonnegligible fractions of the age of the universe. But the emission spectra of many quasars are remarkably similar to those from active galactic nuclei. The interpretation of the line strengths from these objects is extremely difficult because the structure of the emitting region is poorly known. However, predictions based on the models available have shown little evidence for large deviations from solar abundances (Gaskell, Shields, and Wampler 1981, Uomoto 1984).

Osmer (1980) described "the best case yet for abundance anomalies in quasars." From a spectral survey of 125 objects (Osmer and Smith 1980), one was chosen as the most unusual. However, the suggested abundance anomaly necessary to account for the observations of the quasar, designated Q0353-383, was only an enhancement of an order of magnitude in the N/C ratio. It seems that within the domain of active galactic nuclei, the dominant abundance pattern is not far from "cosmic" (SAD).

In addition to emission lines, some quasars show absorption lines (Blades, Turnshek, and Norman 1988). These lines fall into two categories. Some of the absorptions are associated with the quasar itself, that is, the absorption systems contain metallic lines that can be identified, and the wavelength shifts of the absorption lines are very close to those of the quasar emission lines.

York *et al.* (1991) have catalogued a large number of quasar absorption-line systems. Many absorptions are attributed to atomic hydrogen in intervening clouds of material between us and the quasar. The strongest absorption line is then Lyman-α of H I. Some quasars have a rich Lyman-α "forest." It has been thought that most, perhaps all, of the absorption features in quasar spectra *short* of the quasar's Lyman-α emission are due *only* to intergalactic hydrogen. This material might then represent primordial gas with abundances unchanged from the synthesis of the early universe. Lu (1991) searched among quasar Lyman-α clouds for the C IV resonance line, and discusses a detection that would challenge the notion that these clouds have pristine compositions.

Enormous efforts have been made to understand the absorption lines in quasars that *can* be identified with elements heavier than helium. The formidable task of identifying these features with some confidence was attacked in an early work by Bahcall (1968). Current methods are essentially those discussed by Aaronson, McKee, and Weisheit (1975), and modern workers have moved on to the analysis of these features for their abundance systematics.

Steidel (1990) discussed abundances in clouds with sufficient column densities to be optically thick in the Lyman continuum, the so-called Lyman limit systems. He found that the "heavy elements" (C, O, N, and Si) have logarithmic abundances relative to the SAD in the range -3.0 to -1.5. The absorptions may belong to "subgalactic fragments which are thought to have existed during the early evolution of the extended halos of galaxies."

In a small subgroup of absorptions in quasars, the Lyman lines exhibit damping wings, indicating relatively large column densities of hydrogen in the line of sight. It appears that these damped Lyman-α systems systems contain dust (Pei, Fall, and Bechtold 1991), and possibly even molecules (see Donahue and Shull 1991).

We now leave the quasars, and turn to a brief consideration of the diffuse, intergalactic gas that occurs in clusters of galaxies. The existence of this gas has been known since the early days of X-ray astronomy (Gursky *et al.* 1971, Sarazin 1986). The gas is extremely hot, ranging to somewhat more than 10^8 K, and the amount of mass involved can exceed that of the stellar content of the galaxies!

Some of the gas may be primordial, but much of it was at one time associated with the galaxies within the clusters. There are two lines of evidence leading to the latter conclusion. First, the galaxies in the denser clusters are preferentially (the gas-poor) elliptical or "intermediate" types (S0's) which appear to be spiral systems from which the gas has been removed. Second, the gas contains significant abundances of heavy elements.

At one time it was thought that gas was removed from galaxies in clusters as a result of direct interactions such as collisions or tidal stripping. More recently, other mechanisms have been preferred (cf. Dressler 1984), because predicted collision frequencies among galaxies in clusters have been revised downward as a result of new distances to the clusters. David, Forman, and Jones (1990) discuss the effects of supernovae-driven winds on this gas.

Emission lines have been discovered in the X-ray spectra of the *intracluster* gas. The original identifications were made with K-shell transitions in Fe XXV and XXVI, the latter spectrum being that from an iron ion with a single electron. Additional transitions have been identified from lower energy transitions due, for example, to ions of oxygen, sulfur, silicon, and magnesium. Typical observed features are due not to single lines, but to complicated blends, and the standard analytical procedure is to model the emission spectrum of the gas. The results of these studies are generally that the abundances in the intracluster gas range from a few tenths to nearly solar (see the review by Forman and Jones 1991).

We now end our discussion of abundances in the most distant systems. From the point of view of cosmochemistry, the abundances found thus far are surprising in their *similarity* to the composition of stars near the sun. It seems likely that only the faintness of these sources and our relatively crude observational techniques have prevented the detection of objects with truly unusual compositions. To date, nothing has been found at the limits of space to rival the variety of compositions regularly investigated by the stellar spectroscopist and the geochemist.

16.10 Dark Matter in the Universe

Dark matter is a phrase used in modern astronomy and cosmology to describe material that is not directly observed. Its existence is required by our current understanding of the laws of physics – primarily, the law of gravitation. Astronomers routinely infer the existence of "unseen" matter. For example, in many binary stellar systems, only one of the component stars may be detected by its light. The second component is inferred from velocity variations of the observed star. We know that the brightnesses of stars vary enormously relative to their masses, and we expect to find some systems in which one component totally outshines the other.

The modern problem of dark matter cannot be simply understood. It may well be the most critical problem for the physics of our time. Dark matter is inferred with the help of the law of gravitation, just as unseen secondary stars in binary systems are. The problem is that the amount of unseen material is much greater than matter that can be detected by any means *other* than by gravity. And the amount of unseen matter is staggering – one to two orders of magnitude greater than ordinary, or "visible" matter.

The problem in its critical form has been known since the work of Zwicky (1933). He pointed out that the motions of galaxies in clusters were incompatible with stable systems. The amount of matter required for stability was larger by one to two orders of magnitude than could be accounted for from the galaxies themselves.

The physics involved in this inference is very straightforward. It is easy to show that in a gravitationally bound, stable system, the total kinetic energy must be on the average, one half the magnitude of the potential energy. This relation comes from the more general *virial theorem*, which is discussed in many books (cf. Binney and Tremaine 1987).

Zwicky's problem remained on the back burner of astronomical thinking for several decades. Toward the end of the 1970's it was brought into focus by another set of observations of extragalactic systems. Largely through the work of Rubin and her coworkers (Rubin, Ford, and Thonnard 1978), it became generally known that the rotational velocities of spiral galaxies behaved in an extraordinary way. We know that the centrifugal acceleration in circular motion is V^2/R, where V is the velocity, and R the distance from the center. This must be balanced by the central force, and every astronomer expected this force to decline with distance from the center of galaxies. Indeed, we expected the force to decline as $1/R^2$ for large distances, when the majority of the mass of the galaxy was well within the radius R.

Rubin's observations were displayed as "rotation curves," or plots of V as a function of R. The typical curves for spirals showed no sign of the decreased rotation – they were flat. The only way that these curves could be explained dynamically was with the addition of unseen or dark matter. The amount of additional material required depends on the assumptions made about the size of the galaxy itself, and whether the flat rotation curves can be extrapolated. In general, the amount of dark matter that could be inferred from the flat rotation curves was somewhat less than that from the motions of galaxies within clusters. Nevertheless, the striking observations of flat rotation curves were a strong factor in awakening the current keen interest in dark matter in the universe.

We cannot detail here the manifold observational indications of dark matter. Nor can we delve into modern cosmological theories that require sufficient matter to close the universe. We refer to review articles (Trimble 1987, Narlikar and Padmanabhan 1991), and turn briefly to current ideas on the nature of the material itself.

There are a variety of reasons for thinking the dark matter may not be baryonic, that is, in the form of electrons, protons, and neutrons. Baryonic matter usually *radiates*, and by definition, dark matter does not. It is still possible that unusual forms of baryonic matter are responsible. Some discussion has been devoted to brown dwarfs, or Jupiter-like bodies. Earth-like bodies (or bricks!) are generally discounted on the basis of our understanding of cosmical abundances. Such bodies would imply (by complementary patterns) large amounts of hydrogen, which have not been detected.

Some workers favor exotic particles, such as *wimps* or weakly interacting massive particles. If (one of) the neutrinos were to have mass, they would be wimps. Other, still undiscovered particles are often discussed, usually within the context of theories which would unify the fundamental forces of physics. The literature on this field is enormous, and the reader has only to scan the indices of the major journals to find them.

If the solution to dark matter lies in exotic particles or cosmic strings, the topic falls well beyond the scope of processes described in this book. To this writer, a most fascinating aspect of this topic is the attitude of his colleagues toward it. Crises occur in science when observations cannot be explained with current theories. This is surely the present situation with dark matter and the theory of gravitation. Attempts have been made to modify the theory of gravitation, in such a way as to account for the observations that require dark matter. Such work is rarely taken seriously.

In the 1930s observations of β-decay threatened to destroy notions of conservation of energy and momentum. These pillars of physics were saved when Pauli postulated the neutrino. This preserved the underlying structure of the physical theory of that time at the expense of complications in detail. Perhaps the same situation will be true for dark matter, but it is not possible at this time to be certain.

16.11 Problems

1. Show that if the infall parameter γ is equal to the lockup α, then all of the gas is used up ($\mu \to 0$) in the limit as the total mass becomes infinite.
2. Show that an exact solution for the distribution function $p(x)$ in the case of *the* simple model ($\gamma = 0$) is $p(x) = N \cdot 10^x(1-Z_0 10^x)^{(1/y-1)}$, where N is a normalization constant. Find the normalization constant when x is nonzero only between x_1 and x_2.
3. Evaluate the goodness of the fit of Figure 16.6 using a χ^2-test. Read the numerical values off the plot. Can the differences of the observations from the theory be accounted for in terms of root-N fluctuations? At what confidence level can one reject this null hypothesis?
4. Rich (1990) fit the abundance distribution of stars near the galactic center with *the* simple model using an assumed yield of about twice the solar Z. Verify this result with the help of the theory discussed in §16.6 and following. The *maximum* values of Z, or Z_1, for the solar neighborhood and galactic bulge may be read from Figures 16.5 and 16.8. The ratio of these figures will be related to the appropriate yields, and the remaining gas fractions (μ_1). Since numerical estimates have been given for y and μ_1 for the solar neighborhood and the galactic center, you can use the simple model to predict the rato of the Z_1's. The uncertainties in Z_{peak}, Z_1, μ_1, and y, must be about a factor of two. Granted these uncertainties, can one reconcile the observations for the galactic center with the simple model?
5. The globular clusters *may* represent systems where the residual gas was quickly expelled by star formation. Assume that the relevant yields were of the order of

0.02. What must the ratios $r = \gamma/\alpha$ have been to produce abundance maxima in the range $x = -2$ to -1?

6. Use the general form 16.41 or 16.52 to find cases where it is possible that y could be *greater than* the value of Z_{peak} corresponding to x_{peak}. Is infall or gas loss necessary to get $y > Z_{peak}$? Discuss your results in connection with Figure 16.7.

Appendix

Table A1. *Physical Constants*

Cohen and Taylor: Rev. Mod. Phys. **59**, 1121 (1987)

Name	Symbol	Value	cgs exponent
Constant of gravitation	G	$6.673 \cdot 10^{-11}$ m^3 kg^{-1} s^{-2}	$\cdot 10^{-8}$
Speed of light	c	$2.997\,924\,58 \cdot 10^{8}$ m s^{-1}	$\cdot 10^{10}$
Planck's constant	h	$6.626\,08 \cdot 10^{-34}$ J s	$\cdot 10^{-27}$
Elementary charge	e	$1.602\,177 \cdot 10^{-19}$ C 4.803 207 (esu)	$\cdot 10^{-20}$ $\cdot 10^{-10}$
Avogadro's number	$\mathcal{N}_a$	$6.022\,14 \cdot 10^{23}$ mol^{-1}	$\cdot 10^{23}$
Mass unit At. Wt.	$^{12}C/12$	$1.660\,54 \cdot 10^{-27}$ kg 931.494 MeV	$\cdot 10^{-24}$
Electron mass	m_e	$9.109\,39 \cdot 10^{-31}$ kg 0.510 999 MeV	$\cdot 10^{-28}$
Proton mass	m_p	$1.672\,623 \cdot 10^{-27}$ kg 938.272 MeV	$\cdot 10^{-24}$
Neutron mass	m_n	$1.674\,93 \cdot 10^{-27}$ kg 939.565 MeV	$\cdot 10^{-24}$
Boltzmann's constant	k	$1.380\,7 \cdot 10^{-23}$ J K^{-1} $8.617\,4 \cdot 10^{-5}$ eV K^{-1}	$\cdot 10^{-16}$
Gas constant	$\mathcal{R}$	8.314 5 J mol^{-1}K^{-1} 1.987 cal mol^{-1} K^{-1}	$\cdot 10^{+7}$
Bohr radius	a_0	$0.529\,177\,2 \cdot 10^{-10}$ m	$\cdot 10^{-8}$
Stefan-Boltzmann constant	σ	$5.671 \cdot 10^{-8}$ W m^{-2}K^{-4}	$\cdot 10^{-5}$
Radiation constant	$a = 4\sigma/c$	$7.566 \cdot 10^{-16}$ W m^{-3}s	$\cdot 10^{-15}$
Wien's displacement	$\lambda_{\max} T$	$0.289\,78 \cdot 10^{-2}$ m K	$\cdot 10^{0}$

Table A2. *Astronomical and Geophysical Constants*

Name	Symbol	Value (SI units)	cgs exponent
Sun's mass	$M_{\odot}$	$1.9891 \cdot 10^{30}$ kg	$\cdot 10^{33}$
Sun's radius	$R_{\odot}$	$6.959 \cdot 10^{8}$ m	$\cdot 10^{10}$
Sun's luminosity	$L_{\odot}$	$3.84 \cdot 10^{26}$ W s^{-1}	$\cdot 10^{33}$
Sun's surface gravity	g	$2.736 \cdot 10^{2}$ m s^{-2}	$\cdot 10^{4}$
Astronomical Unit	AU	$1.495978 \cdot 10^{11}$ m	$\cdot 10^{13}$
Parsec	pc	$3.084 \cdot 10^{16}$ m	$\cdot 10^{18}$
Light year	ly	$9.460 \cdot 10^{15}$ m	$\cdot 10^{17}$
Mile	mi	$1.6093 \cdot 10^{3}$ m	$\cdot 10^{5}$
Sidereal year	yr	$3.1558 \cdot 10^{7}$ s	$\cdot 10^{7}$
Earth's mass	$M_{\oplus}$	$5.9742 \cdot 10^{24}$ kg	$\cdot 10^{27}$
Earth's mean density	$\bar{\rho}_{\oplus}$	$5.515 \cdot 10^{3}$ kg^{-3}	$\cdot 10^{0}$
Earth's equatorial radius	$R_{\oplus}$	$6.37814 \cdot 10^{6}$ m	$\cdot 10^{8}$
Moon's mass		$7.3483 \cdot 10^{22}$ kg	$\cdot 10^{25}$

Table A3. *Conversion Factors*

Name	Alternate Value
1 eV	$1.602177 \cdot 10^{-19}$ J $\quad \cdot 10^{-12}$ ergs
1 joule	10^{7} ergs
1 cm^{-1} = 1 kayser	$1.2398 \cdot 10^{-4}$ eV
1 calorie	4.1868 J = 1 gm calorie (small calorie)
1 atmosphere	101 325 Pa (pascals, exact)
	$1.01325 \cdot 10^{6}$ dynes/cm^{2}
1 tesla	10^{4} gauss

Table A4. *Elemental Abundances: Earth's Crust and Sad*

(alphabetic order)

Element	Symbol	At. No.	At. Weight	$\log(\epsilon_{Si})=6.00$	
				Crust: $\log(\epsilon)E$	SAD: $\log(\epsilon)$
Actinium	Ac	89	227.03		
Aluminum	Al	13	26.98	5.50	4.93
Antimony	Sb	51	121.76	−.77	−.51
Argon	Ar	18	39.95	−1.23	5.00
Arsenic	As	33	74.92	.39	.82
Astatine	At	85	(210.)		
Barium	Ba	56	137.33	2.47	.65
Beryllium	Be	4	9.01	1.36	−.14
Bismuth	Bi	83	208.98	−2.39	−.84
Boron	B	5	10.81	1.93	1.33
Bromine	Br	35	79.90	.51	1.07
Cadmium	Cd	48	112.41	−.83	.21
Calcium	Ca	20	40.08	5.08	4.79
Carbon	C	6	12.01	3.19	7.00
Cerium	Ce	58	140.12	1.69	.06
Cesium	Cs	55	132.91	.30	−.43
Chlorine	Cl	17	35.45	2.56	3.72
Chromium	Cr	24	52.00	2.38	4.13
Cobalt	Co	27	58.93	1.70	3.35
Copper	Cu	29	63.55	2.04	2.72
Dysprosium	Dy	66	162.50	.28	−.40
Erbium	Er	68	167.26	.33	−.60
Europium	Eu	63	151.96	.16	−1.01
Fluorine	F	9	19.00	3.47	2.93
Francium	Fr	87	(223.)		
Gadolinium	Gd	64	157.25	.60	−.48
Gallium	Ga	31	69.72	1.45	1.58
Germanium	Ge	32	72.61	.33	2.08
Gold	Au	79	196.97	−2.68	−.73
Hafnium	Hf	72	178.49	.21	−.81
Helium	He	2	4.00	−4.46	9.43
Holmium	Ho	67	164.93	−.10	−1.05
Hydrogen	H	1	1.01	5.19	10.45
Indium	In	49	114.82	−.67	−.74
Iodine	I	53	126.90	−.43	−.05
Iridium	Ir	77	192.22	−3.27	−.18
Iron	Fe	26	55.85	5.06	5.95
Krypton	Kr	36	83.80	−5.22	1.65
Lanthanum	La	57	138.91	1.41	−.35
Lead	Pb	82	207.2	.81	.50
Lithium	Li	3	6.94	2.43	1.76
Lutetium	Lu	71	174.97	−.53	−1.44
Magnesium	Mg	12	24.30	5.07	6.03
Manganese	Mn	25	54.94	3.30	3.98
Mercury	Hg	80	200.59	−1.36	−.47
Molybdenum	Mo	42	95.94	.11	.41
Neodymium	Nd	60	144.24	1.45	−.08
Neon	Ne	10	20.18	−3.92	6.54
Nickel	Ni	28	58.69	2.23	4.69
Niobium	Nb	41	92.91	1.35	−.16
Nitrogen	N	7	14.01	2.14	6.50
Osmium	Os	76	190.23	−2.57	−.17
Oxygen	O	8	16.00	6.47	7.38

Table A4. *(continued)*

Element	Symbol	At.No.	At. Weight	$\log(\epsilon_{Si})$=6.00	
				Crust: $\log(\epsilon)E$	SAD: $\log(\epsilon)$
Palladium	Pd	46	106.42	−1.84	.14
Phosphorus	P	15	30.97	3.57	4.02
Platinum	Pt	78	195.08	−2.28	.13
Polonium	Po	84	(209.)		
Potassium	K	19	39.10	4.69	3.58
Praseodymium	Pr	59	140.91	.82	−.78
Promethium	Pm	61	(145.)		
Protactinium	Pa	91	231.04		
Radium	Ra	88	226.02		−1.45
Radon	Rn	86	(222.)		
Rhenium	Re	75	186.21	−3.41	−1.29
Rhodium	Rh	45	102.91	−2.30	−.46
Rubidium	Rb	37	85.47	1.97	.85
Ruthenium	Ru	44	101.07	−1.99	.27
Samarium	Sm	62	150.36	.68	−.59
Scandium	Sc	21	44.96	1.76	1.53
Selenium	Se	34	78.96	−1.19	1.79
Silicon	Si	14	28.09	6.00	6.00
Silver	Ag	47	107.87	−1.12	−.31
Sodium	Na	11	22.99	5.01	4.76
Strontium	Sr	38	87.62	2.65	1.37
Sulfur	S	16	32.07	3.04	5.71
Tantalum	Ta	73	180.95	−.01	−1.68
Technetium	Tc	43	(98.)		
Tellurium	Te	52	127.60	−2.09	.68
Terbium	Tb	65	158.93	−.12	−1.22
Thallium	Tl	81	204.38	−.44	−.74
Thorium	Th	90	232.04	.56	
Thulium	Tm	69	168.93	−.52	−1.42
Tin	Sn	50	118.71	.26	.58
Titanium	Ti	22	47.88	4.13	3.38
Tungsten	W	74	183.85	−.17	−.88
Uranium	U	92	238.03	.00	−2.05
Vanadium	V	23	50.94	2.44	2.47
Xenon	Xe	54	131.29	−6.33	.67
Ytterbium	Yb	70	173.04	.27	−.61
Yttrium	Y	39	88.91	1.55	.67
Zinc	Zn	30	65.39	2.08	3.10
Zirconium	Zr	40	91.22	2.26	1.06

Notes: The crustal abundances are from Carmichael (1982, Table 99), with some additions from Mason (1966, Tables 3.3 and 8.2). Entries for noble gases are from the latter table and apply to the whole earth rather than the crust. SAD abundances are from Anders and Grevesse (1989). For the SAD, the total number is $1.098 \cdot 10^{12}$. The mass fractions of hydrogen, helium, and all other species are 0.708, 0.273, and 0.019. Atomic Weights from IUPAC (1992) and *HCP*.

Table A5. *SAD Elemental Abundances: Astronomical scale*

(order: atomic number)

Element	Symbol	At. No.	At. Weight	$\log(\epsilon_H)=12.00$[1]	
				ϵ	$\log(\epsilon)$
Hydrogen	H	1	1.01	9.991(+11)	12.00
Helium	He	2	4.00	9.740(+10)	10.99
Lithium	Li	3	6.94	2.045(+03)	3.31
Beryllium	Be	4	9.01	2.614(+01)	1.42
Boron	B	5	10.81	7.592(+02)	2.88
Carbon	C	6	12.01	3.617(+08)	8.56
Nitrogen	N	7	14.01	1.121(+08)	8.05
Oxygen	O	8	16.00	8.523(+08)	8.93
Fluorine	F	9	19.00	3.019(+04)	4.48
Neon	Ne	10	20.18	1.232(+08)	8.09
Sodium	Na	11	22.99	2.055(+06)	6.31
Magnesium	Mg	12	24.30	3.846(+07)	7.59
Aluminum	Al	13	26.98	3.040(+06)	6.48
Silicon	Si	14	28.09	3.581(+07)	7.55
Phosphorus	P	15	30.97	3.724(+05)	5.57
Sulfur	S	16	32.07	1.844(+07)	7.27
Chlorine	Cl	17	35.45	1.876(+05)	5.27
Argon	Ar	18	39.95	3.617(+06)	6.56
Potassium	K	19	39.10	1.350(+05)	5.13
Calcium	Ca	20	40.08	2.188(+06)	6.34
Scandium	Sc	21	44.96	1.225(+03)	3.09
Titanium	Ti	22	47.88	8.594(+04)	4.93
Vanadium	V	23	50.94	1.049(+04)	4.02
Chromium	Cr	24	52.00	4.834(+05)	5.68
Manganese	Mn	25	54.94	3.420(+05)	5.53
Iron	Fe	26	55.85	3.223(+07)	7.51
Cobalt	Co	27	58.93	8.057(+04)	4.91
Nickel	Ni	28	58.69	1.765(+06)	6.25
Copper	Cu	29	63.55	1.869(+04)	4.27
Zinc	Zn	30	65.39	4.512(+04)	4.65
Gallium	Ga	31	69.72	1.354(+03)	3.13
Germanium	Ge	32	72.61	4.261(+03)	3.63
Arsenic	As	33	74.92	2.349(+02)	2.37
Selenium	Se	34	78.96	2.224(+03)	3.35
Bromine	Br	35	79.90	4.226(+02)	2.63
Krypton	Kr	36	83.80	1.611(+03)	3.21
Rubidium	Rb	37	85.47	2.539(+02)	2.40
Strontium	Sr	38	87.62	8.415(+02)	2.93
Yttrium	Y	39	88.91	1.662(+02)	2.22
Zirconium	Zr	40	91.22	4.082(+02)	2.61
Neobium	Nb	41	92.91	2.500(+01)	1.40
Molybdenum	Mo	42	95.94	9.131(+01)	1.96
Technetium	Tc	43	(98.)		
Ruthenium	Ru	44	101.07	6.661(+01)	1.82
Rhodium	Rh	45	102.91	1.232(+01)	1.09
Palladium	Pd	46	106.42	4.978(+01)	1.70
Silver	Ag	47	107.87	1.740(+01)	1.24
Cadmium	Cd	48	112.41	5.765(+01)	1.76
Indium	In	49	114.82	6.589(+00)	.82
Tin	Sn	50	118.71	1.368(+02)	2.14
Antimony	Sb	51	121.76	1.107(+01)	1.04
Tellurium	Te	52	127.60	1.722(+02)	2.24

Table A5. *(continued)*

Element	Symbol	At. No.	At. Weight	$\log(\epsilon_H) = 12.00$[a]	
				ϵ	$\log(\epsilon)$
Iodine	I	53	126.90	3.223(+01)	1.51
Xenon	Xe	54	131.29	1.683(+02)	2.23
Cesium	Cs	55	132.91	1.332(+01)	1.12
Barium	Ba	56	137.33	1.608(+02)	2.21
Lanthanum	La	57	138.91	1.597(+01)	1.20
Cerium	Ce	58	140.12	4.068(+01)	1.61
Praseodymium	Pr	59	140.91	5.977(+00)	.78
Neodymium	Nd	60	144.24	2.965(+01)	1.47
Promethium	Pm	61	(145.)		
Samarium	Sm	62	150.36	9.246(+00)	.97
Europium	Eu	63	151.96	3.484(+00)	.54
Gadolinium	Gd	64	157.25	1.182(+01)	1.07
Terbium	Tb	65	158.93	2.159(+00)	.33
Dysprosium	Dy	66	162.50	1.412(+01)	1.15
Holmium	Ho	67	164.93	3.183(+00)	.50
Erbium	Er	68	167.26	8.981(+00)	.95
Thulium	Tm	69	168.93	1.354(+00)	.13
Ytterbium	Yb	70	173.04	8.877(+00)	.95
Lutetium	Lu	71	174.97	1.314(+00)	.12
Hafnium	Hf	72	178.49	5.515(+00)	.74
Tantalum	Ta	73	180.95	7.413(-01)	−.13
Tungsten	W	74	183.84	4.763(+00)	.68
Rhenium	Re	75	186.21	1.851(+00)	.27
Osmium	Os	76	190.23	2.417(+01)	1.38
Iridium	Ir	77	192.22	2.367(+01)	1.37
Platinum	Pt	78	195.08	4.798(+01)	1.68
Gold	Au	79	196.97	6.696(+00)	.83
Mercury	Hg	80	200.59	1.218(+01)	1.09
Thallium	Tl	81	204.38	6.589(+00)	.82
Lead	Pb	82	207.2	1.128(+02)	2.05
Bismuth	Bi	83	208.98	5.157(+00)	.71
Polonium	Po	84	(209.)		
Astatine	At	85	(210.)		
Radon	Rn	86	(222.)		
Francium	Fr	87	(223.)		
Radium	Ra	88	226.02		
Actinium	Ac	89	227.03		
Thorium	Th	90	232.04	1.200(+00)	.08
Protactinium	Pa	91	231.04		
Uranium	U	92	238.03	3.223(−01)	−.49

[a]We adopt Anders and Grevesse's (1989) transformation from the meteoritic to the astronomical scale by multiplying all former abundances by 35.81. The transformation is based on an average of well determined abundances in the meteorites and the sun. The hydrogen abundance then does not come out *(exactly)* 10^{12}.

For reference, we give some additional properties of the SAD. The abundances sum to $1.098 \cdot 10^{12}$, and the mean molecular weight is 1.298. The fraction of hydrogen by mass, X = 0.70782, and that of helium, Y = 0.27329. The mass fraction of all of the remaining elements is 0.01888.

Table A6. *Isotopic Abundances and Mass Excesses* ($\Delta = M - A$)

Z	El	A	Abundance (%) half-life	Process[a]	Δ
0	n		10.3m		8.0714
1	H	1	~ 100		7.2890
		2	~.002	U, ?	13.1358
		3	12.2y		14.9499
2	He	3	0.018	U, ?	14.9313
		4	~ 100	U, H	2.4249
3	Li	6	7.42	X	14.0874
		7	92.58	X, H, U	14.9083
4	Be	9	100	X	11.3480
5	B	10	19.64	X	12.0518
		11	80.36	X	8.6679
6	C	12	98.89	He	0.0000
		13	1.11	H	3.1250
		14	5715. y		3.0199
7	N	14	99.63	H	2.8634
		15	0.37	H	0.1015
8	O	16	99.76	He	−4.7370
		17	0.037	H	−0.8100
		18	0.204	He, N	−0.7830
9	F	19	100	N	−1.4874
10	Ne	20	88.9 :[b]	C	−7.0430
		21	0.27 :	He, N	−5.7331
		22	10.8 :	He, N	−8.0262
11	Na	23	100	C	−9.5297
12	Mg	24	78.7	C	−13.9307
		25	10.1	C	−13.1908
		26	11.2	C	−16.2125
13	Al	26	$7.2 \cdot 10^5$y		−12.2076
		27	100	C	−17.1944
14	Si	28	92.21	O, Si	−21.4913
		29	4.70	O	−21.8938
		30	3.09	O	−24.4318
15	P	31	100	O	−24.4396
16	S	32	95.0	O, Si	−26.0152
		33	0.76	O, Si	−26.5860
		34	4.22	O, Si	−29.9313
		36	0.01	NSi	−30.6659
17	Cl	35	75.53	O, Si	−29.0137
		37	24.47	O, Si	−31.7618
18	Ar	36	84.2	O, Si	−30.2313
		38	15.8	O, Si	−34.7151
		40		s	−35.0403
19	K	39	93.10	O, Si	−33.8063
		40	0.02	O, Si	−33.5353
		41	6.88	O, Si	−35.5598

Table A6. *(continued)*

Z	El	A	Abundance half-life	Process[a]	Δ
20	Ca	40	96.97	O, Si	−34.8969
		42	0.64	Si, s	−38.5439
20	Ca	43	0.15	Si, s	−38.4054
		44	2.06	Si, s	−41.4660
		46	0.0033	NSi	−43.1382
		48	0.19	NSi	−44.216
21	Sc	45	100	Si, E	−41.0665
22	Ti	46	7.93	E	−44.1227
		47	7.28	E	−44.9310
		48	73.94	E	−48.4877
		49	5.51	E	−48.5587
		50	5.34	E, NSi	−51.4321
23	V	50	0.24	E	−49.2193
		51	99.76	E	−52.1991
24	Cr	50	4.31	E	−50.2580
		52	83.7	E	−55.4153
		53	9.55	E	−55.2837
		54	2.38	E	−56.9313
25	Mn	55	100	E	−57.7100
26	Fe	54	5.82	E	−56.2514
		56	91.66	E	−60.6041
		57	2.19	E	−60.1790
		58	0.33	E	−62.1518
27	Co	59	100	E	−62.2264
28	Ni	58	67.88	E	−60.2243
		60	26.23	E	−64.4702
		61	1.19	E	−64.2191
		62	3.66	E	−66.7454
		64	1.08	E	−67.0979
29	Cu	63	69.09	E	−65.5785
		65	30.91	E	−67.2615
30	Zn	64	48.89	E	−66.0012
		66	27.81	E	−68.8983
		67	4.11	E, s	−67.8796
		68	18.57	E, s	−70.0063
		70	0.62	E, s	−69.5599
31	Ga	69	60.4	E, s	−69.3215
		71	39.6	E, s	−70.1415
32	Ge	70	20.52	E, s	−70.5614
		72	27.43	E, s	−72.5826
		73	7.76	E, s	−71.2935
		74	36.54	E, s	−73.4221
		76	7.76	E, s	−73.2135
33	As	75	100	s, r	−73.0339
34	Se	74	0.87	p	−72.2127
		76	9.02	s	−75.2592

Table A6. *(continued)*

Z	El	A	Abundance half-life	Process[a]	Δ
34	Se	77	7.58	s, r	−74.6061
		78	23.52	s, r	−77.0315
		80	49.82	s, r	−77.7613
		82	9.19	r	−77.586
35	Br	79	50.54	s, r	−76.0700
		81	49.46	s, r	−77.977
36	Kr	78	0.35	p	−74.151
		80	2.27	s, p	−77.897
		82	11.56	s	−80.592
		83	11.55	s, r	−79.9846
		84	56.90	s, r	−82.4319
		86	17.37	r	−83.264
37	Rb	85	72.15	s, r	−82.1588
		87	$4.88 \cdot 10^{10} y$	r	−84.5957
38	Sr	84	0.56	p	−80.641
		86	9.86	s	−84.5121
		87		s	−84.8689
		88	82.56	s, r	−87.9106
39	Y	89	100	s, r	−87.6953
40	Zr	90	51.46	s, r	−88.7646
		91	11.23	s, r	−87.8925
		92	17.11	s, r	−88.4561
		94	17.40	s, r	−87.2639
		96	2.80	r	−85.4447
41	Nb	93	100	s, r	−87.2090
42	Mo	92	15.84	p	−86.807
		94	9.04	p	−88.4123
		95	15.72	s, r	−87.7121
		96	16.53	s	−88.7949
		97	9.46	s, r	−87.5445
		98	23.78	s, r	−88.1154
		100	9.63	r	−86.190
44	Ru	96	5.51	p	−86.076
		98	1.87	p	−88.227
		99	12.72	s, r	−87.6198
		100	12.62	s	−89.2216
		101	17.07	s, r	−87.9516
		102	31.61	s, r	−89.1005
		104	18.58	r	−88.100
45	Rh	103	100	s, r	−88.024
46	Pd	102	0.96	p	−87.926
		104	10.97	s	−89.400
		105	22.23	s, r	−88.422
		106	27.33	s, r	−89.914
		108	26.71	s, r	−89.524
		110	11.81	r	−88.335

Table A6. *(continued)*

Z	El	A	Abundance half-life	Process[a]	Δ
47	Ag	107	51.35	s, r	−88.405
		109	48.65	s, r	−88.722
48	Cd	106	1.22	p	−87.132
		108	0.88	p	−89.252
		110	12.39	s	−90.349
		111	12.75	s, r	−89.254
		112	24.07	s, r	−90.5779
		113	12.26	s, r	−89.0503
		114	28.86	s, r	−90.0196
		116	7.58	r	−88.7176
49	In	113	4.28	p, s	−89.372
		115	95.72	s, r	−89.542
50	Sn	112	0.96	p	−88.659
		114	0.66	p	−90.560
		115	0.35	p, s, r	−90.0351
		116	14.30	s	−91.5261
		117	7.61	s, r	−90.3989
		118	24.03	s, r	−91.6536
		119	8.58	s, r	−90.0667
		120	32.85	s, r	−91.1018
		122	4.72	r	−89.946
		124	5.94	r	−88.240
51	Sb	121	57.25	s, r	−89.5884
		123	42.75	s, r	−89.2175
52	Te	120	0.09	p	−89.404
		122	2.46	s	−90.304
		123	0.87	s	−89.1655
		124	4.61	s	−90.5183
		125	6.99	s, r	−89.019
		126	18.71	s, r	−90.066
		128	31.79	r	−88.9923
		130	34.48	r	−87.348
53	I	127	100	s, r	−88.980
54	Xe	124	0.13 :	p	−87.45
		126	0.12 :	p	−89.163
		128	2.17 :	s	−89.8612
		129	27.5 :	s, r	−88.6975
		130	4.26 :	s	−89.8811
		131	21.4 :	s, r	−88.421
		132	26.0 :	s, r	−89.286
		134	10.17 :	r	−88.126
		136	8.39 :	r	−86.426
55	Cs	133	100	s, r	−88.090
56	Ba	130	0.10	p	−87.303
		132	0.10	p	−88.453
		134	2.42	s	−88.969
		135	6.59	s, r	−87.871

Table A6. *(continued)*

Z	El	A	Abundance half-life	Process[a]	Δ
56	Ba	136	7.81	s	−88.907
		137	11.32	s, r	−87.734
		138	71.66	s, r	−88.274
57	La	138		p	−86.525
		139	99.91	s, r	−87.232
58	Ce	136	0.19	p	−86.50
		138	0.25	p	−87.565
		140	88.48	s, r	−88.082
		142	11.07	r	−84.536
59	Pr	141	100	s, r	−86.019
60	Nd	142	27.11	s	−85.950
		143	12.17	s, r	−84.000
		144	23.85	s, r	−83.746
		145	8.30	s, r	−81.430
		146	17.22	s, r	−80.924
		148	5.73	r	−77.408
		150	5.62	r	−73.683
62	Sm	144	3.09	p	−81.965
		147	$1.06 \cdot 10^{11} y$	s,r	−79.266
		148	11.24	s	−79.336
		149	13.83	s, r	−77.136
		150	7.44	s	−77.050
		152	26.72	r	−74.762
		154	22.71	r	−72.454
63	Eu	151	47.82	s, r	−74.650
		153	52.18	s, r	−73.364
64	Gd	152	0.20	p	−74.704
		154	2.15	s	−73.704
		155	14.73	s, r	−72.072
		156	20.47	s, r	−72.536
		157	15.68	s, r	−70.826
		158	24.87	s, r	−70.692
		160	21.90	r	−67.944
65	Tb	159	100	s, r	−69.537
66	Dy	156	0.05	p	−70.528
		158	0.09	p	−70.411
		160	2.29	s	−69.675
		161	18.88	s, r	−68.057
		162	25.53	s, r	−68.182
		163	24.97	s, r	−66.383
		164	28.18	s, r	−65.968
67	Ho	165	100	s, r	−64.897
68	Er	162	0.14	p	−66.336
		164	1.56	p, s	−65.941
		166	33.41	s, r	−64.922
		167	22.94	s, r	−63.287
		168	27.07	s, r	−62.986

Table A6. *(continued)*

Z	El	A	Abundance half-life	Process[a]	Δ
68	Er	170	14.88	r	−60.105
69	Tm	169	100	s, r	−61.270
70	Yb	168	0.14	p	−61.566
		170	3.03	s	−60.760
		171	14.31	s, r	−59.303
		172	21.82	s, r	−59.251
		173	16.13	s, r	−57.547
		174	31.84	s, r	−56.941
		176	12.73	r	−53.491
71	Lu	175	97.41	s, r	−55.160
		176		s	−53.382
72	Hf	174	0.18	p	−55.831
		176	5.20	s	−54.568
		177	18.50	s, r	−52.880
		178	27.14	s, r	−52.435
		179	13.75	s, r	−50.463
		180	35.24	s, r	−49.780
73	Ta	180	0.01	p	−48.939
		181	99.99	s, r	−48.426
74	W	180	0.14	p	−49.625
		182	26.41	s, r	−48.229
		183	14.40	s, r	−46.348
		184	30.64	s, r	−45.688
		186	28.41	r	−42.499
75	Re	185	37.07	s, r	−43.803
		187	$4.2 \cdot 10^{10} y$	s, r	−41.206
76	Os	184	0.02	p	−44.234
		186	1.29	s	−42.988
		187		s	−41.209
		188	13.3	s, r	−41.126
		189	16.1	s, r	−38.979
		190	26.4	s, r	−38.700
		192	41.0	r	−35.876
77	Ir	191	37.3	s, r	−36.699
		193	62.7	s, r	−34.520
78	Pt	190	0.01	p	−37.318
		192	0.78	s	−36.284
		194	32.9	s, r	−34.766
		195	33.8	s, r	−32.803
		196	25.3	s, r	−32.653
		198	7.21	r	−29.921
79	Au	197	100	s, r	−31.151
80	Hg	196	0.15	p	−31.846
		198	10.2	s	−30.965
		199	16.84	s, r	−29.558
		200	23.13	s, r	−29.515

Table A6. *(continued)*

Z	El	A	Abundance half-life	Process[a]	Δ
		201	13.22	s, r	−27.673
		202	29.80	s, r	−27.357
		204	6.85	r	−24.704
81	Tl	203	29.50	s, r	−25.770
		205	70.50	s, r	−23.838
82	Pb	204	1.97	s	−25.118
		206	18.83	s, r	−23.796
		207	20.60	s, r	−22.464
		208	58.55	s, r	−21.760
83	Bi	209	100	s, r	−18.268
90	Th	232	100	r	−35.4472
92	U	235	~ 24.	r	−40.9164
		238	~ 76.	r	−47.3070

Notes: Cameron's isotopic abundances are for the time of solar system formation some 4.6 eons ago. Missing abundance entries indicate uncertainties due to radioactivities; in many cases a rough estimate is possible by adding that percentage necessary to bring the total to 100.

[a]Cameron has assigned nuclides to various named processes. U: Cosmological nucleosynthesis; H: Hydrogen burning; N: Hot hydrogen burning, possibly synthesized in novae; He: Helium burning; C: Explosive carbon burning; Si: explosive silicon burning; NSi: neutron-rich silicon burning; E: nuclear statistical equilibrium; s-, r-, and p-processes.

[b]Uncertain (§8.7).

Sources: Cameron (1982), Lederer and Shirley (1978).

Table A7. *First Four Ionization Energies of the Elements*

Element	Symbol	Electron Volts (eV)			
		χI	χII	χIII	χIV
Hydrogen	H	13.60			
Helium	He	24.59	54.42		
Lithium	Li	5.39	75.64	122.46	
Beryllium	Be	9.32	18.21	153.90	217.72
Boron	B	8.30	25.16	37.93	259.37
Carbon	C	11.26	24.38	47.89	64.49
Nitrogen	N	14.53	29.60	47.45	77.47
Oxygen	O	13.62	35.12	54.94	77.41
Fluorine	F	17.42	34.97	62.71	87.14
Neon	Ne	21.57	40.96	63.46	97.12
Sodium	Na	5.14	47.29	71.62	98.92
Magnesium	Mg	7.65	15.03	80.14	109.27
Aluminium	Al	5.99	18.83	28.45	119.99
Silicon	Si	8.15	16.35	33.49	45.14
Phosphorus	P	10.49	19.73	30.20	51.44
Sulfur	S	10.36	23.33	34.83	47.31
Chlorine	Cl	12.97	23.81	39.61	53.47
Argon	Ar	15.76	27.63	40.74	59.81
Potassium	K	4.34	31.63	45.73	60.91
Calcium	Ca	6.11	11.87	50.91	67.10
Scandium	Sc	6.54	12.80	24.76	73.67
Titanium	Ti	6.82	13.58	27.49	43.27
Vanadium	V	6.74	14.66	29.31	46.71
Chromium	Cr	6.77	16.50	30.96	49.10
Manganese	Mn	7.44	15.64	33.67	51.20
Iron	Fe	7.87	16.18	30.65	54.80
Cobalt	Co	7.86	17.08	33.50	51.30
Nickel	Ni	7.64	18.17	35.17	54.90
Copper	Cu	7.73	20.30	36.84	55.20
Zinc	Zn	9.39	17.97	39.72	59.40
Gallium	Ga	6.00	20.51	30.71	64.00
Germanium	Ge	7.90	15.94	34.22	45.71
Arsenic	As	9.81	18.59	28.35	50.14
Selenium	Se	9.75	21.19	30.82	42.94
Bromine	Br	11.81	21.80	36.00	47.30
Krypton	Kr	14.00	24.36	36.95	52.50
Rubidium	Rb	4.18	26.05	39.02	52.60
Strontium	Sr	5.70	11.03	42.88	56.28
Yttrium	Y	6.38	12.24	20.52	60.60
Zirconium	Zr	6.84	13.13	22.99	34.41
Neobium	Nb	6.88	14.32	25.04	38.30
Molybdenum	Mo	7.10	16.15	27.16	46.40
Technetium	Tc	7.28	15.26	29.55	44.00
Ruthenium	Ru	7.37	16.76	28.47	50.00
Rhodium	Rh	7.46	18.08	31.06	53.00
Palladium	Pd	8.34	19.43	32.93	61.00
Silver	Ag	7.58	21.49	34.83	61.00
Cadmium	Cd	8.99	16.91	37.48	58.00
Indium	In	5.79	18.87	28.04	54.00

Table A7. *(continued)*

Element	Symbol	Electron Volts (eV)			
		χI	χII	χIII	χIV
Tin	Sn	7.34	14.63	30.50	40.74
Antimony	Sb	8.64	16.53	25.30	44.20
Tellurium	Te	9.01	18.60	27.96	37.42
Iodine	I	10.45	19.13	33.00	44.00
Xenon	Xe	12.13	21.21	32.10	46.70
Cesium	Cs	3.89	25.14	33.38	48.00
Barium	Ba	5.21	10.00	35.84	47.10
Lanthanum	La	5.58	11.06	19.18	49.95
Cerium	Ce	5.54	10.85	20.20	36.76
Praseodymium	Pr	5.47	10.55	21.64	38.98
Neodymium	Nd	5.53	10.73	22.10	40.41
Promethium	Pm	5.58	10.90	22.30	41.10
Samarium	Sm	5.64	11.07	23.40	41.40
Europium	Eu	5.67	11.24	24.92	42.60
Gadolinium	Gd	6.15	12.09	20.63	44.00
Terbium	Tb	5.86	11.52	21.91	39.79
Dysprosium	Dy	5.94	11.67	22.80	41.47
Holmium	Ho	6.02	11.80	22.84	42.50
Erbium	Er	6.11	11.93	22.74	42.65
Thulium	Tm	6.18	12.05	23.68	42.69
Ytterbium	Yb	6.25	12.18	25.05	43.74
Lutetium	Lu	5.43	13.90	20.95	42.25
Hafnium	Hf	6.65	14.90	23.30	33.37
Tantalum	Ta	7.89	16.20	22.00	33.00
Tungsten	W	7.98	17.70	24.00	35.00
Rhenium	Re	7.88	16.00	26.00	18.00
Osmium	Os	8.70	16.00	28.00	43.00
Iridium	Ir	9.10	17.00	30.00	45.00
Platinum	Pt	9.00	18.56	35.00	51.00
Gold	Au	9.23	20.50	37.00	54.00
Mercury	Hg	10.44	18.76	34.20	53.00
Thallium	Tl	6.11	20.43	29.83	50.00
Lead	Pb	7.42	15.03	31.94	42.32
Bismuth	Bi	7.29	16.69	25.56	45.30
Polonium	Po	8.42	17.00	29.00	40.00
Astatine	At	9.00	19.00	29.00	42.00
Radon	Rn	10.75	21.00	32.00	42.00
Francium	Fr	3.00	20.00	32.00	43.00
Radium	Ra	5.28	10.15	31.00	43.00
Actinium	Ac	5.17	11.85	17.00	43.00
Thorium	Th	6.08	11.88	20.00	28.80
Protactinium	Pa	5.89	11.60	18.00	29.00
Uranium	U	6.05	11.90	18.90	32.10

Sources: Moore (1970), Martin, Zalubas, and Hagan (1978), Cowan (1981).

Table A8. Ion Radii in Angstrom Units for Valences Indicated

H							He		
Li +1 0.68	Be +2 0.35	B +3 0.23	C +4 0.16	N +5 0.13	O −2 1.32	F −1 1.33	Ne		
Na +1 0.97	Mg +2 0.66	Al +3 0.51	Si +4 0.42	P +5 0.35	S +6 0.30 −2 1.84	Cl −1 1.81	Ar		
K +1 1.33	Ca +2 0.99	Sc +3 0.73	Ti +4 0.68	V +5 0.59 +3 0.74	Cr +6 0.52 +3 0.63	Mn +4 0.60 +2 0.80	Fe +3 0.64 +2 0.74	Co +3 0.63 +2 0.72	Ni +2 0.69
Cu +2 0.72 +1 0.96	Zn +2 0.74	Ga +3 0.62	Ge +4 0.53	As +5 0.46	Se +6 0.42	Br −1 1.96	Kr		
Rb +1 1.47	Sr +2 1.12	Y +3 0.89	Zr +4 0.79	Nb +5 0.69	Mo +6 0.62	Tc +7 0.98	Ru +4 0.67	Rh +3 0.68	Pd +2 0.80
Ag +1 1.26	Cd +2 0.97	In +3 0.81	Sn +4 0.71	Sb +5 0.62 +3 0.76	Te +6 0.56	I −1 2.20	Xe		
Cs +1 1.67	Ba +2 1.34	La +3 1.06 Lu +3 0.85	Hf +4 0.78	Ta +5 0.68	W +6 0.62	Re +4 0.72	Os +6 0.69 +4 0.88	Ir +4 0.68	Pt +2 0.80
Au +1 1.37	Hg +2 1.10	Tl +1 1.47	Pb +4 0.84 +2 1.20	Bi +3 0.96	Po +6 0.67	At +7 0.62	Rn		

References

Aaronson, M., McKee, C. F., and Weisheit, J. C. 1975, "The Identification of Absorption Redshift Systems in Quasar Spectra," *Ap. J.*, **198**, 13.

Abell, G. O. 1958, "The Distribution of Rich Clusters of Galaxies," *Ap. J. Suppl.*, **3**, 211.

Abramowitz, M., and Stegun, I. A. 1964, *Handbook of Mathematical Functions* (Washington DC: Government Printing Off.).

Abt, H. A., Gomez, A. E., and Levy, S. G. 1990, "The Frequency and Formation Mechanism of B2-B5 Main-Sequence Binaries," *Ap. J. Suppl.*, **74**, 551.

Acker, A., Marcout, J., Ochsenbein, F., Stenholm, B., and Tylenda, R. 1992, *The Strasbourg-ESO Catalogue of Galactic Planetary Nebulae*, in 2 vols. (Garching bei München: European Southern Observatory).

Adams, S., Seaton, M. J., and Howarth, I. D. 1984, "K648, The Planetary Nebula in the Globular Cluster M15," *Mon. Not. Roy. Astron. Soc.*, **207**, 471.

Adamson, A. W. 1990, *Physical Chemistry of Surfaces*, 5th ed. (New York: Wiley).

Adelman, S. J., Cowley, C. R., and Hill, G. 1988, "An Introduction to the Coadded Spectrograms of Phi Herculis and Omicron Pegasi," in *Elemental Abundance Analyses*, ed. S. J. Adelman and T. Lanz (Lausanne: Institut d'Astronomie), p. 15.

Aitken, D. K. 1981, "Spectrophotometry of Dust," in *Infrared Astronomy*, ed. G. C. Wynn-Williams and D. P. Cruikshank, IAU Symposium **96** (Dordrecht: Reidel), p. 207.

Aitken, D. K., Roche, P. F., Smith, C. H., James, S. D., and Hough, J. H. 1988, "Infrared Spectropolarimetry of AFGL 2591: Evidence for an Annealed Grain Component," *Mon. Not. Roy. Astron. Soc.*, **230**, 629.

Alcock, C., and Illarionov, A. 1980, "The Surface Chemistry of Stars. II. Fractionated Accretion of Interstellar Matter," *Ap. J.*, **235**, 541.

Alfvén, H., and Arrhenius, G. 1975, *Structure and Evolutionary History of the Solar System* (Dordrecht: Reidel).

Allamandola, L. J., and Tielens, A. G. G. M. (ed.) 1989, *Interstellar Dust*, IAU Symposium **135** (Dordrecht: Kluwer).

Allamandola, L. J., Tielens, A. G. G. M., and Barker, J. R. 1989, "Interstellar Polycyclic Aromatic Hydrocarbons: The Infrared Emission Bands, the Excitation/Emission Mechanism, and the Astrophysical Implications," *Ap. J. Suppl.*, **71**, 733.

Allen, C. W. 1973, *Astrophysical Quantities*, 3rd ed. (London: Athlone Press). Revision edited by A. N. Cox in preparation.

Aller, L. H. 1961, *The Abundance of the Elements*, (New York: Interscience).

Aller, L. H. 1963, *The Atmospheres of the Sun and Stars*, 2nd ed. (New York: Ronald Press).

Aller, L. H. 1984, *Physics of Thermal Gaseous Nebulae* (Dordrecht: Reidel).

Aller, L. H. 1989, "Chemical Compositions of Highly Evolved Stars and their Ejecta," in

Cosmic Abundances of Matter, ed. C. J. Waddington (New York: American Institute of Physics), p. 224.
Aller, L. H., and Keyes, C. D. 1987, "A Spectroscopic Survey of 51 Planetary Nebulae," *Ap. J. Suppl.*, **65**, 405.
Alpher, R. A., Bethe, H. A., and Gamow, G. 1948, "The Origin of the Chemical Elements," *Phys. Rev.*, **73**, 803.
Alpher, R. A., and Herman, R. 1972, "Reflections on 'Big Bang' Cosmology," in *Cosmology, Fusion, and Other Matters*, George Gamow Memorial Volume, ed. F. Reines (Boulder: Colorado Associated University Press), p. 1.
Alter, G., Balázs, B. and Ruprecht, J. 1970, *Catalogue of Star Clusters and Associations* (Budapest: Akadémiai Kiadó).
American Astronomical Society 1993 *Membership Directory* (published by the American Astronomical Society).
Amnuel, P. R. 1993, "Chemical Abundances in Galactic Planetary Nebulae," *Mon. Not. Roy. Astron. Soc.*, **261**, 263.
Anders, E. 1971, "Meteorites and the Early Solar System," *Ann. Rev. Astron. Ap.*, **9**, 1.
Anders, E. 1988, "Circumstellar Material in Meteorites: Noble Gases, Carbon and Nitrogen," in *Meteorites and the Early Solar System*, ed. J. F. Kerridge and M. S. Matthews (Tucson: University of Arizona Press), p. 927.
Anders, E., and Ebihara, M. 1982, "Solar System Abundances of the Elements," *Geochim. Cosmochim. Acta*, **46**, 2363.
Anders, E., and Grevesse, N. 1989, "Abundances of the Elements: Meteoritic and Solar," *Geochim. Cosmochim. Acta.*, **53**, 197.
Anderson, D. L., Miller, W. F., Latham, G. V., Nakamura, Y., Toksöz, M. N., Dainty, A. M., Duennebier, F. K., Lazarewicz, A. R., Kovach, R. L., and Knight, T. C. D. 1977, "Seismology on Mars," *J. Geophys. Res.*, **82**, 4524.
Anderson, L. S. 1989, "Line Blanketing without Local Thermodynamic Equilibrium. II. A Solar-Type Model in Radiative Equilibrium," *Ap. J.*, **339**, 558.
Anderson, O. L. 1988, "Simple Solid-State Equations for Materials of Terrestrial Planetary Interiors," in *The Physics of the Planets*, ed. S. K. Runcorn (Chichester: Wiley), p. 27.
Arfken, G. 1985, *Mathematical Methods for Physicists*, 3rd ed. (Orlando, Fla.: Academic Press).
Arnett, W. D. 1969, "A Possible Model of Supernovae: Detonation of ^{12}C," *Ap. Sp. Sci.*, **5**, 180.
Arnett, W. D. 1971, "Galactic Evolution and Nucleosynthesis," *Ap. J.*, **166**, 153.
Arnett, W. D., Bahcall, J. N., Kirshner, R. P., and Woosley, S. E. 1989, "Supernova 1987A," *Ann. Rev. Astron. Ap.*, **27**, 629.
Arnett, W. D., Fryxell, B., and Müller, E. 1989, "Instabilities and Nonradial Motion in SN 1987A," *Ap. J.*, **341**, L63.
Arpigny, C. 1991, "Comets," in *Reports on Astronomy*, ed. D. McNally, *Transactions of the International Astronomical Union*, **XXIA** (Dordrecht: Kluwer), p. 137.
Aten, A. H. W. 1948, "The Isotopic Composition of Xenon," *Phys. Rev.*, **73**, 1206.
Atkins, P. W. 1990, *Physical Chemistry*, 4th ed., (New York: Freeman).
Atreya, S. K. 1986, *Atmospheres and Ionospheres of the Outer Planets and their Satellites* (Berlin: Springer).
Audoze, J., and Reeves, H. 1982, "The Origin of the Light Elements," in *Essays in Nuclear Astrophysics*, ed. C. A. Barnes, D. D. Clayton, and D. N. Schramm (Cambridge: University Press), p. 355.
Audoze, J., and Tinsley, B. M. 1976, "Chemical Evolution of Galaxies," *Ann. Rev. Astron. Ap.*, **14**, 43.

Baade, W. 1944, "The Resolution of Messier 32, NGC 205, and the Central Region of the Andromeda Nebula," *Ap. J.*, **100**, 137.

Baade, W. 1963, *Evolution of Stars and Galaxies*, (Cambridge, Mass.: Harvard University Press).

Bahcall, J. N. 1968, "A Systematic Method for Identifying Absorption Lines as Applied to PKS 0237–23," *Ap. J.*, **153**, 679.

Bahcall, J. N. 1989, *Neutrino Astrophysics* (Cambridge: University Press)

Baker, J. G., and Menzel, D. H. 1938, "Physical Processes in Gaseous Nebulae III. The Balmer Decrement," *Ap. J.*, **88**, 52.

Bania, T. M., Rood, R. T., and Wilson, T. L. 1987, "Measurements of the ^{3}He Abundance in the Interstellar Medium," *Ap. J.*, **323**, 30.

Bao, Z. Y., and Käppeler, F. 1987, "Neutron Capture Cross Sections for *s*-Process Studies," *At. Data Nuc. Data Tables*, **36**, 411.

Barbuy, B., and Renzini, A. (ed.) 1992, *The Stellar Populations of Galaxies*, IAU Symposium **149** (Dordrecht: Kluwer).

Barin, I. 1989, *Thermochemical Data of Pure Substances* (Weinheim, Germany: VCH Publishers), in two volumes.

Barnes, J. E., and Hernquist, L. E. 1991, "Fueling Starburst Gala with Gas-Rich Mergers," *Ap. J.*, **370**, L65.

Barrow, G. M., 1988, *Physical Chemistry*, 5th ed. (New York: McGraw-Hill).

Barshay, S. S., and Lewis, J. S. 1976, "Chemistry of Primitive Solar Material," *Ann. Rev. Astron. Ap.*, **14**, 81.

Batten, A. H. 1973, *Binary and Multiple Systems of Stars* (Oxford: Pergamon).

Beers, T. C., Preston, G. W., and Shectman, S. A. 1985, "A Search for Stars of Very Low Metal Abundances. I.," *Astron. J.*, **90**, 2089.

Beers, T. C., Preston, G. W., and Shectman, S. A. 1988, "A Catalog of Candidate Field Horizontal-Branch Stars. I.," *Ap. J. Suppl.*, **67**, 461.

Beichman, C. A. 1987, "The IRAS View of the Galaxy and the Solar System," *Ann. Rev. Astron. Ap.*, **25**, 521.

Bel, N., Viala, Y. P., and Guidi, I. 1986, "Chemical Equilibrium from Diffuse to Dense clouds," *Astron. Ap.*, **160**, 301.

Bell, R. A. 1988, "The Overall Abundances of Globular Clusters," in *The Harlow-Shapley Symposium on Globular Cluster Systems in Galaxies*, ed. J. E. Grindlay and A. G. D. Philip (Dordrecht: Kluwer), p. 79.

Bell, R. A., and Gustafsson, B. 1978, "The Colours of G and K Type Giant Stars. II.," *Astron. Ap. Suppl.*, **34**, 229.

Bell, R. A., and Oke, J. B. 1986, "Analysis of Four F Subdwarfs Used as Spectrophotometric Standards, *Ap. J.*, **307**, 253.

Berry, L. G., and Mason, B. 1959, *Mineralogy: Concepts, Descriptions, Determinations* (San Francisco: Freeman).

Berry, L. G., and Mason, B. 1983, *Mineralogy: Concepts, Descriptions, Determinations*, 2nd ed. revised by R. V. Dietrich (San Francisco: Freeman).

Bertout, C. 1989, "T Tauri Stars: Wild as Dust," *Ann. Rev. Astron. Ap.*, **27**, 351.

Bessell, M. S., and Norris, J. 1984, "The Ultra-Metal-Deficient (Population III?) Red Giant CD −38° 245," *Ap. J.*, **285**, 622.

Bethe, H. 1939, "Energy Production in Stars," *Phys. Rev.*, **55**, 434. (See also letter on p. 103, same volume.)

Betz, A. 1987, "Infrared Observations of Circumstellar Molecules," in *Astrochemistry*, ed. M. S. Vardya and S. P. Tarafdar, IAU Symposium **120** (Dordrecht: Reidel), p. 327.

Bignell, R. C. 1983, "High Resolution Maps with the VLA," in *Planetary Nebulae*, ed. D. R. Flower, IAU Symposium **103** (Dordrecht: Reidel), p. 69.

Binney, J., and Tremaine, S. 1987, *Galactic Dynamics* (Princeton: University Press).

Black, D. C., and Matthews, M. S. (ed.) 1985, *Protostars and Planets II*, (Tucson: University of Arizona Press).

Black, J. H., and van Dishoeck, E. F. 1987, "Fluorescent Excitation of Interstellar H_2," *Ap. J.*, **322**, 412.

Blades, J. C., Turnshek, D., and Norman, C. A. 1988, *QSO Absorption Lines: Probing the Universe*, Proc. Sp. Sci. Tel. Inst. Symp. **2** (Cambridge: University Press).

Bodansky, D., Clayton, D. D., and Fowler, W. A. 1968, "Nuclear Quasi-Equilibrium During Silicon Burning," *Ap. J. Suppl.*, **16**, 299.

Boesgaard, A. M., and Steigman, G. 1985, "Big Bang Nucleosynthesis: Theories and Observations," *Ann. Rev. Astron. Ap.*, **23**, 319.

Böhm-Vitense, E. 1992, *Introduction to Stellar Astrophysics:* Volume 3. *Stellar Structure and Evolution* (Cambridge: University Press).

Bolton, C. T. 1971, "Spectral Synthesis of Low-Dispersion Luminosity Criteria in A and F Type Stars," *Astron. Ap.*, **14**, 233.

Bond, H. E., Luck, R. E., and Newman, M. J. 1979, "The Extraordinary Composition of U Aquarii," *Ap. J.*, **233**, 205.

Booth, R. S. 1991, "Extragalactic Molecules," in *Molecular Clouds*, ed. R. A. James and T. J. Millar (Cambridge: University Press), p. 157.

Bowen, I. S. 1928, "The Origin of the Nebular Lines and the Structure of the Planetary Nebulae," *Ap. J.*, **67**, 1.

Bowen, N. L. 1928, *The Evolution of Igneous Rocks* (Princeton: University Press) (1956 Dover reprint).

Bowen, N. L., and Tuttle, O. F. 1950, "The system $NaAlSi_3O_8$–$KAlSi_3O_8$–H_2O," *J. Geol.*, **58**, 489.

Boynton, W. V. 1985, "Meteoritic Evidence Concerning Conditions in the Solar Nebula," in *Protostars and Planets II*, ed. D. C. Black and M. S. Matthews (Tucson: University of Arizona Press), p. 772.

Branch, D. 1990, "Spectra of Supernovae," in *Supernovae*, ed. A. G. Petchek (Berlin: Springer), p. 30.

Brenner, J. D., and Barker, J. R. 1992, "Infrared Emission Spectra of Benzene and Naphthalene: Implications for the Interstellar Polycyclic Aromatic Hydrocarbon Hypothesis," *Ap. J.*, **388**, L39.

Bregman, J. N., Hogg, D. E., and Roberts, M. S. 1992, "Interstellar Matter in Early-Type Galaxies. II. The Relationship Between Gaseous Components and Galaxy Types," *Ap. J.*, **387**, 484.

Brocklehurst, M. 1971, "Calculations of the Level Populations for the Low Levels of Hydrogenic Ions in Gaseous Nebulae," *Mon. Not. Roy. Astron. Soc.*, **153**, 471.

Broecker, W. S., and Oversby, V. M. 1971, *Chemical Equilibria in the Earth* (New York: McGraw-Hill).

Brotzman, L. E. 1992, *ADC Electronic News*, newsletter distributed by e-mail (Goddard, Md: Astronomical Data Center).

Brownlee, D. E. 1985, "Cosmic Dust: Collection and Research," *Ann. Rev. Earth Plan. Sci.*, **13**, 147.

Burbidge, E. M., Burbidge, G. R., Fowler, W. A., and Hoyle, F. 1957, "Synthesis of the Elements in Stars," *Rev. Mod. Phys.*, **29**, 547, (B^2FH).

Burington, R. S., and May, D. C., Jr. 1970, *Handbook of Probability and Statistics With Tables*, 2nd ed. (New York: McGraw-Hill).

Burkhart, C., and Coupry, M. F. 1991, "The A and Am-Fm Stars I. The Abundances of Li, Al, Si, and Fe," *Astron. Ap.*, **249**, 205.

Burrows, A. 1990, "Neutrinos from Supernova Explosions," *Ann. Rev. Nuc. Part. Sci.*, **40**, 181.

Butler, D. 1975, "Metal Abundances of RR Lyrae Stars in Galactic Globular Clusters," *Ap. J.*, **200**, 68.

Butler, D., Dickens, R. J., and Epps, E. 1978, "Studies of RR Lyrae Variables in the Unusual Globular Cluster Omega Centauri. I. Spectroscopic Observations," *Ap. J.*, **225**, 148.

Cameron, A. G. W. 1957, "Nuclear Reactions in Stars and Nucleogenesis," *Pub. Astron. Soc. Pac.*, **169**, 201.

Cameron, A. G. W. 1973, "Abundances of the Elements in the Solar System," *Sp. Sci. Rev.*, **15**, 121.

Cameron, A. G. W. 1979, "The Neutron-Rich Silicon-Burning and Equilibrium Processes of Nucleosynthesis," *Ap. J.*, **230**, L53.

Cameron, A. G. W. 1982, "Elemental and Nuclidic Abundances in the Solar System," in *Essays in Nuclear Astrophysics*, ed. C. A. Barnes, D. D. Clayton, and D. N. Schramm (Cambridge: University Press), p. 23.

Cameron, A. G. W. 1985, "Formation and Evolution of the Primitive Solar Nebula," in *Protostars and Planets II*, ed. D. C. Black and M. S. Matthews (Tucson: University of Arizona Press), p. 1073.

Cameron, A. G. W., Cowan, J. J., Klapdor, H. V., Metzinger, J., Oda, T., and Truran, J. W. 1983, "Steady Flow Approximations to the Helium *R*-Process," *Ap. Sp. Sci.*, **91**, 221.

Cameron, L. M. 1985, "Metallicities and Distances of Galactic Clusters as Determined from UBV-Data: I. The Effects of Metallicity and Reddening on the Colors of Main-Sequence Stars," *Astron. Ap.*, **146**, 59.

Cameron, M., Collerson, K. D., Compston, W., and Morton, R. 1981, "The Statistical Analysis and Interpretation of Imperfectly-Fitted Rb–Sr Isochrons from Polymetamorphic Terrains," *Geochim. Cosmochim. Acta.*, **45**, 1087.

Cannon, A. J., and Pickering, E. C. 1918–1924, "The Henry Draper Catalogue," *Harvard Annals*, **91–99**.

Carmichael, I. S. E., Turner, F. J., and Verhoogen, J. 1974, *Igneous Petrology* (New York: McGraw-Hill).

Carmichael, R. S. 1982, *Handbook of Physical Properties of Rocks*, Vol. I (Boca Raton, Fla.: CRC Press), Table 16.

Carr, M. H. 1981, *The Surface of Mars* (New Haven, Conn.: Yale University Press).

Castellan, G. W. 1983, *Physical Chemistry*, 3rd ed. (Menlo Park, Calif.: Benjamin/Cummings).

Caughlan, G. R., and Fowler, W. A. 1988, "Thermonuclear Reaction Rates V," *At. Data Nuc. Data Tables*, **40**, 283.

Cayrel, R., and Cayrel, G. 1963, "A Detailed Analysis of the Spectrum of Epsilon Virginis," *Ap. J.*, **137**, 431.

Cayrel de Strobel, G., and Spite, M. 1988, *The Impact of Very High S/N Spectroscopy on Stellar Physics*, IAU Symposium **132** (Dordrecht: Kluwer).

Champagne, A. E., and Wiescher, M. 1992, "Explosive Hydrogen Burning," *Ann. Rev. Nucl. Part. Sci.*, **42**, 39.

Chandrasekhar, S. 1949, "Turbulence – A Physical Theory of Astrophysical Interest," *Ap. J.*, **110**, 329.

Chandrasekhar, S. 1983, *Eddington, The Most Distinguished Astrophysicist of His Time* (Cambridge: University Press).

Chase, M. W., Jr., Davies, C. A., Downey, J. R., Jr., Frurip, D. J., McDonald, R. A., and Syverud, A. N. 1986, "JANAF Thermochemical Tables, 3rd ed.," *J. Phys. Chem. Ref. Data.*, **14**, Suppl. **1**, Part I, Al–Co, Part II, Cr–Zr.

Cherchneff, I., and Barker, J. R. 1989, "Infrared Emission from a Polycyclic Aromatic Hydrocarbon (PAH) Excited by Ultraviolet Laser," *Ap. J.*, **341**, L21.

Clark, C. W. 1984, "Isotope Shifts of Some Ultraviolet Transitions of First Row Elements," *Ap. J.*, **285**, 322.

Clary, D. C. 1990, "Fast Chemical Reactions: Theory Challenges Experiment," *Ann. Rev. Phys. Chem.*, **41**, 61.

Clayton, D. D. 1968, *Principles of Stellar Evolution and Nucleosynthesis* (New York: McGraw-Hill). Reprinted with updated references 1983, University of Chicago Press.

Clayton, D. D., Fowler, W. A., Hull, T. E., and Zimmerman, B. A., 1961, "Neutron Capture Chains in Heavy Element Synthesis," *Ann. Phys.*, **12**, 331.

Clayton, D. D., and Leising, M. D. 1987, "^{26}Al in the Interstellar Medium," *Phys. Rept.*, **144**, 1.

Clayton, D. D., and Ward, R. A. 1974, "*s*-Process Studies: Exact Evaluation of an Exponential Distribution of Exposures," *Ap. J.*, **193**, 397.

Clayton, R. N., Grossman, L., and Mayeda, T. K. 1973, "A Component of Primitive Nuclear Composition in Carbonaceous Meteorites," *Science*, **182**, 485.

Clayton, R. N., and Mayeda, T. K. 1977, "Correlated Oxygen and Magnesium Isotopic Anomalies in Allende Inclusions: I Oxygen," *Geophys. Res. Letters*, **4**, 295.

Clegg, R. E. S. 1987, "Collisional Effects in He I Lines and Helium Abundances in Planetary Nebulae," *Mon. Not. Roy. Astron. Soc.*, **229**, 31P.

Climenhaga, J. L. 1960, "Curve of Growth of C_2 Absorption Bands Applied to the Problem of the C^{12}/C^{13} Abundance Ratio," *Pub. Dom. Ap. Obs.*, **XI**, 307.

Cohen, E. R., and Taylor, B. N. 1987, "The 1986 Adjustment of the Fundamental Physical Constants," *Rev. Mod. Phys.*, **59**, 1121.

Condon, E. U., and Shortley, G. H. 1935, *The Theory of Atomic Spectra* (Cambridge: University Press).

Cord, M. S., Lojko, M. S., and Peterson, J. D. 1968, *Microwave Spectral Tables V. Spectral Line Listing* (Washington DC: Government Printing Off.).

Cord, M. S., Peterson, J. D., Lojko, M. S., and Haas, R. H. 1968, *Microwave Spectral Tables IV. Polyatomic Molecules without Internal Rotation* (Washington DC: Government Printing Off.).

Cowan, J. J., Thielemann, F.-K., and Truran, J. W. 1991, "The R-Process and Nucleochronology," *Phys. Rep.*, **208**, 267.

Cowan, R. D. 1981, *The Theory of Atomic Structure and Spectra* (Berkeley: University of California Press).

Cowley, C. R. 1970, *The Theory of Stellar Spectra* (New York: Gordon and Breach).

Cowley, C. R., and Adelman, S. J. 1983, "Abundances in Normal and Chemically Peculiar Stars," *Quart. J. Roy. Astron. Soc.*, **24**, 393.

Cowley, C. R., and Aikman, G. C. L. 1975, "A Study of the λ3984 Feature in the Mercury–Manganese Stars," *Pub. Astron. Soc. Pac.*, **87**, 513.

Cowley, C. R., and Cowley, A. P. 1964, "A New Solar Curve of Growth," *Ap. J.*, **140**, 713.

Cowley, C. R., and Downs, P. L. 1980, "Barium Stars and the *s*-Process," *Ap. J.*, **236**, 648.

Cowley, C. R., and Frey, M. 1989, "Hyperfine Structure in the Barium Resonance Line: Curves of Growth for Solar and *r*-Process Isotopic Mixtures," *Ap. J.*, **346**, 1030.

Cowley, C. R., and Hensberge, H. 1981, "R Andromedae and the Method of Wavelength Coincidence Statistics," *Ap. J.*, **244**, 252.

Crawford, D. L. 1987, *Kitt Peak National Observatory Facilities Book* (Tucson, Ariz.: Kitt Peak National Observatory/NOAO).

Dalgarno, A. 1991, "Interstellar Chemistry," in *Chemistry in Space*, ed. J. M. Greenberg and V. Pirronello (Dordrecht: Kluwer), p. 71.

Dana, J. D. 1985, *Manual of Mineralogy (after James D. Dana)*, 20th ed. by C. Klein and C. S. Hurlbut, Jr. (New York: Wiley).

Danziger, I. J. 1989, "Abundances in Supernova Remnants," in *Cosmic Abundances of Matter*, ed. C. J. Waddington (New York: American Institute of Physics), p. 239.

Danziger, I. J., Lucy, L. B., Bouchet, P., and Gouiffes, C. 1991, "Molecules, Dust, and Ionic Abundances in SN 1987A," in *Supernovae*, ed. S. E. Woosley (Berlin: Springer), p. 69.

David, L. P., Forman, W., and Jones, C. 1990, "Enrichment and Heating of the Intracluster Medium Through Galactic Winds," *Ap. J.*, **380**, 39.

Davidson, K. 1972, "Photoionization and the Emission Line Spectra of Quasi-Stellar Objects," *Ap. J.*, **171**, 213.

Davies, R. L., Sadler, E. M., and Peletier, R. F. 1993, "Line-Strength Gradients in Elliptical Galaxies," *Mon. Not. Roy. Astron. Soc.*, **262**, 650.

Debye, P. 1909, "Der Lichtdruck auf Kugeln von beliebigem Material," *Ann. Physik*, **30**, 57.

Deer, W. A., Howie, R. A., and Zussman, J. 1966, *An Introduction to the Rock-Forming Minerals* (London: Longman).

de Lapparent, V., Geller, M. J., and Huchra, J. P. 1991, "Measures of Large-Scale Structure in the CfA Redshift Survey Slices," *Ap. J.*, **369**, 273.

de Muizon, M., Strom, R. G., Oort, M. J. A., Claas, J. J., and Braun, R. 1988, "G70.7+1.2: Supernova, Nova, or Stellar Shell?" *Astron. Ap.*, **193**, 248.

Deubner, F.-L., and Gough, D. 1984, "Helioseismology: Oscillations as a Diagnostic of the Solar Interior," *Ann. Rev. Astron. Ap.*, **22**, 593.

d'Hendecourt, L. B., and Allamandola, L. J. 1986, "Time Dependent Chemistry in Dense Molecular Clouds. III. Infrared Band Cross Sections of Molecules in the Solid State at 10 K," *Astron. Ap. Suppl.*, **64**, 453.

d'Hendecourt, L. B., and Jourdain de Muizon, M. 1989, "The Discovery of Intersetllar Carbon Dioxide," *Astron. Ap.*, **223**, L5.

Dodd, R. T. 1986, *Thunderstones and Shooting Stars* (Cambridge, Mass.: Harvard University Press).

Dohnanyi, J. S. 1970, "On the Origin and Distribution of Meteoroids," *J. Geophys. Res..* **75**, 3468.

Donahue, M., and Shull, J. M. 1991, "New Photoionization Models of Intergalactic Clouds," *Ap. J.*, **383**, 511.

Donahue, T. M. 1991, private communication.

Donn, B. 1986, "Experimental Investigations Relating to the Properties and Formation of Cosmic Grains," in *Interrelationships Among Circumstellar, Interstellar, and Interplanetary Dust*, ed. J. A. Nuth III and R. E. Stencel, NASA Conference Publication **2403** (NASA), p. 109.

Draine, B. T. 1979, "Time-Dependent Nucleation Theory and the Formation of Interstellar Grains," *Ap. Sp. Sci.*, **65**, 313.

Draine, B. T. 1985, "Tabulated Optical Properties of Graphite and Silicate Grains," *Ap. J. Suppl.*, **57**, 587.

Draine, B. T. 1989, "On the Interpretation of the λ2175 Å Feature," in *Interstellar Dust*, IAU Symposium **135**, ed. L. J. Allamandola and A. G. G. M. Tielens (Dordrecht: Kluwer).

Draine, B. T., and Salpeter, E. E. 1977, "Time-Dependent Nucleation Theory," *J. Chem. Phys.*, **67**, 2230.

Dressler, A. 1984, "The Evolution of Galaxies in Clusters," *Ann. Rev. Astron. Ap.*, **22**, 185.

Durrance, E. M. 1986, *Radioactivity in Geology: Principles and Applications* (New York: Halsted Press).

Dwight, H. B. 1961, *Tables of Integrals and Other Mathematical Data*, 4th ed. (New York: Macmillan).

Dworetsky, M. M. 1969, "Identification of Pt II in the Spectra of Some Peculiar Stars," *Ap. J.*, **156**, L101.

Dworetsky, M. M. 1986, "Non-Magnetic Intermediate-Temperature Stars: A Review," in *Upper Main Sequence Stars with Anomalous Abundances*, ed. C. R. Cowley, M. M. Dworetsky, and C. Megessier, IAU Colloquium **90** (Dordrecht: Reidel), p. 397.

Dworetsky, M. M., Castelli, F., and Faraggiana, R. 1993, *Peculiar Versus Normal Phenomena in A-Type and Related Stars*, Proceedings of IAU Colloquium **138** (San Francisco: Astron. Soc. Pac).

Eddington, A. S. 1927, *Stars and Atoms* (New Haven, Conn.: Yale University Press) (see p. 102).

Edmonds, A. R. 1960, *Angular Momentum in Quantum Mechanics*, Revised printing 1968 (Princeton: University Press).

Edmunds, M. G. 1989, "The Determination and Interpretation of Chemical Abundances from H II Regions," in *Evolutionary Phenomena in Galaxies*, ed. J. E. Beckman and B. E. J. Pagel (Cambridge: University Press), p. 356.

Edmunds, M. G., and Solomon, P. M. 1980, "Introduction," in *Giant Molecular Clouds in the Galaxy*," ed. P. M. Solomon and M. G. Edmunds (Oxford: Pergamon).

Eğe, S. N. 1989, *Organic Chemistry*, 2nd ed. (Lexington, Mass.: D. C. Heath).

Ehlers, E. G. 1987, *The Interpretation of Geological Phase Diagrams* (New York: Dover).

Engleman, R., Jr. 1989, "The Structure and Wavelength of Some Pt II Lines of Astrophysical Interest," *Ap. J.*, **340**, 1140.

Evensen, N. M., Murthy, V. R., and Coscio, M. R., Jr. 1973, "Rb–Sr Ages of Some Mare Basalts and the Isotopic and Trace Element Systematics of Lunar Fines," *Proc. Fourth Lunar. Sci. Conf., Geochim. Cosmochim. Acta*, Suppl. **4**, Vol. **2**, p. 1707.

Eyring, H., Hirshfelder, J. O., and Taylor, H. S. 1936, "The Theoretical Treatment of Chemical Reactions Produced by Ionization Processes. Part I. Ortho–Para Hydrogen Conversion by α-Particles," *J. Chem. Phys.*, **4**, 479.

Faber, S. M. 1973, "Variations in Spectral-Energy Distributions and Absorption-Line Strengths Among Elliptical Galaxies," *Ap. J.*, **179**, 731.

Faraggiana, R., Gerbaldi, M., Castelli, F., and Floquet, M. 1986, "High Resolution Spectrum of the Li I 6708 Region for a Sample of CP 2 and Reference Stars," *Astron. Ap.*, **158**, 200.

Faure, G. 1986, *Principles of Isotope Geology*, 2nd ed. (New York: Wiley).

Ferland, G. J. 1991, *Hazy: A brief Introduction to CLOUDY*, Ohio State University Astronomy Department Internal Report 91-01.

Ferland, G. J., Williams, R. E., Lambert, D. L., Shields, G. A., Slovak, M., Gondhalekar, P. M., and Truran, J. W. 1984, "*IUE* Observations of DQ Herculis and its Nebula and the Nature of the Cold Nova Shells," *Ap. J.*, **281**, 194.

Fesen, R. A., Blair, W. P., and Kirshner, R. P. 1982, "Spectrophotometry of the Cygnus Loop," *Ap. J.*, **262**, 171.

Fesen, R. A., Hamilton, A. J. S., and Saken, J. M. 1989, "Discovery of the Remnant of S Andromedae (SN 1885) in M31," *Ap. J.*, **341**, L55.

Fesen, R. A. and Kirshner, R. P. 1982, "The Crab Nebula. I. Spectrophotometry of the Filaments," *Ap. J.*, **258**, 1.

Fesen, R. A., Wu, C.-C., Leventhal, M., and Hamilton, A. J. S. 1988, "High-Velocity Ultraviolet Iron, Silicon, Oxygen, and Sulfur Absorption Features Associated with the Remnant of SN 1006," *Ap. J.*, **327**, 164.

Fitzpatrick, E. L. 1989, "Interstellar Extinction in External Galaxies," in *Interstellar Dust*, IAU Symposium **135**, ed. L. J. Allamandola and A. G. G. M. Tielens (Dordrecht: Kluwer), p. 37.

Fitzpatrick, E. L., and Massa, D. 1986, "An Analysis of the Shapes of Ultraviolet Extinction Curves. I. The 2175 Å Bump," *Ap. J.*, **307**, 286.

Fitzpatrick, E. L., and Massa, D. 1988, "An Analysis of the Shapes of Ultraviolet Extinction Curves. II. The Far-UV Extinction," *Ap. J.*, **328**, 734.

Foing, B. H. 1991, *Helioseismology from Space* (Oxford: Plenum).

Forman, W., and Jones, C. 1991, "Hot Gas in Clusters of Galaxies," in *Clusters of Galaxies*, ed. W. R. Oegerle, M. J. Fitchett, and L. Danly (Cambridge: University Press), 257.

Foukal, P. 1990, *Solar Astrophysics* (New York: Wiley).

Fowler, W. A. 1972, "New Observations & Old Nucleocosmochronologies," in *Cosmology, Fusion, & Other Matters*, George Gamow Memorial Volume, ed. F. Reines (Boulder: Colorado Associated University Press), p. 67.

Fowler, W. A. 1984, "Experimental and Theoretical Nuclear Astrophysics: the Quest for the Origin of the Elements," *Rev. Mod. Phys.*, **56**, 149 (Nobel Prize Lecture).

Fowler, W. A. 1987, "The Age of the Observable Universe," *Quart. J. Roy. Astron. Soc.*, **28**, 87.

Fowler, W. A., Engelbrecht, C. A., and Woosley, S. E. 1978, "Nuclear Partition Functions," *Ap. J.*, **226**, 984.

Fred, M., Tomkins, F. S., Brody, J. K., and Hamermesh, M. 1951, "The Spectrum of He^3 I," *Phys. Rev.*, **82**, 406.

Frondel, J. W. 1975, *Lunar Mineralogy* (New York: Wiley).

Fukuda, I. 1982, "A Statistical Study of Rotational Velocities of the Stars," *Pub. Astron. Soc. Pac.*, **94**, 271.

Gardiner, W. C., Jr. 1969, *Rates and Mechanisms of Chemical Reactions* (Menlo Park, Calif.: Benjamin).

Gaskell, C. M., Cappellaro, E., Dinerstein, H. L., Garnett, D. R., Harkness, R. P., and Wheeler, J. C. 1986, "Type Ib Supernovae 1983n and 1985f: Oxygen-Rich Late Time Spectra," *Ap. J.*, **306**, L77.

Gaskell, C. M., Shields, G. A., and Wampler, E. J. 1981, "Abundances of Refractory Elements in Quasars," *Ap. J.*, **249**, 443.

Gautier, D., and Owen, T. 1989, "The Composition of Outer Planet Atmospheres," in *Origin and Evolution of Planetary and Satellite Atmospheres*, ed. S. K. Atreya, J. B. Pollack, and M. S. Matthews (Tucson: University of Arizona Press), p. 487.

Gautier, T. N. III, Fink, U., Treffers, R. R., and Larson, H. P. 1976, "Detection of Molecular Hydrogen Quadrupole Emission in the Orion Nebula," *Ap. J.*, **207**, L129.

Genzel, R., and Stutzki, J. 1989, The Orion Molecular Cloud and Star-Forming Region," *Ann. Rev. Astron. Ap.*, **27**, 41.

Gezari, D. Y., Schmitz, M., Pitts, P. S., and Mead, J. M. 1993, *Catalog of Infrared Observations*, 3rd ed., NASA Publication **1294**, (Washington DC: NASA).

Ghoshal, S. N. 1950, "An Experimental Verification of the Theory of Compound Nucleus," *Phys. Rev.*, **80**, 939.

Gilra, D. P. 1971, "Composition of Interstellar Grains," *Nature*, **229**, 237.

Golay, M. 1974, *Introduction to Astronomical Photometry* (Dordrecht: Reidel).

Goldschmidt, V. M. 1937, "Geochemische Verteilungsgesetze der Elemente IX Die Mengenverhältnisse der Elemente und der Atom-Arten," *Skrifter Utgitt av Det Norske Videnskaps-Akademi I Oslo I. Mat.-Naturv. Klasse*, No. 4

Goldschmidt, V. M. 1954, *Geochemistry*, ed. A. Muir (Oxford: Clarendon Press).

Goldsmith, P. F. 1987, "Molecular Clouds: An Overview," in *Interstellar Processes,* ed. D. J. Hollenbach and H. A. Thronson, Jr. (Dordrecht: Kluwer), p. 51.
Goldstein, H. 1980, *Classical Mechanics,* 2nd ed. (Reading, Mass.: Addison-Wesley).
Gordon, M. A. 1988, "H II Regions and Radio Recombination Lines," in *Galactic and Extragalactic Radio Astronomy,* 2nd ed., ed. G. L. Verschuur and K. I. Kellerman (Berlin: Springer Verlag), p. 37.
Gordy, W., and Cook, R. L. 1984, *Microwave Molecular Spectra,* 3rd ed. (New York: Wiley).
Graham, A. L., Bevan, A. W. R., and Hutchison, R. 1985, *Catalogue of Meteorites with Special Reference to those Represented in the British Museum (Natural History)* (Tucson: University of Arizona Press).
Gray, D. F. 1976, *The Observation and Analysis of Stellar Photospheres* (New York: Wiley).
Gray, D. F. 1988, *Lectures on Spectral-Line Analysis: F, G, and K Stars* (Arva, Ontario: The Publisher).
Gray, D. F. 1991, Report of Commission 36: Theory of Stellar Atmospheres in *Reports on Astronomy,* ed. D. McNally, *Transactions of the International Astronomical Union,* **XXIA** (Dordrecht: Kluwer), p. 439.
Greenberg, J. M. 1991, "Physical, Chemical and Optical Interactions with Interstellar Dust," in *Chemistry in Space,* ed. J. M. Greenberg and V. Pirronello (Dordrecht: Kluwer), p. 227.
Grindlay, J. E., and Philip, A. G. D. (ed.) 1986, *The Harlow-Shapley Symposium on Globular Cluster Systems in Galaxies,* IAU Symposium **126** (Dordrecht: Kluwer).
Grossman, L. 1980, "Refractory Inclusions in the Allende Meteorite," *Ann. Rev. Earth Plan. Sci.,* **8**, 559.
Grossman, L., Ganapathy, R., and Davis, A. M. 1977, "Trace Elements in the Allende Meteorite – III. Coarse-Grained Inclusions Revisited," *Geochim. Cosmochim. Acta,* **41**, 1647.
Grossman, L., and Larimer, J. W. 1974, "Early Chemical History of the Solar System," *Rev. Geophys. Space Phys.,* **12**, 71.
Gruen, E. 1991, "Dust Characteristics," in *Reports on Astronomy,* ed. D. McNally, *Transactions of the International Astronomical Union,* **XXIA** (Dordrecht: Kluwer), p. 225.
Gulliver, A. F., and Stadel, J. G. 1990, "Automated Spectral Line Identification," *Pub. Astron. Soc. Pac.,* **102**, 587.
Gursky, H., Kellogg, E., Murray, S., Leong, C., Tananbaum, H., and Giacconi, R. 1971, "A Strong X-ray Source in the Coma Cluster Observed by *Uhuru,*" *Ap. J.,* **167**, L81.
Gustafsson, B. 1989, "Chemical Analyses of Cool Stars," *Ann. Rev. Astron. Ap.,* **27**, 701.
Gustafsson, B., and Bell, R. A. 1978, "The Colours of G and K Type Giant Stars. II.," *Astron. Ap. Suppl.,* **34**, 229.
Hagen, J. B., and Boksenberg, A. 1991, *The Astronomical Almanac for the Year 1992* (Washington DC: Government Printing Off.).
Hall, D. N. B. 1982, "The Space Telescope Observatory," *NASA,* **CP-2244** (Washington DC: NASA).
Hall, J. S. 1949, "Observations of the Polarized Light from Stars," *Science,* **109**, 166.
Halpern, I. 1959, "Nuclear Fission," *Ann. Rev. Nuc. Sci.,* **9**, 245.
Hamilton, A. J. S., and Fesen, R. A. 1988, "The Reionization of Unshocked Ejecta in SN 1006," *Ap. J.,* **327**, 178.
Hammermesh, M. 1962, *Group Theory and its Application to Physical Problems* (Reading, Mass.: Addison-Wesley).
Harkness, R. P., and Wheeler, J. C. 1990, "Classification of Supernovae," in *Supernovae,* ed. A. G. Petchek (Berlin: Springer), p. 1.
Harris, M. J., Fowler, W. A., Caughlan, G. R. and Zimmerman, B. A. 1983, "Thermonuclear Reaction Rates, III," *Ann. Rev. Astron. Ap.,* **21**, 165.

Harrison, G. R. 1939, *MIT Wavelength Tables* (New York: Wiley).
Hartmann, W. K. 1993, *Moons and Planets*, 3rd ed., (Belmont, Calif.: Wadsworth Pub. Co.)
Hartmann, W. K., Phillips, R. J., and Taylor, G. J. (ed.) 1986, *Origin of the Moon* (Houston: Lunar and Planetary Institute).
Hartoog, M. R., and Cowley, A. P. 1979, "The Helium-3 Stars," *Ap. J.*, **228**, 229.
Hartwick, F. D. A. 1976, "The Chemical Evolution of the Galactic Halo," *Ap. J.*, **209**, 418.
Haustein, P. E. 1988, "An Overview of the 1986–1987 Mass Predictions," *At. Data Nuc. Data Tables*, **39**, 185.
Havnes, O. 1974, "Magnetic Accretion and Atmospheric Peculiarities in Early Type Stars," *Astron. Ap.*, **32**, 161.
Haynes, R., and Milne, D. (ed.) 1991, *The Magellanic Clouds*, IAU Symposium **148** (Dordrecht: Kluwer).
Hecht, E., with Zajac, A. 1989, *Optics, 2nd ed.*, (Reading, Mass.: Addison-Wesley).
Heiken, G., Vaniman, D., and French, B. M. 1991, *Lunar Sourcebook* (Cambridge: University Press).
Heintz, W. D. 1978, *Double Stars* (Dordrecht: Reidel).
Hemley, R. J., and Cohen, R. E. 1992, "Silicate Perovskite," *Ann. Rev. Earth Plan. Sci.*, **20**, 553.
Henchman, M. 1972, "Rate Constants and Cross Sections," in *Ion–Molecule Reactions*, Vol. I, ed. J. L. Franklin (New York: Plenum), p. 101.
Herbig, G. H. 1965, "Lithium Abundances in F5-G8 Dwarfs," *Ap. J.*, **141**, 588. (Erratum, *Ap. J.*, **141**, 1592.)
Herbig, G. H. 1975, "The Diffuse Interstellar Bands. IV. The Region 4400–6850 Å," *Ap. J.*, **196**, 129.
Herbig, G. H., and Leka, K. D. 1991, "The Diffuse Interstellar Bands. VIII. New Features Between 6000 and 8650 Å," *Ap. J.*, **382**, 193.
Herbst, E. and Millar, T. J. 1991, "Interstellar Chemistry," in *Molecular Clouds*, ed. R. A. James and T. J. Millar (Cambridge: University Press), p. 209.
Herzberg, G. 1944, *Atomic Spectra and Atomic Structure*, tr. J. W. T. Spinks (New York: Dover).
Herzberg, G. 1945, *Molecular Spectra and Molecular Structure II. Infrared and Raman Spectra of Polyatomic Molecules* (Princeton: Van Nostrand).
Herzberg, G. 1950, *Molecular Spectra and Molecular Structure I. Spectra of Diatomic Molecules*, 2nd ed. (Princeton: Van Nostrand).
Herzberg, G. 1966, *Molecular Spectra and Molecular Structure III: Electronic Spectra and Electronic Structure of Polyatomic Molecules* (New York: Van Nostrand).
Herzberg, G. 1971, *The Spectra and Structures of Simple Free Radicals* (Ithaca, NY: Cornell University Press).
Hesser, J. E. (ed.) 1980, *Star Clusters*, IAU Symposium **85** (Dordrecht: Reidel).
Hewitt, A., and Burbidge, G. 1987, "A New Optical Catalog of Quasi-Stellar Objects," *Ap. J. Suppl.*, **63**, 1.
Hewitt, A., and Burbidge, G. 1989, "The First Addition to the New Optical Catalog of Quasi-Stellar Objects," *Ap. J. Suppl.*, **69**, 1.
Hewitt, A., and Burbidge, G. 1991, "An Optical Catalog of Extragalactic Emission-Line Objects Similar to Quasi-Stellar Objects," *Ap. J. Suppl.*, **75**, 297.
Hiltner, W. A. 1949, "Polarization of Light from Distant Stars by Interstellar Medium," *Science*, **109**, 165.
Hirshfelder, J. O., Curtiss, C. F., and Bird, R. B. 1954, *Molecular Theory of Gases and Liquids* (New York: Wiley).
Hobbs, L. M. 1985, "The Lithium Isotope Ratio in Five F or G Dwarfs," *Ap. J.*, **290**, 284.

Hobbs, L. M., Welty, D. E., Morton, D. C., Spitzer, L., and York, D. G. 1993, "The Interstellar Abundances of Tin and Four Other Heavy Elements," *Ap. J.*, **411**, 750.
Hodge, P. W., and Kennicutt, R. C. 1983, *An Atlas of H II Regions in 125 Galaxies,* (PAPS Document ANJOA88-296-300) (New York: AIP Auxiliary Pub. Serv.)
Hoefs, J. 1987, *Stable Isotope Geochemistry,* 3rd ed. (Berlin: Springer).
Hoffleit, D. 1982, *The Bright Star Catalogue*, 4th ed. (New Haven, Conn.: Yale University Observatory).
Hollenbach, D. J., and Thronson, H. A., Jr. (ed.) 1987, *Interstellar Processes* (Dordrecht: Reidel).
Holweger, H. 1979, "Abundances of the Elements in the Sun," in *Les Eléments et leurs isotopes dans l'univers*, Communications presented at the 22nd Liège International Astrophysical Colloquium (Liège: Institut D'Astrophysique), p. 117.
Holweger, H., and Stürenburg, S. 1993, "Carbon and Silicon in Normal A Stars and in Lambda Bootis Stars," in *Peculiar Versus Normal Phenomena in A-Type and Related Stars*, ed. M. M Dworetsky, F. Castelli, and R. Faraggiana, Proc. IAU Colloq. **138**. (San Francisco: Astron. Soc Pac.), p. 356.
Houk, N. 1982, *Michigan Catalogue of Two-Dimensional Spectral Types for the HD Stars*, Volume 3 (Ann Arbor: Department of Astronomy, University of Michigan).
Houk, N. 1992, material from Volumes 1–4, assembled in advance of publication of Volume 4.
Hoyle, F. 1955, *Frontiers of Astronomy* (New York: Harper & Brothers).
Hoyle, F. 1982, "Two Decades of Collaboration with Willy Fowler," in *Essays in Nuclear Astrophysics*, ed. C. A. Barnes, D. D. Clayton, and D. N. Schramm, (Cambridge: University Press), p. 1.
Hoyle, F., and Wickramasinghe, N. C. 1991, *The Theory of Cosmic Grains* (Dordrecht: Kluwer).
Hubbard, W. B. 1984, *Planetary Interiors* (New York: Van Nostrand).
Hubble, E. 1936, *The Realm of the Nebulae*, (New Haven, Conn.: Yale University Press), Silliman Foundation Lectures, 1982 reprint with foreword by J. E. Gunn.
Huber, K. P., and Herzberg, G. 1979, *Molecular Spectra and Molecular Structure IV. Constants of Diatomic Molecules* (New York: Van Nostrand).
Hughes, P. (ed.) 1991, *Beams and Jets in Astrophysics* (Cambridge: University Press).
Hughes, R. H. 1955, "Isotope Shift in the First Spectrum of Atomic Lithium," *Phys. Rev.*, **99**, 1837.
Hunger, K. 1956, "Zur Theorie der Wachstumskurven," *Zs. f. Astrophys.*, **39**, 36.
Hunten, D. M., Donahue, T. M., Walker, J. C. G., and Kasting, J. F. 1989, "Escape of Atmospheres and Loss of Water," in *Origin and Evolution of Planetary and Satellite Atmospheres*, ed. S. K. Atreya, J. B. Pollack, and M. S. Matthews (Tucson: University of Arizona Press), p. 386.
Huntress, W. T. 1977, "Laboratory Studies of Bimolecular Reactions of Positive Ions in Interstellar Clouds, in Comets, and in Planetary Atmospheres of Reducing Composition," *Ap. J. Suppl.*, **33**, 495.
Icke, V., Preston, H. L., and Balick, B. 1989, "The Evolution of Planetary Nebulae. III. Position–Velocity Images of Butterfly-Type Nebulae," *Astron. J.*, **97**, 462.
Ikezoe, Y. 1987, *Gas Phase Ion–Molecule Reaction Rate Constants through 1986* (Tokyo: Ion Reaction Research Group of the Mass Spectroscopy Society of Japan).
IRAS Science Team 1986, "IRAS Catalogues and Atlases. Atlas of Low Resolution Spectra," *Astron. Ap. Suppl.*, **65**, 607.
Irvine, W. M. 1991, "The Molecular Composition of Interstellar Clouds," in *Chemistry in Space*, ed. J. M. Greenberg and V. Pirronello (Dordrecht: Kluwer), p. 89.

Irvine, W. M. 1992, "Chemistry in the Cosmos," in *Frontiers of Life*, ed. J. T. V. Trân, K. T. V. Trân, J. C. Mounolou, J. Schneider, and C. McKay (Editions Frontières), p. 263.

IUPAC 1989, "Isotopic Compositions of the Elements," *Pure Appl. Chem.,*, **63**, 991.

IUPAC 1992, "Atomic Weights of the Elements 1991," *Pure Appl. Chem.,*, **64**, 1519.

Jackson, J. D. 1975, *Classical Electrodynamics*, 2nd ed. (New York: Wiley)

Jaschek, C. 1989, *Data in Astronomy* (Cambridge: University Press).

Jaschek, C., and Jaschek, M. 1987, *The Classification of Stars* (Cambridge: University Press).

Jaschek, C., and Jaschek, M. 1995, *The Behaviour of Chemical Elements in Stars* (Cambridge: University Press), in press.

Jaschek, M., Andrillat, Y., and Jaschek, C. 1990, "Infrared Observations of υ Sagitarii," *Astron. Ap.*, **232**, 126.

Jaschek, M., and Keenan, P. C. (ed.) 1985, *Cool Stars with Excesses of Heavy Elements* (Dordrecht: Reidel).

Jeanloz, R. 1990, "The Nature of the Earth's Core," *Ann. Rev. Earth Plan. Sci.*, **18**, 357.

Jeans, J. H. 1935, *An Elementary Treatise on Theoretical Mechanics* (Boston, Mass.: Ginn and Co.)

Jeffery, C. S. 1989, "The Analysis of Astronomical Spectra," *Quart. J. Roy. Astron. Soc.*, **30**, 195.

Jeffery, C. S. (ed.) 1993, *Newsletter on Analysis of Astronomical Spectra*, **19** (St. Andrews: Daresbury Laboratory), see also other issues.

Jeffery, D. J., and Branch, D. 1990, "Analysis of Supernova Spectra," in *Supernovae*, ed. J. C. Wheeler, T. Piran, and S. Weinberg (Jerusalem Winter School for Theor. Phys. Vol. 6) (Singapore: World Scientific), p. 149.

Jeffreys, H. 1976, *The Earth, Its Origin, History, and Physical Constitution*, 6th ed. (Cambridge: University Press).

Jenkins, E. B. 1987, "Element Abundances in the Interstellar Atomic Material, in *Interstellar Processes*, ed. D. J. Hollenbach and H. A. Thronson, Jr. (Dordrecht: Kluwer), p. 533.

Jenkins, F. A. 1953, "Report of Subcommittee f (Notation for the Spectra of Diatomic Molecules", *J. Opt. Soc. Am.*, **43**, 425.

Johansson, S., and Cowley, C. R. 1988, "Complex Atoms in Astrophysical Spectra," *J. Opt. Soc. Am.*, **5B**, 2264.

Johnson, H. L., and Morgan, W. W. 1953, "Fundamental Stellar Photometry for Standards of Spectral Type on the Revised System of the Yerkes Spectral Atlas," *Ap. J.*, **117**, 313.

Jorissen, A., Smith, V. V., and Lambert, D. L. 1992, "Fluorine in Red Giant Stars: Evidence for Nucleosynthesis," *Astron. Ap.*, **261**, 164.

Kaler, J. B. 1976, "A Catalog of Relative Emission Line Intensities Observed in Planetary and Diffuse Nebulae," *Ap. J. Suppl.*, **31**, 517.

Kaler, J. B. 1985, "Planetary Nebulae and Their Central Stars", *Ann. Rev. Astron. Ap.*, **23**, 89.

Kaler, J. B., and Jacoby, G. H. 1990, "The Relation Between Chemical Enrichment and Core Mass in Planetary Nebulae," *Ap. J.*, **362**, 491.

Kallemeyn, G. W., Rubin, A. E., and Wasson, J. T. 1991, "The Compositional Classification of Chondrites: V. The Karoonda (CK) Group of Carbonaceous Chondrites," *Geochim. Cosmochim. Act.*, **55**, 881.

Kaplan, W. 1991, *Advanced Calculus*, 4th ed. (Redwood City, Calif.: Addison-Wesley).

Käppeler, F., Beer, H., and Wisshak, K. 1989, "*s*-Process Nucleosynthesis – Nuclear Physics and the Classical Model," *Rep. Prog. Phys.*, **52**, 945.

Kaula, W. M. 1968, *An Introduction to Planetary Physics* (New York: Wiley).

Kaula, W. M. 1986, "The Interiors of the Terrestrial Planets: Their Structure and Evolution," in *The Solar System*, ed. M. G. Kivelson (Englewood Cliffs, NJ: Prentice-Hall), p. 78.

Kennett, B. L. N., and Engdahl, E. R. 1991, "Traveltimes for Global Earthquake Location and Phase Identification," *Geophys. J. Int.*, **105**, 429.

Kennicutt, R. C. 1988, "Properties of H II Region Populations in Galaxies. I. The First-Ranked H II Regions," *Ap. J.*, **334**, 144.

Kerridge, J. F., and Matthews, M. S. (ed.) 1988, *Meteorites and the Early Solar System* (Tucson: University of Arizona Press).

King, E. A. 1983, *Chondrules and their Origin* (Houston: Lunar and Planetary Institute).

Kirshner, R. P., and Chevalier, R. A. 1977, "Spectra of Cassiopeia A. I. Observations," 1977, *Ap. J.*, **218**, 142.

Kirshner, R. P., Oemler, A., Schechter, P. L., and Shectman, S. A. 1981, "A Million Cubic Megaparsec Void in Boötes?," *Ap. J. Lett.*, **248**, L57.

Kittel, C. 1971, *Introduction to Solid State Physics*, 4th ed. (New York: Wiley).

Koeberl, C., and Cassidy, W. A. 1991, "Differences Between Antarctic and non-Antarctic Meteorites: An Assessment," *Geochim. Cosmochim. Acta*, **55**, 3.

Kollatschny, W. 1989, "Aktiv Galaxien," *Sterne u. Weltraum*, **4/89**, 215.

Krauskopf, K. B. 1979, *Introduction to Geochemistry*, 2nd ed. (New York: McGraw-Hill).

Krauss, L. M., and Romanelli, P. 1990, "Big Bang Nucleosynthesis: Predictions and Uncertainties," *Ap. J.*, **358**, 47.

Kudritzki, R. P., and Hummer, D. G. 1990, "Quantitative Spectroscopy of Hot Stars," *Ann. Rev. Astron. Ap.*, **28**, 303.

Kuiper, G. P. 1951, "On the Origin of the Solar System," in *Astrophysics – A Topical Symposium*, ed. J. A. Hynek (New York: McGraw-Hill), p. 357.

Kunth, D., and Sargent, W. L. W. 1983, "Spectrophotometry of 12 Metal-Poor Galaxies: Implications for the Primordial Helium Abundance," *Ap. J.*, **273**, 81.

Kuroda, P. K. 1960, "Nuclear Fission in the Early History of the Earth," *Nature*, **187**, 36.

Kuroda, P. K. 1982, *The Origin of the Chemical Elements and the Oklo Phenomenon* (Berlin: Springer-Verlag).

Kurucz, R. L. 1991a, "South American Tour Summer 1991," Harvard-Smithsonian Preprint No. 3348.

Kurucz, R. L. 1991b, "New Lines, New Models, New Colors," in *Precision Photometry: Astrophysics of the Galaxy*, Van Vleck Observatory Contribution No. 11, ed. A. G. Davis Phillip, A. R. Upgren, and K. A. Janes (Schenectady, NY: L. Davis Press).

Kurucz, R. L. 1993, "A New Opacity-Sampling Model Atmosphere Program for Arbitrary Abundances," in *Peculiar Versus Normal Phenomena in A-Type and Related Stars*, ed. M. M Dworetsky, F. Castelli, and R. Faraggiana, Proc. IAU Colloq. **138**, p. 87.

Kurucz, R. L., and Peytremann, E. 1975, "A Table of Semiempirical gf Values," *Smithsonian Ap. Obs. Spec. Rep.*, **362**, in 3 parts.

Kurucz, R. L., Peytremann, E., and Avrett, E. H. 1974, *Blanketed Model Atmospheres for Early-Type Stars* (Washington DC: Smithsonian Inst.).

Küstner, W. 1921, "Der kugelförmige Sternhaufen Messier 15," *Veröf. Univ. Sternwarte Bonn*, **15**.

Kutner, M. L., Evans, N. J. II, and Tucker, K. D. 1976, "A Dense Molecular Cloud in the OMC-1/OMC-2 Region," *Ap. J.*, **209**, 452.

Kwano, L., Schramm, D. and Steigman, G. 1988, "Primordial Lithium: New Reaction Rates, New Abundances, New Constraints," *Ap. J.*, **327**, 750.

Laird, J. B. 1985, "Abundances in Field Dwarf Stars. II. Carbon and Nitrogen Abundances," *Ap. J.*, **289**, 556.

Laird, J. B., Rupen, M. P., Carney, B. W., and Latham, D. W. 1988, "A Survey of Proper-Motion Stars. VII. The Halo Metallicity Distribution Function." *Astron. J.*, **96**, 1908.

Lambert, D. L. 1988a, "The Chemical Composition of Main Sequence Stars," in *Cosmic Abundances of Matter*, ed. C. J. Waddington (New York: American Institute of Physics), p. 168.

Lambert, D. L. 1988b, "M, S, C Stars," in *Reports on Astronomy, Transactions of the International Astronomical Union*, **XXA**, ed. J.-P. Swings (Dordrecht: Kluwer), p. 342.

Lambert, D. L. 1992, "The *p*-nuclei: Abundances and Origins," *Astron. Ap. Rev.*, **3**, 201.

Landau, L. D., and Lifshitz, E. M. 1958, *Statistical Physics* (Reading, Mass.: Addison-Wesley).

Landau, L. D., and Lifshitz, E. M. 1970, *Theory of Elasticity*, 2nd ed. (Oxford: Pergamon Press).

Lang, K. R. 1980, *Astrophysical Formulae*, 2nd ed. (Berlin: Springer).

Lang, K. R. 1992, *Astrophysical Data: Planets and Stars*, (Berlin: Springer).

Langevin, P. 1905, "Une formule fondamentale de théorie cinétique," *Ann. Chim. et Phys.*, **5**, 8th Ser, 245.

Larson, R. B. 1969, "Numerical Calculations of the Dynamics of a Collapsing Proto-star," *Mon. Not. Roy. Astron. Soc.*, **145**, 271.

Lattanzio, J. C. 1984, "The Effect of Grain Sedimentation on Stellar Evolution", *Mon. Not. Roy. Astron. Soc.*, **207**, 309.

Lattimer, J. M., Schramm, D. N., and Grossman, L. 1978, "Condensation in Supernova Ejecta and Isotopic Anomalies," *Ap. J.*, **219**, 230.

Lebreton, Y., and Maeder, A. 1987, "Stellar Evolution with Turbulent Mixing, VI. The Solar Model, Surface ^{7}Li and ^{3}He Abundances, Solar Neutrinos and Oscillations," *Astron. Ap.*, **175**, 99.

Leckrone, D. S. 1984, "The Resonance Line of Hg II in *IUE* Spectra of Chemically Peculiar Stars," *Ap. J.*, **286**, 725.

Leckrone, D. S., Johansson, S. G., and Wahlgren, G. M. 1991, "High Resolution UV Spectroscopy of the Chemically Peculiar B-Star Chi Lupi," in *The First Year of HST Observations* ed. A. L. Kinney and J. C. Blades (Baltimore, Md.: Space Telescope Inst.), p. 83.

Lederer, C. M., and Shirley, V. S. (ed.) 1978, *Table of Isotopes*, 7th ed. (New York: Wiley)

Lee, T. 1988, "Implications of Isotopic Anomalies for Nucleosynthesis," in *Meteorites and the Early Solar System*, ed. J. F. Kerridge and M. S. Matthews (Tucson: University of Arizona Press), p. 1063.

Levy, E. H., and Lunine, J. I. (ed.) 1993, *Protostars and Planets III* (Tucson: University of Arizona Press).

Lewis, G. N., and Randall, M. 1961, *Thermodynamics*, 2nd ed. revised by K. S. Pitzer and L. Brewer (New York: McGraw-Hill).

Lewis, J. S. 1988, "Origin and Composition of Mercury," in *Mercury*, ed. F. Vilas, C. R. Chapman, and M. S. Matthews (Tucson: University of Arizona Press), p. 651.

Lewis, J. S., and Prinn, R. G. 1984, *Planets and their Atmospheres: Origin and Evolution* (Orlando, Fla.: Academic Press).

Lide, D. R. (ed.) 1992, *Handbook of Chemistry and Physics, 72nd Ed.*, (Boca Raton, Fla.: CRC Press).

Lin-Vien, D., Colthup, N. B., Fateley, W. G., and Grasselli, J. G. 1991, *The Handbook of Infrared and Raman Characteristic Frequencies of Organic Molecules* (Boston, Mass.: Academic Press).

Loeb, L. B. 1961, *The Kinetic Theory of Gases*, 3rd ed. (New York: Dover).

Longhi, J. and Boudreau, A. E. 1979, "Complex Igneous Processes and the Formation of Primitive Lunar Crustal Rocks," *Proc. Lunar Plan. Sci. Conf.*, **10**, 2085.

Lovas, F. J. 1992, "Recommended Rest Frequencies for Observed Interstellar Molecular Microwave Transitions – 1991 Revision," *J. Phys. Chem. Ref. Data*, **21**, 181.

Lu, L. 1991, "The Carbon Abundance of the Lyman-Alpha Clouds," *Ap. J.*, **379**, 99.

Lutz, J. H., and Kaler, J. B. 1983, "Misclassified and Misidentified Planetary Nebulae and Nuclei," *Pub. Astron. Soc. Pac.*, **95**, 739.

Lynden-Bell, D. 1975, "The Chemical Evolution of Galaxies," *Vistas in Astron.*, **19**, 299.

Lynds, B. T. 1962, "Catalogue of Dark Nebulae," *Ap. J. Suppl.*, **7**, 1.

Lynds, B. T. 1965, "Catalogue of Bright Nebulae," *Ap. J. Suppl.*, **12**, 163.

MacAlpine, G. M. 1971, "Static Photoionization Models for Emission-Line Regions of Quasi-Stellar Objects and Seyfert Galaxies," Ph.D. thesis, University of Wisconsin.

MacAlpine, G. M., McGaugh, S. S., and Mazzarella, J. 1989, "The Geometry, Composition, and Mass of the Crab Nebula," *Ap. J.*, **342**, 364.

Maeder, A. 1991, "Chemical Abundances and Models of WR Stars and Blue Supergiants," in *Evolution of Stars: The Photospheric Abundance Connection*, ed. G. Michaud and A. Tutukov (Dordrecht: Kluwer), p. 221.

Magazzù, A., Rebolo, R., and Pavlenko, Ya.V. 1992, "Lithium Abundances in Classical and Weak T Tauri Stars," *Ap. J.*, **392**, 159.

Magisos, M. 1985, "Glossary," in *Protostars and Planets*, ed. D. C. Black and M. S. Matthews (Tucson: University of Arizona Press), p. 1157.

Malaney, R. A., and Lambert, D. L. 1988, "Neutron Source, Neutron Density, and the Origin of Barium Stars," *Mon. Not. Roy. Astron. Soc.*, **235**, 695.

Mallas, J. H., and Kreimer, E. 1978, *The Messier Album* (Cambridge, Mass.: Sky Publishing Corp.).

Margenau, H., and Murphy, G. M. 1956, *The Mathematics of Physics and Chemistry*, 2nd ed. (Princeton, NJ: Van Nostrand)

Martin, G. A., Fuhr, J. R., and Wiese, W. L. 1988, "Atomic Transition Probabilities Scandium Through Manganese,"*J. Phys. Chem. Ref. Data*, **17**, Suppl. 3.

Martin, W. C. 1992, "Sources of Atomic Spectroscopic Data for Astrophysics," in *Atomic and Molecular Data for Space Astronomy: Needs, Analysis and Availability*, ed. P. L. Smith and W. L. Wiese (Berlin: Springer).

Martin, W. C., Zalubas, R., and Hagan, L. 1978, "Atomic Energy Levels – The Rare Earth Elements", *NSRDS-NBS*, **60**.

Mason, B. 1966, *Principles of Geochemistry*, 3rd ed. (New York: Wiley).

Mason, B. (ed.) 1971, *Handbook of Elemental Abundances in Meteorites* (New York: Gordon and Breach).

Mason, B. 1979, "Data of Geochemistry, Sixth Edition, Chapter B. Cosmochemistry Part 1. Meteorites," *Geol. Sur. Prof. Paper*, **440**, B-1.

Mason, B., and Moore, C. B. 1982, *Principles of Geochemistry*, 4th ed. (New York: Wiley)

Mathews, G. J., and Cowan, J. J. 1990, "New Insights Into the Astrophysical r-Process," *Nature*, **345**, 491.

Mathews, G. J., and Ward, R. A. 1985, "Neutron Capture Processes in Astrophysics," *Rep. Prog. Phys.*, **48**, 1371.

Mathis, J. S. 1986, "Observations and Theories of Interstellar Dust," in *Interrelationships Among Circumstellar, Interstellar, and Interplanetary Dust*, ed. J. A. Nuth and R. E. Stencel, NASA Conference Publication 2403, p. 29.

Mathis, J. S. 1990, "Interstellar Dust and Extinction," *Ann. Rev. Astron. Ap.*, **28**, 37.

Mathis, J. S., Rumpl, W., and Nordsieck, K. H. 1977, "The Size Distribution of Interstellar Grains," *Ap. J.*, **217**, 425.

Matteucci, F. 1991, "Chemical Enrichment of the Interstellar Medium," in *Chemistry in Space*, ed. J. M. Greenberg and V. Pirronello (Dordrecht: Kluwer).

Matteucci, F., and François, P. 1992, "Oxygen Abundances in Halo Stars as Tests of Galaxy Formation," *Astron. Ap.*, **262**, L1.

Matthews, T. A., and Sandage, A. R. 1963, "Optical Identification of 3C 48, 3C 196, and 3C 286 with Stellar Objects," *Ap. J.*, **138**, 30.

Matz, S. M., Share, G. H., Leising, M. D., Chupp, E. L., Vestrand, W. T., Purcell, W. R., Strickman, M. S., and Reppin, C. 1988, "Gamma-Ray Line Emission from SN1987A," *Nature*, **331**, 416.

Mayer, M., and Jensen, J. H. D. 1955, *Elementary Theory of Nuclear Shell Structure* (New York: Wiley).

Mazzali, P. A. 1987, "The Ultraviolet Spectrum of Beta Lyrae," *Ap. J. Suppl.*, **65**, 695.

McAlister, H. A. 1991, Report of Commission 26: Double and Multiple Stars, in *Reports on Astronomy*, ed. D. McNally, *Transactions of the International Astronomical Union*, **XXIA** (Dordrecht: Kluwer), p. 243.

McCall, M. L., Rybski, P. M., and Shields, G. A. 1985, "The Chemistry of Galaxies. I. The Nature of Giant Extragalactic H II Regions," *Ap. J. Suppl.*, **57**, 1.

McClure, R. D. 1976, "Standard stars for DDO photometry," *Astron. J.*, **81**, 182.

McClure, R. D. 1984, "The Barium Stars," *Pub. Astron. Soc. Pac.*, **96**, 117.

McCullough, P. R. 1992, "The Interstellar Deuterium-to-Hydrogen Ratio: A Reevaluation of Lyman Absorption-Line Measurements," *Ap. J.*, **390**, 213.

McGaugh, S. S. 1991, "H II Region Abundances: Model Oxygen Line Ratios," *Ap. J.*, **380**, 140.

McNally, D. (ed.) 1991, *Reports on Astronomy, Transactions of the International Astronomical Union*, **XXIA** (Dordrecht: Kluwer).

McSween, H. Y., Jr. 1987, *Meteorites and their Parent Planets* (Cambridge: University Press).

McSween, H. Y., Jr. 1989, "Achondrites and Igneous Processes on Asteroids," *Ann. Rev. Earth Plan. Sci.*, **17**, 119.

Meggers, W. F., Corliss, C. H., and Scribner, B. F. 1975, *Tables of Spectral-Line Intensities*, NBS Monograph **145** (Washington DC: Government Printing Off.).

Meissner, R., and Janle, P. 1984, in *Landolt–Börnstein, New Series: Vol. 2a. Geophysics of the Solid Earth, the Moon, and the Planets*, ed. K.-H. Hellwege and O. Madelung (Berlin: Springer), see p. 410.

Mendoza, C. 1983, "Recent Advances in Atomic Calculations and Experiments of Interest in the Study of Planetary Nebulae," in *Planetary Nebulae*, ed. D. R. Flower (Dordrecht: Reidel), p. 143 (see the appendix).

Mendoza, C., and Zeippen, C. J. 1982, "Transition Probabilities for Forbidden Lines in the $3p^3$ Configuration," *Mon. Not. Roy. Astron. Soc.*, **198**, 127.

Menzel, D. H. (ed.) 1962, *Selected Papers on Physical Processes in Ionized Plasmas* (New York: Dover).

Menzel, D. H., Aller, L. H., and Hebb, M. H. 1941, "Physical Processes in Gaseous Nebulae XIII. The Electron Temperatures of Some Typical Planetary Nebulae," *Ap. J.*, **93**, 230.

Merrill, K. M., and Ridgway, S. T. 1979, "Infrared Spectroscopy of Stars," *Ann. Rev. Astron. Ap.*, **17**, 9.

Merrill, P. W. 1952, "Technetium in the Stars," *Science*, **115**, 484.

Merrill, P. W. 1956, *Lines of the Chemical Elements in Astronomical Spectra*, (Washington DC: Carnegie Institution).

Merrill, P. W., and Greenstein, J. G. 1956, "Revised List of Absorption Lines in the Spectrum of R Andromedae," *Ap. J. Suppl.*, **2**, 225.

Meyer, J.-P. 1989, "Elemental Abundances in the Interstellar Medium...and Elsewhere," in *Cosmic Abundances of Matter*, ed. C. J. Waddington (New York: American Institute of Physics), p.245.

Meyer, P., Ramaty, R., and Webber, W. R. 1974, "Cosmic Rays – Astronomy With Energetic Particles," *Phys. Today*, **27**, 23.

Mezger, P. G. 1991, "Galactic Molecular Masers," in *Reports on Astronomy*, ed. D. McNally, *Transactions of the International Astronomical Union*, **XXIA** (Dordrecht: Kluwer), p. 459.

Michaud, G. 1970, "Diffusion Processes in Peculiar A Stars," *Ap. J.*, **160**, 641.

Michaud, G., and Tutukov, A. (ed.) 1991, *Evolution of Stars: The Photospheric Abundance Connection*, IAU Symposium **145** (Dordrecht: Kluwer).

Mie, G. 1908, "Beitrag zur Optik trüber Medien, speziell kolloidaler Metallösungen," *Ann. Physik*, **25**, 377.

Mihalas, D. 1978, *Stellar Atmospheres*, 2nd ed. (San Francisco: Freeman).

Mihalas, D., and Binney, J. 1981, *Galactic Astronomy*, 2nd ed. (San Francisco: Freeman).

Millar, T. J., Rawlings, J. M. C., Bennett, A., Brown, P. D., and Charnley, S. B. 1991, "Gas Phase Reactions and Rate Coefficients for use in Astrochemistry. The UMIST Ratefile," *Astron. Ap. Suppl.*, **87**, 585.

Millar, T. J., and Williams, D. A. (ed.) 1988, *Rate Coefficients in Astrochemistry* (Dordrecht: Kluwer).

Minkowski, R. 1946, "New Emission Nebulae," *Pub. Astron. Soc. Pac.*, **58**, 305.

Minnaert, M., and Mulders, G. F. W. 1931, "Dopplereffekt und Dämpfung bei den Fraunhoferschen Linien," *Z. f. Astrophys.*, **2**, 165.

Mitchell, D. N., and Le Roy, D. J. 1973, "Rate Constants for the Reaction $D + H_2 = DH + H$ at Low Temperatures Using ESR Detection," *J. Chem. Phys.*, **58**, 3449.

Moore, C. E. 1949, *Atomic Energy Levels*, Volume I, NBS Circular, **467** (Washington DC: Government Printing Off.). See also Vol. II, 1952, and Vol. III, 1958.

Moore, C. E. 1970, "Ionization Potentials and Ionization Limits Derived from the Analyses of Optical Spectra," *NSRDS-NBS*, **34**.

Moore, C. E. 1972, "A Multiplet Table of Astrophysical Interest, Revised Edition," *NSRDS-NBS*, **40**, reprinted from *Contributions from the Princeton University Observatory*, No. 20, 1945.

Moore, C. E., Minnaert, M. G. J., and Houtgast, J. 1966, *The Solar Spectrum* 2935 Å *to* 8770 Å, NBS Monograph **61**.

Morgan, W. W., and Keenan, P. C. 1973, "Spectral Classification," *Ann. Rev. Astron. Ap.*, **11**, 29.

Murnaghan, F. D. 1951, *Finite Deformation of an Elastic Solid* (New York: Wiley).

Myers, W. D. 1977, *Droplet Model of Atomic Nuclei* (New York: IFI/Plenum).

Narlikar, J. V., and Padmanabhan, T. 1991, "Inflation for Astronomers," *Ann. Rev. Astron. Ap.*, **29**, 325.

Newburn, R. L., Jr., Neugebauer, M., and Rahe, J. (ed.) 1991, *Comets in the Post-Halley Era* (Dordrecht: Kluwer).

Nickel, E. H., and Nichols, M. C. 1991, *Mineral Reference Manual* (New York: Van Nostrand).

Nomoto, K., Shigeyama, T., and Tsujimoto, T. 1991, "Supernova Abundance Generation," in *Evolution of Stars: The Photospheric Abundance Connection*, ed. G. Michaud and A. Tutukov (Dordrecht: Kluwer), p. 21.

O'Neill, H. St.C. 1991a, "The Origin of the Moon and the Early History of the Earth – A Chemical Model. Part 1: The Moon," *Geochim. Cosmochim. Acta*, **55**, 1135.

O'Neill, H. St.C. 1991b, "The Origin of the Moon and the Early History of the Earth – A Chemical Model. Part 2: The Earth," *Geochim. Cosmochim. Acta*, **55**, 1159.

O'Nions, R. K., Carter, S. R., Evensen, N. M., Hamilton, P. J. 1979, "Geochemical and Cosmochemical Applications of Nd Isotope Analysis," *Ann. Rev. Earth Plan. Sci.*, **7**, 11.

Oort, J. H. 1977, "The Galactic Center," *Ann. Rev. Astron. Ap.*, **15**, 295.

Oort, J. H. 1983, "Superclusters," *Ann. Rev. Astron. Ap.*, **21**, 373.

Osmer, P. S. 1980, "Q0353-383: The Best Case Yet for Abundance Anomalies in Quasars," *Ap. J.*, **237**, 666.

Osmer, P. S., and Smith, M. G. 1980, "Discovery and Spectrophotometry of the −40° Zone of the CTIO Curtis Schmidt Survey," *Ap. J. Suppl.*, **42**, 333.

Osterbrock, D. E. 1989, *Astrophysics of Gaseous Nebulae and Active Galactic Nuclei* (Mill Valley, Calif.: University Science Books).

Ott, U. 1993, "Physical and Isotopic Properties of Surviving Interstellar Carbon Phases," in *Protostars and Planets III*, ed. E. H. Levy, and J. I. Lunine (Tucson: University of Arizona Press), p. 883.

Page, L. 1935, *Introduction to Theoretical Physics*, 2nd ed. (New York: Van Nostrand).

Pagel, B. E. J. 1979, "Abundances in Unevolved Cool Stars," in *Les Eléments et leurs isotopes dans l'univers*, 22nd Liège International Astrophysical Colloquium (Liège: Institut d'Astrophysique), p.261.

Pagel, B. E. J. 1989, "The G-Dwarf Problem and Radio-active Cosmochronology," in *Evolutionary Phenomena in Galaxies*, ed. J. E. Beckman and B. E. J. Pagel (Cambridge: University Press), p. 201.

Pagel, B. E. J. 1990, "Abundances in H II Regions," in *Chemical and Dynamical Evolution of Galaxies*, ed. F. Ferrini, J. J. Franco, and F. Matteucci.

Pagel, B. E. J. 1991, "Big-Bang Nucleosynthesis – Observational Aspects," *Phys. Scr.*, **T36**, 7.

Pagel, B. E. J., and Edmunds, M. G. 1981, "Abundances in Stellar Populations and the Interstellar Medium in Galaxies," *Ann. Rev. Astron. Ap.*, **19**, 77.

Pagel, B. E. J., and Patchett, B. E. 1975, "Metal Abundances in Nearby Stars and the Chemical History of the Solar Neighbourhood," *Mon. Not. Roy. Astron. Soc.*, **172**, 13.

Palme, H., and Boynton, W. V. 1993, "Meteoritic Constraints on Conditions in the Solar Nebula," in *Protostars and Planets III*, ed. E. H. Levy, and J. I. Lunine (Tucson: University of Arizona Press), p. 979.

Papanastassiou, D. A., and Wasserburg, G. J. 1969, "Initial Strontium Isotopic Abundances and the Resolution of Small Time Differences in the Formation of Planetary Objects," *Earth Plan. Sci. Let.*, **5**, 361.

Papike, J., Taylor, L., and Simon, S. 1991, "Lunar Minerals," in *Lunar Sourcebook*, ed. G. H. Heiken, D. T. Vaniman, and B. French (Cambridge: University Press), p. 121.

Parker, B. 1990, *Colliding Galaxies: the Universe in Turmoil* (New York: Plenum).

Pasteris, J. D. 1984, "Kimberlites: Complex Mantle Melts," *Ann. Rev. Earth Plan. Sci.*, **12**, 133.

Pauling, L., and Wilson, E. B. 1935, *Introduction to Quantum Mechanics* (New York: McGraw-Hill).

Pearse, R. W. B., and Gaydon, A. G. 1976, *The Identification of Molecular Spectra*, 4th ed. (New York: Wiley).

Pease, F. G. 1928, "A Planetary Nebula in the Globular Cluster Messier 15," *Pub. Astron. Soc. Pac.*, **40**, 342.

Peebles, P. J. E. 1993, *Principles of Physical Cosmology* (Princeton: University Press).

Pei, Y. C., Fall, S. M., and Bechtold, J. 1991, "Confirmation of Dust in Damped Lyman-Alpha Systems," *Ap. J.*, **378**, 6.

Penzias, A. A. 1972, "Cosmology and Microwave Astronomy," in *Cosmology, Fusion, and Other Matters*, George Gamow Memorial Volume, ed. F. Reines (Boulder: Colorado Associated University Press), p. 29.

Pepin, R. O. 1989, "Atmospheric Compositions: Key Similarities and Differences," in *Origin and Evolution of Planetary and Satellite Atmospheres*, ed. S. K. Atreya, J. B. Pollack, and M. S. Matthews (Tucson: University of Arizona Press), p. 291.

Pepin, R. O. 1992, "Origin of Noble Gases in the Terrestrial Planets," *Ann. Rev. Earth Plan. Sci.*, **20**, 389.

Perek, L., and Kohoutek, L. 1967, *Catalogue of Galactic Planetary Nebulae* (Prague: Czechoslovak Academy of Sciences).

Petschek, A. G. (ed.) 1990, *Supernovae* (Berlin: Springer).

Phillips, T. G., Huggins, P. J., Wannier, P. G., and Scoville, N. Z. 1979, "Observations of CO (J = 2-1) Emission From Molecular Clouds," *Ap. J.*, **231**, 720.

Pickles, A. 1987, "Population Synthesis of Composite Systems," in *Structure and Dynamics of Elliptical Galaxies*, IAU Symposium **127**, ed. T. de Zeeuw (Dordrecht: Reidel), 203.

Podosek, F. A. 1978, "Isotopic Structures in Solar System Materials," *Ann. Rev. Astron. Ap.*, **16**, 293.

Podosek, F. A., and Swindle, T. D. 1988, "Extinct Radionuclides," in *Meteorites and the Early Solar System*, ed. J. F. Kerridge and M. S. Matthews (Tucson: University of Arizona Press), p. 1093.

Pollack, J. B. 1985, "Formation of the Giant Planets and their Satellite-Ring Systems: An Overview," in *Protostars and Planets II*, ed. D. C. Black and M. S. Williams (Tucson: University of Arizona Press), p. 791.

Pouchert, C. J. 1989, *The Aldrich Library of FT-IR Spectra*, (Milwaukee, Wis.: Aldrich Chemical Co.).

Prantzos, N., Hashimoto, M., Rayet, M., and Arnould, M. 1991, "The p-Process in a Realistic Supernova Model," in *Supernovae*, ed. S. E. Woosley (Berlin: Springer), p. 622.

Preston, G. W. 1959, "A Spectroscopic Study of the RR Lyrae Stars," *Ap. J.*, **130**, 507.

Preston, G. W. 1974, "The Chemically Peculiar Stars of the Upper Main Sequence," *Ann. Rev. Astron. Ap.*, **12**, 257.

Preston, M. A. 1962, *Physics of the Nucleus* (Reading, Mass.: Addison-Wesley).

Prévot, M. L., Lequeux, J., Maurice, E., Prévot, L., and Rocca-Volmerange, B. 1984, "The Typical Interstellar Extinction in the Small Magellanic Cloud," *Astron. Ap.*, **132**, 389.

Prinn, R. G. 1993, "Chemistry and Evolution of Gaseous Circumstellar Disks," in *Protostars and Planets III*, ed. E. H. Levy, and J. I. Lunine (Tucson: University of Arizona Press), p. 1005.

Prior, G. T. 1916, "On the genetic Relationship and Classification of Meteorites," *Mineral. Mag.*, **18**, 26.

Proffitt, C. R., and Michaud, G. 1989, "Abundance Anomalies in A and B Stars and the Accretion of Nuclear Processed Material From Supernovae and Evolved Giants," *Ap. J.*, **345**, 998.

Prutton, C. F., and Maron, S. H., 1965, *Fundamental Principles of Physical Chemistry*, 4th ed. (New York: Macmillan).

Przybylski, A. 1961, "HD 101065—a G0 Star with High Metal Content," *Nature*, **189**, 739.

Przybylski, A. 1966, "Abundance Analysis of the Peculiar Star HD 101065," *Nature*, **210**, 20.

Przybylski, A. 1977, "Is Iron Present in the Atmosphere of HD 101065?" *Mon. Not. Roy. Ast. Soc*, **178**, 71.

Przybylski, A., and Kennedy, P. M. 1963, "The Spectrum of HD 101065," *Pub. Astron. Soc. Pac.*, **75**, 349.

Puget, J. L, and Léger, A. 1989, "A New Component of the Interstellar Matter: Small Grains and Large Aromatic Molecules," *Ann. Rev. Astron. Ap.*, **27**, 161.

Purcell, E. M., and Pennypacker, C. R. 1973, "Scattering and Absorption of Light by Nonspherical Dielectric Grains," *Ap. J.*, **186**, 705.

Rana, N. C. 1991, "Chemical Evolution of the Galaxy," *Ann. Rev. Astron. Ap.*, **29**, 129.

Reader, J., and Corliss, C. H. (ed.) 1991, "Line Spectra of the Elements," in *Handbook of Chemistry and Physics*, 72nd ed., ed. D. R. Lide (Boca Raton, Fla.: CRC Press), p. **10**-1.

Rebolo, R. 1990, "Surface Abundances of Light Elements in Stars," in *New Windows to the Universe*, Vol. I, ed. F. Sanchez and M. Vazquez (Cambridge: University Press), p. 301.

Redfors, A. 1991, "Oscillator Strengths for Y III and Zr III in the IUE Region," *Astron. Ap.*, **249**, 589.

Reeves, H. 1991, "Elements in the Cosmos: Origin of the Light Elements," in *Evolution of Stars: The Photospheric Abundance Connection*, ed. G. Michaud and A. Tutukov (Dordrecht: Kluwer), p. 3.

Reid, M. J., and Moran, J. M. 1981, "Masers," *Ann. Rev. Astron Ap.*, **19**, 231.

Reif, F. 1965, *Fundamentals of Statistical and Thermal Physics* (New York: McGraw-Hill).

Renson, P. 1967, "Abondances anormales d'éléments dans les étoiles Ap," *Ann. d'Ap.*, **30**, 697.

Reynolds, J. H. 1967, "Isotopic Anomalies in the Solar System," *Ann. Rev. Nucl. Sci.*, **17**, 253.

Rich, R. M. 1990, "Chemical Evolution of the Galactic Bulge," in *ESO/CTIO Workshop on Bulges of Galaxies* (Garching bei München: European Southern Observatory), p. 65.

Richardson, S. M., and McSween, H. 1989, *Geochemistry: Pathways and Processes* (Englewood Cliffs, NJ: Prentice-Hall).

Ringwood, A. E. 1975, *Composition and Petrology of the Earth's Mantle* (New York: McGraw-Hill).

Ringwood, A. E. 1979, *Origin of the Earth and Moon* (New York: Springer).

Robie, R. A., Hemingway, B. S., and Fisher, J. R. 1978, "Thermodynamic Properties of Minerals and Related Substances at 298.15 K and 1 Bar (10^5 Pascals) Pressure and at Higher Temperatures," *Geol. Survey Bull.*, **1452**.

Robie, R. A., and Waldbaum, D. R. 1968, "Thermodynamic Properties of Minerals and Related Substances at 298.15 °K (25.0 °C) and One Atmosphere (1.013 Bar) Pressure and at Higher Temperatures," *Geol. Survey Bull.*, **1259**.

Rodney, W. S., and Rolfs, C. E. 1982, "Hydrogen Burning in Massive Stars," in *Essays in Nuclear Astrophysics*, ed. C. A. Barnes, D. D. Clayton, and D. N. Schramm, (Cambridge: University Press), p. 171.

Rolfs, C. E., and Rodney, W. S. 1988, *Cauldrons in the Cosmos* (Chicago: University Press).

Rood, H. J., and Sastry, G. N. 1971, "'Tuning-Fork' Classification of Rich Clusters of Galaxies," *Pub. Astron. Soc. Pac.*, **83**, 313.

Rood, R. T., Bania, T. M., and Wilson, T. L. 1992, "Detection of Helium-3 in a Planetary Nebula," *Nature*, **355**, 618.

Rosen, B. (ed.) 1970, *Spectroscopic Data Relative to Diatomic Molecules,* Vol. 17 of International Tables of Selected Constants, ed. S. Bourcier (Oxford: Pergamon).

Rubin, V. C., Ford, W. K., and Thonnard, N. 1978, "Extended Rotation Curves of High-Luminosity Spiral Galaxies: IV. Systematic Dynamical Properties," *Ap. J.*, **225**, L107.

Russell, H. N. 1934, "Molecules in the Sun and Stars," *Ap. J.*, **79**, 317.

Russell, H. N., Dugan, R. S., and Stuart, J. Q. 1938, *Astronomy: A Revision of Young's Manual of Astronomy. II. Astrophysics and Stellar Astronomy* (Boston, Mass.: Ginn and Co.), see pp. 819 ff.

Russell, R. W., Soifer, B. T., and Willner, S. P. 1977, "The 4 to 8 Micron Spectrum of NGC 7027," *Ap. J.*, **217**, L149.

Russell, S. C., and Bessell, M. S. 1989, "Abundances of Heavy Elements in the Magellanic Clouds. I. Metal Abundances of F-Type Supergiants," *Ap. J. Supp.*, **70**, 865.

Sadakane, K., Jugaku, J., and Takada-Hidai, M. 1985, "The Resonance Lines of B II and Be II in Hg–Mn Stars," *Ap. J.*, **297**, 240.

Salpeter, E. E. 1952, "Nuclear Reactions in Stars Without Hydrogen," *Ap. J.*, **115**, 326.

Salpeter, E. E. 1977, "Formation and Destruction of Dust Grains," *Ann. Rev. Astron. Ap.*, **15**, 267.

Sandage, A. 1969, "New Subdwarfs. II. Radial Velocities, Photometry, and Preliminary Space Motions for 112 Stars with Large Proper Motion," *Ap. J.*, **158**, 1115.

Sandage, A., Sandage, M., and Kristian, J. (ed.) 1975, *Galaxies and the Universe*, vol. IX of *Stars and Stellar Systems*, ed. G. Kuiper and B. Middlehurst (Chicago: University Press).

Sandford, S. A., Allamandola, L. J., Tielens, A. G. G. M., Sellgren, K., Tapia, M., and Pendleton, Y. 1991, "The Interstellar C–H Stretching Band Near 3.4 Microns: Constraints on the Composition of Organic Material in the Diffuse Interstellar Medium," *Ap. J.*, **371**, 607.

Sandford, S. A., and Walker, R. M. 1985, "Laboratory Infrared Transmission Spectra of Individual Interplanetary Dust Particles from 2.5 to 25 Microns," *Ap. J.*, **291**, 838.

Sarazin, C. L. 1986, "X-ray Emission from Clusters of Galaxies," *Rev. Mod. Phys.*, **58**, 1

Sargent, W. L., and Jugaku, J. 1961, "The Existence of He^3 in 3 Centauri A," *Ap. J.*, **134**, 777.

Sato, M. 1976, "Oxygen fugacity and Other Thermochemical Parameters of Apollo 17 High-Ti Basalts and their Implications on the Reduction Mechanism," *Proc Lunar Sci. Conf.*, **7**, 1323.

Savage, B. D., and Mathis, J. S. 1979, "Observed Properties of Interstellar Dust," *Ann. Rev. Astron. Ap.*, **17**, 73.

Scalo, J. M. 1976, "A Composite Hertzsprung–Russell Diagram for the Peculiar Red Giants," *Ap. J.*, **206**, 474.

Scalo, J. M. 1986, "The Stellar Initial Mass Function," *Fund. Cos. Phys.*, **11**, 1.

Schadee, A. 1964, "The Formation of Molecular Lines in the Solar Spectrum," *Bull. Astron. Soc. Netherlands*, **17**, 311.

Scheffler, H. 1982, "Structure of the Galaxy," in *Landolt-Börnstein, New Series*, ed. K. Schaifers and H. H. Voigt, Vol. 2c, see p. 205.

Schmalberger, D. C. 1963, "Depth Dependence of Turbulence in the Solar Photosphere," *Ap. J.*, **138**, 693.

Schmidt, M. 1963, "The Rate of Star Formation. II. The Rate of Formation of Stars of Different Mass," *Ap. J.*, **137**, 758.

Schmidt-Kaler, Th. 1982, in *Landolt–Börnstein, New Series*, ed. K. Schaifers and H. H. Voigt, Vol. 2b, see p. 19.

Schramm, D. N. 1982, "The r-Process and Nucleocosmochronology," in *Essays in Nuclear Astrophysics*, ed. C. A. Barnes, D. D. Clayton, and D. N. Schramm (Cambridge: University Press), p. 325.

Schramm, D. N., and Fowler, W. A. 1971, "Synthesis of Superheavy Elements in the *r*-Process," *Nature*, **231**, 103.

Schwartz, R. D., Cohen, M., and Williams, P. M. 1987, "Near-Infrared H_2 Emission from Herbig-Harrow Objects. I. A Survey of Low-Excitation Objects," *Ap. J.*, **322**, 403.

Schwarz, H. E., and Melnick, J. 1989, *The ESO User's Manual*, (Garching bei München: European Southern Observatory).

Schwarzschild, M. 1958, *Structure and Evolution of the Stars* (Princeton: University Press) (Dover Reprint 1965).

Schweizer, F., Seitzer, P., Faber, S. M., Burstein, D., Dalle Ore, C. M. and Gonzalez, J. J. 1990, "Correlations Between Line Strengths and Fine Structure in Elliptical Galaxies," *Ap. J.*, **364**, L33.

Searle, L., and Sargent, W. L. W. 1972, "Inferences from the Composition of Two Dwarf Blue Galaxies," *Ap. J.*, **173**, 25.

Seaton, M. J. 1979, "Interstellar Extinction in the UV," *Mon. Not. Roy. Astron. Soc.*, **187**, 73P.

Secchi, P. A. 1878, *Die Sterne* (Leipzig: F. A. Brockhaus).

Seeger, P. A., Fowler, W. A., and Clayton, D. D. 1965, "Nucleosynthesis of Heavy Elements by Neutron Capture," *Ap. J. Suppl.*, **11**, 121.

Sellgren, K., Allamandola, L. J., Bregman, J. D., Werner, M. W., and Wooden, D. H. 1985, "Emission Features in the 4–13 Micron Spectra of the Reflection Nebulae NGC 7023 and NGC 2023," *Ap. J.*, **299**, 416.

Sharp, C. M. 1988, "Condensation Calculations in Circumstellar Shells for Different C/O Ratios," in *Rate Coefficients in Astrochemistry*, ed. T. J. Millar and D. A. Williams (Dordrecht: Kluwer), 309.

Sharp, C. M., and Huebner, W. F. 1990, "Molecular Equilibrium with Condensation," *Ap. J. Suppl.*, **72**, 417.

Shaver, P. A., McGee, R. X., Newton, L. M., Danks, A. C., and Pottasch, S. R. 1983, "The Galactic Abundance Gradient," *Mon. Not. Roy. Astron. Soc.*, **204**, 53.

Shimanouchi, T. 1972, *Tables of Molecular Vibrational Frequencies, Consolidated Volume I, NSRDS-NBS* **39**, (Washington DC: Government Printing Off.).

Shimanouchi, T. 1977, "Tables of Molecular Vibrational Frequencies, Consolidated Volume II," *J. Phys. Chem. Ref. Data*, **6**, 993.

Shklovsky, I. S. 1960, *Cosmic Radio Waves* (Cambridge, Mass.: Harvard University Press).

Shull, J. M., and Beckwith, S. 1982, "Interstellar Molecular Hydrogen," *Ann. Rev. Astron. Ap.*, **20**, 163.

Signer, P., and Suess, H. 1963, "Rare gases in the Sun, in the Atmosphere, and in Meteorites," in *Earth Science and Meteoritics*, ed. J. Geiss and E. D. Goldberg (Amsterdam: North-Holland), p. 241.

Silk, J. 1991, "The Cosmic Microwave Background," *Phys. Scr.*, **T36**, 16.

Silva, D. R. 1991, "Empirical Population Synthesis: New Directions," PhD thesis, University of Michigan.

Silvergate, P. R. 1984, "Observations of Heavy Element Radio Recombination Lines from C II Regions," *Ap. J.*, **278**, 604.

Simpson, J. A. 1983, "Elemental And Isotopic Composition of the Galactic Cosmic Rays," *Annual. Rev. Nucl. Part. Science*, **33**, 323.

Smith, H. A. 1984, "Metal Abundances of RR Lyrae Stars: Results From Delta-S Spectroscopy," *Pub. Astron. Soc. Pac.*, **96**, 505.

Smith, P. L., and Wiese, W. L. 1992, *Atomic and Molecular Data for Space Astronomy*, (Berlin: Springer).

Smith, V. V. 1993, "The Surface Composition of Chemically Peculiar Stars," in *Nuclei in the Cosmos*, ed. F. Käppeler and K. Wisshak (Bristol: IOP Publishing), p. 17.

Smith, V. V., and Lambert, D. L. 1990, "The Chemical Composition of Red Giants. III. Further CNO Isotopic and s-Process Abundances in Thermally Pulsing Asymptotic Giant Branch Stars," *Ap. J. Suppl.*, **72**, 387.

Sneden, C., Kraft, R. P., Prosser, C. F., and Langer, G. E. 1991, *Astron. J.*, **102**, 2001.

Sobelman, I. I. 1992, *Atomic Spectra and Radiative Transitions*, 2nd ed. (Berlin: Springer).

Soifer, B. T., and Pipher, J. L. 1978, "Instrumentation for Infrared Astronomy," *Ann. Rev. Astron. Ap.*, **16**, 335.

Solomon, P. M., and Edmunds, M. G. 1980, *Giant Molecular Clouds in the Galaxy* (New York: Pergamon).

Spite, F. 1983, "Chemical Composition of Halo Field Stars and the Chemical Evolution of the Halo," in *Highlights of Astronomy*, Vol. **6**, ed. Richard M. West, (Dordrecht: Kluwer Academic), 119.

Spite, F., and Spite, M. 1982, "Abundance of Lithium in Unevolved Halo Stars and Old Disk Stars: Interpretation and Consequences," *Astron. Ap.*, **115**, 357.

Spite, M., Spite, F., Peterson, R. C., and Chaffee, F. H., Jr. 1987, "Lithium Abundance in Two Extreme High-Velocity Metal-Poor Halo Dwarfs," *Astron. Ap.*, **172**, L9.

Spitzer, L., Jr. 1978, *Physical Processes in the Interstellar Medium* (New York: Wiley), see p. 158.

Spitzer, L., Jr. 1982, *Searching Between the Stars* (New Haven, Conn.: Yale University Press).

Steidel, C. C. 1990, "The Properties of Lyman Limit Absorbing Clouds at $z = 3$: Physical Conditions in the Extended Gaseous Halos of High-Redshift Galaxies," *Ap. J. Suppl.*, **74**, 37.

Steinberg, E. P., and Wilkins, B. D. 1978, "Implications of Fission Mass Distributions for the Astrophysical *r*-Process," *Ap. J.*, **223**, 1000.

Stratton, J. A. 1941, *Electromagnetic Theory* (New York: McGraw-Hill).

Strömgren, B. 1966, "Spectral Classification Through Photoelectric Narrow-Band Photometry," *Ann. Rev. Astron. Ap.*, **4**, 433.

Stull, D. R., and Prophet, H. 1971, *JANAF Thermochemical Tables*, 2nd ed. *NSRDS-NBS*, **37** (Washington DC: Government Printing Office).

Suess, H. E. 1965, "Chemical Evidence Bearing on the Origin of the Solar System," *Ann. Rev. Astron. Ap.*, **3**, 217.

Suess, H. E. 1987, *Chemistry of the Solar System: An Elementary Introduction to Cosmochemistry* (New York: Wiley).

Suess, H. E., and Urey, H. C. 1956, "Abundances of the Elements," *Rev. Mod. Phys.*, **28**, 53.

Suess, H. E., and Zeh, H. D. 1973, "The Abundances of the Heavy Elements," *Ap. Sp. Sci.*, **23**, 173.

Sugar, J., and Corliss, C. H. 1985, "Atomic Energy Levels of the Iron-Period Elements: Potassium through Nickel," *J. Phys. Chem. Ref. Data*, **14**, Suppl. 2.

Sulentic, J. W., and Tifft, W. G. 1973, *The Revised New General Catalogue of Nonstellar Astronomical Objects* (Tucson: University of Arizona Press).

Swindle, T. D. 1988, "Trapped Noble Gases in Meteorites," in *Meteorites and the Early Solar System*, ed. J. F. Kerridge and M. S. Matthews (Tucson: University of Arizona Press), p. 535.

Talbot, R. J., and Newman, M. J. 1977, "Encounters Between Stars and Dense Interstellar Clouds," *Ap. J. Suppl.*, **34**, 295.

Tammann, G. A. 1982, "Cosmology," in *Landolt–Börnstein Series*, ed. K. Schaifers and H. H. Voigt, Vol. 2c, see p. 361.

Taylor, S. R. 1982, *Planetary Science: A Lunar Perspective* (Houston: Lunar and Planetary Inst.).

Taylor, S. R. 1988, "Planetary Compositions," in *Meteorites and the Early Solar System*, ed. J. F. Kerridge and M. S. Matthews (Tucson: University of Arizona Press), p. 512

Taylor, S. R. 1992, *Solar System Evolution* (Cambridge: University Press).

Taylor, S. R., and McLennan, S. M. 1985, *The Continental Crust: Its Composition and Evolution*, (Oxford: Blackwell).

Thaddeus, P. 1991, "Large-Scale CO Observations of our Galaxy and Its Nearest Neighbors," in *Molecular Clouds*, ed. R. A. James and T. J. Millar (Cambridge: University Press), p. 1.

Thielemann, F.-K., Arnould, M., and Truran, J. W. 1986, "Thermonuclear Reaction Rates from Statistical Model Calculations," in *Advances in Nuclear Astrophysics*, ed. E. Vangioni-Flam, J. Audoze, M. Casse, J. P. Chieze, and J. T. T. Van. (Gif sur Yvette, France: Frontières), p. 525.

Timmes, F. X. 1991, "On Supernovae Rates, Oxygen, and Iron Abundances," in *Supernovae*, ed. S. E. Woosley (Berlin: Springer).

Tinsley, B. M. 1980, "Evolution of the Stars and Gas in Galaxies," *Fund. Cos. Phys.*, **5**, 287.

Tolman, R. C. 1934, *Relativity, Thermodynamics, and Cosmology* (Oxford: Clarendon Press).

Tomkin, J., and Lambert, D. L. 1983, "Heavy-Element Abundances in the Classical Barium Star HR 774," *Ap. J.*, **273**, 722.

Torres-Peimbert, S., Peimbert, M., and Daltabuit, E. 1980, "*IUE* and Visual Observations of the Orion Nebula and IC 418: The Carbon Abundance," *Ap. J.*, **238**, 133.

Townes, C. H., and Schawlow, A. L. 1975, *Microwave Spectroscopy* (New York: Dover). Reprint of 1955 edition.

Trimble, V. 1975, "The Origin and Abundances of the Chemical Elements," *Rev. Mod. Phys.*, **47**, 877.

Trimble, V. 1982, "Supernovae. Part I: The Events," *Rev. Mod. Phys.*, **54**, 1183.

Trimble, V. 1983, "Supernovae. Part II: The Aftermath," *Rev. Mod. Phys.*, **55**, 511.

Trimble, V. 1987, "Existence and Nature of Dark Matter in the Universe," *Ann. Rev. Astron. Ap.*, **25**, 425.

Trimble, V. 1991, "The Origin and Abundances of the Chemical Elements Revisited," *Astron. Ap. Rev.*, **3**, 1.

Truran, J. W. 1991, "Chemical Evolution of Galaxies: Abundance Trends and Implications," in *Evolution of Stars: The Photospheric Abundance Connection*, ed. G. Michaud and A. Tutukov (Dordrecht: Kluwer), p. 13.

Truran, J. W., Cameron, A. G. W., and Gilbert, A. 1966, "The Approach to Nuclear Statistical Equilibrium," *Can. J. Phys.*, **44**, 563.

Truran, J. W., and Livio, M. 1986, "On the Frequency of Occurrence of Oxygen–Neon–Magnesium White Dwarfs in Classical Nova Systems," *Ap. J.*, **308**, 721.

Tsuji, T. 1973, "Molecular Abundances in Stellar Atmospheres. II.," *Astron. Ap.*, **23**, 411.

Tsuruta, S., and Cameron, A. G. W. 1965, "Composition of Matter in Nuclear Statistical Equilibrium at High Densities," *Can. J. Phys.*, **43**, 2056.

Tuan, T. X., Montmerle, T., and Van, J. T. T. 1987, *Starbursts and Galaxy Evolution* (Gif Sur Yvette, France: Frontières).

Turner, B. E., and Gammon, R. H. 1975, "Interstellar CN at Radio Wavelengths," *Ap. J.*, **198**, 71.

Ulrich, R. K. 1982, "The s-Process," in *Essays in Nuclear Astrophysics*, ed. C. A. Barnes, D. D. Clayton, and D. N. Schramm (Cambridge: University Press), p. 301.

Unsöld, A. 1955, *Physik der Sternatmosphären*, 2nd ed. (Berlin: Springer).

Unsöld, A. 1976, "Abundance Distributions and Origin of the Elements," *Die Naturwissenschaften*, **63**, 443.

Uomoto, A. 1984, "Spectrophotometry of Intermediate-Redshift Quasars," *Ap. J.*, **284**, 497.

Urey, H. C. 1952, *The Planets: Their Origin and Development* (New Haven, Conn.: Yale University Press).

van de Hulst, H. C. 1957, *Light Scattering by Small Particles* (New York: Wiley). Dover Reprint, 1981.

van den Bergh, S. 1957, "Interstellar Gas and Star Creation," *Zs. f. Ap.*, **43**, 236.

van den Bergh, S., Marscher, A. P., and Terzian, Y. 1973, "An Optical Atlas of Galactic Supernovae Remnants," *Ap. J. Suppl.*, **26**, 19.

van den Heuvel, E. P. J. 1967, "On the Origin of Peculiar and Metallic-Line Stars," *Bull. Astron. Soc. Netherlands*, **19**, 11.

van den Heuvel, E. P. J. 1968, "A Study of Stellar Rotation II. The Origin of Ap and Am Stars and Other Slowly Rotating A- and B-Type Main Sequence Stars," *Bull. Astron. Soc. Netherlands*, **19**, 326.

van der Hucht, K. A., and Hidayat, B. 1991, *Wolf–Rayet Stars and Interrelations with Other Massive Stars in Galaxies*, IAU Symposium **143** (Dordrecht: Kluwer).

Vaniman, D., Dietrich, J., Taylor, G. J., and Heiken, G. 1991, "Exploration, Samples, and Recent Concepts of the Moon," in *Lunar Sourcebook*, ed. G. H. Heiken, D. T. Vaniman, and B. French (Cambridge: University Press), p. 6.

van Schmus, W. R., and Wood, J. A. 1967, "A Chemical–Petrologic Classification of Chondritic Meteorites," *Geochim. Cosmochim. Acta*, **31**, 747.

Vauclair, S., and Vauclair, G. 1982, "Element Segregation in Stellar Outer Layers," *Ann. Rev. Astron. Ap.*, **20**, 37.

Venn, K. A., and Lambert, D. L. 1990, "The Chemical Composition of Three Lambda Bootis Stars," *Ap. J.*, **363**, 234.

Verhoogen, J., Turner, F. J., Weiss, L. E., Wahrhaftig, C., and Fyfe, W. S. 1970, *The Earth, An Introduction to Physical Geology*, (New York: Holt, Rinehart, and Winston).

Véron-Cetty, M.-P., and Véron, P. 1991, "A Catalogue of Quasars and Active Nuclei," 5th ed., *European South. Obs. Sci. Rep.*, **10**.

Viala, Y. P. 1986, "Chemical Equilibrium from Diffuse to Dense Interstellar Clouds. I. Galactic Molecular Clouds," *Astron. Ap. Suppl.*, **64**, 391.

Viala, Y. P. 1990, "Chemical Processes in Interstellar Clouds," in *New Windows to the Universe*, Vol. II, ed. F. Sanchez and M. Vazquez (Cambridge: University Press), p. 373.

Vidal, G., Cooper, J., and Smith, E. 1971, "Unified Theory Calculations of Stark Broadened Hydrogen Lines Including Lower State Interactions," NBS Monograph **120**.

Vílchez, J. M., Pagel, B. E. J., Díaz, A. I., Terlevich, E., and Edmunds, M. G. 1988, "The Chemical Composition Gradient Across M33," *Mon. Not. Roy. Astron. Soc.*, **235**, 633.

von Zahn, U., Kumar, S., Niemann, H., and Prinn, R. 1983, "Composition of the Venus Atmosphere," in *Venus*, ed. D. M. Hunten, L. Colin, T. M. Donahue, and V. I. Moroz (Tucson: University of Arizona Press).

Vorontsov-Vel'yaminov, B. A., Kostyakova, E. B., Dokuchaeva, O. D., and Arkhipova, V. P. 1965, "A Redetermination of the Absolute Emission-Line Intensities of 25 Planetary Nebulae," *Soviet Astron. J.*, **9**, 364.

Wacker, P. F., Cord, M. S., Burkhard, D. G., Peterson, J. D., and Kukol, R. F. 1969, *Microwave Spectral Tables III. Polyatomic Molecules with Internal Rotation* (Washington DC: Government Printing Off.).

Wacker, P. F., Mizushima, M., Peterson, J. D., and Ballard, J. R. 1964, *Microwave Spectral Tables I. Diatomic Molecules* (Washington DC: Government Printing Off.).

Wacker, P. F., and Pratto, M. R. 1964, *Microwave Spectral Tables II. Line Strengths of Asymmetric Rotors* (Washington D.C.: Government Printing Off.).

Wagoner, R. V. 1973, "Big Bang Nucleosynthesis Revisited," *Ap. J.*, **179**, 343.

Wagoner, R. V., Fowler, W. A., and Hoyle, F. 1967, "On the Synthesis of the Elements at Very High Temperatures," *Ap. J.*, **148**, 3.

Wallerstein, G. 1962, "Abundances in G Dwarfs. VI. A Survey of Field Stars," *Ap. J. Suppl.*, **6**, 407.

Wannier, P. G. 1980, "Nuclear Abundances and Evolution of the Interstellar Medium," *Ann. Rev. Astron. Ap.*, **18**, 399

Wapstra, A. H., Audi, G. and Hoekstra, R. 1988, "Atomic Masses from (Mainly) Experimental Data," *At. Data Nuc. Data Tables*, **39**, 281.

Ward, R. A., and Fowler, W. A. 1980, "Thermalization of Long-Lived Nuclear Isomeric States Under Stellar Conditions," *Ap. J.*, **238**, 266.

Wasserburg, G. J., and Papanastassiou, D. A. 1982, "Some Short-Lived Nuclides in the Early Solar System – A Connection with the Placental ISM," in *Essays in Nuclear Astrophysics*, ed. C. A. Barnes, D. D. Clayton, and D. N. Schramm (Cambridge: University Press), p. 77.

Wasserburg, G. J., Papanastassiou, D. A., Nenow, E. V., and Bauman, C. A. 1969, "A Programmable Magnetic Field Mass Spectrometer with On-Line Data Processing," *Rev. Sci. Inst.*, **40**, 288.

Wasson, J. T. 1974, *Meteorites: Classification and Properties* (Berlin: Springer).

Wasson, J. T. 1985, *Meteorites: Their Record of Early Solar System History,* (New York: Freeman).

Watson, W. D. 1976, "Interstellar Molecule Reactions," *Rev. Mod. Phys.*, **48**, 513.

Weast, R. C. (ed.) 1980, *Handbook of Chemistry and Physics*, 61st ed. (Boca Raton, Fla.: CRC Press).

Wegner, G., Cummins, D. J., Byrne, P. B., and Stickland, D. J. 1983, "Element Identifications in the Ultraviolet Spectrum of HD 101065," *Ap. J.*, **272**, 646.

Wetherill, G. W. 1988, "Accumulation of Mercury from Planetesimals," in *Mercury*, ed. F. Vilas, C. R. Chapman, and M. S. Matthews (Tucson: University of Arizona Press), p. 670.

Weizsäcker, C. F. von 1935, "Zur Theorie der Kernmassen," *Zs. f. Physik*, **96**, 431.

Wheeler, J. C. 1991, "SN 1987A," in *Reports on Astronomy*, ed. D. McNally, *Transactions of the International Astronomical Union*, **XXIA** (Dordrecht: Kluwer), p. 430.

Wheeler, J. C., Sneden, C., and Truran, J. W., Jr. 1989, "Abundance Ratios as a Function of Metallicity," *Ann. Rev. Astron. Ap.*, **27**, 279.

White, R. E., Vaughan, A. H., Jr., Preston, G. W., and Swings, J. P. 1976, "Isotopic Abundances of Hg in Mercury Stars Inferred From Hg II λ3984," *Ap. J.*, **204**, 131.

Whitney, B. A., Soker, N., and Clayton, G. C. 1991, "Model for R Coronae Borealis Stars," *Astron. J.*, **102**, 284.

Whitten, D. G. A., and Brooks, J. R. V. 1972, *The Penguin Dictionary of Geology*, (Harmondsworth, Middx.: Penguin).

Wickramasinghe, N. C. 1973, *Light Scattering Functions for Small Particles* (New York: Wiley).

Wildey, R. L., Burbidge, E. M., Sandage, A. R., and Burbidge, G. R. 1962, "On the Effect of Fraunhofer Lines on *U*, *B*, *V*, Measurements," *Ap. J.*, **135**, 94.

Wilkins, G. A. "Documentation and Astronomical Data," in *Reports on Astronomy*, ed. D. McNally, *Transactions of the International Astronomical Union*, **XXIA** (Dordrecht: Kluwer), p. 7.

Williams, R. E., Woolf, N. J., Hege, E. K., Moore, R. L., and Kopriva, D. A. 1978, "The Shell Around Nova DQ Herculis 1934," *Ap. J.*, **224**, 171.

Witt, A. N. 1988, "Visual and Ultraviolet Observations of Interstellar Dust," in *Dust in the Universe*, ed. M. E. Baily and D. A. Williams (Cambridge: University Press), p. 1.

Witt, A. N. 1989, "Visible/UV Scattering of Interstellar Dust," in *Interstellar Dust*, IAU Symposium **135**, ed. L. J. Allamandola and A. G. G. M. Tielens (Dordrecht: Kluwer), p. 87.

Witt, A. N., Bohlin, R. C., and Stecher, T. P. 1984, "The Variation of Galactic Interstellar Extinction in the Ultraviolet," *Ap. J.*, **279**, 698.

Wolfendale, A. W. 1991, "The Mass of Molecular Gas in the Galaxy," in *Molecular Clouds*, ed. R. A. James and T. J. Millar (Cambridge: University Press), p. 41.

Wolff, S. C. 1983, "The A-Type Stars: Problems and Perspectives," *NASA-SP* **463**, National Tech. Inf. Service: Springfield, Va. 22161.

Wood, J. A. 1979, *The Solar System* (Englewood Cliffs, NJ: Prentice-Hall).

Woodgate, B. E., Tandberg-Hansen, E. A., Bruner, E. C., Beckers, J. M., Brandt, J. C., Henze, W., Hyder, C. L., Kalet, M. W., Kenny, P. J., Knox, E. D., Michalitsianos, A. G., Rehse, R., Shine, R. A., and Tinsley, H. D. 1980, "The Ultraviolet Spectrometer and Polarimeter on the Solar Maximum Mission," *Solar Phys.*, **65**, 73.

Woods, R. D., and Saxon, D. S. 1954, "Diffuse-Surface Optical Model for Nucleon–Nuclei Scattering," *Phys. Rev.*, **95**, 577.

Woosley, S. E. (ed) 1991, *Supernovae* (Berlin: Springer).

Woosley, S. E., Arnett, W. D., and Clayton, D. D. 1973, "The Explosive Burning of Oxygen and Silicon," *Ap. J. Suppl.*, **26**, 231.

Woosley, S. E., and Howard, W. M. 1990, "^{146}Sm Production by the Gamma Process," *Ap. J.*, **354**, L21.
Woosley, S. E., Taam, R. E., and Weaver, T. A. 1986, "Models for Type I Supernova. I. Detonations in White Dwarfs," *Ap. J.*, **301**, 601.
Woosley, S. E., and Weaver, T. A. 1982, "Nucleosynthesis in Two $25M_\odot$ Stars of Different Population," in *Essays in Nuclear Astrophysics*, ed. C. A. Barns, D. D. Clayton, and D. N. Schramm (Cambridge: University Press), p. 377.
Woosley, S. E., and Weaver, T. A. 1985, "Theoretical Models for Type I and Type II Supernovae," in *Nucleosynthesis and its Implications on Nuclear and Particle Physics*, ed. J. Audoze and N. Mathieu (Dordrecht: Reidel).
Woosley, S. E., and Weaver, T. A. 1986, "The Physics of Supernova Explosions," *Ann. Rev. Astron. Ap.*, **24**, 205.
Wynne, J. J. 1984, *Current Trends in Atomic Spectroscopy* (Washington DC: National Academy Press).
York, D. G., Yanny, B., Crotts, A., Carilli, C., and Garrison, E. 1991, "An Inhomogeneous Reference Catalogue of Identified Intervening Heavy Element Systems in the Spectra of QSOs," *Mon. Not. Roy. Astron. Soc.*, **250**, 24.
Yorka, S. J. (ed.) 1991, *Newsletter of Chemically Peculiar Red Giant Stars*, Newsletter of Working Group on Peculiar Red Giants, IAU Commissions 29 and 45.
Young, J. S. and Scoville, N. Z. 1991, "Molecular Gas in Galaxies," *Ann. Rev. Astron. Ap.*, **29**, 581.
Zahnle, K. 1993, "Planetary Noble Gases," in *Protostars and Planets III*, ed. E. H. Levy, and J. I. Lunine (Tucson: University of Arizona Press), p. 1305.
Zappala, V. 1991, "Minor Planets," in *Reports on Astronomy*, ed. D. McNally, *Transactions of the International Astronomical Union*, **XXIA** (Dordrecht: Kluwer), p. 150.
Zarkov, V. N. 1986, *Interior Structure of the Earth and Planets* (Chur, Switz.: Harwood Academic Publishers).
Zinner, E. 1988, "Interstellar Cloud Material in Meteorites," in *Meteorites and the Early Solar System*, ed. J. F. Kerridge and M. S. Matthews (Tucson: University of Arizona Press), p. 956.
Zinner, E., Sachiko, A., Anders, E., and Lewis, R. 1991, "Large Amounts of Extinct ^{26}Al in Interstellar Trains from the Murchisson Meteorite," *Nature*, **349**, 51.
Zwicky, F. 1933, "Die Rotverschiebung von extragalaktischen Nebel," *Helv. Phys. Acta*, **6**, 110.

Index

Printed in the United States
By Bookmasters